T0338538

WORKING WITH NETWORK DATA

Drawing examples from real-world networks, this essential book traces the methods behind network analysis and explains how network data is first gathered, then processed and interpreted. The text will equip you with a toolbox of diverse methods and data modelling approaches, allowing you to quickly start making your own calculations on a huge variety of networked systems. This book sets you up to succeed, addressing the questions of what you need to know and what to do with it, when beginning to work with network data. The hands-on approach adopted throughout means that beginners quickly become capable practitioners, guided by a wealth of interesting examples that demonstrate key concepts. Exercises using real-world data extend and deepen your understanding, and develop effective working patterns in network calculations and analysis. Suitable for both graduate students and researchers across a range of disciplines, this novel text provides a fast-track to network data expertise.

JAMES BAGROW is Associate Professor in Mathematics & Statistics at the University of Vermont. He works at the intersection of data science, complex systems and applied mathematics, using cutting-edge methods, mathematical models and large-scale data to explore and understand complex networks and systems.

YONG-YEOL AHN is Professor at Indiana University and a former Visiting Professor at the Massachusetts Institute of Technology. He specializes in network and data science and machine learning, and his research on complex social and biological systems has been recognized by many awards, including the Microsoft Research Faculty Fellowship.

WORKING WITH NETWORK DATA

A Data Science Perspective

JAMES BAGROW
University of Vermont

YONG-YEOL AHN
Indiana University, Bloomington

CAMBRIDGE
UNIVERSITY PRESS

Shaftesbury Road, Cambridge CB2 8EA, United Kingdom

One Liberty Plaza, 20th Floor, New York, NY 10006, USA

477 Williamstown Road, Port Melbourne, VIC 3207, Australia

314–321, 3rd Floor, Plot 3, Splendor Forum, Jasola District Centre, New Delhi – 110025, India

103 Penang Road, #05–06/07, Visioncrest Commercial, Singapore 238467

Cambridge University Press is part of Cambridge University Press & Assessment,
a department of the University of Cambridge.

We share the University's mission to contribute to society through the pursuit of
education, learning and research at the highest international levels of excellence.

www.cambridge.org
Information on this title: www.cambridge.org/9781009212595

DOI: 10.1017/9781009212601

First published 2024

A catalogue record for this publication is available from the British Library

ISBN   978-1-009-21259-5   Hardback

Additional resources for this publication at www.cambridge.org/network-data.

Contents

Preface

Why should we care about networks and network data? For one, they are *everywhere*. We are living in networks, we are using networks, and we *are* networks.

Too philosophical? Do you use Google? Did you browse some online news or social media today? They are all built on the web, a giant network of webpages. Whenever we click a link on the web, we are navigating the web. Google was one of the first companies that recognized the value of understanding the web's network structure, which led to the development of "PageRank," the famous algorithm behind their success. In fact, look at any large technological company. You will see that most of them are now built on some type of network and are extracting massive value from those networks.

Whenever we see *relationships*, there are networks. Do you want to build a massive online marketplace? Your chance of success may increase if you understand and leverage the network between customers and products. Do you want to understand how our cells and brains work? It boils down to understanding the cellular network (interactions between biological components) or the brain network.

If your work calls for understanding a complex system, chances are, you will be examining some kind of network. You will be working with network data!

Why this book?

There are great textbooks on network science. We complement these with a focus on the practical side of network science—working with network data. The purpose of this book is to provide a more practical guide for data scientists to use network science.

For instance, think about the process of *defining* a network. Although it is usually not emphasized, the process of defining nodes and edges from data is often critical but non-trivial! What about visual, exploratory analyses of networks? While an integral part of network science research—how you visualize networks can make or break your project and determine whether you discover salient patterns—it is often not discussed or taught in enough detail.

So, here we are. First, we hope that this book can help researchers in day-to-day tasks, starting from the very act of conceptualizing networks through to sophisticated network analysis, from exploratory analysis to statistical modeling and machine learning. We believe that this book can be a useful resource for simple yet critical questions that researchers and practitioners face everyday, such as: How should I define a network from this data? Should I conceptualize this network as a weighted, directed network, or should we ignore those properties? How can I reduce or simplify this network? How can I visualize this network? How can I interpret the results of community analysis? And so on and so forth.

Second, we aim to give data scientists a foundational understanding of the tools, both mathematical and computational, at their disposal. The breadth and depth of statistical methods we can now use on network data is dizzying. We wish to take the prepared data scientist from their base knowledge of mathematics and statistics forward on a journey through the fundamentals of network data.

We hope to help students and researchers navigate both common yet critical questions and empower them with cutting-edge tools and the understanding to bend them to address their work.

How to use this book

This book is organized into two major parts—*practice* and *fundamentals*—connected by an interlude about good computing practices. As we emphasize the *practice* of network science, we present it first. But that does not mean that a course, or a self-study, should follow the same structure. Depending on the needs, goals, and level of preparation, one may pick and choose either the practical or the theoretical part. For instance, a network science course may use chapters from both parts in parallel to cover theoretical and practical aspects of each module's topic (e.g., see Chs. 11 and 12 for a practical guides to data and Ch. 22 for theoretical models of that data).

These two parts are preceded by an introduction and background that explains the basic concepts of network science as well as mathematical primers for the theory and data ethics. They may be used as a quick review of the basics or an introductory material.

Throughout the text we use *boxes* to emphasize certain points:

 "Info boxes" provide definitions and contexts.

 "Good boxes" point out good practices and steps to take.

 "Warning boxes" highlight dangers and concerns to be mindful of.

Prerequisites   Although not required, you probably want to understand the basics of network science as may be conveyed by a one-semester introductory undergraduate course. Familiarity with probability, statistics, and linear algebra will be necessary, with stronger familiarity needed for Part III. Some experience with basic programming as well as basic knowledge of algorithms and data structures is also assumed and we will on occasion present code snippets, either in pseudocode or in Python. We review many of these topics in Ch. 4.

A website accompanying this book is available online at cambridge.org/network-data.

Acknowledgements

We are grateful to the many students, collaborators, and mentors who inspired us to write this book, enriched our understanding of this material, and helped us improve our text. We are indebted to David Hemsley and Nicholas Gibbons, Stephanie Windows, and Jane Chan at Cambridge University Press for shepherding our book from creation to production. Above all, we thank our loved ones for their support and encouragement during the writing of this book.

James Bagrow
Williston, Vermont

Yong-Yeol Ahn
Bloomington, Indiana

February, 2024

Part I

Background

Chapter 1

A whirlwind tour of network science

Network science has exploded in popularity since the late 1990s. But it flows from a long and rich tradition of mathematical and scientific understanding of complex systems. In this chapter, we set the stage by highlighting network science's ancestry and the exciting scientific approaches that networks have enabled, followed by a tour of the basic concepts and properties of networks.

1.1 Networks as a powerful analogical framework

System thinking

To understand the power of network thinking, it's worth taking a brief detour to the emergence of *system thinking* in the nineteenth century. The nineteenth century was a time of great advancements in science. As the industrial revolution upended how societies and economies function, science became a profession; physics, chemistry, biology, and so many other fields were established and advanced rapidly.

An important perspective that emerged during the nineteenth century was the recognition of *complex systems*—although the name came later—across domains. The idea is that our body, society, and *everything* consists of numerous individual elements. For instance, matter consists of "atoms" and "molecules." Although the idea of *basic elements* was suggested by ancient Greeks, this philosophical atomism turned into a concrete scientific theory—the *atomic theory*, propelled by the discovery of chemical laws through precise measurements of the chemical reactions (see Fig. 1.1). In other words, people began to look at *everything* as a system of atoms and molecules and realized that not only *what* constitutes the matter *matters*, but also *how it is arranged* matters.

Similar theories emerged across many fields around the same time. In biology, new instruments like lenses and microscopes led scientists to look at the structure of cells and organisms on a finer scale. This gave us *cell theory*, which postulates that every

3

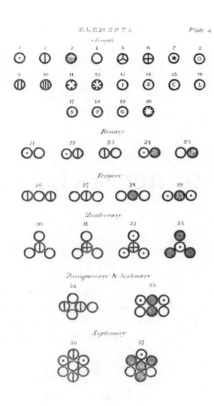

Figure 1.1   Atoms and molecules depicted in John Dalton's *A New System of Chemical Philosophy* (1808) [120].

living organism is made up of cells—the atoms of life. This is also the beginning of the realization that the rich biological phenomena originate not only from the nature of an individual cell, but also from *how* those cells interact with each other.

It was called "neuron doctrine" for the brain. With the ability to examine individual neurons and their dendrites, most notably in the work of Santiago Ramón y Cajal (1852–1934), scientists began to realize that a brain is essentially a giant, networked system of neurons. The realization began to take shape that the connections and interactions—rather than the make-up of different cells—are what the brain is all about.

Interestingly, the establishment of *social statistics* also happened around the same time. As cities grew and countries established themselves, there was strong need to get detailed information about the population within, and this led to the rapid development of *statistics*.[1] The same story again: the strong recognition of a *system* (cities, states, countries, etc.) as a collection of individuals and the ability to concretely think about and quantitatively measure this collection as a whole.

All these revolutions were about recognizing the *systems*, and the *elements* that make up the systems. Living organisms are made of *cells*; the brain consists of *neurons*; everything is made of *atoms*, and cities and countries are made of individual *people*!

---

[1] The term *statistics* originated from a German word "Statistik," which came from "statisticum" (Latin) or "statistica" (Italian). The term "Statistik" was introduced by Gottfried Cornwall (1719–1772) in 1749 to refer to the *analysis of data about the state*.

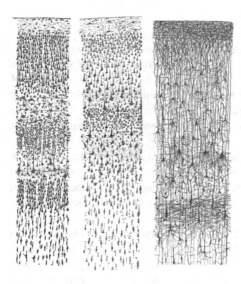

Figure 1.2 Ramón y Cajal's drawings of the human cortex, from *Comparative Study of the Sensory Areas of the Human Cortex* (1899). Figure from [386].

And every one of these revolutions was propelled by the development of new technologies, instruments, and capacities. Being able to see finer, measure more precisely, and calculate more accurately was the primary driving force behind these breakthroughs.

At the same time, although recognition was dawning of the importance of interactions in the system—for example, cells are remarkably similar to each other even when they belong to completely different systems and brains are different because of how they are wired—there was no way to accurately capture, at that time, the *interactions*.

Ramón y Cajal's drawings (see Fig. 1.2) beautifully illustrate how neurons produce incredibly complex branches (dendrites), but he could not see how these dendrites and axons are wired together. It took a hundred years until we could accurately connect the dots and measure the precise connections between individual neurons. Although scientists could *see* individual cells, it took a long time to be able to measure cellular interactions happening at the molecular level.

Yet, even without the science of *interactions*, system thinking could provide a powerful framework with which scientists crossed disciplinary boundaries. There was a plethora of analogy-making between disciplines and systems, unified by system thinking. Working with social *statistics* facilitated the development of probabilistic and statistical thinking and it had a huge influence on the way scientists think about the world. For instance, physicist Ludwig Boltzmann (1844–1906), one of the founders of statistical mechanics, drew inspiration from the social census, making an analogy between molecules and individuals. Boltzmann wrote (emphasis ours):

> *The molecules are like to many individuals*, having the most various states of motion, and the properties of gases only remain unaltered because the number of these molecules which on the average have a given state of motion is constant. (p. 69, *Critical Mass* by Philip Ball [33])

Network thinking

The difficulty and trickiness of measuring interactions delayed recognizing and appreciating the importance of interactions in complex systems. It also limited scientists to the paradigm of *reductionism*—the approach that we need to understand each part of a system to understand the whole. Needless to say, reductionism is not inherently bad; it is indeed true that we should know the parts to understand the whole. The issue is that it is *not enough*, especially when we are dealing with complex systems that exhibit *emergent behaviors*.

It is not enough to understand how an individual neuron works to understand how cognition works; it is not enough to understand the chemical properties of individual molecules to understand how our body produces energy and renews itself; it is not enough to understand how individuals behave to understand large-scale social phenomena.

Network science arrived as it became clear that understanding the elements of a system is not enough, and as technology advanced to be able to measure interactions. But even without new scientific tools, there is a system where we could already measure interactions fairly easily—by asking the elements of the system directly. People can talk, and that is awfully convenient for network measurement. Because it is much easier to measure social networks, social network research has a much longer history than other fields of network science.

Jacob L. Moreno (1889–1974), a psychiatrist and social psychologist who pioneered psychodrama and group psychotherapy, was fascinated by the connections between inter-personal relationships (the social network!) and psychology. In the 1930s, he created a series of remarkable diagrams—which he called *sociograms*—by examining relationships between students at the New York Training School for Girls, a training institute for delinquent girls (see Fig. 1.3). The primary reason that he mapped the social relationships was to investigate why so many girls ran away from the school.[2]

To understand this phenomenon, Moreno proposed and conducted "*sociometry*"—a comprehensive measurement of the social interactions between students—with which he argued that the social network is what influences the eventual outcome of the students. In other words, he argued that the driving force of social dynamics is in the *network* of interactions.

The recognition of network interactions greatly affected sociology and related fields, allowing them to recognize the importance of *social structure* and the "*social forces*" that the social network can exert on the individuals. Network thinking arrived in other fields a bit later, mostly due to the difficulty of actually *seeing* and *recording* the connections. The aforementioned technology was still needed.

This difficulty has been gradually overcome with better engineering and new tools, especially combined with computers and digital technologies. With computers and the Internet, numerous databases of network data emerged across fields. Although our knowledge of chemical reactions is as old as human civilization, it was in the 1990s when the first computerized and shared databases of chemical reactions were created. Similar databases for protein interactions were also created. The Internet itself—a pretty useful

---

[2] Ella Fitzgerald, legendary jazz singer, attended the school and also ran away after about a year. She might be in one of the sociograms that Jacob Moreno created!

EVOLUTION OF GROUPS

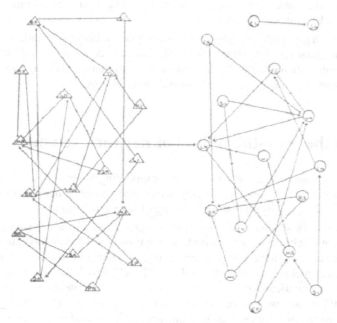

CLASS STRUCTURE, 4TH GRADE
CRITERION: STUDYING IN PROXIMITY, ACTUALLY SITTING BESIDE THE
PUPILS CHOSEN; 2 CHOICES
17 boys and 16 girls. *Unchosen* 6, EP, RY, EL, FA, SI, CF; *Pairs* 17, GR-SI,
GR-LI, MR-LN, LN-SM, YL-KN, AB-BA, BA-BR, KI-KN, AB-PN, FC-VN,
BU-CV, LN-WI, LN-MR, BR-MC, BR-RS, WI-MR, MC-RS; *Stars* 2, LN, VN;
Chains 0; *Triangles* 2, BR-RS-MC; LN-WI-MR; *Inter-sexual Attractions* 1;
Not Choosing 1, SH.

Figure 1.3　Jacob L. Moreno's sociogram of 4th grade students at the New York Training School for Girls. Figure from [318]; originally published in [317].

network—was mapped and compiled into a network dataset. Online social networking services emerged and provided a completely new way to map social interactions at a societal scale.

The fact that network datasets and maps began to emerge *across many fields* also nudged scientists to think more analogically; network thinking flourished. In particular, statistical physicists who are used to the idea of complex systems and computer scientists who are used to computer networks and data structures emerged as the primary driving force to look for universal patterns in these network data.

By identifying and abstracting the *network structure* in all kinds of systems, many universal or prevalent characteristics of these networks have been discovered. Networks now serve as an analogical framework to translate insights and methods from one system to another. For instance, based on our intuition and understanding of the major travel hubs in the network of airports, we can ask whether major "hubs" appear between our proteins or among people. Or, we can apply an algorithm developed to discover groups of people, or "communities," in a social network to instead find "communities" of proteins in the network of interacting proteins.

Network thinking often gives us a superpower to cross rigid disciplinary boundaries. As Boltzmann used the *system analogy* between people and molecules to develop statistical mechanics, network scientists use the *network analogy* to transcend disciplinary boundaries and understand the universal patterns of complex systems.

1.2   Data and theory—the pillars of network science

Our ability to *measure* interactions has been critical to the blooming of network science. Network science, as an empirical science, is largely driven by measurements and data. New *network data* have always driven the understanding of networks around us.

The explosion of network science coincided with the explosion of our capacity to systematically collect data. In biology, the Human Genome Project opened up molecular understanding of human biology in the 1990s; scientists began to assemble the *metabolic network* (the widely used metabolic network database KEGG[3] was initially released in 1995); methods for systematically probing interactions between proteins were first developed in the late 1980s and early 1990s. In neuroscience, techniques like *diffusion tensor imaging* (DTI) were also proposed around this time [42, 43] and computerized measurements of brain networks began. In social science, although network research had a long history since Moreno, the emergence of the Internet and the "social web" allowed us to collect unprecedented, high-resolution, societal-scale social interaction data.[4]

This sudden deluge of network data drew a lot of attention from scientists, leading to the discovery of universalities and commonalities across real-world networks, which then sparked the emergence of network science theories. This pattern continues today, whenever there is new network data from important complex systems, it pushes network science forward by forcing scientists to perform new measurements, develop new methods, and invent new theories.

However, despite being our focus, ***data alone cannot paint a full picture***. Big data without any insight or useful theoretical framework is expensive junk. We need coherent theories that equip us to understand the data and make predictions. For instance, network growth models allow us to ponder the mechanisms *behind* the growth of networks; the theory of random graphs can tell us what should we expect to see in a network under certain assumptions about the system; statistical inference can help us identify large-scale patterns in the networks. Finally, theories make *predictions*, which are powerful directives that lead to useful measurements and data collection.

Network science, like all science, stands upon the dual pillars of data and theory.

---

[3] Kyoto Encyclopedia of Genes and Genomes.

[4] Whether social web data can be a good proxy of social relationships is another interesting question that we will discuss more later.

1.3 Networks are everywhere

One of the primary reasons why networks can serve as such a powerful analogical framework to study complex systems is that networks are *everywhere*. Just look at ourselves. Society is a giant network of people. It is intriguing that so many of the "Big Tech" companies are built on the power of social networks. Google's secret sauce was recognizing the value of the network between web pages and developing the algorithm, *PageRank*, that can harness it. Facebook (Meta) was literally built as a platform to share and communicate with people through the social network. Once it reached a dominant position—pulled enough people into the platform—it became extremely *sticky*. Because so many of our friends use Facebook, it is difficult for any one person to leave and this is true for everyone on the platform. The more people use an online social networking service, the harder it is to quit due to this *network effect*.

If we look inside ourselves, we can find that we (and every living organism) rely on many levels of networks, from the network of biochemical reactions to the network of neurons in the brain. All biological phenomena (and diseases) arise from the interactions between molecules, cells, and organs.

You can also easily find networks in unexpected places. For instance, it has been suggested that there is an interesting network in forests called the "*mycorrhizal network*," which is a network between plants, underground, formed by fungi.[5] This fungal network has been described as the "Internet" or postal network between plants, through which plants can exchange nutrients and communicate with each other. "Mycorrhizal network" doesn't exactly roll off the tongue, but fortunately there is a more memorable name for this network; it's the "*wood wide web*"!

Speaking of plants and fungi, we can construct another interesting network between the foods we eat, based on their flavor similarity. This network—the "*flavor network*"[6]— was inspired by the "food pairing hypothesis," which argues that two ingredients go well together if they share the same flavor-producing chemical compounds because the flavor gets enhanced by the combination (see Fig. 1.4). You can see why this hypothesis calls for a network—look at all those intricate patterns!

When we inhale and exhale these flavor molecules, *olfactory receptors* (proteins) in our nose need to do their job to recognize the flavors. Because they bind with the flavor molecules (like a lock and a key), it is critical to know the exact shapes of the proteins that constitute the receptors. Although the sequences of proteins are well-known, deducing three-dimensional (3D) structure from them is far from trivial; rather, it is the opposite of trivial. This is called the *protein folding* problem and has been a notoriously difficult open problem.

The biggest challenge is that pretty much any pair of amino acids, regardless of their distance in the genetic sequence, might be right next to each other in the final, folded protein. In other words, the amino acids in a protein have a lot of non-trivial long-range *interactions* and each protein can be thought of as a *network* of amino acids that are eventually *connected* in the final 3D shape. Surprisingly, this elusive problem was effectively solved by the *AlphaFold* team at Google's Deep Mind in 2021, and this

---

[5] Note that there is a controversy over the existence of this network. Some scientists criticize that it is unproven and over-hyped.

[6] It was constructed and studied in part by yours truly, see Ahn et al. [6].

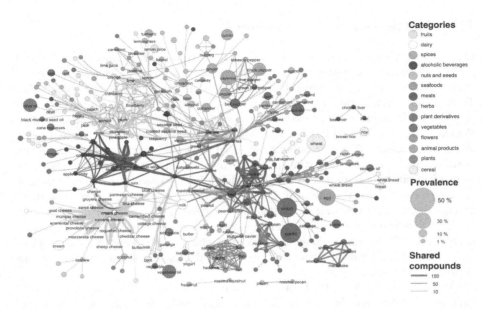

Figure 1.4   The flavor network of ingredients that shared flavor compounds. This visualization (Ch. 13) displays the structure of the *network backbone* (Ch. 10), the most prominent connections. Figure from [6].

achievement was just recently selected by *Science* as the "Breakthrough of the Year." You can probably guess what we want to say here: AlphaFold's machinery indeed uses this network perspective to represent the interaction between amino acids and an important breakthrough was their ability to accurately predict this network structure!

Here is another fun example, now at the scale of the galaxy and universe. Have you ever imagined traveling through the galaxy to reach other stars? We can formulate this problem more concretely by imagining the *maximum distance* that humanity may be able to travel and connect the stars that are reachable with this distance. Not only is this an entertaining musing on life, the universe, and everything, but it also is a curious case of the *percolation* problem that we will talk about later in the book.[7] Can we humans percolate through the galaxy like our coffee percolates through a filter?

OK, we have traveled from our society to the forest, and from our dinner plates to the universe. Still, we would argue that these are just small tips of the network iceberg. Once you fully embrace network thinking, you will start to see network structure and network data *everywhere*!

---

[7] Which, again, is an excellent example that supports our point that network thinking is a powerful analogical framework to translate theories and tools across domains.

1.4 Basic terminology

Here is a brief overview of some basic network terminology. If you have already been exposed to these, it may be a useful review; if you haven't, don't worry. We will return to them in more detail as we progress through the book.

1.4.1 Basic concepts

Mathematically, a network is represented by an object called a *graph*, which we denote $G = (V, E)$. Graphs contain at least two sets: nodes (or vertices) and edges (or links). Here V is the set of nodes in the network, while E is the set of edges. Each edge connects a pair of nodes. We denote the number of nodes in G with N and the number of edges M. You've seen pictures of nodes and edges already, in Figs. 1.3 and 1.4, where nodes were drawn as circles or other shapes and edges were lines or arrows. We use the terms *edges* and *links*—as well as *nodes* and *vertices*—interchangeably throughout this book. We also use the terms *network elements* or *elements* to refer to either nodes or links or both.

> **ⓘ** A network consists of two kinds of elements:
>
> **Nodes** also known as vertices.
>
> **Edges** also known as links.
>
> Understanding what the nodes represent, and what relationship(s) between the nodes are represented by the links, is critical for understanding any network data.

We illustrate some basic network concepts in Fig. 1.5.

Two nodes are connected when a link exists between them, making them *neighbors*. The set of neighbors connected to a node is its *neighborhood*. We denote the set of neighbors of a node i with N_i. The number of neighbors a node has is called its *degree* and we are often interested in the *degree distribution* $\Pr(k)$, the probability that a randomly chosen node from our network has degree k. This distribution is fundamental.

A *walk* is a sequence of edges that join a sequence of nodes; a *trail* is a walk where all edges are distinct; a *path* is a trail where all nodes are distinct. For instance, the edges (i, j), (j, u) and (u, v) form a path (assuming $i \neq j \neq u \neq v$) i–j–u–v that goes from node i to node v through nodes j and u. A path that begins and ends at the same node is called a *cycle*. Although there are other types of walks and paths, we are often interested in *shortest paths*, the paths connecting nodes using as few links as possible. The lengths of shortest paths are often used to define *distances* over a network. Information spreads over paths. These paths are fundamental.

If a path exists from every node to every other node, then we say the network is *connected*; otherwise, it is *disconnected*. A disconnected network will consist of two or more *connected components* (sometimes just called *components*). A connected component is a subset of nodes where paths exist between every pair of nodes in that subset. The connected component containing the largest number of nodes is called the *giant connected component* (GCC) or *giant component*. Often, a network is disconnected but the giant component contains the large majority of nodes.

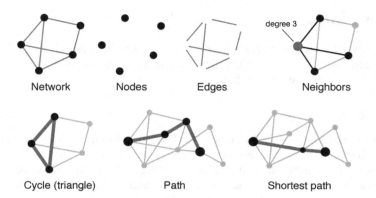

Figure 1.5 Nodes and edges form the basic elements of networks. Neighborhoods, cycles, paths, and shortest paths are some of the structures we examine in network data.

For a connected network, the *eccentricity* of a node is the length of the *longest* shortest path beginning at that node. The longest shortest path between any two nodes in the network is the *diameter* of the network.[8]

1.4.2   Operationalizing a network

The key decision we need to make when working with network data is determining what the elements should represent (to *operationalize* the network). **What are the nodes?** Is it a social network where each node represents a person? Or perhaps each node represents a group of people? Along those lines, **what are the links**? If nodes are people, do links exist between two people when they are friends on a first-name basis? Or is it based on when they have exchanged messages through a website? If nodes are groups of people, what are the links? Perhaps links exist when groups overlap, sharing one or more people? Or perhaps instead two groups are linked if they share a common context, perhaps they are teams that work in the same branch of an organization? Often these questions are not made explicit. Yet, appropriately defining nodes and links between nodes is the key to extracting the right network—or networks!—from data.

1.4.3   Basic types of networks

The "zoo" of networks contains many network animals, depending on what properties they have or definitions they meet (Fig. 1.6). An *undirected* network is one where links are symmetric and have no direction. In an undirected network, if node i is connected to node j then node j is also connected to i. A *directed* network, on the other hand, is one where links are not necessarily symmetric. The link $i \rightarrow j$ may exist without the corresponding link $j \rightarrow i$.[9]

---

[8] If the network is not connected, we usually treat it as having a diameter of infinity.

[9] Directionality brings with it some complexity. Instead of node degree, we now need to consider *in-degree* and *out-degree*. Links point from a *source node* to a *target node*. Likewise, shortest paths that exist between

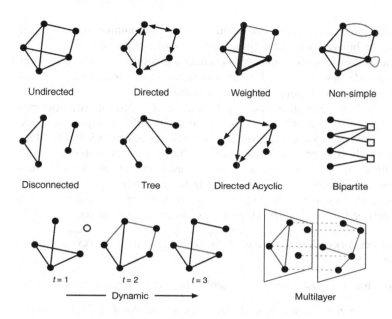

Figure 1.6 A sample of the network "zoo."

Networks can also be considered *weighted* or *unweighted*. A weighted network is one where each link, denoted i, j has an associated edge weight w_{ij}, typically a scalar, that captures how "strong" the link is. Usually, larger weights denote stronger links. And of course, a network can be both (un)directed and (un)weighted.

A network may, or may not, contain *multi-edges* and *self-loops*. A multi-edge is one where multiple edges occur between the same two nodes while a self-loop is an edge between a node and itself. When these are forbidden the network is called *simple* because it can be represented by a simple graph. When multi-edges are present, the network is represented by a *multigraph*. Likewise, when self-loops are present, the network is represented by a *graph with loops* or a *loopy graph*. [10]

Trees are networks where no cycles exist. Between any two nodes there is at most one path connecting them. A disconnected tree is sometimes called a *forest*. More complicated than a tree, in directed networks, a *directed acyclic graph* (DAG) is one where cycles exist but no directed cycles are present. One cannot leave a node and, following directed edges, follow a path back to the start.

Bipartite networks are networks that have two distinct sets of nodes, where each link (usually) lands only between nodes of different sets. Surprisingly, many network datasets contain bipartite networks or are derived from a bipartite network via a projection. A *bipartite projection* is a network derived from one of the two node sets in a bipartite

---

nodes ignoring link direction may no longer exist if we only consider paths that follow the directions of links. So we distinguish between *weakly connected* and *strongly connected* networks, where the former ignores direction when considering connectedness while the latter does not.

[10] Not to be confused with cycles.

network by adding edges between nodes who had neighbors in common in the original bipartite network; the bipartite edges are "projected" onto one of the sets.

Dynamic networks, often called *temporal networks* are networks that change over time. The links, or nodes and links, are functions of time. (Occasionally, for weighted temporal networks, it is the link weights that vary in time.)

Multilayer networks are networks where nodes can be arranged into multiple categories or layers, and edges can connect between nodes in different layers or in the same layer. A *multiplex network* is a special case of a multilayer network where edges between layers can only connect nodes that represent the same entity. It may be easier to think of a multiplex network as a network without layers but where edges fall into categories that convey the same information as the multiplex layers.

Hypernetworks, also known as hypergraphs, are networks where edges are not limited to being between pairs of nodes. Hyperedges can contain two or more nodes. [11]

ⓘ The zoology of networks. Here are a few of the basic kinds of networks we tend to encounter in data.

Undirected If a link i, j exists then i is a neighbor of j and j is a neighbor of i.

Directed If $i \rightarrow j$ then j is a neighbor of i but i is not a neighbor of j unless the link $j \rightarrow i$ is also present.

Weighted Associated with each link i, j is a weight w_{ij}, usually a non-negative scalar, representing how "strong" or "important" the link is. Weighted networks can be directed or undirected.

Simple A network where each link can exist at most once and where self-loops (links beginning and ending with the same node) are not allowed.

Bipartite A network whose nodes fall into two disjoint sets and links only exist between nodes of different sets. While it may sound esoteric, bipartite networks are surprisingly common in data and have many real-world applications.

Trees Networks without loops. At most one path exists between any pair of nodes.

Dynamic also known as temporal networks. The links and possibly nodes are functions of time.

Multilayer Links and possibly nodes fall into difference categories called layers.

Hypernetworks also called hypergraphs or higher-order networks, are networks where edges may contain more than two nodes.

And networks are often combinations of these categories, such as a weighted directed network or a dynamic multilayer network.

---

[11] There is a one-to-one correspondence between a hypernetwork and a bipartite node–hyperedge graph.

1.5 Common properties of real networks

As network science progresses, our view of real networks and their features has sharpened. Networks are complex, with many fascinating and important properties. (We'll encounter many of these properties in this book.) Numerous research tools, both new and found, have been leveraged to describe these properties. Here are a few of the major properties scientists have uncovered.

Sparsity Most real networks are sparse, most pairs of nodes are not connected by an edge. The average degree is much less than the number of nodes.

Small world Many real networks are small in terms of distances, meaning that it only takes a few hops along links to navigate between any pair of nodes, even in a network with a massive number of nodes [309, 486]. This is the source of the famous "six degrees of separation" aphorism in social networks [235].

Transitivity, homophily, and mixing patterns Links are often transitive, meaning that if two nodes have a common neighbor they are likely to be neighbors themselves. The process forming such links, *triadic closure*, has fascinating effects and is often driven by *homophily*, the idea that similar nodes are more likely to be connected. More generally, different kinds of networks can exhibit different mixing patterns, where edges fall not at random but between related nodes.

Hubs and heterogeneity The number of neighbors that a node has can vary wildly across nodes in many real networks. Most nodes have only a modest neighborhood, while a few, called hubs, have absolutely massive numbers of connections. Strong degree heterogeneity and the unexpected existence of these hubs in many networks was one of the early surprises of network science.

Density variation Many real networks exhibit regions that have relatively more connections and other regions that are empty of links. Community structure, where networks have densely connected groups with bottlenecks between them, is one example of such density variation. Core–periphery structure, nestedness, the "rich club," structural holes [85], and more, are all organizing principles connected to variation in density.

Many of these properties interrelate. For example, transitivity can drive the dense communities of a network, and mixing patterns can lead to hubs and degree heterogeneity. These diverse properties, often all at play simultaneously, make networks both exciting and challenging to study. We will focus on many of these issues throughout this book.

1.6 Summary

As our view of nature developed and more data became available, we began to appreciate complexity and interconnectedness all around us. System thinking flourished. In the century since, researchers have codified a field of study dedicated to networks, with its associated language and interesting problems. This field, which came to be called network science, has blossomed due to the ubiquity of networks across science and technology, from the social networks of Jacob Moreno and the neurons of Ramón y Cajal to the Internet and World Wide Web of the late twentieth century. We can no longer imagine the world without evoking networks. And network data is at the heart of it.

Bibliographic remarks

Jacob L. Moreno's pioneering work on social groups, including what are likely the first drawings of a network (Fig. 1.3), is documented in his work, *Who Shall Survive* [317]. His daughter, Regina Moreno, recently published a memoir on her life and family, *Words of the Daughter* [319], including bringing to light the sadly overlooked, critical contributions her mother, Florence Bridge Moreno, an accomplished counselor and teacher, made to Jacob's groundbreaking work.

Santiago Ramón y Cajal, a Nobel laureate who elucidated the shape and structure of neurons and neuronal networks, has been the subject of much interest. Most recently, *The Brain in Search of Itself*, by Benjamin Ehrlich [143] is a fascinating read. Readers interested in more direct sources are encouraged to read Ramón y Cajal's own memoir, *Recollections of my Life* [388] as well as his inspiring guide *Advice for a Young Investigator* [387]. You can also enjoy his mesmerizing drawings of neurons from a beautifully compiled book *The Beautiful Brain* [330].

Readers eager for more on networks and history should consider *The Square and the Tower* by Niall Ferguson [157], a fascinating general audience book on the role of networks throughout world history.

A number of general texts on network science are worth the reader's attention including Newman [342], Barabási [35], and Menczer et al. [305].

Exercises

1.1 Can you find an interesting network that was not mentioned in this chapter? Describe what the nodes and edges represent, and then identify the type of your network. Is it a directed network? Or a bipartite network? Finally, identify a couple of concrete scientific questions and uses for the network.

1.2 Can you find a specific example where a network method or theory crossed a disciplinary boundary? For instance, a model that was originally developed to describe social systems was later applied to explain the dynamics of brain.

1.3 Let's say we want to model the social network in our classroom. How would you operationalize the network? What would be your nodes and edges? Would you introduce edge weights or directions? Why or why not? Please describe the type of your network, your operationalization choices, and rationales behind your operationalization.

Chapter 2

Network data across fields

We seek to understand network data in part because *networks are everywhere*. Whenever a system has interacting elements, a network may be a useful representation of it. Even when there is no obvious network structure, often network structure emerges from the interactions in the system. As a result, most complex systems can be described as networks. The network framework makes itself useful.

For instance, consider again the problem from Ch. 1 of predicting the 3D structure of a protein. Structure prediction has been a famous outstanding challenge for biologists until recently, when DeepMind's AlphaFold made an incredible breakthrough [226]. AlphaFold has been touted[1] as the most impactful contribution of deep learning because finding the 3D structure of a protein has been a bottleneck in the investigation of protein functions and interactions. At first glance, there is no obvious network structure we can see in this problem—it's about understanding the complex energy landscape of various possible 3D molecular structures, right? Indeed, at the basic level it is about finding a 3D configuration of a long chain of amino acids. But at the same time, the problem is about effectively predicting the *long-range* connections between amino acids. A pair of amino acids can be separated by hundreds of other acids in the chain, yet be neighbors in the folded protein. So, how should we represent this "neighboring" relationship? Yes, you can represent it as a weighted network between amino acid residues! One of the key components of AlphaFold's machinery is exactly predicting this network structure.

Beyond proteins, networks are pervasive throughout biology and in fact all fields of research. In this chapter we'll discuss examples of networks and the data used to measure them through biology, neuroscience, sociology, economics, and more. We'll also identify several *focal points*—network data we will take with us throughout this book as working examples and use cases. These networks were chosen because they collectively capture most of the variations that we want to cover and because they are easy to grasp and understand.

---

[1] AlphaFold's success surprised many. Scientists have struggled with the structure prediction problem for nearly 50 years. "I never thought I'd see this in my lifetime," said John Moult, cofounder of the protein structure prediction competition that AlphaFold won [258]. In 2022, DeepMind announced that AlphaFold had solved the structures of over 200 million proteins, nearly every protein known to exist [90]. And as successful as it has been and promises to be, AlphaFold is not the end of the story, but the beginning [316].

2.1   Biology

Networks come into play at all levels of biological research. Organismal processes are fundamentally governed by network processes, and a major research effort of biologists has been gathering, at scale, the data necessary to describe such networks. Here are some examples.

Networks in the cell

We can find many networks across a wide range of scales in biological systems, from interaction networks between microscopic molecules to ecological networks between living organisms. Let's talk about some of these networks.

First, consider the essential parts of biology's *central dogma*. All of the most critical molecules in living organisms—DNAs, RNAs, and proteins—interact with each other. As mentioned earlier, each protein molecule can be thought of as a weighted network between amino acids, connected by the long-range connections created when folding the protein. These proteins are *translated* from RNA molecules, which are the result of the *transcription* process from DNA molecules. All of these processes are executed by tiny protein machines called protein complexes.

Now we already have many choices on how to operationalize the network between these molecules. We can focus only on the physical interactions between individual proteins (e.g., will these two proteins stick together or not?), or we can focus on the underlying *genes* by considering interactions between genes *mediated* by the proteins. The former is called the *protein–protein interaction network* (PPI) and one example of the latter is the *genetic regulatory network* (GRN). A gene can either promote or inhibit the expression of another gene, depending on what the protein that it generates can do.

Proteins perform many different tasks in living organisms. Some proteins regulate the production of other proteins. Some act as catalysts for the synthesis or breakdown of molecules, governing whether a certain metabolic reaction is possible or not. Some proteins are for the structure of the cells. Some proteins carry messages.

Protein networks capture interactions in a very real sense: molecules fitting together to make complexes. But networks can be more phenomenological as well. Genetic interaction networks are one example. In these networks, two genes are connected based on how the deletion of one or the other (or both) affects the organism. If knocking out both genes produces a surprisingly devastating impact on the organism, then we can say that the two genes are *interacting*, maybe through a process of compensation.

Other molecules synthesized and controlled by the activities of the organism's genes and proteins are worth studying as networks. The *metabolic network* captures how metabolites are transformed along synthesis pathways. Here nodes represent metabolites and links represent the transformation of one metabolite into another as part of a synthesis pathway. Often times diseases manifest as dysfunction of these pathways, and comparing the networks with and without disease can guide us to understand the disease's effects and even inform possible drug targets or other medical treatments.

In all these networks, we are reducing the complex interactions that involve DNA, RNA, and proteins and other molecules by focusing only on one type of element at a time because it simplifies the picture a lot. It is also possible to model different types of elements together, which may be closest to reality, but this also introduces a lot of complexity to our data. And of course, with so many possible interactions, the data we need is vast and, as rich as current data are, we are still just beginning to see the larger picture of the cell.

Network neuroscience

Going up the scale, we can examine interactions between cells. One of the most obvious examples would be the neuronal network of the brain. A brain is a dynamic network of neurons that communicate with each other using chemical and electrical signals.

The brain can be modeled in many different ways. At the base level, there is an actual network of neurons that are connected by synapses. This is of course very difficult to measure because the synapses are microscopic. Yet surprisingly, scientists have brain network data at the level of individual neurons, using methods to collect data about individual connections between neurons. For instance, the connectome project takes pictures of extremely thin slices of a brain, which are computationally aligned to map out connection structure from across the slices. Other methods use multi-electrode arrays and calcium imaging to show the spiking of neurons over time. And scientists have developed methods to infer the network structure based on the timing of spikes across different neurons.

Perhaps the most famous example of a complete neuron-level brain network is that of *C. elegans*, a small (about 1 mm) nematode. It does not have many cells and neurons, and nematodes with identical genes develop the same neuronal network. In heroic effort, scientists have mapped every cell in the nematode, including the neurons and the connections between the neurons. They froze the worm and created very thin slices across the body and painstakingly followed every neuron across slices. This work, in combination with other mapping of *C. elegans* genetics, development, and more, led to the 2002 Nobel Prize for Physiology or Medicine.

Other methods can map the brain network but only at a larger scale. Instead of working at individual neurons, we can think of brain regions that roughly correspond to certain brain functions as nodes. Now each node can be millions or billions of neurons. But what are the edges? Again the edges can be defined in many ways, based on the data available. One way to define and measure edges is from physical connections between regions in the brain. Methods like *diffusion tensor imaging* (DTI) allow us to follow large bundles of axons that wire different parts of the brain together. By inferring these axon bundles from measurements, we can "connect" brain regions. This network is often called the "structural connectome."

But there is also a completely different way to measure connections, called the "functional connectome." Here, instead of examining physical connections, we measure the activities of brain regions with *functional magnetic resonance imaging* (fMRI). If two regions fire together often, it probably means that they are functionally connected. This is the idea behind the functional brain network.

Network neuroscience continues to expand its data. Gene expression data is now often incorporated into network studies, and brain networks are often compared to sociodemographic information using large-population brain imaging studies such as the massive ABCD (Adolescent Brain Cognitive Development) study [97].

Networked ecology

Networks enter into ecology in a variety of ways. One example is *trophic networks*, more commonly known as food webs. In an ecosystem's trophic network, nodes represent species and links represent the "exchange of carbon"—a pleasant euphemism that often means members of one species *eat* members of the other. Food webs have an intriguing hierarchical structure starting from apex predators at the top and flowing down to organisms that survive on detritus—leftover, usually decaying, organic matter—or on photosynthesis. Capturing data on these trophic interactions is challenging: ecologists must often conduct long-running surveys to determine if one species indeed feeds off another. This labor-intensive research leads to precious data, especially when compounded by the critical need to understand and maintain the health of ecosystems[2] in the face of ongoing climate change.

Another network commonly studied is one of plants and plant pollinators, for example flowers and bees. Here nodes represent organisms (species), either plants or pollinators, and a link exists from a plant to a pollinator if that pollinator is known to pollinate that plant.[3] Such a network is often called a *mutualistic network*. This is fundamentally an experimental network: field observations are made tracking how often different pollinators were caught *in flagrante delicto* with different plants. Combining these observations, perhaps with some data cleaning or other processing, and a plant–pollinator network is made. Studying these networks in different regions and over time tells us about the health of those ecosystems, including their resilience to shocks such as climate change.

2.2   Socioeconomic systems

Our society is also a huge network—our social network. Broadly speaking, nodes in the network are the people in the society and edges are social relationships. But there are countless ways to think about and define specific social networks, especially the social ties. Maybe we are interested in whether two people *know* each other or not; sometimes we want to map the network of *close friends*; or perhaps we wish to understand proximity, the network of individuals who live geographically close to one another. How about family ties? How about shared social groups such as clubs? There are many ways to define a network and map social ties to edges. What definition or definitions we choose should be carefully considered and strongly motivated by the research question at hand.

How are social networks commonly measured? Historically, sociologists relied on surveys, questionnaires, and interviews. Visiting with individuals, they would ask, "Are you friends with X?" or perhaps, "List your 10 closest friends." Already we can see the power of the surveyor: the form of the question can strongly dictate the final network. If they ask, "List the 10 people you spend the most time with" you may get a very different network than asking about close friends. Likewise, why stop at 10 social ties? This will truncate the network in many ways, filtering out casual acquaintances or weak ties. Yet, one of the most famous results in social networks tells us weak ties are

---

[2] For one, much of our food supply is at stake.
[3] An example of a bipartite network.

important to understand: job seekers often find employment opportunities not from their close, or strong ties, who likely have similar social circles and thus access to the same information, but weak ties [190]. The manual methods for surveying social networks were the backbone of social network research, but the scale of social network makes it difficult to gather sufficient volumes of data.

The rise of both computing and the Internet has changed studies of social networks. Survey data no longer need to be processed by hand but can be analyzed automatically with computers. More importantly, many new sources of data that do not require laborious surveys are available. Mobile phones are very popular devices, and billing records managed by telecommunications companies, tracking who-calls-whom, are a valuable source of communication interactions. And the advent in recent decades of online social networks, platforms where users sign in to share and consume information, give even richer sources of social network data.[4] The social ties are essentially collected automatically from building the friends and "followers" lists of users. Massive amounts of data are now available.

While online social networks are a boon for researchers, they also change the situation in difficult ways. Online social interactions can be quite different from interactions in the real world. A close tie online may not be a close tie in real life, and vice versa. The behaviors people display online may be quite different from real life, and vice versa. The set of users of an online platform may not be representative of all people, often favoring wealthier and more tech-adept people. It may not be possible to tell if one person is using different accounts, or if different accounts actually belong to the same person. Users of the platform may not even be people at all—bots, automated accounts, are quite common. Both measuring people, the nodes, and measuring social interactions, the links, can be fraught. Any inferences about the "real" social network that a researcher draws from the online data may not hold.

Beyond social networks, many other networks play roles in socioeconomic systems. Economic and financial systems are driven by networks at all scales. From the social side, we can study the social relationships between individuals associated with different companies. One way is to build a network where nodes represent company board members and people are connected if they serve on at least one board. Many strategic relationships between companies are associated with shared board members. Most companies are required to publish the memberships of their boards, making this data publicly available.

Another example of a socioeconomic network is a labor flow network. Here nodes consist of job seekers and companies while links exist when a job seeker is employed by a given company. This network evolves in time, tracking the movements between companies as people switch jobs, and can reveal interesting structure among different sectors of the larger labor market [358]. Although not publicly available, data for this network is now tracked by online employment platforms where companies post job openings and job seekers submit applications.

Lastly, the stock market and other financial industries are ripe for network analysis. Using data on worldwide trade, we can build a network between countries based on what the countries are trading and to whom, giving us network insights into global

---

[4] There can also be some serious privacy concerns with such data (Ch. 3).

trade. Networks can also be extracted from stock market data, where nodes represent traded companies and links exist between companies whose trading prices are, in some manner, heavily associated or correlated over time. (Extracting an underlying network from time series data is a very common task.) This network can reveal connections between companies when their stock prices move together. The banking sector provides even more opportunities for networks. A network between banks, for instance, tracking who holds assets in what, lets us study the stability or robustness of the banking sectors [47]. This became important when recovering from the 2008 financial crisis because it can inform, based on the risk of different asset portfolios, where and how far economic shocks can propagate.

2.3   Other fun networks

Networks arise in countless other contexts. Here are just a few possibilities.

One example is the flavor network (Fig. 1.4). Here nodes represent the ingredients that go into foods, and links exist between ingredients that share chemical compounds. Exploring the structure of this network and how it relates to recipes, sets of foods, may help us discover novel food pairings, new and under-explored recipes [6].

Another network comes from the design of electronic circuit boards and integrated circuits. Here nodes represent electrical components such as resistors and capacitors and links exist between nodes that are electrically connected. Circuit design uses such networks to calculate current flows, voltages, and such, but the network is also spatial in that it must be laid out on a circuit board, and laying out the components so that the electrical connections (the edges) are as short as possible is a challenging design problem. This design is made even more difficult when you realize that edges cannot cross, which may be impossible on a planar circuit board. To lay out such a circuit requires using multiple boards stacked in layers. But using many layers makes the device more expensive to build, and connections between layers are more costly than connections within a layer. This points us to designs that minimize the number of layers and maximize the number of edges within layers.

At the very opposite end of the size range from such microelectronics are infrastructure networks such as the power grid. For the power grid, nodes represent power generators and consumers while links represent electric transmission (load flow) between nodes, often in the form of high-voltage long-distance transmission lines. The modern power grid is exceptionally reliable, but blackouts do occur. The advent of renewable resources such as solar panels and wind farms promises to make controlling the grid more difficult, as these power sources are not controllable like coal or nuclear power plants.

In an entirely different domain, language and linguistics provides fertile ground for networks. Consider networks where the nodes are words. The thesaurus: edges denote words that are synonyms. Word association: directed edges denote words that people respond to when prompted by, "what's the first word you think of when I say the word X"? Word co-occurrence: edges denote words that appear next to one another in written documents. This last network has been a major data contributor when constructing *large language models*, machine learning models that can respond to and create convincingly natural written language.

On the subject of machine learning, *neural networks* have become the dominant method of making predictions. The nodes of a neural network represent areas where data (numbers) are aggregated (summed) and transformed using some type of (nonlinear) *activation function*, in analogy with the "integrate-and-fire" model of biological neurons. Links exist when the output of one node serves as an input to another. These inputs are often modified using a weighted sum, with the weights being parameters that we learn by "training" the network: passing data with known output through the network and examining the network's final output, learning algorithms can adjust the weights to guide the output to match known results. The overall organization of the network is called its *architecture*, and neural networks can be designed to solve many problems by the right choice of architecture. Neural networks can be studied using network science tools and neural networks can be used to study other networks, which we'll explore later in this book.

Of course, the sky is the limit when it comes to networks. Their ubiquity is yet another reason why they are such valuable, important objects of study.

2.4   *Focal Points*: networks used throughout this book

Let's pick some networks for our journey. They should be interesting, representative of certain domains and characteristics, and manageable. These focal networks will be referred to throughout the text, grounding our discussion of real-world issues and practices.

Data for each network is available online at cambridge.org/network-data. Later chapters will work through how to use and study these data.

Zachary's Karate Club

Almost no treatment of networks is complete without some reference to the famous Karate Club [505]. This small network was gathered through surveys by Wayne Zachary during the early 1970s. It captures members of a university martial arts club who interacted heavily outside the club, according to Zachary's data. What's interesting about this network, and what has driven its long-running popularity, is that the members of this club had a disagreement and split into two groups, one focused on the club president and the other focused on the club's karate instructor. These groups are visible in the network's structure prior to the split, making the network a test case for group identification methods.

Plant–pollinator network

Our second focal point is a plant–pollinator network [40]. Here the nodes fall into two groups: pollen-spreading organisms such as bees, and plants who are pollinated by those organisms. Links in the network connect only plant to pollinator, capturing field observations of that pollinator acting to pollinate that plant. This condition, where nodes fall into two groups and links exist only between—not within—the groups, is the definition of a **bipartite network**. Bipartite networks such as this one are often studied in ecology. One type of study is to examine differences in the network over time due to climate change, invasive species, and so forth. This particular network was collected from field observations conducted in Spain, and the data includes metadata: the species names associated with each node in the network.

Developer collaboration network

This focal point is a network representing software developers contributing to open source projects hosted by IBM on the GitHub online development platform [27]. Nodes represent developers (identified by their GitHub usernames) and links connect developers who have edited one or more source code files in common, a simple measure of collaboration. We treat the network as weighted by associating with each link a weight counting the number of files commonly edited by the two developers. This makes the network a "projection"[5] of a bipartite network between developers and source code files. This network is also dynamic, evolving over the years 2013–2017.

Flavor network

Mentioned before, this network is derived from a reference text describing what flavor molecules are present in different food ingredients [6]. Food chemists use this reference when devising new flavor additives. But we can use the network to understand better the quality and nature of different recipes (combinations of food ingredients). While cooking is a highly multidimensional process, with preparatory steps, cooking temperature, aroma, and other factors playing important roles in taste, these flavor molecules provide a quantitative starting point to understanding flavor. Indeed, the *pairing hypothesis* states that foods that share many flavor molecules are more likely to taste well together than foods that share few molecules. Testing this hypothesis using a large set of recipes, Ahn et al. [6] found that indeed the hypothesis holds, but more for Western cuisine. East Asian cuisines tend to avoid pairing foods that share compounds. With these data, network analysis can help drive the study of systems gastronomy.

Human Reference Interactome

Our next focal point is HuRI: the Human Reference Interactome [283]. Here nodes in the network represent proteins and links exist between proteins that were observed to interact, according to high-throughout assay experiments. HuRI is the result of a decades-long effort to map out the human *proteome*, the interaction network of human proteins. At the time of this writing, HuRI is the most complete protein–protein interaction (PPI) network to date. Nodes in the network are represented by standardized IDs. A researcher interested in these data can enrich their study with node *metadata*, in

---

[5] A *projection* of a bipartite network is one where two nodes in the projected network were connected to the same node in the bipartite graph, that common node being absent in the projection. A bipartite network can be projected onto either "side," either set of nodes. For example, a network of film actors where two actors are connected if they costarred in any movies is the projection of an actor–movie network onto the actors.

this case using standard GENCODE gene annotations, a "controlled vocabulary" that biologists use to describe information about the protein.

Malawi Sociometer Network

This network came from a study that asked individuals in a village in Malawi to wear small proximity sensors on their chests as they went about their day-to-day business [353]. These proximity sensors can detect and record the presence of other sensors worn by study participants. Tracking what sensors are near one another and when leads to a contact network between participants that changes over time. Here nodes in the network are study participants (sensor wearers) and a link is noted when the two corresponding sensors have detected one another in close proximity. We treat this network both as a static and a dynamic network (Ch. 15), with the static network made by summing the number of contacts observed between participants over all time. (In other words, an edge in the static network represents the total number of contacts between two individuals.) This focal point also illustrates how gathering and studying network data gives rise to *ethical concerns* (Ch. 3): the authors of the original study took care to acquire informed consent from study participants.

 We will use a "bolt" symbol (⚡) in the margin when discussing a focal network.

2.5 Summary

All fields of science benefit from gathering and analyzing network data. This chapter has summarized only a small portion of the ways networks are found in research fields thanks to increasing volumes of data and the computing resources needed to work with that data. Epidemiology, dynamical systems, materials science, and many more fields than we can discuss here, use networks and network data. We'll encounter many more examples during the rest of this book.

Bibliographic remarks

Networks pervade biology. For a influential review in the context of cell biology, see Barabási and Oltvai [37]. In the area of neuroscience, readers may be interested in Bassett and Sporns [44] for a review of network neuroscience, or the more expansive *Networks of the brain* [442]. For those interested in ecological studies, consider Pascual and Dunne [361], Proulx et al. [380], and Bascompte [41].

Networks have been a part of sociology from the very beginning, dating all the way back to Jacob Moreno's pioneering work [317]. In many ways, the standard text for social network analysis remains Wasserman and Faust [485]. With the rise of the Internet and new data sources, sociology has kept up, with the new field of computational social science arising [264]. For a exciting general audience overview of social science and these new data, consider Salganik [412].

Readers interested in other areas may wish to consult Ahn et al. [6] for the flavor network study; Baker [31] or the now classic work of Mead and Conway [303] for an overview of circuit board design, known as VLSI (very large scale integration); Chu and Iu [105] for a review of the power grid (and the "smart grid") from a network perspective; or Cong and Liu [114] for a review of human language as a network.

Exercises

2.1 Collecting data on networks is costly, which was especially limiting before computers and computerized data collection. Suppose you are surveying a group of 100 students to learn about the social network of their school. It costs $X = \$10$ to interview each student, during which you ask them to list their 10 closest friends. Later, it costs, $Y = \$2$ to validate each reported social link.

    (a) How much will it cost to collect and validate the data? Do interviews or link validations contribute more to the total survey cost?

    (b) One student may list another as a friend but the other student may disagree, leading to a social link that is not *reciprocated*. If a link only needs to be checked once, regardless of whether student i listed j as a tie or j listed i, how will the survey's cost change based on how often friendships are reciprocated?

2.2 (**Focal network**) The flavor network captures whether food ingredients share chemical compounds. We could also define a network based on recipes, where nodes represent recipes and links exist between recipes that have common ingredients. While the flavor network itself is not *multilayer* (Sec. 1.4), if a recipe network were brought in, we could think of it as such.

    (a) How can we connect the layers together, meaning how can we place links from nodes in one network to nodes in the other?

    (b) More generally, would a combined flavor–recipe network be worth studying? Speculate on some ways the second "layer" of the network may relate to the first. What scientific questions can we investigate with this combined network?

2.3 (**Focal network**) The plant–pollinator network is a bipartite network. The developer collaboration network *comes from* a bipartite network. That the two networks share similarities in how they are defined, despite coming from entirely different research areas, is intriguing. Speculate on some similarities and differences between the two networks, think of ways to compare them directly, and describe some hypotheses that may come from drawing a kind of "analogy," broadly speaking, between one network and the other.

Chapter 3

Data ethics

Scientists must be ethical and conscientious, always. Data bring with them much promise to improve our understanding of the world around us, and improve our lives within it. But there are risks as well. Scientists must understand the potential harms of their work, and follow norms and standards of conduct to mitigate those concerns. But network data are different. As we discuss in this chapter, network data are some of the most important but also most sensitive data. Before we dive into the data and methods in later chapters, here we discuss the ethics of data science in general and network data in specific.

3.1   Introduction to data ethics

When working with network data, we must be keenly aware of a multitude of ethical issues. Although complex ethical issues can exist for *any* data, network data poses additional challenges because individual data points are not isolated from each other. So let's walk through the common challenges as well as more network-specific issues.[1]

The very first thing we need to understand is that a dataset may contain deeply private data and the privacy may be due to the relationships contained in the data. How would you feel if all your social media activities were sold and shared? How would you feel if your personal genome or health records were accidentally published? When dealing with data about social relationships and activities, privacy issues become even more thorny because it is not only about the information and people captured in the dataset, but it extends even to the information or people who are *not in the dataset*. For instance, capturing communication activities of my friends may well include lots of information about myself. So what are the important privacy risks? What should we do to mitigate those risks?

Another important fact is that *a dataset is never an object reality*. Because it is not possible to capture *everything*, data collection always forces choices of what to

---

[1] If you have not done so already, we strongly encourage you to familiarize yourself with human subjects research ethics and practices, such as informed consent and institutional review boards (IRB).

collect, what to ignore, and how. Such choices are often unintentional but their knock-on effects can introduce biases and harms. For instance, consider genetic data. Because the data collection is primarily done, particularly during the early days of genetics, by elite research universities and prestigious hospitals that are engaging in cutting-edge research, the samples tend to come from *rich areas of rich countries*, making the composition of the sample largely focused on those with European ancestry [241, 434]. What could be the implications of such biased omissions and inclusions?

Finally, ethical considerations extend beyond what is in the data itself to include how the data is processed, used, shared, and published. Data can be misused in many ways, ranging from carelessness to outright research misconduct. The misuse can start simply from carelessness—not being aware of the inclusion bias, privacy risks, and other ethical issues. Even without any bad intentions, such carelessness still can produce social harm. For instance, not fully understanding the data may lead to erroneous analysis and biased conclusions, which can then lead to problematic social policies or medical practices that have lasting impacts on society. We also hear about data breaches all the time; it is not trivial to protect the privacy of people captured in the dataset. Finally, data can also be maliciously manipulated—for example, due to the pressure towards "better" or "cleaner" results in academia—by researchers. What should we do to prevent such misuses, errors, and misconducts?

3.2   Biases in the dataset

Inclusion bias

Creating a dataset requires making choices and those choices can introduce systematic biases. At the first glance, this may not sound too bad. Why is it a problem to have fewer people from a certain ethnic group or gender? Datasets always capture only a small sample of the population anyway, right?

Here is the problem: because models, insights, and policies are derived from data, and if the data do not represent a certain population, those models, insights, and policies will not represent the population either. Here are some stories.

Did you know that when drugs are withdrawn from the market due to health risks, those health risks tend to be greater for women [209]? Although it is difficult to pinpoint the exact reasons why this is the case, an important context, and a likely culprit, is that there has been a severe sex-inclusion bias in biomedical research. Although there are numerous sex-based differences from cellular to physiological functions, testing both sexes has not been a consistent practice. If experiments are performed only on male cells, male animals, and male humans, it is only natural to expect that we cannot exactly know how a drug will work for female bodies. Although this issue was recognized and policies have been implemented requiring researchers to include both sexes in their studies, the practice is still far from perfect [449].

Another example is facial recognition. A team of researchers found that commercial facial recognition engines are terrible at detecting faces of Black people [83]. In a stirring presentation, the lead author Joy Buolamwini showed a scene where her face was not recognized by a computer *until she puts on a white mask*. This is again rooted in the inclusion bias of the dataset. Facial recognition models are trained on large databases of human faces. Because it is much easier to find portrait photos of those who live in rich countries (Who have more cameras? Who have more websites?), these image datasets tend to include more "white" images than other ethnic groups, particularly Black people.[2] Guess what kinds of faces are well represented in such databases? Guess what kinds of faces are most accurately detected by the existing facial recognition models? What will happen if self-driving cars, drones, or other machines cannot properly recognize Black human faces as people?

Let's return to the example of genetic data. Inclusion bias in genetic data goes back to the most fundamental genomic data—the human genome. After the Human Genome Project was completed, some researchers and doctors began to discover that they couldn't match some of their patients' (especially those without European ancestry) genetic sequences to the reference genome, which hampered their diagnoses and treatments. Why did it happen? Well, you know the answer: biased sampling.

The African population is collectively much more diverse than any other sub-population and this biased sampling of human subjects ended up creating a systematic bias in the reference human genome. A study examined 910 people and found that 300 million letters of DNA were missing in the reference genome [426]. Using one or a few individuals as genetic reference for the genome simply cannot represent the breadth of the entire human population.

We see the same pattern over and over again across domains. In addition to the cases of computer vision (facial recognition) and genomics, natural language processing (NLP) and understanding (NLU) have been criticized for inclusion bias as well [50]. Because the web is the primary source of natural language examples, the training data is heavily biased towards the dominant web platforms, languages, and populations who control the web—the same bias again.

In fact, from psychology, we already have a nice name for these sample biases: "WEIRD"—"Western, Educated, Industrialized, Rich, and Democratic." This acronym was coined to capture the most common biases in psychological research. Because a huge fraction of psychological research has been conducted in the universities of Western countries, the easiest subject to recruit for university researchers are students in those same universities, who tend to be WEIRD. In other words, a lot of our "understanding" of human psychology may be about the human psychology of the "WEIRD" population, which may or may not generalize to the rest of the world.

Any algorithms or methods trained on biased data will also be biased. Increasing the size of the dataset does not necessarily solve the issue as long as the source of the data is already biased. For instance, let's say we are training a machine learning model based on various personalized genetic and biological interaction data. Just like a facial recognition model failing on Black faces that were not in the training data, our precision health model will not know what to do with the types of data that were not in the training dataset. This will likely lead to worse performance on the underrepresented population in the dataset, which will directly lead to ineffective or even dangerous results.

---

[2] Even basic photographic technology has a historical bias against dark skin tones [272]!

Data reflect systematic biases in reality

Suppose our dataset does not suffer from any inclusion bias (not realistic). Is it then free from ethical issues? Unfortunately, the answer is still no. Simply reflecting reality is still problematic because society suffers from systematic biases. If a machine learning model learns and reflects exactly what is in society, it is completely natural to see the same societal biases emerging out of the model, *even if the data does not have inclusion bias*. For instance, it was discovered that translation services such as older versions of Google Translate produced highly stereotypical results when asked to translate sentences that involve gender stereotypes. In one case, when asked to translate the ungendered Turkish phrases "O bir doktor. O bir hemşire" (He/She is a doctor. He/She is a nurse) into English, Google happily transformed this un-gendered sentence into a gendered sentence: "He is a doctor. She is a nurse."[3]

One may argue that this is not a biased model because it simply reflects reality. The problem is that a biased model can further strengthen societal biases and stereotypes. Furthermore, even a small bias can be amplified by the model; if a model is set up to return the "best" answer, even a slight tilt (e.g., 51% male vs. 49% female) can lead to the case where the model returns "male" every time!

Another issue is about biased measurements. Even without inclusion bias, the measurement itself can be biased due to systematic biases that we have in our society. A recent study demonstrated that a widely used, commercial healthcare decision-making algorithm exhibits a serious bias against Black populations [351]. This algorithm is trained using the *amount of healthcare cost* that a patient will incur as the outcome measure. This sounds reasonable because the total healthcare cost should reflect how bad a patient's health is. If they are healthy, they don't need to incur much cost; if they are seriously ill, it will naturally involve more visits, tests, drugs, and procedures, which will lead to a higher cost. The problem is that healthcare exhibits systematic biases against Black patients. Black patients have been historically marginalized, abused, and neglected by the healthcare system in the USA and that means they may incur, ironically, lower cost—by not being properly treated—than white patients given the same condition. And this is exactly what the algorithm picked up. Given a similar condition (how sick a patient is), the algorithm produced lower risk scores for Black patients (because they are likely to incur lower cost in the future), and thus recommended weaker interventions and treatments. When this type of biased algorithm is widely used (it *was* already), it will exacerbate the existing systematic bias.

Thinking about biases in network data

But you may ask, "but these examples are not about network data, aren't they?" Well, although they may not seem directly related to network data, remember that many network datasets are built on primary data such as genomic data, social data, and so on. Any bias in the base data can creep into the network data. Even worse, we may overlook the serious biases because they are not obviously visible in the network data. Once we

---

[3] Google's translation service has since improved its handling of gender-specific alternatives. Other machine translation services still reflect this bias.

get the "abstract" network data, we often pay little or no attention to how the network was obtained.

Working on biological networks? Any biological networks that we construct and analyze rely on, in one way or another, the human genome, or some biological samples that are quite likely biased. Are you studying protein–protein interaction networks? From whom were those interactions captured? What were the biases in the subjects who donated their blood or tissues? If they are heavily biased, can you guarantee that your methods safely generalize to other underrepresented ethnic groups? How about brain networks? Who were the subjects that your data were built on? Are you making claims about human social networks? What kinds of data are you basing your claims upon? Are the networks that you are considering heavily biased towards a specific population of European–North American subjects? Is your data coming from only the "WEIRD" population? What can you say about the rest of the world?

What should we do?

Unfortunately there is no easy solution to these issues. Although these biases can have a profound impact on subsequent analyses, it is not always easy to mitigate—or even detect—them. Furthermore, it has not been a common practice to examine, investigate, and disclose potential biases when a dataset is being collected and released. However, there is growing consensus and systematic efforts to mitigate biases in the data.

One important initiative, and a critical step, is about better documentation (see also Ch. 18). For instance, establishing a norm to always have a *datasheet* that explicitly details crucial information about the data and data collection process, including the potential biases in the data [177], will be a meaningful step. Completing a formal statement on the biases in the data, ideally from the very beginning when the data collection is planned, can help researchers to recognize biases and mitigate them, or at least let users be aware of the biases in the data.

Another important movement is the call for methods and policies for *algorithmic audits*. Algorithmic audits are the practice of systematically probing for biases in the data (and algorithms that use those data) just like auditing financial books. Good examples are the stories mentioned above, such as the facial recognition case and the case of algorithmic bias in healthcare [351]. Although an audit is not guaranteed to identify problems in the data or algorithms, it can ensure better transparency and accountability.

Finally, the development of models that can de-bias or mitigate existing biases is another active area of research [429, 16, 457].

In sum, we need to think about data bias all throughout the life cycle of datasets and their usage. It should start from careful planning that considers potential biases drive by data collection—for example, we need to ask, "what kinds of biases are we introducing by recruiting people locally from our town or university?" Further questions should be asked about measurements. Does this measurement reflect systematic biases in our society? These considerations should then be clearly documented in a datasheet that accompanies the dataset. Even if data collection is imperfect (which it will be for most cases), clearly documenting and communicating why and how can prevent downstream misuses and misinterpretations of the dataset. We should also ensure that any models trained on the dataset should provide, especially when used for important decision making, transparency and interpretability, by using more interpretable models, by allowing independent external audits by third parties, and by publishing necessary details of the models. These steps can only be realized when we are keenly aware of the mechanisms and implications of the aforementioned ethical issues.

3.3   Privacy and surveillance

Some call the era we are living in "the age of surveillance capitalism" [512], where personal data is obsessively collected, commercialized, weaponized. Unfortunately, network data is at the heart of these practices. Let us talk about the issues of privacy and surveillance regarding network data.

What is special about network data?

Just like other data, network data poses privacy risks. But network data can pose even more risks because of the *connections* that it contain. For instance, imagine a dataset that contains the network of sexual relationships, which was collected to study the sexual behavior of people as well as the spread of sexually transmitted infections. If you are one of the "nodes" in the dataset, would you be OK if your identity was revealed? Probably not. OK, assume that your identity will *not* be directly revealed. Are you OK with the information of other people being revealed? How about the identity of those who are *connected to you* in this network? It can be pretty easy to identify you, once we know the other nodes that are connected to you.

As you can see in this example, in the case of network data, privacy is not just about *yourself*. Due to the connected nature of the network data, it is also about the data of other people, especially those who are connected with you, in the network. Even if *you* did not agree to share your "data" about your past sexual relationships, your previous partners may be willing to share, disclosing highly sensitive, revealing information about you.

Public information should be OK, right?

You may say this is an extreme example that deals with intimate personal information. How about public information? Are you OK with sharing all of your information that is "public"? If you say yes, you may want to rethink your answer after reading about a study conducted on Facebook's public "like" data. Researchers found that, if they have access to a large set of *public* data about what people *liked* on Facebook, they can reliably predict all kinds of *private* attributes of people in the data, including their sexual orientation [251].

How about some stupid, sarcastic tweets that you wrote many years ago and completely forgot about? Are you sure that you have *never* offended anyone—who were *wrong* of course—on the Internet? Even if the information is *in principle* public, it carries a different weight when aggregated into a dataset that can be systematically analyzed, searched, and scrutinized. Furthermore, once combined with other data, even public information can have a devastating power to ruin one's life. Digital footprints, albeit public in principle, can haunt us after many years.

Limits of anonymization

Then, you may say, "OK, but it is fine as long as the data is anonymized, right?" Unfortunately, it has been demonstrated that safe anonymization is extremely challenging when social data is involved. Here is a story. The IJCNN (The International Joint Conference on Neural Networks) set up a challenge using a social network dataset, by partnering with Kaggle, a well-known platform for machine learning competitions. The challenge was to predict missing links (Ch. 10) in a social network of users of the Flickr online photo-sharing service. The data were prepared by *anonymizing* it, stripping away all user identity information, and then removing a fraction of the links between users which competitors then tried to predict.

Can you guess how the competition was won?

The winning solution *did not perform "link prediction"* per se [327]. But then how could they accurately predict links without predicting links? Instead of predicting links, they crawled the Flickr social network data *themselves* and then matched it to the competition data, effectively *deanonymizing the competition data.* They used a common property of real networks—a highly heterogeneous degree distribution. Because there are only a few nodes with large degree, they are fairly unique and identifiable. Once you match those *hub* nodes, they can act as reference points to match other nodes. In doing so, the winners could reliably match the networks even without any identifying information, and from there simply *identify links missing from the competition data* that appeared in the network that they had crawled. You don't need to predict the links if you already know them!

This "solution" was an important lesson on just how easy it is to *de-anonymize* social network data. Even without any identifying information attached, it is still possible (and fairly easy) to recover the identity of the users in the dataset if you have access to the full dataset. That means, if a network dataset with sensitive, private information is shared, *even if it is anonymized*, it may still be possible to reidentify the users in the data by cross-referencing publicly available data and then linking private information to the identified users.

Also, as mentioned before, when the data is about the social network, it is not always possible or easy for an individual to opt-out from the data release. An interesting example is the search for Saddam Hussein after the 2003 invasion of Iraq. US military intelligence collected detailed data about the social network around Hussein, to identify the most likely connections that he might have among his kins and close allies, directly contributing to his capture in December 2003 [390]. This case illustrates the power and danger of social network data and analysis. Even without any direct consent from you, the data *surrounding* you can be collected and leveraged against you.

Major features of social networks make it difficult to anonymize any social network data. First, as mentioned, the network around us is not *homogeneous*. Some people have far more social connections than others. This heterogeneity makes a social network extremely easy to "identify." Second, social networks exhibits strong *homophily*, the phenomenon that people tend to share attributes with or be similar to their social connections. This is due to a host of mechanisms, from genetic inheritance to social contagion and population sorting. In other words, people around you may share similar genes, become similar with you because you talk with them, or you may be close to them because you are similar to each other. All of this contributes to privacy or the lack of it: if someone knows a lot about your friends, they also can infer a lot about you.

Surveillance capitalism

These privacy issues are difficult to regulate and mitigate. Companies, with strong pressure to squeeze out more profits, will collect as much data as regulations allow. And the more individual, private information the company has, the more money they can make from it, because they can more precisely target the population. Similar pressure exists for governments. The need to provide citizens with safety and security in the face of threats, likely or not, will lead to more surveillance. With better technologies and online-based communication, governments around the world have increasingly stronger power to monitor their citizens. NSA's PRISM program may be the most famous example, although similar programs exist across many countries. These circumstances and systematic pressures are often captured by the term "surveillance capitalism" [512]. The more surveillance a company does, the better they understand and predict human behaviors, which leads to more ways to make money out of them.

Open science and privacy

One area where privacy issues directly collide with another principle is in scientific research. Faced with the widespread replication crisis and other problems such as scientific misconduct, scientists and funding bodies are compelled to push for open science, sharing data and methods publicly so that other researchers can more easily reproduce and build on published studies. But then what should we do with sensitive, private data? Sharing can pose a serious privacy risk; *not* sharing violates the best practice of open science.

This again, just like all other ethical issues, does not have an easy solution. It is impossible to have strict guidelines that can be applied to every case because each case may have completely different risks and benefits when sharing the data. When the risk of deanonymization and leaking sensitive data is great, we may have to forego the open science principle, for instance by only allowing restricted access for reproducing the results. On the other hand, if the risk is acceptable, it may be more important to stick to the open science principles as closely as possible.

What should we do?

In sum, we need to understand that—while any personal data can pose serious risks regarding privacy—network data may pose even more serious risks because the data is *connected with each other*. Network data may be much easier to de-identify and carries a lot of revealing information about those who are in the network.

In other words, when working with network data, we need to be considerate and careful about potential harm that can be inflicted on the people in the dataset. If the data is sensitive, we should, of course, be extremely careful about privacy leaks. But even if the data is *in principle* public, we should still think about the potential risks and potentially damaging implications of the dataset. Can someone lose their job because of this dataset and my data analysis? Can someone be publicly shamed or harmed by others because of this?

At the same time, we need to understand the broad societal context regarding social data collection and be cognizant about possible complex conflicts between individual privacy, the scientific value of research, the principles of open science, and so on. Many of these problems are difficult, having no clear answers, and call for thoughtful and nuanced approaches.

3.4 Mistakes, misconduct, and how to prevent them

Even if we address all the issues discussed above, ethical issues can still arise from the process of research and data analysis. Even with the best of intentions, researchers can always make mistakes and some can be highly damaging. And what if the researchers are the *baddies*? What if *you* are tempted to manipulate the data or exploit and distort the data to manufacture the conclusion that you want? How can we ensure that, even in the presence of bad actors, we collectively produce auditable and reproducible results free of ethical issues?

Types of misconduct

What are the types of misconduct and ethical issues that can arise during the research process? First of all, a researcher can flat out falsify data. Instead of following the normal procedures of obtaining data (e.g., surveys, web-crawling, etc.), one can potentially manufacture their own data so that it fits their preconceived conclusion. This is difficult to detect because, unlike the derived datasets where we can track its *provenance* (see Ch. 17), raw data do not have ways to check the provenance. That said, falsified data can still be detected through independent re-collection of data (and comparison) as well as discovering artifacts and anomalous patterns in the data—false data often exhibit artificial patterns. Nevertheless, falsifying the primary data is difficult to detect and can inflict lasting damages if the data gets used in many other research projects.

Perhaps worst of all, such dangers lurk at every step throughout the life cycle of data-driven research. Even if the raw data is legitimate, the processed data can still be falsified or messed up, and even with good data, faulty or misused analysis can produce erroneous results or fabrications. In fields with mostly computational research, this type of problem is easier to mitigate because others can replicate the analysis to identify problems. A more tricky issue is so-called p-hacking, sometimes known as inflation bias, which is a malpractice of trying out many different analysis (or even collecting more data) until finding a "significant" result [207]. All statistical results are suspect at best when p-hacking occurs.

And there are always errors. Throughout the process, even without any bad intentions, the researcher can be sloppy and produce erroneous results. Such mistakes can have terrible outcomes. One famous example is the so-called "Reinhart–Rogoff error" committed by Harvard Economists Carmen Reinhart and Kenneth Rogoff, in their paper that bolstered the arguments for austerity measures in EU countries [393]. Their results were wrong simply because they failed to select some cells in their data spreadsheet [68]—affecting the lives of millions.

Why does misconduct happen?

We can think about two primary drivers: the *incentive* and the *probability of getting caught*, and if we include sloppy errors, *bad research practices*. The higher the incentive to commit a misconduct and the lower the changes of being caught, the more likely a researcher may attempt it. In academia, there is a strong incentive for publication and prestige. Decisions for hiring, promotion, and tenure are all made primarily based on publication records. Publishing in the most prestigious journals—which are hungry for surprising and strong results—can guarantee not only *survival*, but fame and funding as well. Therefore, there is a substantial incentive for misconduct—falsifying the data, fabricating results, and so on. If you can simply make up your data or force your analysis to fit your predetermined story, it becomes *much* easier to write a paper, especially one with fascinating, surprising results.

Then there is the risk of being caught. If one does not have to share the data and code, it is difficult for someone else to identify the issues. For this reason, it is increasingly common for journals to require authors to publish their data and code alongside their publication.

In industry or other organizations, the nature of the incentive may be different but it is still present. Companies may have strong reasons to exaggerate the performance of their methods to attract investments or to sell their products. At the same time, because results tends to be directly applied to real-world applications in industrial setting, it may be easier to catch any issues or errors.

Finally, let's talk about mistakes. A lot of mistakes and errors occur due to bad research practices. For instance, using software like Excel makes researchers highly prone to simple errors [356, 511]. Not performing code reviews or other poor software engineering practices are also culprits that produce errors.

What should we do?

Trusting researchers is not a solution. We need to build systems and processes that assume any researcher involved can, sadly, be a bad actor, or simply a human being who is capable of making mistakes. We have to create robust safeguards by ensuring data provenance, replicability of results, and open science. As the whole process of analysis becomes more transparent, auditable, and replicable, there will be stronger deterrence for misconduct. Fortunately, norms and policies are moving towards this direction. Increasing numbers of journals require the publication of replication data and code so that other researchers can directly replicate the results. Even when not published with the paper, sharing code and data with other researchers is becoming a strong norm across many fields. And more software engineering best practices are being embraced by researchers (see Ch. 19). While *publish-or-perish* remains a strong incentive, things are starting to improve.

3.5 Summary

Here we reviewed ethical concerns surrounding network data and network analysis. The ethical issues that we discussed often do not have clear solutions but require thoughtful approaches and understanding complex contexts and difficult circumstances. It is critical to understand how biases creeps into datasets and what kinds of negative implications they can have. It is also critical to understand the trickiness of handling private data, especially in the network context. The first step to mitigate ethical issues working with data is to be clearly aware of them.

Then we need to promote and exercise the best practices. A dataset is not complete without detailed documentation about the data collection process, potential biases, and other contexts surrounding the dataset. We must pay close attention to the bias in the data as well as privacy issues. Carefully processing the data is also important for ensuring data provenance and the correctness of the analysis. Embrace best practices in software engineering to minimize errors in computerized analysis. Ensure replicability of the data processing and results.

Bibliographic remarks

All scientists should be ethical and responsible in how they conduct their research. For general introductions to ethical and responsible research, we recommend Resnik [394], Seebauer and Barry [422] and, especially for the social sciences, Israel and Hay [228].

Our increasingly data-rich, digital world creates new perils. For an engaging, general audience treatment of how data and data algorithms lead to negative societal outcomes, we highly recommend O'Neil [352].

Network data in particular introduce special concerns. Staiano et al. [443] explore ways to infer personality profiles of users through their social network data. Garcia [173] discuss how online platforms can build latent or "shadow" profiles of their users. Sarigol et al. [414] describe how privacy in a networked world ceases to be an individual choice but instead depends on the group. Bagrow et al. [30] demonstrate how actionable, predictive information can be extracted about individuals using only their social ties. The privacy implications of network data remain an active area of research.

Exercises

3.1 (**Focal network**) Consider the Malawi Sociometer Network (Sec. 2.4). The data underlying this network were gathered as part of an experiment where participants wore small devices that could detect who they were in contact with. This is an example of a *human subjects* research project.

   (a) Read the primary publication describing this dataset [353]. Summarize their discussion of the study ethics.

   (b) Are there other ethical concerns? Were you to conduct a similar study, what ethical concerns might you need to address?

3.2 A company wishes to hire you as chief data scientist. They seek to distribute a free, location-based smartphone game where players gain points by exploring their surroundings, promoting exercise and enjoying the outdoors. Their business model is to collect data from the phone sensors, such as Bluetooth and wireless Internet, allowing them to track users' whereabouts and what other devices connect with their phones, then resell this data to businesses seeking advertising profiles, shopper analytics, and more.

   (a) What ethical concerns are posed by this company's plan?

   (b) What options might the company have to address ethical concerns?

3.3 (**Advanced**) A social media provider falls victim to a cyberattack. Hackers leak user details, including email addresses and the usernames of their online friends/followers.

   (a) Describe some ways that users can be negatively affected by revealing this information.

   (b) Suppose at some earlier point the social media provider asked users to provide access to their address books, giving them a list of known contacts. Not all these contacts are users of the platform, giving the platform a new source of potential users. If *those* details are also found by the hackers, how might non-users be negatively affected by the leak?

   (c) Suppose the leak included information describing what devices were used by users to access their account. This information may be cross-referenced, allowing someone to "join together" a user's multiple accounts. Those users may have very important reasons for those separate accounts. Describe some ways users can be negatively affected if it becomes public knowledge that they have multiple accounts. How serious could this be compared to other concerns?

Chapter 4

Primer

Network data calls for—and is analyzed with—many computational and mathematical tools. One needs good working knowledge in programming, including *data structures and algorithms* to effectively analyze networks. In addition to graph theory (Ch. 1), *probability theory* is the foundation for any statistical modeling and data analysis. *Linear algebra* provides another foundation for network analysis and modeling because matrices are often the most natural way to represent graphs. Although this book assumes that readers are familiar with the basics of these topics, here we review the computational and mathematical concepts and notation that will be used throughout the book. You can use this chapter as a starting point for catching up on the basics, or as reference while delving into the book.

4.1 Coding and computation

Studying network data requires creating and using computer code. We assume the reader is familiar with writing code using basic programming concepts, including loops, conditional statements, and functions, as well as basic data structures and algorithms. Here we provide some additional background and notation.

ⓘ For vs. foreach.

One programming concept worth discussing is the *foreach loop*. A traditional for loop uses a numeric index variable that increments between a start and end value in constant steps. Here is a for loop written with a C-style code:

```
for (int i=0; i<10; i+=){
    [...]
}
```

The loop (or index) variable i begins at 0 and increases by 1 while it is less than 10.

By comparison, a foreach loop generalizes a for loop by defining a sequence of values, not necessarily a range of numbers, that are iterated over. For example:

```
A = ['a', 'b', 'c']
foreach x in A:
    [...]
done
```

Here A is a list of characters, and our variable x will be equal to 'a', then 'b', then 'c', as the loop progresses. In practice, foreach loops are very useful, as we can focus on what we are looping over and not have to deal with maintaining and using index variables.

The concepts are simple but confusion can set in when syntax is succinct: often a foreach loop is clear from the context and only "for" is written. Indeed, some programming languages, chiefly Python, use foreach loops exclusively. When first using such a language, think of the for as really being "foreach" when reading or writing code.

Working with files

Working with data computationally necessarily requires working with data files.[1] We assume familiarity with basic file read and write operations in your programming language of choice. Usually files are "opened" for reading and writing and a file "handle" is created that provides access to the file from your code. In Python, for instance, a file can be opened for writing with fout = open("filename", 'w') and we can use the fout *handle* to send data to the file:

```
values = ['a', 'b', 'c']
for s in values:
    fout.write(s)
    fout.write("\n")
fout.close()
```

In the example, we explicitly add newline characters ("\n") after writing each string s and we "closed" the file handle when finished. Python, and most modern languages, provide great functionality to handle such bookkeeping automatically.

Working with files also requires specifying the locations of those files. For that, *paths* are used, which specify where the file is located within your computer's file structure by describing the sequence of folders (directories) and sub-folders we would follow to find the file. Paths can be *global* (or *absolute*), specified from the very beginning of your file system, or *relative*, specified in relation to your program's *working directory*. Managing a working directory can be a chore for beginners, but by allowing your code to use relative paths, your code will be portable, and work without being changed on other computers with different file structures.

 Debugging tip: find out how to print your code's working directory and add this to the top of your script if you are struggling with file errors. In Python:

```
import os
print(os.getcwd())
```

---

[1] And databases too!

and in R:

```
print(getwd())
```

Algorithms and pseudocode

An *algorithm* is a sequence of instructions to solve a problem or perform a computational task. Many network calculations boil down to a combination of algorithms. For instance, the problem of calculating the shortest paths between two nodes in a network can be achieved by the *breadth-first search* (BFS) algorithm or *Dijkstra algorithm.*

To describe algorithms, we use *pseudocode*, which is a simplified notation of instructions that resemble spoken language.[2] Readers who are familiar with programming in any common programming language should be able to read pseudocode, regardless of what language they know. Algorithm 4.1 is pseudocode of the BFS algorithm that returns all nodes that can be reached along paths starting from node s.

Algorithm 4.1 Reachable nodes with breadth-first search (BFS). With a few modifications, this algorithm can also compute the distances (or shortest path lengths) between nodes in G.

1: Input: Graph $G = (V, E)$, starting node s
2: Define Q as a new queue with s as its only element
3: Define `visited` as a new set with s as its only element
4: **while** Q is not empty **do**
5:     current = Q.dequeue()   ▷ *Get a node to visit*
6:     **for** $u \in$ neighbor(current) **do**   ▷ *Get the neighbors*
7:         **if** $u \notin$ `visited` **then**
8:             `visited`.add(u)
9:             Q.enqueue(u)
10: Return `visited`

Notice that BFS as written in Alg. 4.1 depends on a *queue*, a data structure that makes it easy and fast to retrieve items (in this case nodes) in a particular order—algorithms and data structures go hand-in-hand. An alternative to BFS is called DFS—Depth-First Search. Based on the names and Alg. 4.1, can you judge what makes them different?

Computational complexity

Often there are multiple ways to solve a computational problem and the solutions (algorithms) can wildly vary in terms of their efficiency, which is concretely defined and studied as *computational complexity*. For instance, imagine that we are sorting an array of numbers. Among numerous sorting algorithms, one particularly interesting—and incredibly simple—algorithm is called "Bogosort," which can be expressed with just two lines of pseudocode (see Alg. 4.2).

---
[2] There is an old joke saying that "Python is executable pseudocode" given its syntax resembling pseudocode and natural language.

Algorithm 4.2 Bogosort

---

while not sorted(deck) **do**
    shuffle(deck)

If the shuffle[3] is random and all elements are distinct, given n total elements, the expected number of comparisons and swaps that should be performed (in the average case) roughly scales with $n!$. In other words, if we have an array with just one million items, we are looking at about $10^{5,565,709}$ operations. Even if a supercomputer can perform 10^{20} operations per second, we'll need to wait... for some time.

Yes, this is indeed a ridiculous example. But, if you are not equipped with at least a basic understanding of computational complexity, you *will* accidentally implement your own bogo-algorithms. Pretty much everyone who codes had the experience of waiting for a program to finish, realizing only later that it will not finish within their lifetime (or until the end of our solar system). The difference between an efficient algorithm and an inefficient one can determine—not just *how long* we should wait—but whether the computation is even *possible*.

Can't we just use faster programming languages or more powerful computers? This is a valid point but only to some extent. In most cases, the *fundamental* efficiency of the algorithm easily tops the power of the hardware and programming language.

Time complexity and space complexity The two major components of computational complexity are *time complexity* and *space complexity*. Time complexity is about how the amount of operations increases (scales) with the size of problem (e.g., the length of an array in a sorting problem or the number of nodes and edges in a network calculation). In other words, the question is: would we do 10 times more computation if our array becomes 10 times longer? Would we do 100 times more computation? Space complexity is similar, but about how the amount of memory scales with the size of the problem. There are often—not always—tradeoffs between these two; we can sometimes speed up a computation by putting more data into memory or save space by doing more computation.

Big-O Since an algorithm's performance matters most when the size of input is large, we usually focus on the limiting behavior of algorithms. (This also abstracts away pesky implementation details like how fast the CPU is and how efficient the programming language is.) These are conceptualized based on the "Big-O notation" that captures the limiting behavior of mathematical functions. As $x \to \infty$,

$$f(x) = O(g(x)) \tag{4.1}$$

if there exists a positive real number M and a real number x_0 such that

$$|f(x)| \leq Mg(x), \text{ for all } x \geq x_0. \tag{4.2}$$

---

[3] By "shuffle" we mean to randomly permute all the elements of a sequence.

For instance, if the time complexity of an algorithm, given the size of input n, can be written as $T(n) = 2n^2 + 5n + 100$, we say $T(n) = O(n^2)$ Because, for any $n > 1$,

$$2n^2 + 5n + 100 < 2n^2 + 5n^2 + 100n^2 = 107n^2, \tag{4.3}$$

where $M = 107$ and $g(n) = n^2$.

In most cases, the big-O complexity can be calculated by simply keeping only the fastest-growing term in the full complexity expression. Note that big-O notation specifies an upper bound, not a precise scaling relationship. In other words, if $f(n) = O(n)$, then we could also write $f(n) = O(n^2)$. However, in the context of algorithm analysis, big-O notation usually denotes the tighter bounds.

Another crucial consideration is that algorithms' performance can vary immensely based on the input. Even a terrible sorting algorithm may work very well if the input is already almost sorted. Therefore, it is critical to consider multiple scenarios, especially worst cases. It is customary to report both the average and worst case complexity when they differ. For instance, the worst-case time complexity of the Quicksort algorithm is $O(n^2)$ although its average-case time complexity is $O(n \log n)$.

A sneak peek into the complexity zoo $O(1)$ refers to the case where the algorithm does not depend on the size of input data at all. Whether it's $n = 10$ items or $n = 10{,}000{,}000$, an $O(1)$ algorithm returns its result within a constant time that does not scale with n. For instance, the operation of obtaining the size of an array or a set is usually $O(1)$ because the data structure usually keeps track of the number of items.

$O(N)$ algorithms tend to be those that traverse the data at least once. For instance, if we want to identify the maximum value of an unsorted array, we need to scan the entire array and examine every element at least once. $O(N^2)$ algorithms tend to require repeat traversals. For instance, in terms of time complexity, bubble sort is a well-known example of an $O(N^2)$ algorithm. Given N elements, we go through every item but the last one and compare that item to every subsequent item in the array; the operations scale as N^2. $O(\log N)$ algorithms tend to appear when the data are "nicely organized" and we can eliminate major portions of the data at every step of the computation. For instance, if we have an already sorted array, finding the location of a target number within the array is easy: we can recursively bisect the array. If the value at the middle of the array is, say, larger than the target value, then we can safely ignore the second half of the array because all those values will be larger than the target value. This bisection reduces the computation drastically and we need only the order of $O(\log N)$ computation. Having a sublinear algorithm for a problem is fantastic![4]

Data structures

Algorithms (and their complexity) are tightly coupled with *data structures*—the ways to computationally organize and manipulate collections of data. For instance, sorting algorithms like *heap sort* depend on a clever binary-tree data structure called a *heap* that allows us to access the smallest (or the largest) item in the collection in constant time.

---

[4] On the other hand, an exponential or even *nondetermistic polynomial* (NP) algorithm is very challenging. This is the famous P *vs.* NP question.

Let's describe—mathematically and as "computational objects"—some of the common data structures that we'll rely on throughout this book.

Arrays and lists

In the abstract sense, an *array* or a *list* is a sequence of items, which do not have to be unique, usually denoted with square brackets: $[a, b, \ldots]$. An *array* usually refers to the most primitive type of lists that stores the same type of data in a consecutive block of memory. An array can be indexed; we can obtain the first item or 100th item (e.g., a[0] and a[100]) in $O(1)$. The size of an array is usually fixed when we allocate it. It is not possible to quickly find out whether an item is in the array or not. To do so, either we should scan the whole array (unsorted) or otherwise do a search (e.g., a binary search on a sorted array).

Lists usually refer to a data structure that can be dynamically lengthened, shrunk, or modified. A *linked list* is a data structure constructed by creating a chain of items that each point to the next (and/or previous) item in the list. We can traverse the linked list sequentially by following these *pointers*. Items can be easily added or removed from the list by modifying these pointers. However, a linked list does not allow us to access an item by its index.

The number of items in a list is its *length*. Lists can be homogeneous (every item in the list is the same type) or inhomogeneous, containing multiple types of items. A list can even contain elements that are themselves lists, making a *list-of-lists*.

Tuples, which are usually denoted with parentheses: (x, y, z), are similar to lists. Like a list, a tuple is still a sequentially ordered list of items. Usually the distinction is whether we can change (list) or not (tuple) the data. Some computer languages implement list and tuple data structures slightly differently; lists in Python, for instance, are mutable, while tuples are immutable.

> **i** A *mutable* data structure (or variable) is one whose contents can change after it has been created while an *immutable* data structure cannot be changed once it is created. Using an immutable data structure guarantees that it stays the same once initialized. We can pass that data structure to a function and be sure the function won't have any side effects on the data. While immutable data limits the computational tasks and algorithms that can be pursued, it is safer and, when changes aren't needed, good practice to make data immutable.

Dictionaries

The next data structure worth considering, and one that is used throughout most computer code that works with networks, is the *dictionary*. Dictionaries or "dicts," also called associative arrays, maps, or hashes,[5] allow us to efficiently map *keys* to *values*.

Dictionaries typically require the keys to be unique, distinct from one another. The values need not be unique. The cardinality, size, or length of a dictionary is the number of keys it contains. Dictionaries are denoted with curly braces: $\{k_1 : v_1, k_2 : v_2, \ldots\}$ by

---

[5] You can tell something is important when it has a lot of names!

writing each element of the dictionary in the form key:value, using a colon (:) to denote the mapping from key to value. You may on occasion see a dictionary's key–value pair written equivalently as a tuple: (key, value).

Dictionaries are especially powerful data structures. We can efficiently determine whether a key is a member of a dictionary. (Usually this requires immutable keys.) This means we can efficiently retrieve values when we have a key, giving us a lookup table that works at (approximately) the same speed regardless of how long it is. A use case with networks is defining a dictionary where the nodes in the network are the keys and the values are the sets of neighbors (Ch. 8). Then we can quickly retrieve any node's neighbors regardless of how big the network becomes.

Sets

Once we have a dictionary, we can also implement a *set* (Sec. 4.2.1), which can be considered (and sometimes implemented as) a dictionary with the item as the key.[6] Unlike a list where $[a, b] \neq [b, a]$, a set is unordered, $\{a, b\} = \{b, a\}$, and its elements are unique. The usual implementations of sets guarantee low time complexity ($O(1)$ or $O(\log N)$) operations for item lookup (by using a hashtable, sorted list, or tree-based data structure) and addition and deletion of items. But it does not allow indexing.

Sets are usually mutable as we can insert or remove elements from them but implementations of sets typically forbid using mutable objects as elements because data structures work by identifying each element based on its contents and, if those contents change, the program will lose track of the element. Not only does forbidding mutable set elements avoid these errors, it allows for data structures to work very efficiently, meaning that we can tell if a given item is a member of a set without looking at every element in the set first (i.e., fast item lookup). Regardless of how big the set is, testing for membership requires the same amount of work,[7] This efficient test for set membership is very powerful, and complicated (network) algorithms can often be expressed very easily with set data structures.

4.2 Mathematics

Here we describe some notation we'll use throughout the book and discuss various concepts from linear algebra and probability, both of which are central to studies of network data.

4.2.1 Sets and other notation

A *set* is an unordered collection of unique elements, which can be any mathematical objects, including other sets.[8] A set that contains no elements is called the *empty set*,

---

[6] The value can be `true`, 1, or anything that indicates the existence; the value does not matter much when we use a dictionary as a set.

[7] Strictly speaking this is only approximately true. Sometimes, depending on what algorithm is used to implement the set, there may be more computations needed for bigger sets than smaller sets, but it is usually less than having to look at every element.

[8] As a mathematical object, this is different than the set data structure we just discussed. Some set data structures, depending on how they are implemented, cannot support sets as members of sets.

denoted $\emptyset$ or sometimes $\{\}$. The number of elements in a set is the size or *cardinality* of the set; for a set s, its size is denoted $|s|$. We denote a set when listing out its elements using curly braces: $\{a, b, c\}$ or $\{a_i\}_i^n := \{a_1, a_2, \ldots, a_n\}$. An element x that belongs to a set A is denoted by $x \in A$; likewise, x not in A is denoted $x \notin A$. The union $A \cup B$ of two sets A and B is the set of elements that appear in either or both of the sets. Conversely, the intersection $A \cap B$ is the set of elements that appear in both A and B. The difference $A - B$ is the set of elements in A that are not in B, and likewise, $B - A$ is the set of elements in B that are not in A. One set can be a *subset* of another set: $A \subset B$ means that every element in A is also in B. Conversely, $A \supset B$ means that A is a *superset* of B: every element in B is also in A. A is a *proper subset* of B if and only if A is a subset of B and B contains at least one element not in A, that is, $B - A \neq \emptyset$.

Some sets of *numbers* are common enough that standard symbols are used to denote them. These include the set of integers $\mathbb{Z}$, the set of real numbers $\mathbb{R}$, and the set of complex numbers $\mathbb{C}$. (Note that these sets are infinite and each is a superset of the one that came before: $\mathbb{Z} \subset \mathbb{R} \subset \mathbb{C}$.) Another convention is to use superscripts to denote positive and negative numbers: $\mathbb{R}^+$ represents the positive real numbers, for example.

How can we express sets that have complex definitions? For example, the set of all integers between 0 and 100 that are divisible by 4? We can write this out element-by-element, of course, but the following notation is common:

$$\{a \mid a \in \mathbb{Z}, 0 \leq a \leq 100, a = 0 \bmod 4\}. \tag{4.4}$$

You can read this kind of expression as "the set of all a such that [these conditions hold]". In this case, we'd read Eq. (4.4) as "the set of all values of a such that" ("$|$") "a is an integer" ($a \in \mathbb{Z}$), "a is between 0 and 100 inclusive" ($0 \leq a \leq 100$) and "a is divisible by 4" ($a = 0 \bmod 4$). This notation is called *set-builder notation* or just *set notation*.

On occasion, we will distinguish identities or definitions from statements of equality by using "$:=$" instead of "$=$" ("$\equiv$" is another popular alternative).

4.2.2   Linear algebra

Linear algebra provides a powerful language to organize, represent, and manipulate collections of numbers and variables. As network data are all about collections of nodes and edges, it is natural to use the language of linear algebra.

Basic definitions   A *scalar* x is an individual number. A d-dimensional *vector* $\mathbf{x}$ is a collection of scalars x_i, $i = 1, \ldots, d$, where x_i is the ith *element* of $\mathbf{x}$. (We use boldface variables to distinguish vectors and matrices from scalars.) Vectors of numbers of the same length can be added, subtracted, and multiplied, with a variety of definitions for scalar and vector multiplication. For instance, the *dot product* between two vectors is the sum of the products of their elements: $\mathbf{a} \cdot \mathbf{b} = \sum_{i=1}^{d} a_i b_i$.

A matrix $\mathbf{X}$ of size $n \times m$ is a collection of scalars arranged into an array or grid of n rows and m columns. The elements of a matrix $\mathbf{X}$, typically written[9] X_{ij} and

---

[9] It is sometimes common to use an uppercase letter for the matrix and the corresponding lowercase letter for an element of that matrix. We dispense with that formality.

identified by a pair of indices, represent the row and column of the element: X_{ij} is the element located in row i and column j. A *square matrix* is one where $n = m$; otherwise, the matrix is *rectangular*. Vectors can also be distinguished as row vectors or column vectors. A *row vector* is then a $1 \times d$ matrix and a *column vector* is a $d \times 1$ matrix. Given row and column vectors, a matrix of size $n \times m$ can also be considered as a collection of row vectors of length m arranged in n rows or a collection of column vectors of length n arranged in m columns. Sometimes we need to *transpose* a matrix: $\mathbf{A}^\mathsf{T}$ is the transpose of $\mathbf{A}$, formed by swapping rows and columns, or $A_{ij}^\mathsf{T} = A_{ji}$. Transposing a column vector will create a row vector, and vice versa, which we can use to represent a dot product: $\mathbf{a} \cdot \mathbf{b} = \mathbf{a}^\mathsf{T}\mathbf{b}$ using matrix multiplication (described shortly).

Matrices of numbers are endowed with a variety of mathematical operations. Two matrices $\mathbf{A}$ and $\mathbf{B}$ can be added or subtracted by adding or subtracting their elements element-wise: $[\mathbf{A} \pm \mathbf{B}]_{ij} = A_{ij} \pm B_{ij}$. This requires $\mathbf{A}$ and $\mathbf{B}$ to be the same size (have the same numbers of rows and columns). A matrix can be multiplied by a scalar along the same lines: $[c\mathbf{A}]_{ij} = cA_{ij}$.

Lastly and most importantly is matrix multiplication. A matrix $\mathbf{C} = \mathbf{AB}$ is defined as a collection of dot products between the rows of $\mathbf{A}$ and the columns of $\mathbf{B}$: the element $C_{ij} = \sum_{k=1}^{n} A_{ik}B_{kj}$ of $\mathbf{C}$ is the dot product between the ith row of $\mathbf{A}$ and the jth column of $\mathbf{B}$. This definition has a variety of consequences and is fundamental to all areas of science, engineering, and mathematics. It requires the matrices to be compatible sizes; if $\mathbf{A}$ has n columns, $\mathbf{B}$ should have n rows.

Eigenvalues and eigenvectors For an $n \times n$ square matrix $\mathbf{A}$, when an $n \times 1$ vector $\mathbf{v}$ and scalar λ satisfies the following equation:

$$\mathbf{Av} = \lambda\mathbf{v},$$

we call $\mathbf{v}$ an *eigenvector* of $\mathbf{A}$ and λ an associated *eigenvalue*. When we consider $\mathbf{A}$ as a linear transformation, its eigenvectors are the vectors that do not change direction under this transformation. Eigenvectors and eigenvalues play a role when diagonalizing a matrix. If a square matrix $\mathbf{A}$ is diagonalizable, we can write it as $\mathbf{A} = \mathbf{VDV}^{-1}$, where the columns of $\mathbf{V}$ are the eigenvectors of $\mathbf{A}$ and $\mathbf{D}$ is a diagonal matrix with the eigenvalues of $\mathbf{A}$ on the diagonal ($D_{ii} = \lambda_i$). Often, but not always, we will work with real, symmetric ($\mathbf{A}^\mathsf{T} = \mathbf{A}$) matrices, which is especially convenient because, by the spectral theorem, they can always be diagonalized by an orthonormal basis of real eigenvectors. The eigendecomposition is one of several important *matrix factorizations*; singular value decomposition (SVD), for rectangular matrices, is another we will encounter. As we will see, matrix algebra and spectral analysis play pivotal roles in many problems in network analysis.

4.2.3 Probability

Probability is the fundamental mathematical tool of statistics and data science. Let us introduce the basic probability concepts and notation.

Conceptually, *probability* captures the chances for a random event to occur. An *event A* is when a *random variable* takes on a certain value from its *sample space* Ω, defined as the set of all possible outcomes for the event. For instance, A can be the event where a coin is flipped and takes on the value heads, represented as $A = H$. The sample space of A is $\Omega_A = \{H, T\}$. A single event can encompass multiple acts; event B could be the results from tossing a coin twice in a row, which has a sample space of $\{HH, HT, TH, TT\}$. Two or more events are said to be *mutually exclusive* (or *disjoint*) if at most one of them can occur. We say the events are *collectively exhaustive* if at least one of the events must occur.

An *event A* is said to occur with *probability* $\Pr(A)$ (sometimes written $P(A)$). This $\Pr(A)$, defined for the sample space of A, is a *probability distribution function* if it satisfies:

1. $\Pr(A) \geq 0$ for every A,

2. $\Pr(\Omega_A) = 1$,

3. $\Pr(\cup_{i=1}^{\infty} A_i) = \sum_{i=1}^{\infty} \Pr(A_i)$ for mutually exclusive $A_1, A_2, \ldots$

These axioms of probability—non-negativity, unitarity, and additivity—are the foundation for other properties of probability, such as Bayes' theorem, described below.

The *joint probability* for multiple events $A, B, \ldots$ all occurring is denoted by $\Pr(A, B, \ldots)$. For instance, imagine throwing two dice A and B. Then, $\Pr(A = 1, B = 1)$ is the joint probability that both dice land on 1. The joint probability is the product of the individual probabilities if and only if the random variables are *independent* from each other:

$$\Pr(A, B) = \Pr(A)\Pr(B) \iff A \text{ and } B \text{ are statistically independent.} \tag{4.5}$$

As a notation, the joint distribution is symmetric, meaning $\Pr(A, B) = \Pr(B, A)$.

The *conditional probability* $\Pr(A|B)$ [10] is the probability to observe A given that we observe B. For instance, the probability of rain during the next hour $\Pr(\text{rain})$ can depend heavily on whether there are dark clouds or not. In other words, $\Pr(\text{rain at 4pm} \mid \text{clouds}) > \Pr(\text{rain at 4pm} \mid \text{no clouds})$. Note that the conditional probability does not have to describe a temporal or causal relationship. For instance, it is totally valid to think about $\Pr(\text{clouds at 2pm} \mid \text{rain at 4pm})$ (given a future event, we can look back on past events) or $\Pr(\text{wearing rain boots} \mid \text{carrying an umbrella})$.

The conditional probability and the joint probability are related. If we ask for the probability of both events A and B occurring, it is the same as asking (i) if event B occurs *and* (ii) given B occurred, if event A occurs. Formally, we write this relationship [11] as

$$\Pr(A, B) = \Pr(A \mid B)\Pr(B) \tag{4.6}$$

or

$$\Pr(A \mid B) = \frac{\Pr(A, B)}{\Pr(B)}. \tag{4.7}$$

This is called the *chain rule* (or *general product rule*) of probability. This rule plays an essential role in probability theory by allowing us to reduce any joint probability distribution into a product of conditional probabilities.

---

[10] It can be read as "the probability of A given B."

[11] If A and B are independent, then $\Pr(A \mid B) = \Pr(A)$, which again shows that $\Pr(A, B) = \Pr(A)\Pr(B)$.

What if there are more than two variables? We can iteratively apply the chain rule (Eq. (4.6)) by considering multiple variables as one. For instance, here with three variables, we first group A and B, then apply the chain rule again:

$$\Pr(A, B, C) = \Pr(C \mid A, B) \Pr(A, B)$$
$$= \Pr(C \mid A, B) \Pr(B \mid A) \Pr(A). \qquad (4.8)$$

More generally,

$$\Pr(A_n, \ldots, A_1) = \Pr(A_n \mid A_{n-1}, \ldots, A_1) \cdots \Pr(A_3 \mid A_2, A_1) \Pr(A_2 \mid A_1) \Pr(A_1). \qquad (4.9)$$

When we consider multiple events, the *marginal probability* of a variable is simply the probability of individual events ($\Pr(A)$, $\Pr(B)$, etc.). Using the conditional probability, when we consider two variables, the marginal probability can be written as:

$$\Pr(A) = \sum_{B \in \Omega_B} \Pr(A, B) = \sum_{B \in \Omega_B} \Pr(A \mid B) \Pr(B), \qquad (4.10)$$

which states that we need to consider all possible cases of the *other* variable to calculate the marginal probability. This process of summing/integrating out the other variable or variables (it holds for more than two variables) is called *marginalization*.

Finally, combining the symmetry of the joint distribution with the relationship between joint and conditional distributions reveals *Bayes' theorem*:

$$\Pr(A, B) = \Pr(B, A)$$
$$\Pr(A \mid B) \Pr(B) = \Pr(B \mid A) \Pr(A)$$
$$\Pr(A \mid B) = \frac{\Pr(B \mid A) \Pr(A)}{\Pr(B)}. \qquad (4.11)$$

Equation (4.11), despite its simplicity, is one of humanity's most fundamental mathematical discoveries. It is also connected to many cognitive biases that plague human quantitative reasoning as many of those biases reflect misunderstandings of conditional probabilities.[12] Because Bayes' theorem is a fundamental way to approach *inference problems* and *learning from data* in general, we will be seeing this formula throughout this book.

4.2.4 Random variables and probability distributions

Random variables A *random variable* (RV) X is a variable endowed with a corresponding probability distribution $\Pr(X)$ governing the probability that X randomly takes on one of its *supported values*, denoted x. An event would be the assignment of x to X which would then occur with probability $\Pr(X = x)$. Given a probability distribution $f(x)$, often we use $X \sim f(x)$ to denote that X is *drawn from* the distribution $f(x)$ or equivalently that X is *distributed according to* $f(x)$.

---

[12] Bayes' theorem tells us we can't in general just "flip around" a conditional probability, $P(A \mid B) \neq P(B \mid A)$, something many intuitively wish to do.

For example, a coin flip can be represented by a random variable X that takes on value $x = 0$ when the coin lands tails and $x = 1$ when the coin lands heads. The probability distribution for this X can then be defined as $\Pr(X = 1) = p$, $\Pr(X = 0) = 1 - p$, where $p \in [0, 1]$ is a constant ($p = 1/2$ for an unbiased or fair coin).

Often distributions have *parameters* associated with them, non-random quantities that govern the scale or spread of the RV's values. For the coin flip, p served as a parameter. When discussing parameters generically, we commonly use the symbol θ to represent one or more generic parameters and when defining a function we distinguish between the values of the parameter(s) and the value of the RV with a semicolon: $f(x; \theta)$. (When describing an RV's probability it is also common to use a conditional probability, $\Pr(X \mid \theta)$, to distinguish between values and parameters.)

Lastly, when dealing with multiple variables, we distinguish whether they follow the same or different distributions. Suppose we independently flip a coin ($X \in \{0, 1\}$) and roll a die ($Y \in \{1 \ldots 6\}$). Because they are independent, the joint probability of getting $X = x$ and $Y = y$ is given by

$$\Pr(X = x, Y = y) = f_{\text{coin}}(X = x) f_{\text{die}}(Y = y), \tag{4.12}$$

where we use f_{coin} and f_{die} to distinguish the different distributions the RVs follow. We usually simplify such expressions using the RVs only to distinguish the densities: $\Pr(X = x, Y = y) = \Pr(X = x) \Pr(Y = y)$, with the idea being that the distributions must be different for these different arguments. (When unambiguous, it is also common to drop the values x and y, $\Pr(X, Y) = \Pr(X) \Pr(Y)$.) If there is ambiguity, notation should always make it clear which RV goes with which distribution.

> **ⓘ** If two or more RVs follow the same distribution, they are *identically distributed*. If they are also independent, they are *independent and identically distributed* or **iid**.

Probability mass and probability density For a discrete random variable like the coin flip, the support of X is a finite or countably infinite set. In comparison, a continuous random variable is one with an uncountably infinite support. Typically these RVs are real-valued.

The distribution for a discrete random variable is referred to as a *probability mass function* (**pmf**). The pmf assigns probability to each supported value x such that $\sum_x \Pr(X = x) = 1$ (where the sum runs over all supported values). The probability that X takes on *some* value is 1 so the probabilities must sum to 1 over the supported values.

On the other hand, for a continuous random variable, the probability function is referred to as a *probability density function* (**pdf**), which assigns probability *densities* to values of x. The distinction here is that for a continuous random variable there will be an infinite number of possible values, so each value must be assigned only a infinitesimal probability in order to be normalized. In other words, the values of a pdf are *not probabilities*! Instead, they are probability *densities*. And there is zero *probability mass* at any precise value of the random variable ($\Pr(x) = 0$); the probability mass only exists when we consider an interval—for instance, the probability that x is between a

and b is $\Pr(a < x < b) = \int_a^b \Pr(x)dx$. Defined in this way, normalization is ensured through an integral over the supported values:

$$\int \Pr(x)dx = 1. \tag{4.13}$$

The area under the curve must be equal to 1 for the distribution to be properly normalized.

> ⚠ A further consequence of properly normalized pdfs is that the probability densities do not necessarily remain below 1 as probabilities do. Suppose X is supported for $0 \le x \le 1/2$. The width of this interval is less than 1, so the height of any pdf defined on it must exceed 1 at some point for the area under the curve to equal 1.

Cumulative distributions   For pdfs and pmfs defined on ordinal support, we can derive a *cumulative distribution function* (**CDF**)[13] that also assigns probabilities to values of x. Instead of asking what is the probability or density of a particular value x, the CDF asks what is the *probability of having a value less than or equal to x*. The CDF $F_X(x)$ is defined as:

$$F_X(x) = \Pr(X \le x) = \begin{cases} \sum_{x' \le x} \Pr(x') & \text{for a pmf,} \\ \int_{-\infty}^x \Pr(x')dx' & \text{for a pdf.} \end{cases} \tag{4.14}$$

The CDF is a monotonically increasing function defined over the entire support of X and it connects the percentiles of X to each value of x (for instance, the 50th percentile or median of x is the x where $F_X(x) = 1/2$). In the context of network analysis, we also use the *other* cumulative distribution function, called the *complementary cumulative distribution function* (**CCDF**)[14] and defined as:

$$\bar{F}_X(x) = \Pr(X > x) = 1 - F_X(x). \tag{4.15}$$

While conveying the same information as the CDF, the CCDF is particularly useful when we examine RVs that are strongly *skewed* such that values far larger than the mean and median have some nonvanishing probability of being observed.[15] A logarithmic plot of the CCDF will visually stretch out the small probabilities assigned to the extreme values, whereas those values will be squashed visually as they accumulate near 1 when plotting the CDF.

Both the CDF and CCDF can be obtained from the functional form of the pmf or pdf using Eqs. (4.14) and (4.15), or estimated from the actual data points (creating what are called the "empirical" CDF and CCDF; see Exercises and Ch. 11).

4.2.5   Commonly encountered distributions

There is unlimited variety in probability distributions, both pmfs and pdfs, but some types of distributions are so fundamental and so frequently encountered that they are given specific names. Here we discuss a few.

---

[13] The CDF is sometimes just called the *distribution function*.

[14] The definitions for CDFs and CCDFs are sometimes exchanged.

[15] The degree distribution of networks often exhibits such a skewed or "heavy-tailed" distribution, making this scenario especially important when studying networks.

Bernoulli distribution   A Bernoulli random variable X represents a binary or two-outcome variable ($x = 0$ or $x = 1$) with a constant probability for outcomes:

$$\Pr(X = 1) = p,$$
$$\Pr(X = 0) = 1 - p, \tag{4.16}$$

or, written more compactly,

$$\Pr(X = x; p) = p^x (1 - p)^{1-x} \text{ for } x \in \{0, 1\}. \tag{4.17}$$

The *Bernoulli distribution* is foundational since so many problems can be reduced to binary events, such as true/false or win/lose outcomes. In the network context, whenever we think about random processes such as creating edges at random in a null model, we think about Bernoulli trials.

Binomial distribution   The *binomial distribution* is a companion to the Bernoulli. It describes the number of positive outcomes k from a collection of n independent, identically distributed (*iid*) Bernoulli variables, where each Bernoulli has a positive outcome with probability p:

$$\Pr(X = k; n, p) = \binom{n}{k} p^k (1 - p)^{n-k}, \tag{4.18}$$

for $k = 0, 1, \ldots, n$ and

$$\binom{n}{k} = \frac{n!}{k!\,(n-k)!}$$

is the *binomial coefficient*. Here n and p serve as parameters, $p^k(1 - p)^{n-k}$ is the probability of one particular way to achieve k successes and $n - k$ failures, and $\binom{n}{k}$ captures the number of arrangements of these k successes across n trials. We can write $X \sim \text{Binom}(n, p)$ to denote that X follows a binomial distribution with parameters n and p.

In the context of networks, the binomial can, for instance, describe the degree distribution of a random graph, which is created by performing Bernoulli trials for every possible pair of nodes—an edge exists between the two nodes with probability p or it does not with probability $1 - p$.

Poisson distribution   The binomial distribution can be well approximated, when n is large and p is small, by a *Poisson distribution*:

$$\Pr(X = k; \lambda) = \lambda^k \frac{e^{-\lambda}}{k!}, \tag{4.19}$$

where $\lambda = np$.[16] A Poisson gives the distribution of the number of events in a fixed time (or space) interval when events occur independently from one another and the *rate* of events λ is known.

---

[16] Actually, this is imprecise. The Poisson distribution derives in the limit from the binomial distribution by taking $n \to \infty$ and $p \to 0$ such that the expected value $np \to \lambda > 0$.

The correspondence between the binomial and the Poisson can be understood as follows. If we take a continuous time interval and divide it up into many small intervals, and assume at most one event can occur per segment,[17] the number of segments containing an event can be represented by a binomial with n as the number of segments. If we divide into smaller and smaller segments and reduce the probability for a single event to occur such that the total rate of events occurring is constant (or, $n \to \infty$ and $p \to 0$ such that $np \to \lambda$) then we can show that in the limit the binomial distribution will converge to the Poisson distribution.[18]

Normal (Gaussian) distribution A *normal distribution* or *Gaussian distribution* is one of the classic bell-shaped probability distributions. You are likely already familiar with it. It is ubiquitous due to the *central limit theorem*—the normal distribution is the limiting form for the distribution of sums of *any* iid RVs with finite variances. Thanks to the central limit theorem, both binomial and Poisson distributions can be approximated by a normal distribution when the number of trials is large.

Its pdf is given by

$$f(x; \mu, \sigma) = \frac{1}{\sigma\sqrt{2\pi}} \exp\left(-\frac{1}{2}\left(\frac{x-\mu}{\sigma}\right)^2\right), \tag{4.20}$$

for $x \in \mathbb{R}$, where μ and σ^2 are the mean and variance, respectively, of the random variable. We denote that an RV X is normally distributed with mean μ and variance σ^2 with $X \sim \mathcal{N}(\mu, \sigma^2)$. A *standard normal distribution* is one with $\mu = 0$ and $\sigma^2 = 1$.

Log-normal distribution Sometimes, it is not the "raw" value that is normally distributed, but the *logarithm* of the value:

$$f(x; \mu, \sigma) = \frac{1}{x\sigma\sqrt{2\pi}} \exp\left(-\frac{(\ln x - \mu)^2}{2\sigma^2}\right). \tag{4.21}$$

Just as the normal distribution arises when summing multiple iid RVs (via the central limit theorem), the *log-normal distribution* is produced by a multiplicative process. In other words, when many independent (positive) RVs are multiplied together, the resulting random variable approaches the log-normal distribution (this is also called *Gibrat's law*). For instance, in the distribution of wealth, income, or other financial applications (e.g., the *Black–Scholes model* for option pricing), a log-normal distribution is often a good approximation because the underlying process is *multiplicative*.

The log-normal and many other distributions are said to be *broadly distributed* or *heavy-tailed* in the sense that the distribution spans multiple orders of magnitude and a large value of x can be expected every so often, a value of x that we would never see under a normal distribution with the same mean.

---

[17] Which becomes more valid of an assumption as segments become smaller.

[18] In fact, in Ch. 22 we derive this not using a traditional calculation but to demonstrate a combinatorial tool called *generating functions* that are helpful for mathematically analyzing networks and other problems.

Power-law distribution   Another common "broad" distribution is a *power-law distribution* $p(x)$ where the pdf (x continuous) or pmf (x discrete) follows a power-law functional form:

$$p(x) = Cx^{-\alpha}, \tag{4.22}$$

with exponent $\alpha > 0$ (usually between 2 and 4) and a suitable normalization constant C. Some choices are necessary for the support (since α is positive, we must exclude $x = 0$) to ensure normalization, which dictates the possible values of α; both affect whether the mean, variance, or other moments are finite.

An extension of a *pure* power-law distribution is a power law with an exponential cutoff,

$$p(x) = Cx^{-\alpha}e^{-\lambda x}. \tag{4.23}$$

This introduces a second parameter λ, which governs the value of x when the power law is "overwhelmed" by the exponential. Note that Eq. (4.23) reduces to Eq. (4.22) when $\lambda = 0$.

A useful property of power-law distribution is that it shows up as a straight line in a log–log plot of x and $p(x)$. If we take the logarithm of both sides of the equation, we get

$$\log p(x) = \log C - \alpha \log x. \tag{4.24}$$

In log–log scale, this is a straight line with slope $-\alpha$ and intercept $\log C$. Furthermore, the CCDF of a power-law distribution is *also* a power-law distribution, with exponent $\alpha - 1$ instead of α,

$$\int_x^\infty Cy^{-\alpha}dy = \frac{C}{1 - \alpha}\left[y^{1-\alpha}\right]_x^\infty = C'x^{-(\alpha-1)}. \tag{4.25}$$

Because of this and the abundance of heavy-tailed degree distributions in real-world networks, examining CCDFs in log–log scale is a common exploratory analysis technique for network data. We will revisit power-law distributions again in Chs. 11 and 12.

Note that a power-law probability distribution is distinct from a *power-law scaling relation*. For instance, Newton's law of gravitational attraction in classical mechanics $F = G\frac{m_1 m_2}{r^2} \propto r^{-2}$, an "inverse-square" law, where $A \propto B$ means that there exists some constant c such that $A = cB$, is mathematically a power-law functional form relating F and r but is not describing the distribution of a random variable.

Broadly distributed random variables appear often in network data, especially as degree distributions. The degree distribution of many networks can be explained either by log-normal distribution or the power-law distribution, as we will see later in this book. Because a log-normal distribution exhibits a heavy tail due to its logarithmic nature, it is often not easy to distinguish this distribution from the power-law distribution (or vice versa). In fact, identifying statistically broad distributions from finite data samples is often fiendishly difficult (Ch. 11 and Sec. 22.6)

While not exhaustive, the above distributions are the most commonly encountered by scientists working with (network) data and when analyzing and implementing data analysis methods.

4.3 Statistics

As a subject, statistics is often lumped in with probability, but probability is a fully axiomatic branch of mathematics whereas statistics is a mathematical science that leverages probability (in particular) to understand better the properties of data. Statistical questions often ask about underlying phenomena that give rise to empirical data by comparing that data with probabilistic models. Answering such questions requires care, and often we must thoughtfully consider *uncertainty* when data are noisy or incomplete.

4.3.1 Summary statistics

We often turn to summary or descriptive statistics to quantify our data. Measures of *central tendency* can tell us what are "typical" values, and measures of *dispersion* tell us how tightly packed values tend to be around their centers. The form of our data dictates what mathematical operations, and therefore what statistics, we can use. Categorical variables cannot be added or multiplied, for instance, but we can tell whether two values are equal, so we can only use the mode for a categorical's measure of central tendency. For a numeric variable, however, those operations are permitted, giving us medians or means to measure central tendencies.

When computing (summary) statistics, such as means and variances, throughout this book, we'll use the following notation.

The *mean*, *expectation*, or *expected value* of a random variable X is denoted $\langle x \rangle$ (sometimes $E[X]$) and given by

$$\langle x \rangle = \sum_x x P(x) \quad \text{(pmf)} \qquad \langle x \rangle = \int_{-\infty}^{\infty} x P(x)\, dx \quad \text{(pdf)}, \tag{4.26}$$

where the summation or integration is over X's support. The expectation of a function $f(x)$ of an RV is

$$\langle f(x) \rangle = \sum_x f(x) P(x) \quad \text{(pmf)} \qquad \langle f(x) \rangle = \int_{-\infty}^{\infty} f(x) P(x)\, dx \quad \text{(pdf)}. \tag{4.27}$$

Using this, the nth *moment* (about zero) of a distribution is given by $\langle x^n \rangle$. The mean is the first moment. The *variance* of X compares the first and second moments,

$$\text{Var}(X) = \left\langle (X - \langle X \rangle)^2 \right\rangle = \left\langle X^2 \right\rangle - \langle X \rangle^2. \tag{4.28}$$

(The variance of X is also commonly denoted by σ_X^2 or just σ^2 when the context is clear.) The *standard deviation* is the square root of the variance. The variance is a special case of the *covariance* $\text{Cov}(X, Y)$ for two RVs ($\text{Var}(X) = \text{Cov}(X, X)$):

$$\text{Cov}(X, Y) = \left\langle (X - \langle X \rangle)(Y - \langle Y \rangle) \right\rangle = \langle XY \rangle - \langle X \rangle \langle Y \rangle. \tag{4.29}$$

The covariance tells us how strongly related two variables are, but it's also common to use the (Pearson) *correlation coefficient*,

$$\rho_{XY} = \frac{\text{Cov}(X, Y)}{\sigma_X \sigma_Y} = \frac{\langle XY \rangle - \langle X \rangle \langle Y \rangle}{\sqrt{\langle X^2 \rangle - \langle X \rangle^2} \sqrt{\langle Y^2 \rangle - \langle Y \rangle^2}}, \tag{4.30}$$

instead of the covariance to measure how strongly correlated or linearly related the variables are to each other. Rescaling the covariance in this way ensures $-1 \leq \rho \leq 1$, making for a more interpretable statistic and more easily compared across different datasets.

Useful properties   Expectation is linear:

$$\langle c_1 X + c_2 Y \rangle = c_1 \langle X \rangle + c_2 \langle Y \rangle \quad c_1, c_2 \text{ constants,} \tag{4.31}$$

$$\left\langle \sum_{i=1}^{n} c_i X_i \right\rangle = \sum_{i=1}^{n} c_i \langle X_i \rangle \quad \text{in general.} \tag{4.32}$$

The variance obeys the following (c, c_i constants):

$$\text{Var}(X) \geq 0, \tag{4.33}$$

$$\text{Var}(X + c) = \text{Var}(X), \tag{4.34}$$

$$\text{Var}(cX) = c^2 \text{Var}(X) \quad \text{(that } c \text{ is squared is a common ``gotcha''),} \tag{4.35}$$

$$\text{Var}\left(\sum_{i=1}^{n} c_i X_i\right) = \sum_{i=1}^{n}\sum_{j=1}^{n} c_i c_j \text{Cov}(X_i, X_j) \tag{4.36}$$

$$= \sum_{i=1}^{n} c_i^2 \text{Var}(X_i) + \sum_{i \neq j} c_i c_j \text{Cov}(X_i, X_j), \tag{4.37}$$

$$\text{Var}\left(\sum_{i=1}^{n} c_i X_i\right) = \sum_{i=1}^{n} c_i^2 \text{Var}(X_i) \quad \text{(if the } X_i \text{ are uncorrelated).} \tag{4.38}$$

Sample statistics   Lastly, it can be important to distinguish between the statistics of a random variable defined above and the statistics computed from a data sample. For a collection of n data points $\{x_i\}$ and $\{y_i\}$, define

$$\bar{x} = \frac{1}{n} \sum_{i=1}^{n} x_i \quad \text{(sample mean),} \tag{4.39}$$

$$s_x^2 = \frac{1}{n-1} \sum_{i=1}^{n} (x_i - \bar{x})^2 \quad \text{(sample variance),} \tag{4.40}$$

$$s_{xy}^2 = \frac{1}{n-1} \sum_{i=1}^{n} (x_i - \bar{x})(y_i - \bar{y}) \quad \text{(sample covariance),} \tag{4.41}$$

$$r_{xy} = \frac{\sum_{i=1}^{n} (x_i - \bar{x})(y_i - \bar{y})}{\sqrt{\sum_{i=1}^{n} (x_i - \bar{x})^2} \sqrt{\sum_{i=1}^{n} (y_i - \bar{y})^2}} \quad \text{(sample correlation).} \tag{4.42}$$

The sample covariance is unbiased when using $n - 1$ in the denominator (this is called Bessel's correction).

   While the statistics of a random variable are perfectly defined, they are facts of the distribution, the sample statistics themselves vary depending on the data sample being used. A different sample would lead to a different sample statistic, with this difference decreasing with increasing sample size.

Standardizing   A standard score z is useful when we wish to "correct" for the mean and variance of x:

$$z = \frac{x - \langle x \rangle}{\sigma_x}. \tag{4.43}$$

A standard score of z means that x is z standard deviations above (or below, if negative) the mean of x. This *z-score* lets us compare different data on a common scale and lets us perform *z-tests* using the properties of the standard normal. Note that if the variable we are standardizing is itself the sample mean of a randomly sampled underlying variable, z can be expressed with the underlying variable's standard deviation with a correction in the denominator:

$$z = \frac{\bar{x} - \langle \bar{x} \rangle}{\sigma_X / \sqrt{n}}. \tag{4.44}$$

Intuitively, this makes sense, as the variability in $\bar{x}$ should decrease with increasing n.

4.3.2   Inference

Statistical inference is a process of inferring models (assumptions) that explain the data well. The common approaches to inference are (null) hypothesis testing and Bayesian inference.

Hypothesis testing proceeds by defining a null hypothesis related to a question of interest. For example, if we want to understand how strongly correlated x and y are, we could take as a null hypothesis that "x and y are uncorrelated." We then consider a *test statistic*, something we can measure in our real data and can understand, traditionally mathematically, how it behaves assuming the null is true. We then measure the test statistic in our real data, and ask how probable it is to see a value that large or larger under the null hypothesis. This "tail probability" is given by the CDF of the null distribution of the test statistic. If that probability, called a p-value, is sufficiently small, we can argue that it is unlikely for the null hypothesis to be true. Hypothesis testing is a richly developed set of tools, with many tests for many situations and careful ways to report results. It brings a lot of baggage, however, and is (often, rightfully) criticized. For instance, we are actually only testing the null, not any actual hypothesis. Likewise, we often rely on arbitrary thresholds to determine whether a test is "significant"; commonly, if $p < 0.05$ the test is significant, otherwise it is not. And, since a smaller p-value is a stronger result, researchers often, intentionally or not, find themselves optimizing for p, a misleading practice called *p-hacking* (Ch. 3). Used properly, hypothesis testing has its place, and it can be simple and effective, but these concerns should be front of mind.

Bayesian inference is built on Bayes' theorem. We've seen Bayes' theorem before

(Eq. (4.11)) but let's write it again using two symbols D and θ, and label it:

$$P(\theta \mid D) = \frac{\overbrace{P(D \mid \theta)}^{\text{"likelihood"}} \overbrace{P(\theta)}^{\text{"prior"}}}{\underbrace{P(D)}_{\text{"evidence"}}}.$$

$\underbrace{}_{\text{"posterior"}}$

(4.45)

Here D refers to *data* and θ refers to *parameters*. $P(\theta)$ is called the *prior*, or the prior probability distribution of the parameters. $P(D \mid \theta)$ is called the *likelihood*, or the likelihood that our model with parameters θ generates the data D. $P(D)$ is usually called the *evidence* or *marginal likelihood*.[19] Finally, $P(\theta \mid D)$ is called the *posterior*, or the posterior probability distribution of θ given the data.

Once we write the theorem in this way, we can interpret it as following: we can learn about the conditional probability of parameters given data ("posterior") by knowing the conditional probability of data given parameters ("likelihood"), the probability of parameters ("prior"), and the marginal likelihood of data ("evidence"). In effect, the left-hand side of Eq. (4.45) tells us what statistical models (what parameters) are probable given our data and what models are improbable, but this probability is not easy to compute. The terms on the right-hand side are computable so, thanks to Bayes' theorem, we have a way to address the posterior, the probability we wish to assess.

If you are not familiar with Bayesian inference, you may be wondering how can we even think about the "marginal likelihood of data" or the "prior probability distribution of parameters." Those are great questions! In Bayesian statistics, probability is conceptualized, not as a fixed number that we can estimate by performing many trials, but as the degree of belief that an outcome will occur. That is why we can think of the *prior* probability distribution of parameters. We explicitly state what we believe (or don't believe) about our parameters prior to seeing the data. Data then lets us adjust our initial belief (prior) to a more *informed* belief (posterior).

Let's walk through a toy example. Imagine yourself catching some fish on a fishing boat floating on a lake. Your goal is to create a good model that explains how many fish you can catch (per hour) from the lake. Your *data* are the number of fish you caught in the past several hours, say:

$$D = [4, 1, 2, 10].$$

(4.46)

Then we can come up with some probabilistic models to explain the data. Suppose one model (M_1) assumes a uniform distribution of the number of fish caught with a single parameter N that determines the maximum number of fishes, or

$$\Pr(n; M_1, N) = \begin{cases} \frac{1}{N+1} & n \in \{0, \ldots, N\}, \\ 0 & \text{otherwise.} \end{cases}$$

(4.47)

For a second model (M_2), let's assumes a Poisson distribution, which describes processes that happen with a constant rate:

$$P(n; M_2, \lambda) = \frac{\lambda^n e^{-\lambda}}{n!}.$$

(4.48)

---

[19] This name stems from writing the denominator as $P(D) = \sum_{\theta'} P(D \mid \theta')P(\theta')$.

Once we have our models, we can think about the likelihood, which captures how *likely* it is to have the data given a model. Unlike the probability of a model given the data, the likelihood is straightforward to calculate. For instance, if we assume our datapoints are iid, for the first model with $N = 10$, the likelihood is:

$$\Pr(D \mid M_1; N = 10) = \prod_i \Pr(d_i | M_1; N = 10) \qquad (4.49)$$

$$= \frac{1}{11} \cdot \frac{1}{11} \cdot \frac{1}{11} \cdot \frac{1}{11}. \qquad (4.50)$$

Likewise, we can compute the likelihood of M_2 in the same way (this time without plugging in a parameter value):

$$\Pr(D \mid M_2; \lambda) = \prod_i \Pr(d_i | M_2; \lambda) \qquad (4.51)$$

$$= \frac{\lambda^4 e^{-\lambda}}{4!} \cdot \frac{\lambda^1 e^{-\lambda}}{1!} \cdot \frac{\lambda^2 e^{-\lambda}}{2!} \cdot \frac{\lambda^{10} e^{-\lambda}}{10!}. \qquad (4.52)$$

The ability to calculate the likelihoods lets us compare models and parameters. Let's focus on the first model and think about the ratio of two posterior probabilities given two different parameters ($N = 10$ and $N = 20$):

$$\frac{\Pr(N = 10|D)}{\Pr(N = 20|D)} = \frac{\Pr(D|N = 10)\Pr(N = 10)\Pr(D)}{\Pr(D|N = 20)\Pr(N = 20)\Pr(D)} \qquad (4.53)$$

$$= \frac{\Pr(D|N = 10)\Pr(N = 10)}{\Pr(D|N = 20)\Pr(N = 20)}. \qquad (4.54)$$

If this ratio is larger than 1, that means, given the data, $N = 10$ is more likely than $N = 20$. Starting from here, there are multiple ways to approach the inference—identifying good models and their parameters. In general, the process of comparing and then choosing between models using data is called *model selection*.

4.3.3 Maximum likelihood estimation

We can see from Eq. (4.54) that the ratio of two posterior probabilities is solely determined by the likelihood and the prior. Let's first assume that we do not have any knowledge or prior belief about what the parameters should be. Then we can assume an equal prior for all possible parameters ($\Pr(N_1) = \Pr(N_2)$ for any N_1 and N_2). In this case, the ratio of posteriors is entirely determined by the *likelihood*—the higher the likelihood of a given parameter, the higher the posterior probability of that parameter, which makes sense! In other words, under these conditions,

$$\Pr(\theta|D) \propto \Pr(D|\theta). \qquad (4.55)$$

Then, the best parameter given data can be identified by finding the parameter that maximizes the likelihood:

$$\theta_{\text{MLE}} = \arg \max_{\theta} P(\theta|D) = \arg \max_{\theta} P(D|\theta). \qquad (4.56)$$

(The "arg max" of a function $f(x)$ is the value(s) of x for which $f(x)$ is maximized. A similar definition holds for "arg min.") This is the basic idea of *maximum likelihood estimation* (MLE). It is built on the assumption that, when we don't have any information or prior belief about parameter distribution, it is reasonable to assume that the parameter that maximizes the likelihood is the best parameter. Note that MLE is a *point estimate*— it identifies a single parameter value that maximizes the likelihood but it cannot tell us what is the posterior probability *distribution*.

4.3.4 Maximum a posteriori

What if we have some information or belief about the prior distribution of the parameters? For instance, maybe we have reasons to believe that the fish yield of the lake should be around 5. Maybe lakes of this size and characteristics tend to have similar numbers of fish.

Bayes' theorem provides a straightforward way to incorporate this information through the prior distribution of the parameter $\Pr(\theta)$. We just need to keep the prior:

$$\theta_{\text{MAP}} = \arg\max_{\theta} \Pr(\theta|D) = \arg\max_{\theta} \Pr(D|\theta)\,\Pr(\theta). \qquad (4.57)$$

This is called *maximum a posteriori* (MAP) inference. For many common situations, MAP point estimates tend to be a simple interpolation between the MLE point estimate and the mean of the prior distribution. For more information, read about "conjugate priors" and the exponential family.

4.3.5 Bayesian inference

MLE and MAP obtain *point estimates*—a single point in the parameter space that maximizes the likelihood or posterior. Although a point estimate can be the "solution," it is less useful than inferring the full distribution and sometimes can be misleading. For instance, imagine a hypothetical case where the likelihood function looks like Fig. 4.1.

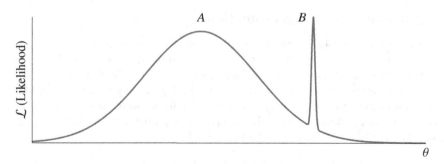

Figure 4.1   A hypothetical likelihood function where most of the probability mass exists (A) far away from the maximum likelihood point (B).

As you can see, MLE will (or may) point us to B, where the sharp peak is located. However, the peak is extremely sharp and the probability mass around the MLE is negligible compared with the broad peak around A. In this case, can we justify the result of MLE, particularly given that most parameter values will be around A if we sample from this distribution?

As shown in this case, a point estimate can be misleading and miss critical information about the distribution. With a single estimated parameter value, it is impossible to know the uncertainty around the estimate and it is also impossible to know whether there exist other parameters that are almost equally likely. When we say "(full) Bayesian inference," we refer to the practice of considering the posterior probability *distribution* rather than the point estimate like MLE or MAP.

Sometimes a full Bayesian inference can be accomplished mathematically, by writing down an expression for the posterior probability. Unfortunately, this tends to be possible only for simple models with tractable likelihoods. In practice, we often resort to computational approaches called *Markov Chain Monte Carlo* (MCMC) that avoid computing the posterior and instead generate large numbers of samples (parameter values) that will be distributed following the posterior.[20] MCMC can be expensive to compute, and may need careful guidance to ensure the posterior is being sampled correctly. This hampered Bayesian inference in the past, but computers are quite fast nowadays, and many new MCMC algorithms make inference both faster and more reliable.

4.4 Summary

Network science is a broadly interdisciplinary field, pulling from computer science, mathematics, statistics, and more. The data scientist working with networks thus needs a broad base of knowledge. Although helpful, we do not expect the reader to know all these areas well. Instead, the reader should be prepared enough that they can learn this material without also needing to learn much of the background that the primer itself depends on. This chapter serves as a primer for the reader on background information that we will use through the rest of the book. When necessary, we will refer back to these materials as we proceed.

Bibliographic remarks

Many resources abound for learning programming in general and for data science in particular. These days, the two major programming languages for data scientists are Python and R. Readers wanting to learn more on practical programming in Python may consider *Python for Scientists* by Stewart [444] or *An Introduction to Python Programming for Scientists and Engineers* by Lin et al. [276], while *R for Data Science* by Wickham and Grolemund [491] is an excellent entry point for those interested in R. *Introduction to Algorithms* by Cormen et al. [117] is one of the classic starting points for studying algorithms and data structures.

Readers seeking to learn more linear algebra are encouraged to begin with the classic *Introduction to Linear Algebra* (Strang [447]) or the more recent *Linear Algebra and*

---

[20] An alternative to MCMC is *variational inference*, which approximates or lower bounds the intractable denominator in Eq. (4.45).

Learning from Data (Strang [446]), which gives a broad introduction to linear algebra using statistics and machine learning as the stage.

For readers interested in brushing up on probability and statistics, we recommend the excellent *All of Statistics* by Wasserman [484], the perennial classic *Probability & Statistics for Engineers & Scientists* by Walpole et al. [481], and *Computer Age Statistical Inference* by Efron and Hastie [142], the latter giving an exciting blend of statistics and computing. For Bayesian statistics, *Doing Bayesian Data Analysis* by Kruschke [256] is a practically focused introduction, while *Bayesian Data Analysis* by Gelman et al. [179] is suitable for those with more background.

Exercises

4.1 A *multiset* acts like a set except duplicates are allowed, so multiset elements need not be unique. Suppose you are working with a programming language that gives you `set`, `list`, and `dict` (dictionary) data structures. Describe how to implement `multiset` using these built-in data structures.

4.2 We have at the ready an infinite number of true/false questions. Assume questions are unrelated and each question has probability $p = 1/20$ to have an answer of T, otherwise the answer is F. We continue to ask questions until the first T answer. How many questions should we *expect* to ask?

4.3 Show that $\text{Var}(X) = 0$ if and only if $\Pr(X = c) = 1$ for some constant c.

4.4 The cumulative distribution function (CDF) $F(x)$ (Eq. (4.14)) can be estimated from data using the empirical CDF (ECDF):

$$F_n(x) = \frac{1}{n} \sum_{i=1}^{n} I(X_i < x), \qquad (4.58)$$

where

$$I(X_i < x) = \begin{cases} 1 & \text{if } X_i \leq x, \\ 0 & \text{otherwise.} \end{cases} \qquad (4.59)$$

An advantage of the ECDF over a typical *histogram* for estimating a distribution is that the ECDF is defined for all n data points, while the histogram is defined at fewer points due to the need to bin the data.

Describe with pseudocode a simple function that computes the ECDF efficiently using a `sort` function.

4.5 We wish to study how a network with N nodes and M edges changes when an edge is removed. We use a method $f(i, j)$ that compares a pair of nodes i, j. We study what removing an edge does by computing f before and after the edge is removed. Computing f has complexity $O(N + M)$ for one pair. For one edge removal, what is the complexity of checking the change in f over every pair of nodes? What is the complexity of checking every edge?

Part II

Applications, tools, and tasks

Chapter 5

The life cycle of a network study

How do we obtain new knowledge? A common distinction is between "*hypothesis*-driven" and "*data*-driven" research. Hypothesis-driven research (or "normal science") begins with a research question or a hypothesis that is built on existing knowledge or "paradigms." Then, driven by the question, researchers seek measurements or data that can answer the question. Einstein's theory of relativity is a good example of such hypothesis-driven research. The idea and mathematical theory was built on several hypotheses about how space and light behaves. The theory then produced several testable hypotheses that were later tested empirically to confirm[1] the theory.

By contrast, in data-driven research, the measurements, observations, or simply *data* are the driver of the research. Researchers may not necessarily have a concrete question in mind at the beginning of the research, but will identify interesting patterns from the data through an iterative, exploratory analysis. Kepler's laws are a good example because Kepler discovered these laws by carefully examining the data that had been collected without any concrete idea or hypothesis about universal laws in mind.

These two approaches are also tied to the nature of data that they deal with. Robert Groves, expert in survey methods and 23rd director of the US Census Bureau, classified research data into *designed* data and *organic* data [194]. (Sociologist Matthew Salganik referred to the latter as *found* data [412].) In traditional social science research, the researcher usually design surveys or data collection methods, which are then used to collect the actual data. Yet in many areas of data science, this model is flipped—usually, the data has already been collected as a byproduct (e.g., a social media company's user logs) without any research questions in mind. Researchers then *find* the usage of such data. Often, the researchers do not form the research question before examining the data. We will talk more about this distinction later in this chapter.

In practice, many research projects contain both aspects. The research questions and hypotheses are often iteratively sharpened by a better understanding of the data obtained through exploratory analyses. Sharper questions can also dictate additional data collection or inspire novel ways to dissect and combine existing datasets.

Although data-driven research is sometimes criticized due to its risk of "researcher-

---

[1] Or, more properly, failed to rule out the theory.

65

degrees-of-freedom" and "p-hacking," it is important to understand that *hypothesis-free observation* is an integral part of the scientific process and plays a foundational role in science. At the same time, it is also crucial to understand the importance of research design and the risks of *p-hacking* and related issues.

In this chapter, we follow the sequence of steps that hypothesis-driven research would follow. This does not mean that this is the only way research can be done. We chose this sequence because the process of hypothesis-driven research is often overlooked in data science, and this process is followed at least during parts of even heavily data-driven research projects. We discuss specific tools and techniques underlying both approaches in later chapters, particularly Ch. 11.

5.1   Network questions

A network study often begins with a *network question*, meaning a question that needs to take connections (edges) or networks into account. Network questions can be questions that are about the structure and dynamics of networks themselves (e.g., *"what are the topological characteristics of this network?"*) or questions about the impact of network structure on network elements (*"can we predict the clinical outcome of an individual by knowing the size and characteristics of their social network?"*).

Note that the problem itself does not necessarily dictate whether it is a network question or not—some problems can either be studied as a network question or as a non-network question. For instance, understanding the spread of an infectious disease like COVID-19 can be studied either by considering the contact network between people and the spread of disease through this network or by ignoring the network. In fact, *network epidemiology*—considering network structure as critical elements to understanding epidemics—is relatively new and still not in the epidemiological mainstream.

5.1.1   Types of network questions

The potential questions can range from micro- (*"can we identify social influence between a pair of students?"*) to macro-level (*"what is the degree distribution of this network?"*), and from descriptive (*"what are the communities in this network?"*) to prescriptive (*"what kinds of interventions can we implement to enhance communication within a company?"*). One data scientist may just be curious about the properties of a new network that no one has seen before. Another may need to create the best link prediction algorithm to minimize user *churn* on their online social media service. The questions can also be defined only loosely and emerge from the exploratory analysis of a network.

In any case, the question will eventually dictate the type of network data that is needed as well as the potential methods of analysis. Even if the analysis began with hypothesis-free exploration, the question should be clarified as quickly as possible to avoid critical mismatch between the question and the data, which can lead to a lot of wasted effort and flawed analysis.

5.1.2 Questions guide operationalization

It often goes under-appreciated that numerous networks can be defined (or "operationalized") from a *single* dataset. For instance, there is no unique "***Twitter network***"; numerous entities can be considered nodes, each of which will result in a different network. Do we consider each user account as a node? How about hashtags, tweets, words, or groups of users? How about the edges? Do we consider the "following" relationship as a directed edge? Should we only consider an edge to be present when the relationship is reciprocated (both i follows j and j follows i), leading to an undirected network? Or, maybe we can construct a bipartite network between users and hashtags? We can keep asking these questions, in social media data and in fact in nearly all data we expect to encounter.

The point is that there are often numerous choices one has to make even to *define* a network, especially when the raw dataset is rich and complex. Consider a ***protein–protein interaction network***. Even if we ignore all the complexities about defining nodes (proteins), several networks can be defined because there exist multiple types of edges. One network can be defined by the edges that have been discovered and studied in the *literature*; another network can be defined by the edges discovered through a systematic *pair-wise* interaction probing (e.g., based on the Y2H (Yeast Two-Hybrid) assay); another can be defined by the AP–MS (Affinity Purification–Mass Spectrometry) method that tends to discover *cliques*[2] rather than binary edges. Depending on which types of edges you include in your network, the network will exhibit very different structural properties as well as biases. Overlooking these biases and characteristics will ensure flawed analyses and results.

Nevertheless, it is your *question* that clarifies what could be the *right* or *ideal* operationalization of your network. For instance, although there are numerous networks that we can identify from Twitter data, once we have a concrete question in mind, it becomes clearer which network fits the question. Do you want to study how social media users are exposed to misinformation? This probably requires you to think about the network structure through which the misinformation can flow, the "following" network on Twitter. Now we have a clue what network to use. Or do you want to study how Twitter users engage in political debates? Then it will be critical to examine the network of "mentions" and "replies." Although these examples are probably still too simplistic, the point is that, in many cases, there are numerous *networks* that coexist in the same dataset and it is crucial to clarify the question and the corresponding, reasonable operationalization of the network.

More often than not, it may be impossible to collect or work with the ideal dataset, but even so, it is still important to clarify what *could* be the ideal network data because it lets us identify the assumptions, compromises, and biases that we introduce when choosing the final network data. For instance, let's assume that we are interested in studying how conspiracy theories about vaccines spread across society.[3] It is of course impossible to work with data that capture *every* possible social communication involving

---

[2] A clique is a completely dense subgraph. It is a set of nodes within the network where every pair of nodes is connected.

[3] Antivaccination campaigns, often rooted in misinformation and fear, have plagued vaccination efforts almost from the very beginning of Jenner's smallpox inoculations [10].

the conspiracy theories. Yet, it is possible to narrow the question to something feasible, such as: *"how did a particular conspiracy theory spread on Twitter?"* or *"are there Facebook groups that promote vaccine conspiracy theories?"* In this process of finding a feasible study, we inevitably introduce biases and assumptions, such as the study population (those who use a particular social media vs. everyone) and the types of communication. It is critical to explicitly identify these biases and assumptions.

Especially when working with existing data, it is easy to overlook these simple, yet fundamental questions. The network data that is currently available may not be right to address the question and it is essential to collect new data or transform the current data. So, *what is your question?*

5.2   Collecting, constructing, and cleaning network data

Once we have a question, we can concretely think about the ideal network that we would like to analyze. Sometimes we can and should collect the network data ourselves; sometimes we need to clarify how to operationalize the network from an existing dataset; in other cases, we should find an existing dataset.

5.2.1   Designed data vs. organic data

As discussed earlier in this chapter, one important distinction about data is whether the data is *designed* or *organic* (or "found") [194, 412]. Designed data refers to datasets that are collected to answer specific research questions. Organic data, on the other hand, refers to datasets that were collected for purposes other than the questions being asked. They may be collected by a company to answer their own questions about products, users, etc., but are later repurposed for research. Designed data tends to be small (because they are collected specifically for the research in question) but high-quality (same reason); organic data tends to be big but may not necessarily contain the exact information that can best answer the question.

Let us talk about an example. If we ask a question about the spread of vaccine conspiracy theories on Twitter, we can either design the data or use organic data. Designed data would most likely mean that we perform a survey that asks people on Twitter about the exact things that we want to know about them. The organic data route would mean that we simply take whatever the Twitter data already contain and try to find with it, as best we can, reasonable operationalizations and proxies.

When we design our datasets, because we are already asking the exact questions that we want to get answers to, the usual barrier is the ability and resources to collect enough data. On the other hand, when we work with existing data, the most critical question is often about finding the best proxies for the certain quantities that we really want to estimate and dealing with all kinds of biases present in the data.

Returning to the misinformation example, Twitter is (presumably) not *designed* to gather vaccine misinformation data, so we will necessarily have organic data. When we use the organic data that Twitter is already collecting, we would have to take data from Twitter and build a network from the data. This process again involves lots of choices. Usually it is not possible to access Twitter's full database (unless you work *at* Twitter) meaning we would need to collect what Twitter makes public. There can be multiple ways to do so (e.g., using the "Streaming API" or the "Search API"[4]) and different data sources may introduce different types of biases. The operationalization of network edges also requires choices. Should we collect who follows whom? Should we simply collect retweets? How about mentions? Shall we combine all of these link types or just pick one of them? These are all tricky questions that need to be carefully evaluated based on the research question and the characteristics of the edge types.

Even after defining a network, questions remain. There are other types of data we can collect. For instance, we need to decide how much information we want to collect about nodes and links. Do we want to collect the profiles of the users? They may change, so do we want to monitor them over time? We can go even further. Do we want to contact those users directly and perform surveys? Directly asking people can lead to invaluable data that lets us peek into their social communication outside Twitter.[5] We can also choose to collect more information about individual edges. Say we want to use only the retweets. Then, should we collect various information *about* those individual retweets, such as the number of retweets, those who retweeted a particular tweet, timestamps of the retweets, and so on? There are numerous metadata (Ch. 9) that can be potentially helpful later. All these decisions affect how the results can be interpreted and how much the study can reveal.

Unfortunately, the importance of these questions is often overlooked. Many studies will simply choose the most convenient or already walked paths, which can create a mismatch between what they are asking and what the data represents.

5.2.2 Exploratory and confirmatory network analysis

Broadly speaking, given an extracted network, analysis can take two paths, exploratory or confirmatory. Both paths can rely on statistical, computational, or mathematical tools, tools which form the bread-and-butter of network science research. And they are not exclusive: the tangled path of real science often follows both.

Exploratory network analysis   Many network questions can be directly answered or illuminated greatly by performing exploratory network analysis. Exploratory network analysis usually involves network visualizations as well as measurements of network statistics. Computational methods, such as community detection or graph embedding, can also be employed for this analysis as well. The goal of exploratory analysis is to understand the overall network structure and to guide further analysis.

Confirmatory network analysis   On the other hand, confirmatory analysis aims to test a concrete hypothesis by employing statistical models. For instance, consider a protein–protein interaction network derived from the Affinity Purification–Mass Spectrometry (AP–MS) technique discussed earlier. AP–MS discovers interactions using a

---

[4] An API or Application Programming Interface is a specification that allows computer processes to communicate with one another and send and receive data. In this case, the API is a specification created by Twitter that says, in essence, "Here is how you can ask us for data and this is what the data will look like if you are allowed to see it and ask for it correctly."

[5] Surveys should always be done following appropriate ethical guidelines; Ch. 3.

"bait protein" designed to capture an entire interacting complex of associated "prey" proteins that can then be identified by subsequent analysis. As discussed above, from the network's point of view, this technique does not sample links independently but instead gathers entire dense subgraphs, groups of nodes called cliques which are completely connected. A network derived from AP–MS will not be representative of the entire structure—for example, we expect triangles, cliques of size 3, to be overrepresented, perhaps heavily so. We can test this hypothesis by building a null model that captures the overall density of our AP–MS network but randomizes away the complexes, then we can compare the observed quantity of triangles in the real data with what we would observe if the null model held true.

We dive deeper into network exploration and confirmatory analysis in Ch. 11; Ch. 13 focuses on visualizing networks, which are often used for presentation but are also useful for network data exploration.

5.3   Iterating on the cycle

Of course, the story does not end after collecting, cleaning, and analyzing the network data. The results of a study are not open and shut. Instead, as the current study addressed your original questions, new questions will emerge. The data were limited to a single time period; does the result hold at other moments? The network only consisted of links from a given layer; what about other layers? The network came from an API that aggregated all activities; do the results hold under a different aggregation? Or when the API changes?

The net result of this is that the outcome of one study will inform the starting point of the next. A new network dataset can be collected and now, armed with the knowledge gleaned from the previous study, we can learn better than before about the system of interest.

5.4   Summary

Network studies follow an explicit form, from framing questions and gathering data, to processing those data and drawing conclusions. And data processing leads to new questions, leading to new data and so forth. Network studies follow a repeating life cycle. Yet along the way, many different choices will confront the researcher, who must be mindful of the choices they are making with their data and the choices of tools and techniques they are using to study their data.

Bibliographic remarks

The tension between open-ended exploration and hypothesis-driven confirmation has been at the foundations of science since the enterprise began. Yanai and Lercher [500] discuss some of the disadvantages of having a hypothesis, that it may lead researchers to fixate their attention on the preexisting question, causing them to miss out on other, critical features within the data. Yet being completely free of hypotheses is also risky, especially if statistical tests intended for confirmatory work are used in an exploratory fashion. Head et al. [207] discuss the prevalence and effects of such practices on scientific research. The long-standing tension between these competing views can only mean that they must be synthesized. As Tukey writes in his seminal book on exploratory data analysis, "[E]xploratory and confirmatory can—and should—proceed side-by-side" [465].

Readers interested in learning more about the dichotomy between designed and organic data are encouraged to consult Groves [194], which includes interesting examples of approaches that use both types of data. Salganik [412] discusses the dichotomy further, in particular in regards to the "big data" era and its influence on social science research.

Readers interested in learning more about using online social network and social media data may wish to consult Russell [409]. Kumar et al. [260] cover further the various approaches and nuances to using Twitter in particular, going beyond the examples we discussed here. Keep in mind that the world of social media is a fast-moving place, and it is likely that these sources will be outmoded in some respects despite still serving as pertinent introductions to the area.

Exercises

5.1 Suppose you have been hired by a large college campus to study how college students spend time together. You realize that college students often spend a lot of time on their computers and smartphones! If you have access to data from the campus wireless network, you should be able to track students, especially if there are a large number of wireless access points.

Write a brief study proposal to give to the college to convince the IT department, which runs the wireless network, to give you access to data that can help you answer your question. In your proposal, map out how a network where nodes are students will be constructed from the data, what it tells you about how students spend time together, what it *doesn't* tell you about how students spend time together, and what kind of next steps may be involved after your study is complete.

5.2 What ethical concerns are there with the data you wish to collect from the previous question? How can the study be modified to address these concerns?

5.3 (**Focal network**) Consider the flavor network. This network was derived from a reference text used by food chemists to give specific flavors to new foods. Suppose a new edition of this book is released. Briefly, describe a study to check the existing flavor network and see if it needs to be updated. Assuming it does need to be updated, describe how to go about doing so.

Chapter 6

Gathering data

In working with network data, data acquisition is often the most basic yet the most important and challenging step. The availability of data and norms around data vary drastically across different areas and types of research. A team of biologists may spend more than a decade running assays to gather a cell's interactome; another team of biologists may only analyze publicly available data. A social scientist may spend years conducting surveys of underrepresented groups. A computational social scientist may examine the entire network of Facebook. An economist may comb through large financial documents to gather tables of data on stakes in corporate holdings.

Often, these data are hard won: suppose you are traveling between villages in southern Cameroon, interviewing residents about their social ties as part of a study on land-usage microgrants. Each link in the social network is the result of significant physical effort on the part of the researcher, and there are thousands of such links. Or imagine you are installing metering devices on a municipal power grid, gathering data on the topology of the grid through system documentation before beginning a project to help a local utility improve their use of renewable energy. Here too we have significant outlays of time and money.

On the other hand, some network data can be gathered with little effort. While building a major social media platform is an enormous undertaking, once they have launched a data API, a researcher can fairly quickly gather potentially massive amounts of data just by writing computer code to automatically send requests to that API.[1] And the growth of open science has made such data availability far more prevalent in research as well.

Yet whether data are available with comparably little effort, or it took a herculean effort, those data must still be gathered. Data gathering precedes network analysis and the resulting data are often not in an apparent network form, except perhaps when you are able to leverage an existing network data source. More generally, you may find yourself faced with gathering "raw" data from which a network (or multiple networks) will need to be extracted (Ch. 7). Here we discuss gathering these data in a variety of potential domains.

---

[1] Consider also the ethical issues if instead of a researcher gathering all that data, it is a nefarious actor (Ch. 3).

6.1   Motives, means and opportunities

Why gather data? We encourage researchers to devise a "mission statement" as they begin to gather data. Ask what you wish to accomplish and how and where the data will address this. A specific research question is often the strongest anchoring point for such a mission statement:

> *We wish to gather data describing the kinships and social ties of individuals participating in [an NGO's] microgrant program. We seek to understand how family ties versus friendships act as support for grant holders.*

Often a data gathering effort will be an ongoing process and such a statement can allow one to remain on track.

Establishing a data gathering effort

How to gather data? This is a broad question that will pull in the core of your field. In anthropology, data sources and best practices are very different than in molecular biology or computer science. One useful lens through which to view mechanisms for collecting data is whether the data is gathered manually, through human work, or automatically, using computer programs.

Manual data gathering

Broadly speaking, manual data gathering falls into four categories:

Data entry Researchers (or hired data coders) manually enter data into a machine-accessible form such as a database or spreadsheet. Reliable data entry often requires a coding standard for how different records are produced, as well as reliability checks. Reliability can be supported by redundancy—having multiple coders enter the same data and then confirm their overlay. This is also useful for data entry that may involve judgment calls on the part of the coders. Statistical methods, such as inter-rater reliability, are helpful here. Data entry can also be a part of other types of manual gathering.

Surveying Some problems require data collection in the field. Gathering soil samples from locations within an animal pasture, for example, or counting plant species at various points through a forest. Surveying may involve gathering physical samples to be returned to a lab, recording the readings taken by a scientific instrument placed in the field, or manually noting observations. The latter cases thus overlap with data entry.

Interviewing Surveys specifically focused on meeting with people and asking questions. While interview data are generally small scale given the manual nature of the data collection and the resources of most researchers, the advent of the Internet allowed for interviews reaching thousands of people. Interview data may be challenging if the interview is free-ranging in the sense that the interviewee's responses to past questions prompt the interviewer to ask different followup questions.

Experimenting Surveying and interviewing are generally examples of observational efforts, data are gathered passively in the sense that the researcher wishes to describe a phenomena of interest as is, without manipulation. Experiments on the other hand, occur when the researcher is able to control or manipulate factors relating to the phenomena, and wishes to design manipulations to test hypotheses. Manual[2] experiments are often the gold standard, and randomized controlled experiments best address specific causal questions.

Automated data gathering

Parallel to manual efforts are automated data gathering, where computer programs are created to collect data for subsequent processing and analysis. For example, you may code up a "crawler" that accesses an online service's API. Starting from an initial record, your crawler retrieves and records all the data fields associated with that record. One such field contains a list of related records. Your crawler adds these to a queue of records to be retrieved and then proceeds to the next record on its "to be crawled" queue.[3]

Coding up a crawler to extract a network structure is straightforward when the records being crawled already contain a field of links. Sometimes, however, the links between records need to be extracted in a less obvious manner. For instance, a *web* crawler records the contents of a web page (HTML data) and other web pages are found by parsing the HTML and extracting the hyperlinks to those pages. Even more challenging is crawling unstructured, natural language documents. Without any structure, you may need to impose rules to extract relationships, perhaps using natural language processing algorithms such as named entity recognition.

Automated data collection has the obvious advantage of labor savings—the computer does most of the work—but it has another useful advantage: you can also automate some of the record-keeping. Suppose you are downloading a collection of social network datasets. You write code that retrieves a list of networks from an online server. Then, for each network, you download the network, save the original file to a specified location, and append a message to a separate "log" file denoting the status of the download (success or failure) and other useful details such as the time when the download occurred. These records are invaluable for documenting the "experiment" (in this case, the data crawl). If you ever need to gather more data, or regather existing data, these records will greatly simplify that task. We discuss record-keeping for research in general in Ch. 17.

Choosing between them—manual vs. automatic

When is manual data gathering appropriate? Perhaps a better question to ask is whether automated gathering is appropriate then see if manual gathering is possible when automated gathering isn't.

---

[2] Of course, in some areas of research, particularly the life sciences, high-throughput experiments are now automated to a fantastic degree, often with robots and complicated machinery. Nowadays, there is not always a clear line dividing automated and manual experiments.

[3] Depending on the ordering of records added to this queue and crawled, this crawler could be following either of the classic depth-first search or breadth-first search (Alg. 4.1) algorithms.

A computer program can help with data gathering when the data are machine-readable and machine-accessible. Nowadays, most scientific instruments, microscopes, telescopes, air quality sensors, and so forth, are electronic and produce and store data electronically. These data are readable by machine.

Data are accessible when they are available from your machine. If your data are spread across various databases or available online through web servers, then you should be able to craft a program that queries these resources and collects the data onto your own machine. But if the data are not reachable from a machine, say they are stored in analog media, or the data are not yet collected, then manual work is the only choice.

The line between machine-readable and machine-unreadable blurs when considering what is being read from the data. Consider natural language data—perhaps you wish to study a large corpus of recorded interviews. The data are stored in an audio file format and are easily listened from any computer. The audio data is encoded and readable by the computer. But the *content* of the audio, the words spoken by the participants, are not really available to the machine. Fortunately, speech recognition, the process of converting spoken audio to text transcriptions of what was spoken, has become far more reliable and practical. Running speech recognition algorithms over the audio files would make text data available and machine-readable, although currently this is still not completely reliable.

Hybrid approaches   Of course, a sufficiently complex (and interesting) data gathering effort may span both manual and automated approaches. One may be interested in data from historical documents that are only available on paper. You begin with a manual effort, scanning or photographing the paper pages to create digital images. For this, you should keep detailed records of who is scanning and when, as well as information such as the document being scanned, if it is not evident from the digital image. Adopting standards for reporting and organizing the manually gathered data, such as clear and consistent filenames, is important.

Once the documents are scanned, it may be possible to switch over to automation. Nowadays, *optical character recognition* (OCR), computer algorithms that can recognize text within images, are increasingly high-quality and reliable. You therefore begin the automated effort by OCRing the digital images, creating a text file corresponding to each image. These text can now be sent to additional programs. For example, *entity extraction* may be able to recognize when entities, people, places, things, are mentioned in the text. These entities now become the focus of your network studies.

Leveraging existing data: network data repositories

The simplest case of network data gathering occurs when network data already exist and your task is to retrieve it. Much of the heavy work has already been accomplished: the format of the data are probably clear, the decisions made to process the network are set,[4] and you merely need to determine the right data to look at and record where and when you retrieved it.

---

[4] Although, there will still be value in reexamining such decisions.

Currently, there are multiple network data repositories online. They are especially useful when the question is about general properties of networks (e.g., do most networks exhibit heavy-tailed degree distribution?), or about general-purpose algorithms that can be applied to a broad class of networks. By examining these properties across many networks over an entire repository, you can develop an understanding of these properties at scale.

Working with a repository of network data will generally require understanding how to download the data, what file format(s) are used, and how to read those file formats with your code. A repository may offer an API allowing you to write code to access the data automatically or it may only offer manual downloads where you must navigate to pages on the repository website and then download the networks there. But you may have some luck after examining the web addresses of these pages: the repository may be organized in a way that the download links are in a predictable format, allowing you to write a program that builds those links then downloads them automatically, in essence converting the repository into a simple API.[5]

In either case, after downloading, you will need to understand the form of the file that you received. (Many repositories offer multiple file formats to choose from.) A network may be stored in a basic text format like a CSV file, or it may be stored in a specialized JSON or XML-based format. Confused about files? We discuss file formats in detail in Ch. 8.

Non-network data as precursors

It's all well and good to have readily available network data to work with, and you should be strongly encouraged to always find out what may already be available when beginning a project. But in reality it can be quite rare to have a network already built for you. Instead you need to extract the network yourself, from non-network data. Such is the life of the pioneer!

Some examples of non-network data include natural language (text) data, where network relations can be drawn out by processing, collections of time series where a network of interactions may be driving correlations in the different time series, and amino acid sequencing data where a network of (non)local interactions can be inferred and used to predict protein structure.

We recommend separating the tasks of non-network data gathering from the tasks of network extraction. Keeping them separated lets you focus on one problem at a time. (Divide-and-conquer!) Further, if you try to combine the steps, extracting the network while gathering the data, and only keep the network as you go, you may be in trouble if you need to revise any aspects of the extraction step. (These revisions are surprisingly common.) Instead, gather and save the non-network data, then, afterwards, you can explore and iterate on the extraction task.

We refer to this problem, extracting the network from precursor data as the "upstream task" and it must occur before any "downstream" network analysis. The upstream task is so important we dedicate an entire chapter, Ch. 7, to discussing it.

---

[5] Keep in mind that the web server providing the data may not have enough resources to serve huge numbers of simultaneous downloads, which may be created by your code. Be kind.

Non-network data as side information   Networks can be highly informative in and of themselves, but they also have great value as a means to organize additional information. Associating non-network information with the nodes, or links, or both, often gives us a better understanding than looking at either alone. Incorporating such non-network metadata as attributes describing the nodes or links is the focus of Ch. 9.

 When additional data are available describing the nodes and links, even more can be learned about the network, and the network can tell us even more about the node and link data.

Recognizing sampling effects and limitations

Lastly, it must be said that you need to recognize, from the beginning of data gathering, that you are deriving a sample of data. How you gather the data, and when, influence the final network you produce. For instance, if you build a crawler that spreads outward from a random data record (which becomes a node in the network), you will have both an incomplete and a biased view (Ch. 21) of the network: the crawler will be more likely to spread to highly connected nodes, and if the network is disconnected, you won't be able to see any data on nodes the crawler cannot reach.

 Do not expect your data to be complete or unbiased. How you gathered the data may drastically change what the final network looks like.

Understanding sampling effects, limitations, and biases is crucial, and Chs. 10, 11, and 24 and Sec. 23.5, all touch upon different aspects of this central issue.

6.2   Data gathering across fields

While it's helpful to discuss data gathering in a generic context, specific research fields will have both opportunities and issues to be addressed. Specific fields may have standard data repositories you can use. Likewise, fields may have established practices for how data are collected and stored. Researchers in those areas should be aware of these specifics. Here we describe some specific points for social networks, biological networks, and other types of networks.

6.2.1   Social networks

Historically, observations or surveys were the primary method of collecting social network data. Indeed, most of the oldest network datasets were collected through surveys or observations. For example, Jacob Moreno's pioneering work on social networks (Ch. 1) featured the social network of a classroom gathered in such a manner [317].

Manually surveying a social network proceeds in a local fashion. Each survey participant is interviewed and asked, essentially, "who are your friends?" (Working one-by-one through the network is an example of *egocentric network sampling*.) This introduces some biases. One, the answer to the survey question largely depends on how one defines "friends" and some participants may have looser or stricter definitions than others. Two, by its nature, the derived network is directed: Amir may claim to be friends with Bob, while Bob may not consider Amir to be a friend.

As opposed to manual collection, digital communication provides a new opportunity to collect more objective (but still biased in many ways) social network data. For instance, the mobile phone communication network leverages call-detail records (CDR) to collect how much one communicates with others. CDR data have been used heavily in social network studies, supporting important claims about how social networks are organized. Yet, they are biased as well. They cannot capture other channels of communication such as online social networks or in-person (face-to-face) communication. If two people are very close but their communication is predominantly in person, their relationship may be invisible in the call data. They also cannot distinguish the context of the calls. One may be on the phone for a long time with someone not because they are friends but for mundane reasons such as service calls about utilities, etc. Call data also tend to be samples taken from a single mobile phone provider. In most cases, a provider only covers a fraction of the entire population. That fraction may not be representative of the population at large and information about how users in the sample communicate with those outside it will be incomplete.

Complementing, and in some ways supplanting, mobile phones, social media and online social networks are a major channel for social communication. Although some social media services have much larger market share than others, no one social media completely dominates our online communication. Data from these sources share similar limitations with the call data: population biases, biases in the type of communication, and limited information on communications outside the social media "layer." Researchers are often unable to distinguish whether users share accounts or users have multiple accounts. And of course, one needs to consider privacy issues with social media data (Ch. 3).

All this is not to say these data are not useful. Quite the opposite. While there are many gaps, biases, and privacy concerns in these forms of social data, used appropriately, acknowledging their limitations, they are still an invaluable way to study social networks and human communication.

6.2.2 Biological networks

Gathering biological network data has been a major scientific effort for decades. Most of these datasets come from experiments, but some networks are actually derived by mining scientific publications. We discuss some examples.

As gene expression and gene sequencing technologies matured, it became cheaper and faster to gather large quantities of genetic data. At the same time, more and more high-throughput experimental techniques developed in other areas of study. For instance, experimental protocols were developed to measure metabolic pathways, protein–protein interactions (PPI), and gene regulatory (gene–protein) interactions. Scientists began exploring these data and building networks. Metabolic networks represent cellular metabolites as nodes and links indicate chemical reactions. PPI networks use nodes to represent proteins and links indicate proteins that interact with one another. For PPI networks, a variety of competing and complementary protocols exist, using different experimental techniques to infer interactions. And of course, interactions that occur in a high-throughput assay may not occur in the cell, and vice versa. Over time, large databases have been built and shared with researchers. One example is the Kyoto Encyclopedia of Genes and Genomes (KEGG), which contains a wealth of data on metabolic pathways, gene sequencings, enzymes, diseases classifications, and more.

In neuroscience, networks have been a driving force from the beginning (Ch. 1). The nematode *C. elegans* was the first organism to have its neuronal network fully mapped out, a painstaking multiyear axon-tracing effort that continues to bear scientific fruit. Brain networks, the "connectome," are increasingly studied at both an anatomical level and through the use of neuroimaging techniques such as MRI machines. Promising population-scale studies are now being conducted, including the Adolescent Brain Cognitive Development (ABCD) study [97], an exciting and massive longitudinal study that includes neuroimaging data. However, the "raw" data coming from MRI machines is far removed from a final network, and considerable research investigates how best to extract a meaningful network from such data. And as with the previous examples, we must be circumspect about what the data are telling us (and not telling us) about the brain.

The scientific literature is now also becoming a source of network data. For example, in a PPI network, establishing an interaction between even a single pair of proteins may warrant a scientific publication, and there are now decades of such publications. Using text mining tools and manual data entry, researchers have gathered *literature curation* (LC) networks, PPI networks where links exist between proteins when interactions have been reported in publications. Such LC networks are used to benchmark against new experimental protocols. As the tsunami of scientific papers continues to grow, it is likely more areas of research will begin extracting networks from the scientific literature, a process made easier and likely more reliable as better natural language processing tools develop.

6.2.3   Other networks

All manner of networks can be derived from all kinds of data. In fact, it's quite rare to see a problem that *doesn't* admit some kind of useful network representation.

Citation networks appear throughout scholarship. These could be scientific or other scholarly publications citing one another, patent filings citing preexisting patents, or legal documents citing case law. Here the data gathered are either the contents of the citing documents themselves, whether written publications, patent text, or legal documents, or else metadata has already been extracted. The latter situation is helpful, as drawing out citations from natural language can be challenging, especially when you need to disambiguate the citations to ensure that citations written in different ways always refer to the same document. (Often, solving this problem is a major undertaking in and of itself. Be wary of what effects may be introduced if this job is done poorly.)

 Can you clearly identify nodes in the network, always being able to determine with certainty whether different references to nodes refer to different nodes or to the same node? If not, you are faced with a ***disambiguation*** task.

Collaboration networks are also common, helping us understand how individuals work together in various organizations. Nodes represent individuals and links between nodes represent individuals who collaborated together. Collaborators could be scientists or scholars who coauthor papers, or they could be actors who appeared in films together, or they could be politicians who cowrote or cosponsored bills. Links between the nodes capture coauthoring, costarring, or cosponsoring between the individuals. Here the data to be gathered will link individuals with the items they collaborated on: coauthors linked with papers, actors linked with movies, or politicians linked with bills.[6] When data are gathered, these links may be easily extracted or there may be challenges to face. Just as citation records may need to be disambiguated, the names of collaborators may also need processing.

Physical infrastructure networks are another class of network data to be gathered. The road network can be extracted from map data or even satellite photos: nodes represent roads and two nodes are connected if one can drive directly from one road onto the other. A power grid is a literal, physical network of transmission and distribution lines connecting power generators and consumers. The network itself can be inferred from the maintenance and construction records of the grid operator but often it is the real-time use of the grid that is most interesting. Phaser Measurement Units (PMUs) can be installed throughout the grid to capture and record the electrical dynamics in near-real time. And another network example is a network of data-transmitting computers, The autonomous systems layer of the Internet, for instance. The Internet, while a built network, is self-organized in a somewhat ad hoc manner, and it remains an interesting inference problem to determine an accurate picture of its topology from routing tables and other data sources, particularly as it constantly changes.

These examples are just the tip of the iceberg when it comes to the possibilities for network data.

6.3 Summary

In this chapter, we have moved one step along the network study life cycle. Data gathering is key to future success—garbage in, garbage out—making it critical to ensure the best quality and most appropriate data is used to power your investigation. Key to data gathering is good record-keeping and data provenance—a point we discuss in detail in the *Interlude*, particularly Ch. 18.

We now turn our attention to the next step in the network study's life cycle, extracting networks from the underlying data.

---

[6] Notice in all these examples that we begin with a network where nodes fall into two categories, collaborators and some kind of artifact (paper, film, bill) and then generate a network containing only collaborators with links between collaborators who have artifacts in common. The collaborator–artifact network is an example of a bipartite network, and the collaborator–collaborator network is a *projection* of the bipartite network onto the collaborators. We could also project onto the artifacts, creating an entirely different network where two artifacts are connected if they shared a collaborator. Keep in mind when working if there is a bipartite network "behind-the-scenes," it may be better to study it than the projection.

Bibliographic remarks

Readers interested in social data gathering are encouraged to check out *Bit by Bit* by Matthew Salganik, a fascinating general account of the emerging area of computational social science [412].

The history surrounding the efforts to study *C. elegans* as a model organism and map out its synaptic connections is riveting. Emmons [146] gives an enjoyable review with a wealth of photographs and evocative imagery. Brenner [74] is a personal account of the early days of *C. elegans* work.

High-throughput screening is a major development in systems biology underlying much modern network or "-omics" data. For a history of its development, see Pereira and Williams [368].

Readers interested in learning more about brain networks and network neuroscience may wish to consult Sporns [442] and Bassett and Sporns [44].

Exercises

6.1 Disambiguation (also known as record linkage) is a common task when processing natural language data. Suppose you are studying a network of actors who costarred in films. The data are the texts of the end credits of each film. Describe some disambiguation challenges you anticipate facing if you try to put the actors into a network from these credits.

6.2 (**Focal network**) A PPI network such as HuRI is generated by an experimental protocol that tests pairs of proteins one at a time to see if they interact. Suppose there are N proteins in total and the experiment has tested $p = 10\%$ of the protein pairs. Approximately 1% of test pairs should show an interaction. How many edges are in the network? How many edges should be in the network if all pairs are tested? What assumptions do you make as you answer these questions? HuRI contains 8272 proteins. If it takes one second to test a pair of nodes for an interaction, how long will it take to exhaustively test for edges.

6.3 (**Focal network**) Read the Zachary Karate Club paper [505]. Draft a brief (500-word) description of the data generating process, how the data were collected, and any processing, filtering, or other manipulation that was performed.

Chapter 7

Extracting networks from data — the "upstream task"

The goal of this chapter is to recognize that, while there are cases where it is straightforward and unambiguous to define a network given data, often a researcher must make choices in how they define the network and that those choices, preceding most of the work on analyzing the network, have outsized consequences for that subsequent analysis.

7.1 What is it?

Sitting between gathering the data and studying the network is the *upstream task*: how to define the network from the underlying or original data. Defining the network precedes all subsequent or "downstream" tasks, tasks we will focus on in later chapters. Often those tasks are the focus of network scientists who take the network as a given and focus their efforts on methods using those data.

> **ⓘ** The *upstream task* is to define the network from the data before you begin to analyze it. Sometimes this is easier said than done: researchers often face choices, sometimes difficult choices, before they can begin to study their network.

The simplest way to visualize the upstream task is to ask yourself two questions: *"What are the nodes?"* and *"What are the links?"* By focusing on these questions, while they seem rather elementary, you can at times reveal important details about why the network was made and even if it should have been defined differently.

Consider these two questions as we discuss several examples below.

Example: social network In social network analysis, a common data source has been social media services like Twitter. Some provide a programming interface (called an API) to retrieve data on users and their activities. But there are multiple ways to generate the network from social media activity data. For instance, from Twitter, one

83

could construct the network using mentions, retweets, and followings. In these cases, the nodes are Twitter users but how to define links may not be clear. Which definition should we choose? Should we simply gather all of them and merge them, defining links between users when any mention, retweet, or following occurs? This can be a critical mistake depending on the research question. It has been shown that each of these connections carry different meaning [72, 115]. In political discourse[1] on Twitter, *retweets* strongly signify agreement with the original author. By contrast, *mentions* tend to cross the political aisle and are used as a channel for fighting and mockery of political opponents. In other words, the retweet network captures in-group relationships while mentions capture both.

Example: protein–protein interaction network   Protein interactions are measured via high-throughput experiments (Ch. 6). When defining a network such as HuRI (Ch. 2), we need to pay attention to our experimental methods because the characteristics and biases in the methods strongly affect downstream tasks and network properties. For instance, the AP–MS and Y2H methods extract different types of links (Ch. 5), leading to potentially very different networks. But often choices need to be made, not just between methods, but within a method, to use it appropriately. To build the HuRI network, for example, the Y2H method was applied to an $N \times N$ search space of proteins nine separate times. And three different versions of Y2H were used [283]. Luck et al. varied the versions and replicated their screenings specifically to enhance the robustness of the discovered interactions. Future studies may vary their protocols further. Overall, we can see how different experimental methods, and choices when applying them, can yield very different pictures of the network being inferred.

Example: brain network   Neuroscientists use imaging experiments to infer the hidden structure and functional dynamics of the living brain (Fig. 7.1). As with protein interaction networks discussed above, experimental protocols will affect the final network being extracted. The brain scanner is part of the upstream task. The field of neuroimaging has taken great pains to understand the most appropriate use of imaging studies, with sometimes great debate as to their ability to yield good descriptions of the brain. From this work has arisen a field of statistical analysis aimed at inferring the nodes and links of brain networks, the connectome. These analyses include methods to pull out signals from time series measurements of blood oxygen levels in the brain, the central focus of functional MRI (fMRI) imaging. Moving from these signals to a network requires many choices of algorithms and parameters, all of which influence the final form of the brain network. The upstream task in network neuroscience is rich and complex.

7.2   Why does it matter?

We emphasize the importance of the upstream task because everything subsequent depends on it. Perhaps you wish to study the community structure in your network but

---

[1] If you can call it that.

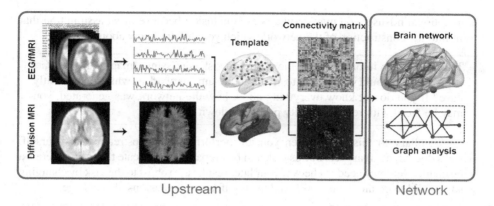

Figure 7.1 An illustration of the upstream task in network neuroscience. Here (left-to-right) functional or structural MRI data are recorded and processed into standardized time series or fiber bundles, respectively, then connectivity matrices are generated which are finally processed to make networks. Many choices are made along the way, from what algorithm to use for standardizing the data to what measure to use for comparing the time series. Network neuroscience is an example where the upstream task is highly visible and well documented due to its complexity, but many other areas feature networks drawn from comparably involved upstream tasks. Figure adapted from [94].

your definition of links depends on a data processing algorithm. If a small change to a parameter of that algorithm leads to a drastic change in the network's structure, then is your discovered community structure fundamental to the data you're investigating or is it simply due to the data algorithm?

Pay close attention to the "data generating process"!

 Always think critically about the upstream task.

When *you* perform the task:
- What would change if you completed the task differently?
- Can you check whether your research results are robust to changes to the task?

When *someone else* performed the task:
- Do you have enough information to understand what they did and how?
- Did they do it in a manner appropriate to your problem?

A further effect of drawing attention to the upstream task is that it shines light on an important aspect of ***data provenance***. As network data are shared, researchers can become fixated on the network and lose track of the preceding work, the extenuating circumstances, that went into creating that data.[2] When this happens there is a risk of

---

[2] Of course, fixating on simple messages while losing track of extenuating circumstances is by no means limited to network studies. Many scientific problems develop a "folk wisdom" where a fact or figure gets passed around, taken for granted and assumed true, all while being supported by one or a few studies or experiments that were limited in scope or even incorrect. One example is the idea that in cold weather you

unintentional misuse, you may draw a poor conclusion because of an assumption that went into the construction of the network which you did not know about.

 Take care when information on the upstream task is missing. Imagine if you start from scratch, would you be able to arrive at the same network as what you currently have? If you do not know everything about how the network was generated, you may be faced with reproducibility problems down the line.

A corollary to this holds when you are performing the upstream task yourself. Always document all aspects of this task and be prepared to replicate the task—there's a good chance you may need to check if your later results are robust to the task by changing what you did to create the network and testing if your conclusions also change.

 Always document your upstream task thoroughly. Ensure this information stays with the network you extract. Be prepared to modify and repeat the upstream task.

7.3 Summary

This brief chapter discussed the upstream task, defining the network by creating a process to extract the network from the gathered data. Envision the upstream task by asking yourself, *"what are the nodes?"* and *"what are the links?,"* with the network following from those definitions. You will find these questions a useful guiding star as you work, and you can learn new insights by re-evaluating their answers from time to time.

Are you satisfied (currently) with the network you've extracted? Good, we can now turn our attention to incorporating non-network data (Ch. 9), further refining the network (Ch. 10), or exploring the network (Chs. 11–13 and beyond).

Bibliographic remarks

Little ink has been spilled on the importance of the upstream task in network science. A notable exception is the excellent perspective piece by Butts [88], which asks such simple—but foundational questions—as "when is a node a node?" and "when is an edge an edge?"

Exercises

7.1 Describe a network where there is one answer to the question, "what is a node?" and from that answer there is really only one answer to the question, "what is an

---

lose most of your heat through your head, based on a single Army study and since called into question [479]. Another, more chilling example is the considerable confusion early in the COVID-19 pandemic that arose over whether the virus spread only over short distances in respiratory droplets or was "airborne" in particles smaller than 5 microns, which spread much farther. But this 5-micron cutoff, well supported by policy, is not well supported by research and a better cutoff for farther spread may in fact be 100 microns [482]!

edge?" Describe another network where, given what nodes represent, there are many answers to the question, "what is an edge?"

7.2 (**Focal network**) Consider the data generating process of HuRI. What biases could be present due to it?

7.3 (**Focal network**) Consider the developer collaboration network. Nodes represent GitHub users, links exist between developers who coedit files. Describe a few other meaningful definitions of links for these nodes.

7.4 (**Focal network**) Write a table summarizing, for each focal network, answers (in your own words) to the questions, "what are the nodes?" and "what are the links?" From these answers, do you see any similarities or differences between the focal networks?

Chapter 8

Implementation: storing and manipulating network data

You've done all this work on scoping out research, gathering data, and defining your nodes and edges, how do you actually get your hands dirty working with the network? Working with network data almost exclusively requires doing so computationally. You need to read the data into the computer's memory, then access and manipulate it in a meaningful manner programmatically.

In this chapter, we'll discuss how to represent network data inside a computer, with some examples of computational tasks and the data structures that enable those computations. We'll also discuss some typical (file) formats for storing network data and recommend some specific network libraries.

8.1 A home for your networks

Mathematically, we represent a network with a graph $G = (V, E)$ where V is the set of nodes and E is the set of edges. But what does this really do for us? What can we use it for? Yes, defining $V = \{a, b, c\}$ and $E = \{(a, b), (b, c), (c, a)\}$ gives us a network (in this case, a triangle) but is that useful? If we want to compute something, for instance the degree of node a, can we do so? Yes, for degree, we can examine each element in E in turn (in code, using a for loop) and determine if a is within it, giving us a count of a's edges, but is that efficient? It turns out there are multiple ways of storing network data and some are better than others!

Network data structures from the perspective of basic operations

Suppose you wish to work with network data using a programming language that you are fond of. The most practical route is to find a suitable network library in that language; most popular programming languages these days come with an extensive ecosystem of libraries, which often include those for working with networks and graphs. Although we will discuss such libraries later in this chapter, and you may never have to implement such a library yourself, it is still useful to understand the inner workings of such libraries and think computationally about network data structure and algorithms.

If we want to work with a network[1] in our code, we need to think about *what we will need to do with it*. What kind of operations will we do and what kind of calculations will we perform? At a basic level, we need a *data structure* that represents the topology of the network. If you have a node, is it within the network or not? If you have two nodes, are they neighbors or not? Given a node, can we obtain the set of its neighbors? Or what is its degree? Those operations, particularly understanding the neighborhoods of nodes, are the most basic operations for examining the structure of the network. For instance, if we can figure out the neighbors of any node, then we can easily implement the breadth-first search (BFS) algorithm (Ch. 4) to identify components in the network or to find the shortest path between two nodes.

On top of the capacity to represent and examine the structure, we also need a way to build, manipulate, and export that structure. We may be reading a data file that contains the network and if we want to put it into a data structure to represent the network and perform operations, we need the ability to insert nodes and edges into the network. To manipulate the network, we will need capabilities to also remove nodes and edges, as well as to change the attributes of nodes and edges. Finally, we may want to export the network to a file or to a database, which requires serializing and formatting the network data into various formats.

Let's discuss potential data structures for representing a network.

 Here is a useful exercise: create a table of basic operations and data structures. Enumerate the most basic operations for analyzing network data (e.g., "testing for connections between two nodes," "obtaining the set of neighbors," etc.) in combination with potential data structures that can be used to represent the network. Then, think about the computational complexity of all the operations for each data structure and compare them to each other. See Ch. 4 for basic data structures and computational complexity.

An array for your network

So, what would be a good data structure for representing a network? Is there one best data structure or are there multiple data structures, each of which is good for a particular purpose? Let's think about some potential data structures.

Node and edge sets   Just as we conceptualize a network mathematically as the sets of nodes and edges, we can implement the network using set data structures for "set of nodes (V)" and "set of edges (E)." They are straightforward to implement and it will be easy to check if a node or an edge exists within the network or not. However, this data structure underwhelms at the most basic operation of finding the neighbors of a node. To find the neighbors of a node, we need to iterate through the entire edge set, checking if the node is within each edge. Thus, the computational complexity of

---

[1] We'll limit this discussion to basic undirected and directed networks, but everything holds for more complex networks, provided a few more data structures and operations are in place.

this basic operation is $O(|E|)$ and running a basic algorithm like BFS will be very inefficient.[2]

Adjacency matrix   Another straightforward way to represent network data is with a two-dimensional (2D) array that stores an *adjacency matrix* **A**, where A_{ij} represents an edge (or edge weight) between i and j,

$$A_{ij} = \begin{cases} 1 \text{ or } w_{ij} & \text{if the } i\text{th and } j\text{th nodes are connected,} \\ 0 & \text{otherwise.} \end{cases} \tag{8.1}$$

Elements in **A** are binary for unweighted networks or they can be edge weights w_{ij} if the network is weighted. Because many network operations can be represented mathematically as matrix operations (Ch. 25), this data structure can be mathematically intuitive and convenient. Directed and undirected networks can both be represented by allowing the matrix to be asymmetric ($A_{ij} \neq A_{ji}$) or symmetric ($A_{ij} = A_{ji}$). And it is trivial to test a connection between two nodes: we just need to check if A_{ij} is zero or not.

But it's not all sunshine and rainbows when it comes to using **A**. The problem is again the efficiency of basic operations. To find out the degree of a node or obtain the set of neighbors of a node, we have to scan the entire row or column of the matrix, which requires $O(|V|)$ operations. Essentially, you count every node in the network, but you really only need to count the neighbors of i.

Furthermore, to store a network, we need to allocate a $|V| \times |V|$ matrix, which is inefficient especially if the network is sparse and large.[3] (We discuss sparse formats below.) In other words, storing the entire adjacency matrix as a 2D array is usually not a good data structure for representing a network.[4]

Lastly, a third reason to avoid an adjacency matrix is simply the lack of flexibility when it comes to representing nodes. A matrix of size $N \times N$ has row and column indices of $1, 2, \ldots, N$.[5] So we can track node 1, node 2, and the like, but we cannot so easily tell which node is 1, which node is 2, etc. If we had an online social network where usernames were used to identify nodes, for instance, we would need a separate data structure tracking which username corresponds to node 1, which corresponds to node 2, and so forth. This is not a particularly burdensome requirement, although it creates opportunities for bugs if node IDs are not mapped correctly, but it just isn't necessary. Alternative data structures that we will cover shortly allow for arbitrary strings and numbers as keys and set elements, and so we need not worry about further tracking the node IDs.[6]

---

[2] Readers familiar with *dataframes* may wish to implement these data structures using them, but we recommend better alternatives for network data shortly.

[3] Bad news: many real-world networks are large and sparse.

[4] If the network is small and dense (e.g., a correlation-based network with almost all-to-all connections), then an adjacency matrix *can* be an excellent data structure; furthermore, there exist more sophisticated matrix data structures that do not require a lot of memory for storing a large, sparse matrix and that allow for some efficient matrix computations.

[5] Or, if it happens to be zero-indexed, $0, 1, \ldots, N - 1$.

[6] One may still choose to use numbers to represent each node to save memory when working with huge networks. If node names are strings, say, and there are very many nodes, you could be using a lot of memory in a data structure where node names need to be stored multiple times. If this becomes prohibitive, using

Adjacency list and adjacency set   Instead of storing all the rows or columns from the adjacency matrix, we can store just the neighbors of each node as a list or a set, and then associate each node with its list or set of neighbors. In other words, we can create an *associative array*[7] where the nodes' IDs are stored as its keys and the neighbors of each node are stored as corresponding values. We'll call this array node2neighbors since, if we plug in node i as a key, we get a set or list of nodes that are i's neighbors in the graph.[8]

An associative array is in many ways the ideal data structure for interacting with a network computationally. As we discussed, to represent the network topology, we need to tell if a node is present in the network and if two nodes are connected. If we use an associative array where nodes are keys, we can immediately, and often efficiently (Sec. 4.1), perform the first operation by asking if a node is present as a key in the array. In pseudocode, we would denote this as "i in Array", for example.

What about the second operation, tracking whether two nodes are connected? This is also easy; we can simply ask if the second node is present in the list (set) of neighbors of the first node, which already involves the operation of obtaining the neighbors of a node.

So what should we use to store the neighbors? A list or a set? The primary difference is that the time complexity of item search in a list is $O(n)$, while the time complexity of item search in a set is $O(1)$ (or at least sub-linear). Although this seems to indicate that the set is the obvious winner and indeed it is a more universal solution, there are some caveats. First, a list (array) is a much simpler data structure than a set, and it is often easier to implement and use. Second, when the number of neighbors is small, searching an item by going through an array can be much faster than searching an item in a set because of the fixed overhead of hashing or other operations happening in a set. If your computational task mostly requires iterating over the neighbors of a node, then a simple array can be more efficient. Still, for a general purpose library, it is more reasonable to use a set due to the worst-case time complexity.

For our (a, b, c) triangle discussed above, using the {key:val, key:val, . . . } notation for an associative array (Ch. 4), we can define node2neighbors as: node2neighbors = {a:{b,c}, b:{c,a}, c:{a,b}}. With this array, we can quickly retrieve the neighbors of, say, a from node2neighbors[a], giving us the set $\{b, c\}$ and from this we can in principle implement any graph algorithms or methods. Want to compute the degree distribution? Loop over each node in the array, get the neighbors and count how many there are:

---

shorter IDs, such as integers or even memory pointers, can make such a data structure less redundant (see also Ch. 27). However, it is best to worry about this only if it becomes a problem, as computer memory continues to increase in abundance.

[7] Python's associative array is called a *dictionary* and it is a critical component of Python, one of the most popular programming languages and a great choice for working with networks.

[8] When we use a list of neighbors instead of a set, it is called an *adjacency list*, although that term is also used for file formats.

```
degrees = []
for node in node2neighbors:
    set_neighbors = node2neighbors[node]
    node_degree = length(set_neighbors)
    degrees.append(node_degree)
```

Here we get the neighbors of node with graph[node] and the degree, the number of those neighbors, is the length of that set, length(graph[node]).[9] With the list of each node's degree we can count how many nodes have degree one, how many have degree two, etc. resulting in the network's degree distribution.

Such an associative array can be used to power any other graph algorithms, such as breadth first search. Further, additional data can be associated with the network (Ch. 9) by using additional arrays (node2attribute, link2weight, etc.).

 An associative array that takes a node as a key and returns a set of nodes as a value is a very good way to represent a network topology. With this "node2neighbors" data structure, we can quickly find any neighbors of a given node.

Operations

Now that we have the node2neighbors data structure, let's turn to some operations using it.

The first operation we discussed, asking whether a node is present in the graph or not, is already in place. This operation simply uses the key membership operation that any associative array already has; how it is implemented, whether you call k in Arr, or Arr.has_key(k) or the like, will of course depend on how your associative array is implemented. For our purposes, we can think of making a simple "wrapper" for this operation to abstract away the specifics and focus on the network structure. Using Python, for instance, we could have a function like this:

```
def has_node(graph, i):
    return (i in graph)
```

(We grouped the return value with parentheses for clarity, but in Python these are not necessary.) Here our has_node function takes a node2neighbors dictionary (Python's name for associative arrays) called graph[10] and a node called i and simply returns the Boolean outcome of the "key in dict" operation, using the built-in in operator.

What about edges? A simple has_edge function can be defined likewise:

```
def has_edge(graph, i, j):
    if not has_node(graph, i):
        return False
    return (j in graph[i])
```

Here, has_edge is similar to has_node: we take the graph array and two nodes, *i* and *j*,

---

[9] This simple code snippet is almost correct Python code, with the exception that for-each loops in Python are denoted with for instead of foreach and the length of a set is given by len not length.

[10] In practice we would want to document what exactly our function expects "graph" to be. A *docstring* is useful for this.

and ask if *j* is a neighbor of *i* by using the in operator, not on the array itself, but on the neighbor set graph[i]. This tells us whether *j* is a neighbor of *i* and thus whether the edge *i*, *j* is present in the network.[11] But there is a wrinkle in complexity too: before we check if *j* is in *i*'s neighborhood set, we first ask if *i* is in the graph. If it's not, we know immediately that the edge must not be present, so we return False. But if we didn't check and went directly to j in graph[i] and *i* is not in graph, we would get an error trying to access an absent key. Hence the need to check for *i* first.[12]

ⓘ Example: **counting triangles**.

Let's use our node2neighbors data structure to count, for a node, how many of its neighbors are connected. Each connected pair of neighbors forms a triangle. Let's call our node2neighbors data structure graph while a is the node whose triangles we wish to count:

```
neighbors = graph[a]
triangles_of_a = 0
for i in neighbors:
    for j in neighbors:
        if i != j and has_edge(graph,i,j):
            triangles_of_a = triangles_of_a + 1
triangles_of_a = triangles_of_a / 2
```

Here, after getting the neighbors of a, we do a double for-loop, comparing every neighbor i to every other neighbor j and checking if i and j are connected. This double for-loop is easy to write but has two drawbacks: it (inefficiently) compares every pair of nodes twice, hence we need to divide our counts by two when finished; and it compares nodes to themselves, hence we skip counts where i == j. A better double for-loop would look at only the $\binom{k_a}{2}$ unordered pairs of neighbors of node a, for instance using the itertools.combinations() function in Python.

Counting triangles in a network is important for many structural measures; see Ch. 12 for more.

What about building the network? The has_* functions assume that graph already exists. We need a way to put nodes and edges into graph. (In principle, we only need to put edges in, but it's useful to put nodes in separately so we can represent nodes that have no neighbors.) Working again with Python, let's first initialize an empty node2neighbors array, then insert a node 'a':

```
graph = {} # empty dict
graph['a'] = set()
```

That second line, where we create the key and value, is all we need for an add_node

---

[11] Here we're focusing on an undirected network and we assume that *j* is only in *i*'s neighbors if *i* is also in *j*'s neighbors. We could check both if we want to, of course. For a directed network we would need to distinguish the order of our query: has_edge(digraph,i,j) is not the same as has_edge(digraph,j,i), assuming we've implemented has_edge for directed networks.

[12] Python specifically has an *"it's better to ask for forgiveness than for permission"* philosophy. In this case, it would be (slightly) better to first try to access the graph[i] entry and then, if it throws an error, *catch* that error and return False. This isn't always the case in other languages.

function, but it's not *safe*. If that node was already present, we would erase its neighbor set from the array! Instead, we first need to check if the node is not in the graph, then add it if it is safe to do so:

```
def add_node(graph, n):
    if not has_node(graph, n):
        graph[n] = set()
```

Here we reuse our has_node function and, if it returns false, add_node modifies graph in place, so it does not need to return anything. Quite simple overall. Now we can initialize our data structure as before, then insert the node:

```
graph = {}
add_node(graph, 'a')
```

> ⚠ Use ***utility functions*** to modify your graph data structure *safely*. Operations such as adding nodes and edges require multiple steps. If something is forgotten, the data structure can become inconsistent. In fact, many errors we discuss in Ch. 10 can be caused by such inconsistent manipulations. Utility functions, implemented correctly, can help prevent these errors.

That's fine for nodes, but of course, we want to add edges too:

```
def add_edge(graph, i, j):
    add_node(graph, i)
    add_node(graph, j)
    graph[i].add(j)
    graph[j].add(i)
```

Here we first call out to our add_node function to make sure i and j are present before we add the edge (and we are safe if either or both are already present). After, we use Python's set.add method to make sure that j is added to i's neighbor set and likewise that i is added to j's neighbor set. (By writing graph[i].add(j) we are referencing the neighbor set graph[i] from within the array and calling its method .add() in a single statement.) We could have used has_edge for a safety check as well, making sure we don't insert j into i's neighbor set if it's already there. In Python this is unnecessary—if we insert an element into a set already containing that element, nothing will happen—but other languages may behave differently.[13]

In principle, a node2neighbors data structure and these four operations (has_node, has_edge, add_node, add_edge) are sufficient to build and study a network, but it's not necessarily convenient. For example, we may want to add multiple nodes (or edges) at once, and an "add_" function that takes a list of nodes (or edges) and performs all the operations for us would be quite handy. More important, however, are "remove" functions, remove_node and remove_edge. These enable *changes*, not just additions, to the graph structure.

---

[13] We may also wish to use a *multiset* for the neighbor set if we wish to represent parallel or multi-edges by allowing copies of a neighbor in the set to represent repeat edges, but we ignore this complexity for now.

8.2   Moving in and out of your home: network and graph data formats and storage systems

Working with network data requires handling and processing that data, dealing with files and file formats. The most common file formats, and the ones we strongly encourage using, are almost always supported by modern programming languages and network libraries (Sec. 8.3).

Perhaps the simplest way to store a network is by saving its adjacency matrix **A** to a file. This option is useful and efficient for some contexts, such as working with small and dense networks (e.g., functional brain network data[14]) in an array-based programming environment like MATLAB. However, it is too space-inefficient for large networks and it is not flexible enough to store other information about the network, such as node attributes. Indeed, even identifiers for nodes cannot be easily captured: if the seventh row and ninth column of this matrix is a 1, how do you know what is the seventh node and the ninth node? You need another file that contains the information.

As mentioned, capturing the sparse form of **A** is a more useful starting point.[15] Instead of storing all $N \times N$ entries, record an $M \times 2$ matrix where each row captures an edge. For example, $A_{79} = 1$ becomes a row "7 9". This is called the *edgelist*.[16] It is more space-efficient unless the network is dense enough that M is similar in magnitude to N^2. More importantly, the matrix can be readily transformed into a two-column table where the columns store the node identifiers, as opposed to the matrix which only stores the indices or locations of nonzero elements of **A**. Row "7 9" becomes row "Amir Bob".

 In most cases, use a sparse edgelist format over a dense adjacency matrix format.

Storing an edgelist to a (text) file requires setting aside specific symbols or characters as *delimiters*. These characters are needed to separate columns within a row and to separate rows. The most common choices are to use commas (",") or tabs (tabstops) to distinguish columns and newlines to distinguish rows:

```
[node],[node]<newline>
[node],[node]<newline>
[node],[node]<newline>
...
```

But you must be careful: if your node identifiers contain the same delimiter characters, you can be in trouble reading your file unambiguously. Commas, in particular, are problematic delimiters as nodes may be names of the form "Surname, Given name" separated by a comma. Even numeric data may contain commas as a thousands or

---

[14] Data coming from neuroimaging devices such as fMRI machines often takes the form (after processing) of a correlation or association matrix **W**. This matrix is dense, meaning it contains nonzero values for *each* W_{ij} that describe the strength of the relationship between brain regions i and j.

[15] The sparse form is sometimes called the *coordinate format* while the dense form is sometimes called the *array format*.

[16] This edgelist format is equivalent to the *coordinate list (COO)* format for storing sparse matrices. Similar formats include *compressed sparse row* and *compressed sparse column* format. Most linear algebra libraries that support sparse matrices implement these formats.

(sometimes) decimal separator. We encourage you to be mindful of *delimiter collisions* when reading and writing network data.

 Be aware of ***delimiter collision***. If a node identifier contains the column separator, for instance, the edgelist file will become ambiguous: you won't be able to tell where columns begin and end.

Of course, we would need some operations to take such a file and populate our graph data structure. Likewise, given a network in that data structure, we need an operation to record the nodes and edges to such a file. Consider in your programming language of choice how to implement such operations.

 Stress test your network-writing code. Good practice when coding is to read back the file you have written and confirm that the network you create from your file is identical to the network you recorded into that file.

 Avoid writing your own string-processing code to deal with delimiters. Instead, use a preexisting library such as Python's csv library. It will be battle-tested and far more robust to data problems, and likely supports *escaping* delimiters and *quoting* fields, both ways to prevent misunderstanding the file due to delimiter collision.

An edgelist format can naturally handle both undirected and directed networks. For a directed network, simply assume one column is the source node and the other column is the destination node for all edges. A bidirectional edge $i \leftrightarrow j$ can be represented by including a row in the file for each direction: $i \rightarrow j$ and $j \rightarrow i$. Be mindful of which column is which; you don't want to end up with a network where every edge is reversed!

 An edgelist file's contents alone can't distinguish a directed vs. an undirected network. We recommend being explicit: use the filename or other documentation to clearly label the network as directed or undirected whenever there is the possibility for confusion.

Sometimes you may encounter a format related to edgelists called an adjacency list or a neighborhood list. This format mimics the node2neighbors array structure we argue for in Sec. 8.1 but as a file and not an in-memory data structure. In this format, each row of the file represents a node and its neighbors:

[node] [delimiter] [neighbor 1] [delimiter] [neighbor 2] ...

Files of this form are "ragged" in that the number of entries per row is not constant (unless every node has the same degree). [17] The advantage of this format over the two-column edgelist format is further space efficiency: by storing every edge $(i, j_1), (i, j_2), \ldots$ separately, you duplicate i each time. The adjacency list format $i, j_1, j_2, \ldots$ eliminates this duplication, leading to a slightly decreased file size. While this is a benefit, it's generally

---

[17] For a hypergraph or higher-order network (Sec. 1.4), this format is identical for storing hyperedges. Each row represents one hyperedge consisting of two or more nodes.

minor: never underestimate growth in file storage capacity in computer hardware. A further benefit of an adjacency format is that you can store node attributes within the file, as there is now one row per node as opposed to edgelists where there is one row per edge. However, storing node attributes brings some complexity as you need a means to distinguish within a row node neighbors from node attributes. This can be done, but for simplicity and portability it may be best to eschew the effort and use a separate format.

Other standard file formats

While we advocate for simple approaches to file formats, an edgelist or adjacency list stored to a CSV file may be limiting. In particular, while it's easy to associate edge attributes in an edgelist, bringing in node attributes will usually require another file. Dealing with multiple files is not the end of the world, but it can be annoying. To address this, many standard formats have emerged that represent the nodes and edges of the network and associated information (Ch. 9) all within a single file. In fact, over 100 such file formats have appeared in the past 25 years [402]. A few of the more prominent:

GraphML A flexible, XML-based[18] format. Uses .graphml file extension. Website: graphml.graphdrawing.org.

GML Graph modeling language. Confusingly similar name to GraphML. Uses a non-XML, non-standard format for storing nodes, edges, and their attributes in a single file. Website: gephi.org/users/supported-graph-formats/gml-format/.

GEXF Graph Exchange XML Format. Another flexible XML-based format. Website: gexf.net.

Another format, not specific to network data, that is worth discussing is *JSON*. Compared to XML, JSON is simpler and can be more directly translated into a programming language's native data structures. There are a number of ways to encode a network as JSON. One form closely mimics that of GML, capturing the nodes and links of the network using associative arrays nested as values within a larger array. Where possible, we encourage using JSON for storing networks as, unless space is a great concern, JSON is portable and easily used within all modern programming languages.

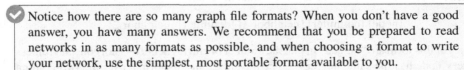 Notice how there are so many graph file formats? When you don't have a good answer, you have many answers. We recommend that you be prepared to read networks in as many formats as possible, and when choosing a format to write your network, use the simplest, most portable format available to you.

---

[18] XML, eXtensible Markup Language, which underpins a number of graph file formats such as GraphML, was quite popular in the 1990s and early 2000s, probably due to its superset relationship with HTML and the rise of the World Wide Web. Since then it has fallen out of fashion, with favor turning over to a simpler format: JSON.

Graph databases

Sometimes, it is worth storing the network in a database instead of a "flat" file. Flat files lack indexing, making it hard to find parts of the network without loading the entire network into your computer's memory.[19] Databases provide indexing features and also manage access, allowing multiple computer processes to read and, more importantly, write data. The database process will manage simultaneous access and prevent collisions and other problems.

> **ⓘ** **File vs. database.** A database is best when you wish to access only a small subset of the network at any one time, for example studying only a single node and its neighbors. A file is best when you expect to examine all of the network. A database is best when many processes want to record changes to the network data at the same time.

The most common databases are relational databases, which store data in tables and use SQL (structured query language) to query the data. In relational databases, although it is possible to have connections between items, each item (row) is often considered independent from each other. By contrast, *graph databases* focus on storing connected items and their relationships so that users can query the data based on the connections.

For instance, consider storing a social network in a relational database to storing it in a graph database. Using a relational database, we would probably choose to store the network data in two tables: one for the nodes and one for the edges. The node table would have a column for the node ID and multiple columns for various node attributes. The edge table would have a column for the edge ID, a column for the source node ID, and a column for the target node ID, linking to the node table. If we want to obtain the friends of a given person, we may run a SQL query like the following:

```
SELECT target_node_id
FROM edge_table
WHERE source_node_id = 123
```

This query will search through the edge table (or its index) and return the target node IDs of all edges that have the source node ID of 123.

In contrast, in a graph database, the network data would be stored as a set of *triples*, which contains *subject*, *predicate*, and *object*. For instance, in a statement "Alice knows Bob," "Alice" is the subject, "knows" is the predicate, and "Bob" is the object. Then, many such triples constitute a graph's data. Graph databases can be queried with *SPARQL*, which is a recursive acronym of "SPARQL Protocol and RDF Query Language." Using SPARQL, one may find the acquaintances of Alice by running a query like the following:

```
PREFIX foaf: <http://xmlns.com/foaf/0.1/>

SELECT ?person
WHERE {
```

---

[19] In fact, many of the issues with using an adjacency matrix that we avoid with our associative array in Sec. 8.1 can also be avoided with a suitable index on top of the adjacency matrix. Essentially, `node2neighbors` *is* an index structure.

```
    <http://example.com/alice> foaf:knows ?person .
}
```

This query[20] uses the "knows" predicate defined by the FOAF (friend of a friend) ontology, foaf:knows, to identify all "?person" that are connected to Alice through the predicate. Unlike the relational database, the query does not search through the whole edge table, but directly accesses the local graph around the given person.

Graph databases are often used for storing large knowledge graphs and online social networks. While less common in research applications, graph databases are commonly used in industry where real-time concurrent access at large-scale is critical; examples include TAO (Facebook), FlockDB (Twitter), JanusGraph (Uber), and Neptune (Amazon).

We discuss graph databases further in Ch. 27.

8.3   Software libraries for network analysis

There are many software packages for networks. Some are full-fledged applications that provide a graphical user interface (GUI) for performing network analysis, while others are libraries that provide a set of functions to use in your code. While pre-built applications are great for getting started, they become limiting, and we strongly encourage data scientists working with networks to take a computational approach, using libraries and writing code to work with their data.

Regarding the choice of programming language, at the time of writing, the two most popular programming languages for data scientists are **Python** and **R**. Both are great choices, with the caveat that interpreted, dynamic languages may be slow relative to compiled languages. However, we generally recognize that it is the efficiency of the researcher in their work more than the efficiency of the computer in running the code that matters. Never discount the speed increases of future computing hardware. Also note that many libraries are written in efficient, lower-level languages such as C, C++, Rust, etc., and then wrapped in a high-level language for ease of use. Thus, even if you are using Python, a relatively slow language, you may still be able to take advantage of the speed of a lower-level language. Other emerging languages, such as Julia, are also worth considering.

As discussed in this chapter, some of the basic operations on networks can be easily implemented. With functionalities to read and write network data, that may be all you need to get the analysis going. Also, "rolling your own" is valuable for learning firsthand the complexity of the data. However, for most cases, leveraging fully featured and thoroughly tested libraries is the way to go. At the time of your reading, some new tools likely have emerged while others have disappeared. So, instead of listing many, we will instead focus on a couple of the most popular and a few notable libraries.

Probably the two most well-known and widely-used libraries for network analysis

---

[20] SPARQL resources are represented with URIs (Uniform Resource Identifiers), which is why we write <http://example.com/alice> instead of alice. These can be abbreviated by declaring "prefixes," as we did in the example for the foaf predicate.

are **NetworkX**[21] and **igraph**.[22] Both NetworkX and igraph are popular, feature-rich open source projects with long histories of development and many developers. NetworkX is written in Python while igraph is written in C. This makes NetworkX generally slower than igraph but easier to install and use, and it is more flexible especially for dynamic networks. On the other hand, igraph is more portable than NetworkX, and can be used from Python, R, or C/C++ itself.

The four "atoms" of a network library—has_node, has_edge, add_node, add_edge—that we discussed in Sec. 8.1, by no coincidence, share names with the corresponding methods in NetworkX. (igraph uses similar names, for instance add_vertex is used instead of add_node.) What is different is that NetworkX implements graph *classes*:

```
import networkx as nx
graph = nx.Graph()
graph.add_edge('a','b')
```

Here graph is an object of the networkx.Graph class and the method .add_edge was used on that object to insert an edge. Objects and methods have advantages where many internal data structures and other methods can be accessed within the object without us knowing about it, a programming concept called *encapsulation*. This lets complex data structure operations occur transparently and it allows for a class to change how it works internally without breaking code using objects from that class. Most network libraries are written in this way and it is a good idea to learn how to use them.

Software moves fast and new libraries and even programming languages arrive all the time. NetworkX and igraph are both worthy starting places, but many hundreds of alternatives exist, from the tried-and-true *Boost Graph Library*, part of the *Boost* C++ libraries,[23] to the up-and-coming JuliaGraphs[24] project. The Julia programming language itself is worth a special mention. Julia, although it reads like high-level programming languages such as Python, is fast—approaching the speed of C. As such, Julia, with its simple syntax and high speed, has been gaining popularity in the data science and scientific computing communities. While Python and R remain the dominant data science languages at the time of writing, keep your eyes peeled on alternatives like Julia. It is an exciting time for scientific computing.

8.4 Summary

Your network data needs to be stored somewhere on the computer. This may take the form of a simple plain text "flat" file, a compact binary file, or a graph database. Flat files can be formatted in many ways, from a basic edgelist saved in a CSV file to a complex XML-flavored GraphML standard format. Flat files are best when networks are small or you need to interact with the entire network at once; databases work best when the network is very large and you only need to examine some of it, and when many computer processes want to interact with the network simultaneously.

---

[21] https://networkx.org
[22] https://igraph.org
[23] https://www.boost.org
[24] https://juliagraphs.org

When working with network data using code, you have many choices of data structures. A simple associative array is usually the most versatile data structure that can store and compute the network efficiently. Writing your own code to process network data can sometimes be straightforward and provide invaluable opportunities to learn network data structures and algorithms. Yet, in most cases, you would use existing libraries that feature extensively tested and efficiently engineered functionalities. Python and R, both excellent programming languages for data science, come well-equipped with third-party libraries for working with network data.

Bibliographic remarks

For a general tour of Python for data science applications, we recommend Vanderplas [473], which is freely available online. Likewise, for readers interested in using R, Wickham and Grolemund [491] is a valuable read. Scopatz and Huff [420] give an excellent, broader tour of good scientific computing practices. See also the interlude of this book.

Exercises

8.1 Create a `remove_node(graph, i)` function that removes i from an undirected network represented by the `node2neighbors` array `graph`. Make sure it does so consistently.

8.2 Describe a function, using either pseudocode or, if you wish, Python, that takes a "node2neighbors" associative array for a network and creates that network's adjacency matrix.

8.3 Produce some pseudocode describing a `write_edgelist` function that takes a graph data structure and filename as input and records the graph to the named file as an edgelist. In your pseudocode solution, manage the writing of "row" and "column" delimiters explicitly.

8.4 *A home for directed networks.* We focused on undirected networks when discussing the `node2neighbors` array structure and introducing the four `has_*`, `add_*` functions. Describe with pseudocode or, if you wish, Python, an implementation of those operations that works specifically for directed networks. Ensure that the directionality of edges is preserved by distinguishing "source" and "target" nodes when needed.

8.5 When reading an edgelist for a directed network, how do you manage the distinction between source and target nodes?

8.6 *Graph intersection.* Suppose you have two networks *G* and *H* and you wish to compute a new network containing only nodes and edges present in both. Describe with pseudocode a function `intersection(G,H)` to accomplish this.

8.7 *Graph union.* Suppose you wish to combine two networks G and H. They may share nodes and edges. Write pseudocode for a union(G,H) function that produces a new graph containing nodes and edges that are present in G or H or both.

Chapter 9

Incorporating node and edge attributes

Much of the power of networks lies in their flexibility. Networks can successfully describe many different kinds of complex systems. These descriptions are useful in part because they allow us to organize data associated with the system in meaningful ways. These associated attributes and their connections to the network are often the key drivers behind new insights.

9.1 Data surround your network

Networks primarily consist of two kinds of objects, nodes and edges. But we can extend this with additional data, using the nodes and edges as "hooks" on which to hang this information. In terms of the network, we can treat this data as "metadata"—attributes that further describe the nodes and edges.[1] See Fig. 9.1.

9.1.1 Node attributes

What kind of data describes nodes? Let's consider some possibilities for different types of networks.

Social networks Consider a social network where each node represents a person belonging to the network. While the links of that node capture their social ties, much more information can be considered. There may be sociodemographic data about the individual such as age, gender, or ethnicity.[2] There may be economic data such as net worth or income. There may even be health information such as disease diagnoses. Such

---

[1] Generally, we'll use the term "network attributes" to refer to either node attributes or edge attributes. However, when working with a large corpus of networks, you could have attributes describing each network in its entirety. We'll call these *network-level attributes*, to avoid confusion. (Our focus here is primarily on single networks.)

[2] And of course, such data are usually sensitive and often protected; Ch. 3.

(a) Node attributes   (b) Edge attributes

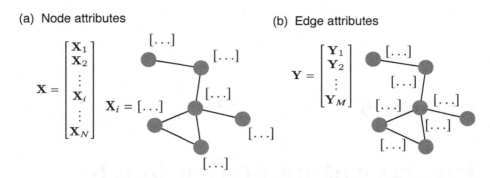

Figure 9.1 Node and edge attributes. For node attributes (**a**), associated with each node is a vector of values called attributes. For edges attributes (**b**), a vector is instead associated with each edge. We can gather these attributes into matrices (Sec. 9.2) whose rows correspond to the nodes or edges. Often the "halo" of data surrounding a network, meaningfully captured as such attributes, gives new insights into the network data.

data may be self-reported, come from a data gathering effort (Ch. 6), or be the result of a predictive or statistical analysis. The latter is especially common in online social networks where interest profiles are generated, often with machine learning, to infer what subjects are most interesting to a person.

Biological networks   Consider a protein–protein interaction network where each node represents a protein of interest and links denote interactions between proteins that are known to occur. These data are usually gathered through experimental procedures that test whether proteins interact. The results of these tests become the links. But the proteins themselves are of extreme interest to both molecular biologists and specialists focused on processes involving those proteins. The net result of this research focus is large bodies of information describing the proteins. This includes the molecular composition of the proteins but also details about how the protein assembles in the cell, what genes are known to encode for the protein, what functions the protein may fulfill, and even what diseases may be related to dysfunction of that protein.

Such protein-specific information is now collected and disseminated using large databases and *ontologies*. For instance, one of the standards adopted in the bioinformatics community are gene ontology (GO) terms. GO terms form a controlled, hierarchically organized "vocabulary" that describe proteins (genes) in terms of their functions and roles in the cell.[3] It is hierarchically organized in the sense that every term, except one term at the top, is a child of another term, forming a *directed acyclic graph* (DAG).[4] For instance, "carbohydrate metabolic process" is a parent term of "carbohydrate biosynthetic process." This hierarchical structure allows researchers to characterize a group of proteins by their GO terms. Discovering, for instance, that a

---

[3] The human reference interactome (HuRI) is one example. The HuRI PPI network has GO terms associated with each of the nodes describing their known roles and functions.

[4] GO terms have three aspects: *biological process*, *cellular component*, and *molecular function*, each of which has its hierarchical structure and terms.

densely interconnected (interacting) set of proteins are all known to play a role in regulation of DNA repair and response to ionizing radiation may give researchers insights when developing new cancer therapeutics. This analysis is called *functional enrichment analysis* or *GO term enrichment analysis* and is a common practice in the field of bioinformatics. This is possible thanks to the fact that GO terms are well-organized *node attributes*.

9.1.2 Edge attributes

Just as nodes can carry associated attributes so too can edges.

The edges in a social network, for instance, describe the social ties between people. In a multi-modal (multilayer) network capturing many possible connections, an edge attribute can categorize the edge, distinguishing social ties from work ties from family ties. Such categories can also capture context—the reason why the edge exists. For a work tie, for example, an edge attribute can tell us where the two individuals worked together, when they worked together, and more.

Or an edge attribute may denote the *strength of the tie*, perhaps the most commonly considered type of edge attribute. Edge weights can encode this numerically, with stronger ties being given higher weight. One way edge weights are gathered is by counting an observable quantity such as the number of times two people interacted in a given period. In fact, this can be extended into a time series denoting when each interaction occurred; an edge attribute can map a static network onto a dynamic representation (Ch. 15)!

And of course, weights are not just for social networks. Another example from neuroimaging data would be using the correlation coefficient (or other measure of association) between different fMRI time series as edge weights when extracting a brain network. Stronger edges exist between more strongly associated time series.

In general, if you are working with a network, see if you can use edge attributes to describe the data. You may very well discover a pathway to interesting new insights.

9.1.3 Using attributes

Node and edge attributes are useful in many ways.

First and foremost, they allow us to justify, validate, or better understand network structure. Many networks exhibit homophily, where similar nodes are more likely to be connected than dissimilar nodes. How to quantify similarity? Attributes! By looking at the sets of attributes of linked nodes, we can measure if homophily is occurring using information related to but distinct from the network structure itself. Likewise, we can cross-reference that structure with the attributes. Look at a random selection of node pairs i and j at a distance d from one another (d is the number of hops in the shortest path between the nodes). How similar do their attributes tend to be? Do closer node pairs (smaller d) tend to be more similar than more distant pairs (larger d)?

Attributes can also justify methods for understanding the network structure. One of the most common problems is community detection (Ch. 12). Community detection methods try to find meaningful groups of nodes in the network, the communities. ("meaningful" here is a loaded word, and can be defined in many different ways [364]).

Suppose you've used a community detection method and found some groups of nodes and they even look like a good group according to the network structure (and as defined by the method you chose to use). But how else can you validate the groups? Attributes! First define a measure for the similarity of attributes, as before. Then compare the similarities for nodes within the same community compared to, perhaps, random pairs of nodes. Alternatively, consider a null model where the attributes are assigned to nodes (or edges) at random. How much more similar are the real attributes within communities compared to the null where attributes are assigned at random? This is an example of the *enrichment analysis* we discussed. It lets us explore how well the community structure, found only using the network, is reflected within the attributes, under the assumption[5] that the attributes relate to the network and community structure.

9.2   Representing attributes

Generically, we can think of storing node and edge attributes in matrices (Fig. 9.1) alongside the nodes V and edges E of the network:[6]

$$G = (V, E, \mathbf{X}, \mathbf{Y}). \tag{9.1}$$

We're augmenting the node set with $\mathbf{X}$, an $N \times p$ matrix of *node attributes*, and the edge set with $\mathbf{Y}$, an $M \times q$ matrix of *edge attributes*. Some networks may possess $\mathbf{X}$ only or $\mathbf{Y}$ only, while other networks possess both.

As an example, suppose we have a social network where nodes, representing individuals, are enriched with demographic attributes. Let's say each node has an age and gender attribute. Then the age and gender can be organized into the $N \times 2$ matrix $\mathbf{X}$ with one numeric column representing age and another column representing gender as a categorical variable. Similarly, suppose each edge in the network has an associated date denoting the first time the two connected individuals were in contact and an associated number describing the total number of times the two individuals were in contact. We can store these edge attributes in an $M \times 2$ matrix $\mathbf{Y}$ where the one column represents the date of first contact and the other represents the number of contacts.

What's particularly nice about this arrangement is that we can easily add attributes simply by appending columns onto the matrices. And sometimes we can bring to bear tools for analyzing matrices, such as singular value decomposition, to learn more about the pattern of the network's attributes.

That said, as literal data structures, these matrices may be poor choices for large networks: if attributes are sparse, meaning only few nodes have attributes, then the matrix will consume more computer memory than necessary. A solution is to use a sparse format for the matrix, a data structure designed to only use memory for nonzero elements.

---

[5] And this is a big assumption. It is entirely possible that the network attributes you are working with have little or nothing to do with the network structure or the communities you are seeking to find. In such a case, poor enrichment results do not imply that the communities you have found are poor.

[6] We use these matrices to describe the data *mathematically*. In practice, matrices may also be useful as data structures within your code. However, many other possible structures may be used, such as an associative array that maps nodes to lists of data. These choices may be better than matrices. See Ch. 8 for details on implementing network data within your code.

Another issue is using these matrices for *multidimensional* attributes such as a time series of contacts for each edge (see also Ch. 15). We could make a single column for the entire time series and let each element of that column be multidimensional, which is fine for many data structures but is more challenging if you wish to use matrix analysis on **Y**. Alternatively, we could expand the time series to be a collection of attributes, essentially making **Y** a matrix of its own, with one row for each link and one column for each time period. It's worth judging the computational aspects as you decide which data organization to choose.

Finally, there is the problem of attributes of different sizes. For instance, the GO terms assigned to proteins discussed above: each node has a set of terms associated with it, but the number of terms per node is not fixed. One can store these in a tabular format where the right side is *ragged*: one row per node where the first entry in the row represents the node while the remaining entries represent the list of terms. One can also make a non-ragged organization where each row represents a (node, term) pairing, and nodes with multiple terms show up multiple times, once per term. Although this format [7] has the disadvantage of spreading a node's attributes across multiple rows, it is an example of *tidy data* and is often a great way to store all kinds of data.

9.2.1 Tidy data and network attributes

Often, a nice way to store attribute data is in a *tidy* format [490]. [8] Tidy datasets (and *tidying* so-called messy data) gives us a simple, but very useful mental model for reasoning about our data (in this specific case, our attributes).

"Tidiness" is now a term-of-art when considering datasets. A dataset is "messy" or "tidy" based on how it is organized. Specifically, a dataset is broken down into four non-disjoint components: *values*, *variables*, *observations*, and *tables* (we really like the abbreviate *VVOT* [9]). The values of the data are organized into variables and observations which are then grouped into tables. A dataset is tidy when the tables are organized such that all values associated with a given variable fall into the same column of a table, all the values associated with a given observation fall into the same row of a table, and all observations of a specific kind (also called an observational unit) are organized in the same table. This mapping of variables to columns, observations to rows, and units to tables is the essence of tidy data.

It also sounds rather self-evident. And, in a sense, it is. However, in practice, one encounters endless amounts of messy data, data that break one or more of these requirements. Data become messy for a variety of reasons and messy data are much, well, messier to deal with [490]. In terms of programs, code written to expect a tidy format [10] will be much more flexible and maintainable over time; messy data often requires one-off, bespoke code that otherwise wouldn't be necessary.

When dealing with network attributes, the matrices **X** and **Y** are naturally tidy. For **X**, each row is a node (each node is an observation) and each column is a node attribute

---

[7] Which we can also interpret as an edgelist for a bipartite graph.

[8] Tidy data can be thought of as a statistical (or data scientific) expression of Codd's 3rd normal form in database theory.

[9] *Vee-vot Las Vegas*.

[10] Tools built specifically for tidy data, such as the "*tidyverse*" in R, are becoming more popular.

(each node attribute is a variable). Likewise, for **Y**, each edge is an observation and each edge attribute is a variable. Thus, we can treat the nodes and edges of the network as our dataset's "observational units."

When dealing with network attributes, it is worth keeping tidiness in mind, both when you are building **X** or **Y**, and when working with them.

9.2.2   Storing attributes

As with storing a network using an $M \times 2$ edgelist or other file format (Ch. 8), we also need files to store attributes associated with the network nodes and edges.

As a format, an edgelist extends quite naturally to storing edge-based attributes. Simply append more columns to the edgelist, one for each variable you wish to store—the first two columns store the nodes of each edge, any subsequent columns then store attributes associated with those edges. The most common example is capturing weighted edges by adding a third column of edge weights. The association with the edge is automatic just from being placed in the same row. And attributes instead of or in addition to edge weights can be captured with further columns. [11]

 An edgelist format makes it easy to store edge attributes: just append them as further columns.

While it's easy to associate edge attributes in an edgelist, bringing in node attributes will usually require another file. Dealing with multiple files is not the end of the world, but it can be annoying. One way to avoid multiple files is to use an extensible file format such as GraphML that is capable of describing nodes, edges, and the attributes of both, all within a single file. However, even with multiple files, as long as network elements are consistent across files, it is easy to combine and process them as necessary, especially so if the files are tidy.

 Consider using tidy data files for node and edge attributes when you wish to leverage the broad array of tidy tools now available. Use network-specific file formats when you wish to keep all attributes in one file.

Remember that downstream tasks, such as network visualization (Ch. 13), may be easier with one format or the other, depending on what tasks you are performing and what tools you will use.

9.3   Patterns and relationships of attributes

As you incorporate attributes into your network, it is important to explore their properties. Seek out patterns and relationships in how attributes relate to each other and how they relate to the network.

Some questions to ask:

---

[11] Again, keep in mind the problem of delimiter collision discussed in Ch. 8. If you have text-based attributes, they may unexpectedly contain the delimiter character even if the node identifiers themselves did not.

Overall distributions What are the overall distributions of attributes? What is the most common value of a categorical attribute? What is the mean or median value of a numeric attribute? What is the range of an attribute, smallest values to largest values? What is the earliest time of a temporal attribute? What is the latest? Treat the attributes as data and apply exploratory data analysis.

Correlations and associations Does one attribute correlate with another attribute? Compute the $p \times p$ correlation matrix for **X** and the $q \times q$ correlation matrix for **Y**.

Attributes and the network What about associations (or correlations) between attributes and the network?

- Correlations between nearest neighbors?
- Correlations within connected components?
- Do correlations decay with graph distance?
- For edge attributes: correlations between colliding edges (i, j and j, k)?

The patterns between neighbors in the network and their node attributes are a good vehicle for using an *assortativity coefficient* [331, 332].[12] Suppose we are working with an undirected network and each node i has an attribute x_i. Suppose also that x_i captures a categorical value, perhaps the self-reported ethnicity of individuals in a social network. Examining all the edges in the network, we can count how often an edge i, j connects a user i with attribute $x_i = u$ and a user j with attribute $x_j = v$. Let e_{uv} be the number of edges connecting one user with attribute value u to another user with attribute value v. (Because the network is undirected, $e_{uv} = e_{vu}$.) The assortativity coefficient for this attribute is then defined as

$$r = \frac{\sum_i e_{ii} - \sum_i a_i b_i}{1 - \sum_i a_i b_i}, \tag{9.2}$$

where $a_i = \sum_j e_{ij}$ and $b_j = \sum_i e_{ij}$ (the sums run over the possible values of x). Defining the assortativity of x with respect to the network in this way will give $r = 0$ if there is no association between edges and x values. If all edges connect to nodes with the same value of x, then $r = 1$ and the network is "perfectly assortative" for that attribute. Likewise, if edges only connect nodes with different values of x, then $r = -1$ and the network–attribute pair is "perfectly dissortative."

By measuring r, we can better understand how nodes "mix" with one another in terms of attribute x. And it can be especially interesting to study r's value in different scenarios, either comparing r computed for different attributes, comparing r over different time periods (for a temporal network) or even comparing r for different data preprocessing steps (Ch. 7).

Equation (9.2) holds for a categorical attribute. What about a numeric quantity? The measure we most naturally turn to is the Pearson correlation coefficient $r(X, Y)$

---

[12] In fact, these associations can measure relationships with *network features* such as degree, not just attributes, which we'll see in Sec. 12.5.

between two numeric variables X and Y (Sec. 4.3):

$$r(X,Y) = \frac{\sum_i X_i Y_i - n\bar{X}\bar{Y}}{\sqrt{\sum_i X_i^2 - n\bar{X}^2}\sqrt{\sum_i Y_i^2 - n\bar{Y}^2}}, \tag{9.3}$$

where $\bar{X}$ and $\bar{Y}$ are the average (sample mean) of X and Y and n is the number of values. Usually, r measures the association between paired observations (x_1, y_1), (x_2, y_2), ... like you would examine visually in a scatter plot. But here we can leverage this expression with our network structure by defining a paired observation for each edge. Specifically, for a node attribute x, for edge $e = (i, j)$, $X_e = x_i$, $Y_e = x_j$, and we use $n = M$, the number of edges, in $r(X, Y)$. In other words, the X variable in Eq. (9.3) is the attribute value of one end of the edge and the Y variable is the attribute value at the other end.

In either case, categorical or numerical attribute x, we can use r to measure whether edges tend to land between nodes with similar values of x ($r > 0$), between nodes with dissimilar values of x ($r < 0$), or whether edges tend to fall without any discernable association with x ($r \approx 0$).

9.4   Connecting attributes and the network—record linkage

You may sometimes find that some attributes are not directly connected or linked to the elements of the network. Meaning, you have node attributes but you're not sure which attribute goes with which node (and likewise with edges). You may need to perform some data manipulations to either draw out attribute values or connect those to nodes or links in the network. Often, this takes the form of a *join*: joining two or more datasets using some common feature or identifier. Sometimes this joining process is straightforward, other times it is challenging.

For example, you may have demographic information for individuals in a social network. But if those attributes are associated with people using a *written name*, while the social network identifies nodes based upon *usernames*, you will have to determine how to link the attribute data with the social network. Perhaps a field exists mapping written name (Johann Smith) to username (jsmith21) but what if this does not provide a one-to-one mapping?[13] In that case, it may be possible to guess the username given the set of nodes (represented by usernames) stored in the social network data and the set of written names stored in the attribute data. Such a process is called, among other things, record linkage, and it is likely to be a serious undertaking.

What about edge attributes? Usually the edges are defined relative to the nodes they connect; an edge ID is identified with a tuple of two node IDs. So if you have a linkage problem with nodes that problem will be compounded when looking at edges.

 Ideally, you can unambiguously connect the nodes in the network with their attributes, and likewise for edges. Sometimes this cannot be done, for example, nodes in a social network identified by username ("jsmith21") but node attributes

---

[13] Most commonly, the usernames will be unique but multiple usernames can have the same written name.

only contain written names ("Johann Smith"). Figuring out accurately which node or edge attribute goes with which node or edge is an example of *record linkage*, a potentially difficult task.

9.5 Missing attributes

It may be that some attributes are only partially observed in your data: entries in the attribute matrix $\mathbf{X}$ or $\mathbf{Y}$ (or both) are absent. For instance, the age of some members of a social network may be missing.[14] Such missingness is a common occurrence in many problems where data are hard to come by.[15]

If you find yourself in such a situation, what should you do? First, ask what and where: what attributes have missing values and where (what nodes or links) are they missing?[16] Is all the missingness concentrated in one particular attribute or does it appear throughout the attribute matrices?

Of course, while you work on "what" and "where," you will surely find yourself asking "why": why are attributes missing? It may be due to mistakes with data processing. But what if the missingness is not due to mis-processing the data but something more serious? What mechanism could be driving the missingness? Could there be a bias driving the missingness making it more likely that certain values of an attribute are missing? Investigating the driving forces underlying missingness patterns is the core of the field of *data imputation* in statistics.

Once you have a handle on how many attribute values are missing, and you've ruled out "obvious" issues such as data processing bugs, you can begin to speculate on missingness mechanisms and then consider courses of action, depending upon your research goals (Ch. 5). Focus on the seriousness of the missing values. If missingness is rare, can the problem be ignored? Do your goals depend on the missingness? Perhaps all the missingness occurs in a particular attribute that you don't feel is necessary to your use of the data. Or you could pivot in your research goal and actually consider addressing the missingness as a research question itself. You may learn a lot just by understanding the root cause of the data's missingness.

You could attempt to tackle the missingness head on. First, of course, is more data gathering. It may be possible that with more time and effort the missing values could be recovered. If so will depend on the circumstances of your data and how it was gathered, of course. Even if the missing values can't be found, simply gathering more data may give you enough observations, even with missingness, to address your original research questions.

---

[14] Worse, it may be present but incorrect.

[15] Missingness in the network itself is also important; Ch. 10.

[16] In principle this is a straightforward search of your data. In practice, however, it can be a challenge depending on how your data are encoded. One issue is when the symbol representing missingness ("n/a," "None," "nan," "." etc.) in the data is not consistent. And are you sure the missing indicator means what you believe it means? This sounds like an obvious question but it is worth taking care: when data are merged it is possible to carry forward different indicators. Some attributes may use "NA" for missingness, others use "-1". If this is not well documented, you may find yourself in a bug hunt.

If further data gathering is not possible, your next resort is likely to be statistical imputation techniques. The best technique, which isn't imputation at all, is **listwise deletion**. Despite the fancy name, listwise deletion simply means to remove any rows in your data matrix (in our case, $\mathbf{X}$ or $\mathbf{Y}$) with any missing values before proceeding on toward further analysis. Listwise deletion is known to work well when there are no patterns to your missingness (this is known as *missing completely at random* or MCAR). However, it throws out data and can introduce problematic biases if missingness is not random.

Imputation methods that try to infer the missing values generally work by building statistical models for the missing value as a function of the observed values.[17] Historically, at first, single models were built, but it was shown these failed to address the greater *uncertainty* we must have in imputed values: after a value was imputed, it was just as good or bad as any other, observed values when it came to later calculations. Current best practice is to perform some form of *multiple imputation*, where each missing value is imputed repeatedly, giving a distribution of values and helping (somewhat) to capture the uncertainty of that value from the spread of the impute distribution.

Imputation is fundamentally difficult, and miraculous recovery of missing values is unlikely. Further, in our case, the network structure will ensure that observations (rows of $\mathbf{X}$ or $\mathbf{Y}$) are related, and not iid. This fundamentally challenges all attempts to impute the missing values.

 Why worry so much about missingness?

The network structure ensures that attribute observations (rows of $\mathbf{X}$ or $\mathbf{Y}$) will be related, and not iid. This fundamentally challenges all attempts to address missingness.

Associations with the network

Section 9.3 asked how the attributes relate to the network structure. When missingness is present, we should ask the same question: how do missing values relate to the network itself?

This relationship can manifest in many ways, and the careful researcher should explore such associations by asking, for example:

1. If a node has a missing attribute, will a neighbor of that node be more likely to also be missing that attribute?

2. Do similar nodes in terms of the network have similar patterns of missingness? For example, do all high-degree nodes miss the same attributes?

3. More generally, do network properties (degree, clustering, centrality) predict missing values? Perhaps low-betweenness edges are more likely to be missing edge attributes?

---

[17] However, missingness may be determined by unobserved variables or even the missing values themselves. In such cases, good imputation is much more difficult.

4. Does the network structure itself exhibit missingness, such as missing links? (For a deep dive on this, see Ch. 10.) If so, does missingness in the network correlate with missingness in the attributes? One simple way to check for this is to investigate if low-degree nodes such as singletons have more missing attributes than other nodes. Although it is difficult to say for certain, those low-degree nodes are most likely to have missing links, and so this check can be a tantalizing clue.

One simple starting point to associate missingness in an attribute with the network is to make a new, binary attribute that indicates the missingness. Suppose x is an attribute with some missing values. Let's build a new attribute $\tilde{x}$ defined with:

$$\tilde{x}_i = \begin{cases} 1 & \text{if } x_i \text{ is missing,} \\ 0 & \text{otherwise.} \end{cases} \tag{9.4}$$

Now we can compute, for instance, the assortativity coefficient (Eq. (9.2)) between nodes and $\tilde{x}$, recycling that mathematical machinery to explore the association between edges and missing values of x. (And of course, we should explore how missingness in one attribute is related to values of other attributes, or even missingness in those attributes.)

Many of these questions are couched in the node attributes, but the same ideas hold for link attributes. And other questions combining nodes and links are worth considering for datasets where both $\mathbf{X}$ and $\mathbf{Y}$ are under study. Do missing node attributes correlate with missing link attributes: if node i or node j (or both) have missing attributes, does link i, j have any missing attributes?

9.6 Summary

Networks rarely exist in isolation. The elements of the network, its nodes and links, often have additional attributes—properties or features—worth describing. In a social network, these may be demographic features, such as the ages and occupations of members of a firm. In a protein interaction network, they may be gene ontology (GO) terms gathered by biologists studying the human genome. Gain insight by gathering data on those features and associating them with the nodes or links of the network. Represent these attributes using node and link matrices kept alongside the network itself. Consider what attributes may already be present in your data and what attributes may be worth adding by joining new data, if possible. Study the association between the network and the attributes: do patterns in the attributes correlate with certain nodes? Note that attribute data may be imperfect: some nodes or links may be missing attributes possessed by others. Can you explain why?

In general, a network is better understood when the halo of data surrounding it is explored and explained.

Bibliographic remarks

For protein–protein interaction networks such as HuRI, the Gene Ontology (GO) knowledge base is a major source for node attributes. The Gene Ontology is the result of a major bioinformatics consortium working since 1998 to describe the functions of genes. GO terms provide a controlled vocabulary to describe these functions and are some of the most important node metadata describing PPI and other biologically relevant networks. For more information, see http://geneontology.org, Ashburner et al. [15] and The Gene Ontology Consortium et al. [456]. The *Gene Ontology Handbook* [128] gives further practical guidance for using these materials.

As discussed in Sec. 9.2, the node and link attribute matrices bring to mind the idea of "tidy data." This general idea, written out in a 2014 paper by Wickham [490], is well worth reading for anyone interested in working with data.

Assortativity and mixing patterns in networks, introduced by Newman [332], have long been a source of descriptive statistics for network data. Here we have considered assortativity in terms of understanding how network attributes relate to the network structure itself. We treat other forms of mixing, such as degree mixing, in Ch. 12.

The example of connecting attributes to the social network based on written names instead of usernames is a special case of the general problem of record linkage, data matching, and reference reconciliation. Closely related problems include fuzzy matching, name disambiguation and deduplication. Readers interested in learning more may wish to consult Christen [104] and Euzenat and Shvaiko [152] for in-depth treatment.

Missing values and their recovery, known as imputation, are a long-standing problem in statistics [407]. A full treatment of modern imputation methods and how missingness is classified (missing completely at random, missing at random, missing not at random) is beyond our scope here. Interested readers are encouraged to study Gelman and Hill [178] and van Buuren [469] for a recent introduction to imputation. We also discuss missing data in Ch. 10.

Exercises

9.1 Suppose we store a network using a node2neighbors array (Ch. 8). Describe a way (data structure and code) to include node attributes.

9.2 Repeat Ex. 9.1 but this time for edge attributes.

9.3 Using your node and edge attribute data structures, describe with pseudocode some methods for updating those data if the network changes, for example if a node or edge is removed.

9.4 Using a specific network of interest that has node attributes, describe a study that will test the "quality" of the network using those attributes. Specifically, explain how node attributes can inform the questions of whether more network data need to be gathered (Ch. 6) or whether the upstream task should be revisited (Ch. 7).

9.5 (**Advanced**) In Sec. 9.3 we defined assortativity for a node attribute (Eq. (9.2)). What if we were working with an edge attribute? Try to define a coefficient r suitable for a categorical edge attribute. Describe the range of values this r can take. Does your coefficient have any bias or structural problems? (*Hint*: consider the effects of very high- vs. very low-degree nodes.)

Chapter 10

Awful errors and how to amend them

Network data, like all data, are imperfect measures of objects of study. There may be missing information or false information. For networks, these measurement errors can lead to missing nodes or links (network elements that exist in reality but are absent from the network data),[1] or spurious nodes or links (nodes or links present in the data but absent in reality). More troubling is that these conditions exist in a continuum, and there is a spectrum of scenarios where nodes or links may exist but not be "meaningful" in some way.

In this chapter, we describe how such errors can appear and affect network data, and introduce some ways these errors can be handled in the data processing steps. (For more on the theory of errors and uncertainty in networks, see Ch. 24.) Such error fixes can lead to different networks, before and after processing, for example. Chapter 14 discusses techniques for comparing networks that may be useful to guide any processing you may wish to do.

> **ⓘ Errors are not mistakes.** Here we use *error* in the sense of scientific uncertainty. Errors in the network come from either measurement uncertainty or processing uncertainty. We distinguish these issues from systematic mistakes or blunders made by the researcher, such as creating buggy computer code. Such blunders are truly awful. (Of course, buggy code used in the past may be *causing* errors in the data you are presently working with.)

10.1 Errors in data: omission and commission

Broadly speaking, and certainly not limited to networks, errors in data fall into three classes:

---

[1] We previously encountered missingness not in the network but in attributes associated with the network; Sec. 9.5.

Missingness where an observation should have been captured in the data but was not.

Spuriousness where an observation appears in the data but did not actually exist.

Uncertainty where an observation was made but its value was recorded incorrectly.

A scientist will always have to contend with the specter of such errors, thinking whether their data are affected and why.

For example, suppose you are gathering data by polling individuals about their political views. You may deal with missing values due to *survey nonresponse*, where some individuals preferentially avoid participating in your poll. Were this nonresponse to be uncorrelated with the data you are gathering, you could deal with it by considering your data as randomly sampled from a larger population. Unfortunately, this is seldom the case. Instead, individuals who avoid your poll are much more likely to hold unusual or non-mainstream views, or to distrust scientists, or the like. Your missingness is now non-random, and the conclusions you will draw about the distribution of political views is likely to be skewed.

Spuriousness and uncertainty can also appear within these polling data. False observations can be recorded due to mistakes or (unfortunately) fraud. Individuals who wish to skew political views may attempt to participate in the poll multiple times or workers canvassing for your poll may accidentally survey the same individuals multiple times. And mixups can occur when the surveyors take notes, leading to uncertainty due to incorrectly recorded survey responses.

This polling example highlights some errors that occur based on the data generating mechanism, how the data are observed and recorded. In this case, people are involved directly, creating the data (by participating in the poll) and recording the data (by conducting the poll). But such problems can still occur, and may even be worse, when people are not in the equation, as automation—automated data gathering and automated processing—can lead to such errors as well.

Take care when addressing errors

Errors in data should be addressed to ensure your conclusions from those data are reliable and well supported. This can lead data scientists toward methods to fix errors. In principle, fixes are great, but often in practice fixes are infeasible and it is best to only identify errors and reason about their sources. In fact, poor fixes can make things worse. Let's discuss some consequences that occur when trying to deal with missing values.

Have you ever started analyzing a data table and discovered there are a bunch of "N/A" or "NaN"[2] values in your table? These symbols are being used to represent missing values and you can't compute even basic summary statistics with them in place.[3] Although it is all too common to immediately drop rows with missing values[4]

---

[2] "NaN" or "Not a Number" represents an ill-defined number in standard floating point numeric convention.

[3] Averaging, for example, a collection of numbers containing a single "NaN" will make the average itself be "NaN."

[4] Dropping any rows in a table of data that contain at least one missing value is sometimes known as *listwise deletion*.

so you can proceed with calculations, in practice, this can be dangerous. The reason is that the mechanism for missingness can severely bias what we can see in the data.

For example, say we are interested in the relationship between income and education by using a survey dataset that contains income and education information for a sample of people. We may be tempted to drop all rows with missing income or education values. What happens then? Let's think about why someone might have not reported their income. It could be that they are embarrassed about their income, or do not have a consistent source of income. If this is the case, then we would expect that people who do not report their income are more likely to have lower income than people who do report their income. In other words, by dropping the datapoints without income values, we are introducing a strong bias in our analysis.

In any serious data science practice, we should never simply drop rows with missing values without investigating it. Instead, we should try to understand the mechanism of missingness and try to address it. The first criterion to think about is whether the missingness is *missing completely at random (MCAR)* or not. If the missingness is MCAR, that means that the probability of missing data is same for all datapoints and therefore it is safe to drop the rows with missing values. Of course, it is very rare that the missingness is MCAR and that we can confirm that the missingness is MCAR.

If the missingness is not MCAR, then there are several possibilities. The first case is *missingness at random (MAR)* (without "completely"). This means that the probability of missingness is not the same for all datapoints, but it only depends on the observed data. If this is the case, we can drop the rows with missing values as long as we can control for the observed characteristics that affect the missingness. Then there can be the case of missingness that depends on unobserved data (or even the missing value itself), or *missing not at random (MNAR)*. One common example is the case of clinical trials, where a patient becomes more likely to drop from the trial if they feel discomfort from the treatment. If the level of discomfort is not measured, then there is no observed variable about the patient that affected the missingness, although the missingness is definitely affected by a specific variable and not random. In this case, the missingness should be explicitly modeled.

While our discussion has centered on missing values, spurious and uncertain values need to be addressed too. In many ways, these are even more challenging, as a spurious value is less obviously spurious than a missing value is missing. Further, attempts to fix these problems can increase uncertainty (a value meant to correct for a noisy observation may be even farther from the truth) or missingness (say, by incorrectly removing a supposedly spurious observation). Once again, the best bet is often to address errors only by identifying them.

10.2 Errors in networks

Our problems compound when confronted with a network structure. Here errors such as missingness or spuriousness can interact through the network structure. Broadly speaking, we can have missing nodes (a node in the network fails to appear in our data), missing links (a link between observed nodes is missing), spurious nodes (a node appears in our data which is not present in reality), and spurious links (a link appears in the data between observed nodes that is not present in reality).

But these categories intersect. A link can be missing because both nodes it connects are missing. A spurious node can be introduced by a duplication error where a node is incorrectly included twice in the network. And if this occurs, what about the links connecting those nodes: are they divvied up somehow between the real and spurious node? Or, are they each duplicated so both nodes have every link?

Missingness in network data

In the context of network data, missingness can arise in multiple contexts. First of all, the missingness can be about the nodes. When we examine a social network obtained from a social media platform, we will miss many of those who do not use the platform. And then we may not see those who have not interacted with anyone in the network or simply were inactive during our observation period. If we examine a protein–protein interaction network, we may miss the proteins that have not been catalogued yet or those that are simply too difficult to test. Missing nodes, especially if they are very central nodes or hubs of high degree, can drastically alter our view of the network's structure.

Missingness can also be about the links. Even if neither node is missing, we may be unaware of the link that connects them. This may come from limitations of the experimental techniques (e.g., a protein–protein interaction network), or limitations of the data collection procedure (e.g., the subjects in a social network survey cannot recall all of their friends or acquaintances). And missing links can also have a profound effect on our view of the network.

Note that there is a difference between missing data (usually links) and measured absence. If the existence of a link was tested or measured, then the lack of the link is not missing data; the absence was measured.

In addition to missing the nodes or links themselves, we can also be missing information about non-missing nodes or links: the missingness can be about the attributes of the nodes or links that we can use for statistical analyses. For instance, missing income can bias our analyses of the relationship between income and social network structure, as described above.

Spuriousness in networks

On the other hand, we may have spurious data in the network. As no experiment or data collection is perfect, spurious nodes or links can find their way into the data. A survey respondent may confuse someone with another; an experiment may yield a false positive; someone on social media may have mistyped their friend's username. Even when there are no such errors in the basic measurements, there still can be spurious links.

One prominent example is *indirect correlation*. When it is difficult to observe interactions or links directly, we often rely on correlational measurements. For instance, if we measure how two genes respond to various perturbations (experimental conditions), we can calculate a correlation matrix between genes. This correlation matrix can then be interpreted as a network. However, this poses a challenge. Not only are we in the situation of multiple hypothesis testing (and therefore some of the *significant* correlation values would be spurious), but also we will have indirect correlation. Let's say gene A strongly affect gene B—B's expression is strongly correlated with A's. At the same time, A also affects another gene C—C's expression is also strongly correlated with A's. In this situation, what do we observe between B and C? Most likely—although not guaranteed—B and C will show significant correlation with each other, even if they do not directly interact. This positive, indirect correlation may be strong enough to even wash out weak, *negative* direct interaction between B and C! This problem exists even if the data collection or experiments are perfect.

What effects can spuriousness have on the network structure? Of course, the predominant issue is links appearing in the network that are either absent in reality or are otherwise meaningless. But compound effects can occur. We may encounter incorrect node *merging*—one node in the data is more than one node in reality. Or we may have the inverse, incorrect node *splitting*—multiple nodes in the data are one node in reality. Such errors arise when nodes are not clearly identified, for example, a social network where users are pseudonymous or only partial identifying information is available can quickly lead to one person appearing multiple times in the network and multiple people appearing as a single node in the network. Node splitting can also lead to link splitting: if a link exists between i and j and j is split into j_1 and j_2, then we may expect link i, j to become two links, i, j_1 and i, j_2. And likewise, node merging can lead to link merging. These node errors can be especially insidious compared to simpler missing or spurious link errors.

10.3 Sources of network errors

Whenever network data are gathered, one must consider whether the gathering mechanism has introduced errors. Often this requires having more than a passing familiarity with the mechanism itself, how it works and how it might fail. Consider link errors. Will false positive links occur, links that appear in your dataset but are not truly present? What about false negatives, where nodes appear disconnected in your data but are linked in reality? Often a given data generating process is more likely to create one type of error than another.

Experimental and observational data will come with errors. Experimental protocols, for instance assays that test protein interactions, will never be perfect, although good methods and good researchers will always put in the effort to understand and if possible mitigate such issues. Observational data, likewise, may fail to capture all facets underlying the network structure and if a computational or statistical method is used to infer links from those data, that method may itself be the source of errors. Many datasets are also generated manually, through surveying or laborious human coding, and such manual efforts are prone to human error.

Based on your understanding of the data you are working with, can you tell what are the main sources of errors? What are the net results of these errors, many false or spurious nodes or links? Or perhaps too few links are present, an error of omission? When many links are included that should not be, you may wish to thin your data. Otherwise, if you suspect much of your network is missing, consider trying to thicken your network. We turn our attention to such fixes now.

10.4   Fixes

Given errors are possible, even likely, what can we do? Is it possible to identify errors in the network structure and correct them? What methods might we use for such *fixes*?

> **!** Fixing errors is non-trivial, sometimes impossible. Often the best we can do or should do is *identify* and *understand* the errors that are likely present. Doing so gives us more confidence in the data because we can anticipate failure modes for problems relying on those data. Unreliable fixes, in comparison, may pollute the data, leaving us worse off than if we had instead done nothing at all.

Having discussed a bit about errors in network data, both the kinds of errors and their possible causes, we now turn our attention to what to do about them.

10.5   Thinning spurious data

At times, the network you extract from the underlying data will be overloaded with too many links, nodes, or both. These elements may be entirely spurious, or they may be deemed extraneous junk to be eliminated.[5] Sometimes, the data are too big to even fit into memory, which makes it tedious to analyze the network. Highly dense networks often appear when considering weighted networks. For instance, networks constructed from co-occurrences or correlations can be extremely dense, even fully connected. When the network is extremely dense, network visualization does not work well and even simple measures like degree can become meaningless.

In these cases, it is worth removing the least important elements of the network in order to facilitate visualization, computation, and analysis. Here we discuss ways to thin out these elements, forming a reduced or "thinned" network.

10.5.1   Thresholding by degree

Probably the simplest way to reduce the size of a network is to remove nodes with degree less than a certain threshold. The implicit assumption here is that the degree of a node reflects its importance and thus removing the least connected nodes is a good way to reduce the network while losing the least amount of information. The simplest case is removing nodes with degree one (also known as "dangling nodes"). Nodes with degree one are nodes that are connected to only one other node and thus it is often reasonable to consider them to belong to the node that they are connected to. For instance, if we imagine community structure in the network, it is natural to assume that all dangling nodes belong to the same community of the node that they are dangling from. Many real-world networks, because they tend to be a sample of a larger population, tend to have numerous dangling nodes that do not contribute a lot of information to the network, and removing them often leads to much smaller networks that are amenable to visualization and computational analyses. And of course, we can and should also consider thresholds other than 1, the challenge being choosing the threshold appropriately.

---

[5] Keep in mind that care should be taken declaring anything "junk" (cf. junk DNA [287, 134]).

Degree thresholding is even more useful in ***directed networks***. In directed networks, either in-degree or out-degree reflects the importance of the node in the network better than the total degree and therefore these filtering methods can be used more effectively. For instance, let's say we have obtained a retweet network about a certain topic from Twitter. In this network, where the nodes represent users and the directed edges represent retweets, the in-degree captures how many times a user's tweets have been retweeted by others, which can be one of the simplest proxy of "influence" on Twitter. Therefore, filtering nodes based on their in-degrees is a good way to remove nodes that are less important to the conversation.

10.5.2 Thresholding by cores

A slightly more sophisticated variation of degree-based thresholding is to use the concept of *k-cores*. A *k*-core is a subnetwork of a network containing every node that is connected to at least *k* other nodes in the subnetwork. For instance, a 2-core subnetwork of a given network is a subnetwork where every node has at least two neighbors. This 2-core subnetwork can be obtained by iteratively removing all nodes with degree less than two. Any *k*-core subnetwork can be obtained similarly. Although not guaranteed, often the cores of the network contain the most important structure to analyze.

10.5.3 Weight thresholding

When the edges are weighted, using the weight of the edge usually provides a higher-resolution proxy of importance. We can not only filter out nodes based on their importance, but also filter out edges based on their weight or other derived measures of importance.

The simplest approach is removing edges based on a weight threshold. We simply set a minimum weight threshold—call it w_{thr}—and ignore all edges that have smaller weight than the weight threshold. The idea behind this method is that more important edges are heavier weight and choosing the right threshold will let us retain those important edges.

Figure 10.1 shows an example of some weight thresholds in practice. Here we have taken the Malawi Sociometer Network,[6] removed nodes outside the largest connected component, and then retained only those edges (i, j) with weight $w_{ij} > w_{thr}$. (After removing edges, any nodes of degree zero were also removed.)

Although this is probably the most common approach, it often fails to reveal meaningful structure and is generally a poor method to reduce the complexity of the network. The reasons that a simple threshold does not work well are (1) real networks often exhibit core–periphery structure or so-called "rich-club" structure (see Ch. 12), and (2) there is seldom a single, universal threshold that, across the entire network, can meaningfully separate edges worth keeping from edges best discarded.

Rich-club structure is characterized by stronger and denser connections among a "rich club" of nodes while the density decreases as we move towards the periphery. If

---

[6] Here we convert the temporal network to a weighted network by summing the number of contacts between participants over time.

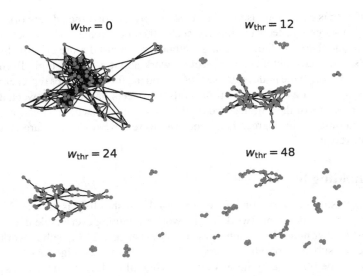

Figure 10.1   Visualizing the Malawi Sociometer Network under different weight thresholds. See Ch. 13 for more on such *network visualizations*.

the network possesses this structure, a simple thresholding will result in the removal of dangling periphery nodes, which does not meaningfully reduce the density at the core. Increasing the threshold value may not help, because it will gradually remove only the most peripheral nodes that are remaining in the network, leaving dense cores unscathed. We see evidence of this happening in Fig. 10.1.

Nevertheless, a single threshold is the simplest way to make use of edge weights and can be useful when a concrete, well-reasoned threshold value can be picked and when the network does not exhibit strong rich-club structure. In biological applications, often the choice of threshold is done by considering the false discovery rate[7] or, for co-occurrence/correlation networks, setting a p-value threshold of the correlation.

10.5.4   Windowing

Windowing is a particular kind of thresholding most useful for dynamic networks (Ch. 15).

Suppose each node or edge of the network has time periods associated with it. Windowing means to retain the subnetwork associated with a particular time period, the data "window," by keeping only those nodes and edges that appear in the network during that window. Windowing the network can allow researchers to focus on the

---

[7] The False Discovery Rate (FDR) is defined based on how often you reject the null hypothesis: $FDR = FP/(FP + TP)$ where FP is the number of false positives (rejecting a null hypothesis which is actually true) and likewise TP is the number of true positives. In the area of hypothesis testing, when repeatedly testing the same data (multiple comparisons) the rate of errors is higher than expected. The FDR was introduced as a better measure to quantify error rates than the significance level α typically used to quantify single hypothesis tests.

most salient time period they wish to study. For example, a social network extracted from online social activity may be most relevant to political scientists during the period leading up to a major national election. Those scientists may want to retain nodes (users) and links (say, retweets) in the period preceding and immediately following the election.

But how to define the start and end of this window of interest? Typically, it will depend on the problem being studied. The political scientists may know that most political activity occurs during the week of an election, or perhaps the preceding month. Likewise, the election under study may have resolved itself quickly and most interest is lost within a few days from the election's conclusion. Conversely, the election may have been highly contested, and interest continued for many weeks, so the scientists should use that information when defining their window's end period.

Window inference. What should be done when a researcher does not have prior information on the time period defining their window of interest? While potentially costly, the best bet is to devise a sweep over time periods. Define a set of window sizes (time durations), slide them over the time span of the data, and examine the network within each period. In other words, suppose your window size is ΔT. Examine the networks extracted from windows $[t_1, t_1 + \Delta T], [t_2, t_2 + \Delta T], \ldots, [t_n, t_n + \Delta T]$, where $t_1, \ldots, t_n$ are a set of time window starting points you have chosen to iterate over.[8] As an example, you may wish to look at a window of size ΔT that you slide over one month of data one day at a time, making $t_1 = $ day 1, $t_2 = $ day 2, and so forth. Then, after iterating over many time periods for a ΔT, choose another value of ΔT and repeat the process. This double-loop inference is expensive, and leads to many networks to compare (Ch. 14) but, in essence, this is a burden you should in general expect to bear when you are operating with data but without prior information.

Equipped with a collection of networks of different windows, which window is most appropriate? Again, you ideally turn to prior domain knowledge of those data. Lacking this knowledge, you may be able to specify a reasonable measure based on the network topology. For example, which window has the densest network or which window has the sparsest network? You may need to control these measures, however. For example, a longer window will probably lead to a denser network regardless of other factors, so you may instead want to look at a normalized density such as number of edges per unit time. Likewise, maximizing sparsity may lead to a period with a (nearly) empty network; instead, consider finding the sparsest network that is still connected. Lastly, if you are equipped with a statistical model for the data (e.g., Ch. 23), you may be able to look at the network with the largest likelihood under that model (again, perhaps controlling for model complexity and window size).

Window robustness. Suppose you have decided on a time period $[t_1, t_2]$ to study. You've extracted the network of nodes and edges active or otherwise associated with that window and you are ready to analyze the network. But doubt creeps in: Do I have the right window of interest or should I consider a different window $[t_1', t_2']$? Should I look at a wider window ($t_1' < t_1, t_2' > t_2$) or a narrower one ($t_1' > t_1, t_2' < t_2$)? Perhaps the start of my window is correct but the end is not, or vice versa?

These questions need to be addressed with a *robustness analysis*. As you vary the window of interest, does the network change? If change occurs, when and by how

---

[8] Can you determine meaningful starting points for your data?

much? Do your conclusions *about* the network change? We strongly encourage such robustness checks: confirm that small or medium changes to either t_1, t_2, or both, do not fundamentally change your results. You can check these changes by comparing (using a suitable measure; Ch. 14) the network taken from window $[t_1, t_2]$ to one taken from a perturbed period $[t_1, t_2 + \delta]$,[9] where δ is a small perturbation, either additive or subtractive, to the end of the time period.

10.5.5   Backbone extraction

Backbone extraction refers to a set of methods that filter edges by computing their "importance"—usually based on their weights and surrounding network structure. Edges deemed less important by some criteria are removed, reducing the density of the network. The remaining edges become the *backbone* of the network that plays the most critical structural role. Numerous methods exist with different operationalizations of the "backbone."

We've already discussed weight thresholding, a basic thinning step for weighted networks where we retain edges i, j whose weights w_{ij} meet or exceed some given threshold: $w_{ij} \geq w_{thr}$. This method uses a global threshold w_{thr}. Introducing a global threshold may sometimes be sufficient (try it out and see; Ch. 11), but other times the final extracted network may be inaccurate or otherwise problematic. In particular, a single, global threshold fails to use structural information about the relationships between different edges at different locations in the network. Some nodes may be surrounded primarily by weak (low weight) edges while other nodes sit among many strong edges; a global threshold will not account for such variations across different portions of the network. Thus, it may be better to define an adaptive threshold that accounts for relative edge strengths across different regions of the network.

Let us discuss one method for backbone extraction called the *disparity filter* [424], which uses a simple and often effective way to define a relative threshold. It assumes that the importance of an edge i, j can be estimated by examining the nodes that are connected to it and their immediate neighbors. Let's imagine a node i with $k_i = 4$ neighbors, each of which is connected to i with a different weight: $\{1, 10, 3, 2\}$. Obviously, the edge with weight of 10 (let's say (i, j)) would be the most important edge to i, but how important? The disparity filter argues that we can think of a simple null model, where we randomly divide the total strength W_i $(1 + 10 + 3 + 2 = 16)$ into four parts, representing what the weights would look like if their total was the same and they were distributed over the same number of edges, but no edge was preferred over another. We can visualize this null model as dropping $k_i - 1$ points on the unit $(0, 1)$ interval uniformly at random to create k_i random intervals that always sum to 1. Then, what is the probability, call it α_{ij}, for one of the intervals created at random to be at least w_{ij}/W_i long? For an interval of size $x = w_{ij}/W_i$ to occur at the leftmost interval (beginning at 0) we have to make sure a point is dropped at x, which occurs with infinitesimal probability dx, and none of the remaining $k_i - 2$ points land at positions less than x, which occurs with probability $(1 - x)^{k_i - 2}$. (Because points are equally likely anywhere, and the entire interval is of length 1, the probability for a point to fall inside a given interval is the length of that

---

[9] Or periods $[t_1 + \delta, t_2]$ or $[t_1 + \delta_1, t_2 + \delta_2]$.

interval.) So the probability for the first interval to be at least length w_{ij}/W_i is given by $\int_{w_{ij}/W_i}^{1} (1-x)^{k_i-2} \, dx$. But there is nothing special about the first interval, created by the leftmost of the $k_i - 1$ points, so the same quantity will hold if we consider the interval created between the leftmost and the second leftmost points, between the second leftmost and third leftmost points, and so on. Thus, the probability we see at least one interval as long or longer than w_{ij}/W_i is one minus the probability we see no intervals longer than w_{ij}/W_i or:

$$\alpha_{ij} = 1 - (k-1) \int_0^{w_{ij}/W_i} (1-x)^{k_i-2} \, dx = \left(1 - \frac{w_{ij}}{W_i}\right)^{k_i-1}. \qquad (10.1)$$

The smaller α, the more unlikely an edge's weight is at random, and therefore the more important that edge is likely to be. The disparity filter works by defining a threshold on the α_{ij}, meaning you keep all edges (i, j) with $\alpha_{ij} < \alpha_{BB}$ for some pre-chosen α_{BB}. Note that the importance of an edge (i, j) can be very different for i and j, meaning, from i's perspective, the link to j may be very important, whereas from j's perspective, the link to i may be unimportant compared to other links from j. The edge's weight may be the largest among i's edges while being the smallest among j's edges. Thus, we expect $\alpha_{ij} \neq \alpha_{ji}$ and in general it is best to keep an edge if it meets the threshold in either direction, meaning either the link to j is important from node i's perspective or the link to i is important from node j's perspective. The power of the disparity filter is that the null model defines a *local* threshold: each node's immediate neighbors are used to determine whether a link is important, unlike a less sophisticated global weight threshold.

A remark on α in the disparity filter method. When proposing the disparity filter, the authors motivated α as a significance threshold (p-value), even using the common choice of $\alpha = 0.05$. Indeed, the derivation of α_{ij} is akin to what one does with hypothesis testing where one defines a test statistic under a null model. However, such an interpretation brings with it a lot of baggage related to significance testing, including multiple comparisons and p-hacking. We have found it instead more fruitful to simply think of α as merely a threshold and not to try to interpret edges as "significant" or not. One effect of this is that a meaningful (i.e., useful) backbone can be had with values of α such as $\alpha = 0.2$ or $\alpha = 0.5$. Such values would be considered too large for hypothesis testing, but here the null model for how link weights are distributed is unlikely to be a plausible generating mechanism for the actual network (and you will surely have more domain knowledge for your network as to why). Were one to desire a proper null hypothesis test, one would need a null model for the network topology and the link weight distribution. The disparity filter works well by sidestepping these interpretive issues.

Example: using the disparity filter   Let's apply the disparity filter to one of our focal networks, the Malawi Sociometer Network.[10] For the weighted version of this network, edge weights w_{ij} represent the number of contacts observed between i's sociometric badge and j's badge over the course of the study. And, while the authors took care to

---

[10] Here again we use the weighted version of the network.

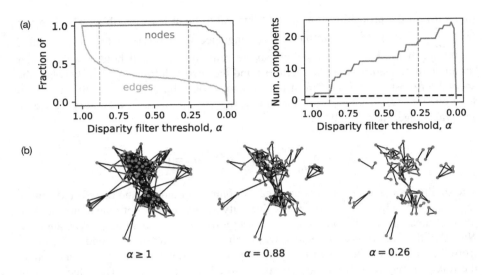

Figure 10.2 Applying the disparity filter to the Malawi Sociometer Network. (**a**) Without guidance on choosing $\alpha_{BB} = \alpha$ for Eq. (10.1), we perform sweeps over a range of values to see where the network structure changes. We've flipped the direction of the x axes to mimic the effect of tuning α *downward* from the original network (at $\alpha = 1$) all the way to removing all edges (at $\alpha = 0$). (**b**) From the α sweeps, we identify a few points of interest and draw the network at those values compared to the original network at $\alpha = 1$.

define time contacts such that brief, likely noisy signals from the badges were excluded, it is still plausible that unimportant edges are captured in the data alongside important edges. So this is a good opportunity to apply the disparity filter using edge weights as inputs to Eq. (10.1).[11] However, doing so requires choosing α, the disparity filter's threshold. What value should it be?

While the authors of the disparity filter advocate for values $\alpha \approx 0.05$, in our experience these are often too small. Indeed, applying such a value to the Malawi Sociometer Network would lead to few retained edges. Instead, let's take a more computationally exhaustive approach by sweeping over a range of α and determining what value (or values) may be appropriate. To guide our search, for each value of α we compute the fractions of nodes and edges that remain in the network (nodes with degree zero after removing edges are removed) as well as the number of connected components. Looking at these quantities as functions of α can guide us towards particular points such as when nodes begin to vanish from the network or the number of components start to rise. We present these α sweeps in Fig. 10.2a.

From the curves in Fig. 10.2a we observe a few possibilities. Edges disappear quickly from the network as we increase the filter (i.e., as we lower α downward from $\alpha = 1$) before their removal rate starts to slow around $\alpha \approx 0.9$. Nodes begin to disappear at $\alpha = 0.26$ and that is a good candidate to examine further. Meanwhile, the number of

---

[11] Of course, being a temporal network, we may wish or need to use methods specific to temporal networks, such as windowing (Sec. 10.5.4), discussed in this chapter. See also Ch. 15.

components makes a sudden jump at $\alpha = 0.88$, meaning that parts of the network begin to break apart and this lines up well with the slowdown in removed edges. Taken together, these statistics point to two possible values of α to examine and we draw in Fig. 10.2b the network at both filtering points alongside the unfiltered network. Decreasing α even further may be useful as well, as many small, disconnected groups begin to emerge, pointing to community groups within the social network. And now, having thinned the network (a little or a lot, depending on our needs), we can continue to investigate the social network of the study participants. Finally, remember that, although backbone extraction allows us to simplify and reveal important structure of the network, many network analyses can be—and often should be—done on the original network rather than in the backbone.

While other backbone methods exist (see remarks below), we find the simple disparity filter to be an excellent starting point in practice, and it is often more than sufficient for our needs.

10.5.6 Using reciprocity

For directed networks, another thinning approach is to look for reciprocated ties, pairs of nodes where both (i, j) and (j, i) edges are present, and retain only those links. Reciprocity is a well-studied feature of social networks in particular, and it can be a good signal of the stronger, more important ties. Non-reciprocated ties, depending on the data, may be less meaningful in comparison. Of course, this depends as always on what network question is being addressed; it may be critically important to retain both reciprocated and non-reciprocated ties.

10.5.7 Using side information

The previous methods have all considered thinning the network using information from the network structure. For networks with attributes (Ch. 9), one can also use those attribute values to decide on a thinning criterion. For instance, suppose each node can be placed into one of three categories. We may have additional information that sampling is much better for one or more of the categories than the others. We can then thin the network simply by removing nodes outside of the better-sampled categories.[12]

Depending on the problem under study and the properties of the node or link attributes, we can devise any manner of other thinning actions. A numerical attribute can be thinned using a global threshold (retain all nodes where x exceeds a threshold we introduce). (If we consider edge weights to be an attribute, then we have actually discussed one example of using side information to thin a network.)

Consider the possibility of combining attribute-based thinning with structural thinning. Building off the example of thinning the network by retaining only nodes from well-sampled categories, imagine combining that thinning with a degree-based thinning. Now nodes are retained if they are high-degree (based on a threshold we define) and they fall in the categories we wish to retain. Combining these filtering conditions

---

[12] Whether this is appropriate of course, is another question. The bias in the sampling process still carries through to the final network.

introduces an order-of-operations complication: if we first filter by node category, the degrees of some nodes will be reduced, changing the results when filtering by degree. (Notice that filtering by attribute is not affected whether we filter first or second by degree.) So is it better to filter first by degree then by attribute? In the end, it will depend on the data at hand; if possible, we encourage you to explore both orders and see what difference—if any!—it makes.

10.6   Thickening missing data

A problem more challenging in many ways than subsetting the non-spurious network nodes and edges, is thickening the data. Here we are trying to grow our network, attempting to infer latent network elements, nodes or links, that were not captured in the data but likely do exist. Such a task is generally far more challenging than taking a known element and deciding whether or not it should be discarded. How can one make such inferences?

10.6.1   Graph union

Suppose you have conducted an experiment to measure protein–protein interactions such as those in HuRI. You have quite a few usable links in your network, but it is still very sparse and you are likely missing many interactions. The experimental procedure you used to measure those links is not the only option, however, and other teams have published networks using other experiments. You decide to thicken your network by *merging it* with other, published networks.[13]

The algorithm underlying this merger is called the graph union. The union of two graphs is defined by taking the unions of their nodes and their edges. Essentially, it defines a new network that contains a node if it exists in either (or both) of the original networks and it defines an edge if that edge exists in either (or both) of the original networks. You can see how this will thicken the network by adding edges where the nodes overlap and by potentially growing the network by adding more nodes.[14] A graph union can represent collating additional data sources into your network, or performing additional experiments.

Computing a graph union is straightforward with one notable but often underappreciated challenge: the nodes need to be *clearly identified*. In other words, given a node i in one graph and a node j in the other graph, you need to be able to tell if they are the same node or not. This is straightforward when nodes have unique, trustworthy identifiers.[15]

---

[13] Indeed, one of the networks studied as part of HuRI is dubbed "HI-union," which is the merger of HuRI with all the research group's previously published PPI screening experiments [283].

[14] Interestingly, the graph *intersection* can be used likewise for thinning spurious data: retain only those edges discovered in multiple networks.

[15] Identifying nodes without ambiguity is often assumed and taken for granted yet it is not nearly so common in many situations. Consider reconstructing a historical social network from contemporaneous records such as written documents and recorded eyewitness accounts. People may refer to one another in different ways, using nicknames; people may even be unsure what parties were involved in particular actions. Cultural effects can make this worse. Consider the difficulties of distinguishing individuals in the data when studying a society where people do not have surnames.

But if not, if you cannot reliably distinguish the nodes in different networks, trouble can ensue.

While computing the union is straightforward, more difficult is to decide whether it is appropriate. Essentially, by merging two networks, you are deciding that the definitions of links[16] need to be extended. Now you are saying a PPI edge, for instance, exists if it was found by one experimental protocol or another. Perhaps this redefinition is fruitful but what if it is not? The sources of the two networks may be fundamentally incompatible and combining their edge sets may simply not make sense.

10.6.2 Link prediction

While the graph union represents the updating of a network with more data, other approaches are more speculative. Link prediction is the problem of inferring links that probably (hopefully!) exist in a network but are not (yet) seen in the data you have—can you predict these absent links with enough accuracy that you can incorporate them into your network data?

Simply put, we can think of a link prediction method as a function that takes a network (optionally including any associated attributes; Ch. 9) and two disconnected nodes i and j and returns either a binary indicator "yes, link i, j exists" or "no, link i, j does not exist" or a weight such as a probability for the link i, j to exist. This function can be simple or quite complicated. The classic Adamic–Adar function [3] is simple but often reasonably effective; we recommend it as a baseline or starting point if you wish to begin applying link prediction to your data. It uses no metadata, simply examines the observed network structure and computes a measure of similarity between i and j based on how many neighbors they have in common. Let N_i be the set of neighbors of node i, such that $|N_i| = k_i$ is the degree of i and $N_i \cap N_j$ is the set of nodes that are neighbors to both i and j. The Adamic–Adar index is then defined as a sum that runs over these common neighbors:

$$AA(i, j) = \sum_{u \in N_i \cap N_j} \frac{1}{\log k_u}. \tag{10.2}$$

This measures how many common neighbors two nodes have, under the assumption that nodes with many neighbors are more likely to be connected than nodes with few neighbors. But it also down-weights those common neighbors based on their degree, assuming a neighbor with many connections is more likely to be common just by chance and therefore should contribute less compared to a neighbor with few connections when deciding if edge i, j exists. The use of $\log(k)$ in the denominator is meant to mitigate the effects of outliers: *very* high-degree nodes will still be down-weighted, but not as much as if k_u was used instead.

By itself, Adamic–Adar's measure, or another scoring function, will not lead to a thicker network. Doing so requires incorporating that function into a decision process that declares "yes, i and j are connected" or "no, i and j are not connected." One decision process is to define a threshold and insert any links that scored above that cutoff. But what cutoff to choose?

---

[16] And nodes, if the two networks have different sets of nodes.

Here again the researcher is confronted with a choice and moving forward without enough information may be difficult. We recommend the following. First, determine if you have access to any information or criteria that can help guide you to a threshold. For instance, you may wish to add at most a given number of edges, say m new edges. Then, use that number instead: insert edges into the graph between the m highest rated pairs of nodes. And if you don't have something as simple as the number of new edges to guide you, perhaps you can look for another property of the newly thickened network. Perhaps you want to ensure the average degree is 5, or the diameter is 4. Then add the highest-scored node pairs until that property is achieved. Of course, it may not be possible to achieve such a desired property by adding in those new edges, so some trial-and-error may be involved.

As always, wherever feasible, perform robustness checks to see how any conclusions you wish to draw from the data are affected by the predictions. Comparing different networks under different prediction parameters is helpful (Ch. 14).

Keep in mind that while much of the literature has studied the problem of link prediction, most studies kept known results on hand for testing purposes. Often for instance, studies do not fully address the choices a researcher needs to make in practice, such as what scoring cutoff to use when deciding whether to predict a new link, instead devising a test scheme that either abstracts away this detail or evaluates different methods averaged over all possible cutoffs. In practice, you should not expect to have access to such ground truth when you're making actual predictions, and the results of these more abstract evaluations may not provide much guidance.

Ideally, a link prediction method will have a reasonable notion of confidence or certainty around its predictions: "We are 95% certain Alice and Bob are friends. We are 50% certain Bob and Carole are friends." Use this information, for example by mapping it onto link weights. Or extract a "high-confidence" backbone (discussed above) that retains observed links and those predicted links of which you are most confident. Without measures of certainty, predicted links may not be worth including, as it becomes too difficult to tell the reliability. This may happen with a machine learning "black box" approach to link prediction, although such methods may still be useful if they are deemed sufficiently trustworthy.

Suppose you have thickened a dataset by applying link prediction. It is good practice to distinguish predicted and observed links. First, for data provenance, you do not want subsequent users of the data to misinterpret a predicted link as an observed link, which may be disastrous. Second, for safety, you probably want to compare any analyses you do on the network with and without the predicted links included. How do your conclusions change due to using the predicted links? Therefore, incorporate a binary observed/predicted link attribute (Ch. 9) and retain it with your data moving forward.

Link prediction is fundamentally a machine learning solution; we discuss machine learning with networks in-depth in Ch. 16.

10.7 Other approaches

Combination methods

We have discussed thinning out spurious network elements and thickening up missing network elements as separate problems. In fact, consider applying both practices together. For example, pairing link prediction with a backbone extraction method. What's interesting to explore, but perhaps daunting itself, is the interactions between the different methods. If you first extract the backbone, the links predicted by a given scoring function may be quite different than if you first apply that scoring function. And likewise, if you first thicken the network with predicted links, then extract a backbone, the final backbone may be quite different. Keep in mind Occam's razor: a simple solution, all else being equal, is superior to one more complex. What is the fundamental problem you are addressing with the data and does it lead you to a complex, composite method? Only with an answer in the affirmative is it likely you will need or want the complexity of combining methods.

Statistical models and uncertainty quantification

In principle, although not always helpful in practice, one can build a statistical model (Ch. 23) for the probabilities of nodes or links to be valid, then query the data to see which network elements are missing or spurious. If the statistical model holds, and is computationally tractable for your network, then you can perform an inferential procedure to estimate high-likelihood links that do not exist in the data (even potentially nodes although this is especially tricky), or low-likelihood links (and nodes) that do exist in the data but perhaps should not. Coupled with a decision rule for when to add or remove links (or, again, nodes), you have now used statistical inference to thin and thicken the data by quantifying elements which are very certain or very uncertain (Ch. 24).

Besides relying on the accuracy and appropriateness of the statistical model, another challenge faced by this approach—and indeed, many approaches—is sparsity and imbalance. Most possible links do not actually appear in the network data, the average degree being much less than the number of nodes. This means it's likely to have a high rate of false positives, predicting nodes or links that do not exist, simply because there are so many more pairs of nodes than there are links. Keep this bias in mind.

10.8 Summary

After gathering data and extracting a network, investigating the network may reveal errors of omission—important structure is missing—or errors of commission—spurious structure is present. They are tricky to handle, but there are some techniques that can help. Use thickening techniques to get insights into omissions: link prediction can point you towards unseen links; graph union can allow you to merge your network with another. Use thinning techniques to address commissions. Thresholding by link weight, time window, or other means, can help you retain the strongest, most certain links;

backbone extraction can gather the critical subset of the network structure based on its properties. But remember that you should be careful; although link prediction may point you towards unseen links, you should not simply analyze the network with those predicted links included as if they are all correct. Backbone extraction methods can be useful but they are also *models* with strong assumptions and therefore should be treated as such.

Even when elements of the network structure are not in error, thinning and thickening techniques can be useful preprocessing for a variety of reasons. Algorithms may simply be challenged by too much density or analysis may be too difficult with so much structure to sift through.

There is not necessarily a bright line separating clearly the data gathering and network extraction tasks from thinning and thickening the data. A backbone extraction algorithm can be an element of a larger, iterated upstream task. The data are gathered, a preliminary network is extracted, spurious edges are removed, and a final network is obtained. This recapitulates in many ways the life cycle of a network study.

Bibliographic remarks

Taylor [455] is a truly lovely introduction to uncertainty and error analysis in scientific measurement. One of us (Bagrow) considers the cover photograph to be one of his favorite scientific illustrations.

Imputation is the field of statistics broadly interested in filling in missing values in data. Imputation has a long history dating back to seminal work by Rubin [406] and Dempster et al. [127] (see Rubin [407] for a history). van Buuren [469] is a more recent textbook on the field. Gelman and Hill [178] also has an excellent chapter about missing data.

Numerous backbone extraction and structural sparsification methods, used to thin networks, have been studied. The disparity filter we discussed here was introduced by Serrano et al. [424]. Other approaches include structural measures, especially for social networks, and spectral methods based on spectral properties of matrices such as the graph Laplacian that capture the network structure. Batson et al. [46] provides an overview of spectral methods (see also Ch. 25). There are also many methods that are specifically designed for networks constructed from associations (e.g., correlation matrices). This is a particularly common problem in systems biology. Soranzo et al. [439] provides an early overview of basic methods such as partial correlation.

Link prediction is a well-studied task in networks. Liben-Nowell and Kleinberg [273] provide an overview and Lü and Zhou [282] give a broader survey of work on the task.

Butts [87] discusses statistical aspects of errors in social networks, along with using Bayesian inference to tackle the errors. Other works of interest along these lines include Guimerà and Sales-Pardo [195], Martin et al. [295], and Newman [337].

Exercises

10.1 *Effects of disambiguation.* Suppose you built a network of actors who costarred in films (Ex. 6.1) as extracted from texts of the end credits of films. Doing so required disambiguating different actors. Suppose you are not confident in how the disambiguation was performed. Describe some errors or issues with the final network that may have occurred due to problems with the disambiguation. How likely or unlikely are the different issues?

10.2 (**Focal network**) Consider the Malawi Sociometer Network. Badges worn by participants (the nodes of the network) could measure proximity to other badges. And we add links between nodes when proximity occurs. Each badge, both its hardware and software, is subject to failure.

   What happens to the network data if:

   (a) A subset of badges occasionally lose memory and no contacts are recorded?

   (b) Some badges use the wrong time when recording contacts?

   (c) Some badges are running different versions of the software and they cannot detect badges with another software version?

   Describe some ways to fix these problems. Does the network being temporal help?

   Can you think of any other problems due to the badges? Problems not due to the badges?

10.3 (**Focal network**) Consider a PPI network such as HuRI. These networks are generated by systematically testing pairs of proteins one at a time to see if an interaction exists. Suppose the network has N nodes and M edges. Those M edges were found by testing M_{test} protein pairs. What is the average degree in the observed network and what do you estimate it to be in the "true" network that would be found if every pair of nodes could be tested?

10.4 Describe the disparity filter (Sec. 10.5.5) in pseudocode using data structures and functions from Ch. 8.

10.5 (**Focal network**) Implement the disparity filter. Apply it to the Malawi Sociometer Network, reproducing Fig. 10.2. Compare these results to extracting the backbone with a global threshold. (See also Ex. 23.6.)

10.6 (**Simulation study**) Write a program that samples a graph's edges uniformly at random. As edges (i, j) are sampled, track how often nodes i and j are observed. Apply this program to two focal networks of interest. (Be sure to average your results over the randomness of sampling.) Use a plot to visualize whether node observations are biased for or against high-degree nodes. How does the sampling rate of nodes relate to their degree?

Chapter 11

Explore and explain: statistics for network data

Data analysis can generally be divided into two main approaches, exploratory and confirmatory. Exploratory approaches seek to understand novel data by iteratively examining its various facets, asking questions and performing calculations along the way to guide the researcher through the data. Often graphical techniques guide the exploration, but descriptive statistics are commonly used as well. Confirmatory analysis, on the other hand, grounds the researcher with specific, preexisting hypotheses or theories, and then seeks to understand whether the given data either support or refute the preexisting knowledge.

In the context of network data, both approaches are valuable and complementary. We discuss both in this chapter. The following chapters, on understanding network structure (Ch. 12) and visualizing networks (Ch. 13), pair well with this chapter, giving further details on measures and techniques to use for exploring and explaining network data.

11.1 Exploratory analysis

Exploratory analysis is a process of examining a dataset with the goal of better understanding its characteristics. For instance, it is immensely useful to visualize the network at hand if the network is fairly small. As we will discuss in a later chapter (Ch. 13), a good network visualization can reveal a lot of details about the network, answering questions including: how dense is the network? Are there clear communities in the network? Is there a core–periphery structure?

But exploratory analysis is not limited to visualization. There are many statistical exploratory analyses as well. The most basic analyses measure the statistical properties of the network structure, by measuring the degree distribution, degree correlation, clustering coefficient, path length distribution, etc. (Ch. 12). Taken together, these measures can tell us about the general properties of the network. However, one should be keenly aware that the statistics cannot capture everything. Exploratory visualization,

137

exploratory statistics, and confirmatory statistics all work hand-in-hand to provide a holistic understanding of a network.

 Exploratory data analysis (EDA) uses summary statistics and graphics to describe the contents of a dataset.

Leveraging EDA tools

A strong understanding of general EDA is valuable. As many networks are large, we often will summarize or quantify the network by computing distributions of (usually) numeric quantities. This summary in effect projects the network information down to non-network data, a great opportunity to leverage existing EDA.

 Use general-purpose EDA methods with network data by extracting (numeric) quantities from the network and exploring them.

One simple example is the degree distribution (Sec. 11.7). For each node in the network, compute its degree. Then ask: what is the mean degree? What is the median degree? Do they differ by a wide margin? How else can we summarize the degrees of nodes? Perhaps we can ask how many nodes have degree 1? What is the largest degree we observe? What is the variance or standard deviation? Moving from summary statistics, we could look at the degree distribution visually using a histogram or cumulative distribution plot.[1] (See below for more on plotting the degree distribution.) Histograms are one of the simplest and yet most important graphical EDA tools and we can leverage them for network data by extracting the right network quantities. Even simple bar charts can be used to visualize categories in effective ways; don't discount the power and expressiveness of simple plots. And in all of this, of course, we are not limited to the degree distribution; many quantities (Ch. 12) are worth examining.

Another major focus of EDA in general is exploring relationships. Are two variables or quantities x and y strongly associated? We can measure this with a correlation coefficient.[2] Other summary statistics can be used to capture the relationship between these variables but we can also use graphics to quickly understand the data. Indeed, the humble scatter plot is a simple, powerful visual tool for understanding pairwise relationships between numeric quantities.

Axes scales and heavy tails   Many numeric network quantities exhibit broad distributions, also known as heavy tails. In such data, there are occasionally extremely large values, far larger than you would expect from a normal (Gaussian) distribution with the same mean and variance. (A handy rule-of-thumb for noticing a heavy tail is when there is a large discrepancy between the mean, which is affected by outlier values, and the

---

[1] The empirical cumulative distribution (ECDF) is a very powerful alternative to histograms. It works by sorting the data and then plotting each value on the horizontal axis against its rank (divided by the number of observations) on the vertical axis. Unlike a histogram, the ECDF is not affected by a choice of binning and without binning it uses every datapoint in its plot. The one disadvantage of an ECDF over a histogram is that it visualizes the cumulative probability, which is an integral of the probability distribution. This integral is less interpretable for some audiences.

[2] This assumes a linear association.

median, which is not.) In such situations, plots need to be explored more thoroughly. We recommend, and it is standard practice, to use *logarithmic scales* on your plots. All major plotting libraries support log-scaled axes. Log axes will stretch out very small values and squash down very large values, and can balance out the broad distribution you may be unable to see with a linear scale. If necessary, try combinations of logarithmic and linear scales for x- and y-axes. But also, keep in mind a logarithmic view is very different from a linear one, and it is not something that comes naturally when humans try to interpret space and distance within a plot. We explore an example in Sec. 11.7.

 Consider plotting network statistics such as the degree distribution. Count how many nodes have degree 1, degree 2, and so forth. Keep in mind a log-scale may be necessary.

Towards network-specific EDA

Converting network quantities of interest into standalone, non-network variables allows us to use general EDA—but at a cost. Necessarily, network information is lost. Yes, we have the degree distribution, and we can tell which nodes have the highest degrees, but are those nodes neighbors in the network or are they far apart from each other? This information has been lost.

 Using only general-purpose EDA tools on data pulled from a network loses much of the network-specific information.

There are two ways to proceed: (i) incorporating the network more directly into the extraction process that generates data for general EDA tools to study, or (ii) move away from general EDA and use network-specific methods. Later, we'll focus on (ii) in greater detail. For now, we turn our attention to (i).

The simplest way to incorporate the network into the data being explored is to track pairs of nodes and whether they are neighbors or not. In the context of the degree, doing so moves us from studying the degree distribution toward studying degree correlations. This helps with the question: are connected nodes of similar degree or not?

We can use graphical tools to explore degree correlations. A scatter plot with a point for each link where the x-axis is the degree of one node and the y-axis is the degree of the other node can be quite handy. One drawback, though, is that the degree is integer-valued and this can lead to a lot of points stacking up in the plot. You can fix this by adding a bit of random noise to each point before plotting or, more systematically, you can produced a heatmap by plotting boxes color-coded by how many links fall into each k_1, k_2 coordinate. The matrix defined by this heatmap is called the degree correlation matrix.

Moving on from a visual of the degree correlations, we can also compute a correlation coefficient between the degrees of connected nodes. This takes a common, general-purpose summary statistic, the Pearson[3] correlation coefficient, and uses network information, only looking at connected nodes, for a numeric network quantity, the

---

[3] The common correlation coefficient is not the only choice. Another good option is the Spearman

degree. Other network quantities than the degree can be quantified based on edges, for instance, node centrality measures. We also discuss degree correlations and other such measures in Ch. 12.

If one wishes, one can move further across the network than just neighbors. Neighbors are one step away from another. What about second-neighbors, also known as next-nearest neighbors? Being farther apart from one another, we would expect them to be less related to each other than directly connected neighbors, but is this the case? It may be worth investigating in your data using, for instance, a shortest path algorithm such as breadth-first search.

 Consider exploring relationships between pairs of nodes as a function of how far apart they are. Nodes very distant from one another in the network are probably less similar than nodes close by, but this is worth testing.

Lastly, for some networks, the ultimate network-specific graphical EDA tool is network visualization. Drawing the network can bring immediate attention to many features. Not all networks are amenable to visualization and some more subtle properties are difficult to notice in a graphic, but in most cases, as long as you are mindful of its limitations, drawing the network is almost always worth a try. Chapter 13 focuses entirely on network visualization.

What to explore and when

As we continue to learn statistical measures and other quantifications, we should ask ourselves when we wish to apply such tools. Here EDA can support both understanding the data we have and informing the data collection (Ch. 6) and network extraction (Ch. 7) processes. Suppose to find a network, we use an algorithm that defines edges based on a similarity function f applied to data gathered for pairs of nodes: nodes i and j are connected if $f(\mathbf{x}_i, \mathbf{x}_j) \geq s$. But what should f be and what value should we set the threshold s? We can test different choices by applying EDA to each resulting network. This might show, for instance, that the network is densely interconnected, but only for a particular function or threshold value. As another example, exploring the network might reveal a bug: self-loops have been introduced because the similarity function was applied not to every pair of distinct nodes i and j, $i \neq j$, but instead to every pair of nodes i and j, *including* $i = j$. This is fine if intended—self-loops are often meaningful—but if unintended and undiscovered, it could be disastrous.

11.2   Network analysis is usually iterative and complementary

It is rare to stick to only exploratory analysis or only confirmatory analysis. Exploratory analysis often raises questions about the network, and confirmatory analysis answers these questions. Also, one can start from a specific question, but then the exploratory

---

correlation coefficient. Spearman measures a monotonicity relation between x and y, whereas Pearson measures the more specific linear relationship.

analysis allows the researcher to revise the question based on a better understanding of the data.

As an example, suppose a researcher wants to study political conversations happening on Twitter and collected a retweet network from Twitter, where nodes are Twitter users and directed edges are retweets. They first quickly visualize the network and see that the network contains multiple blobs. Measuring the *clustering coefficient*[4] also showed that the network may be strongly clustered, although it has not been confirmed with a null model. These exploratory analyses led the researcher to hypothesize that the network is highly polarized and clustered based on users' partisan views. To test this hypothesis, the researcher used methods to identify partisan membership of a Twitter user based on their following or profile information. They also ran community detection algorithms (Ch. 12) to identify large communities in the network. The confirmatory analysis of measuring the correlation between partisan membership and the communities, possibly with appropriate statistical testing, supports the conclusion that the network is clustered based on partisan membership.[5]

As you can see in this (hypothetical) example, exploratory analyses and confirmatory analyses complement each other and most research projects with network data involve both.

11.3 Confirmatory analysis

When we measure a quantity from a network, how do we know whether it is large or small? Is a clustering coefficient of 0.5 large or small? The answer is: we don't know! We cannot say much until we know more details about the network and proper *context*. First, the number that we are measuring will be affected by other properties of the network. If the network is sparse, a clustering coefficient of 0.5 can be almost impossibly large; but if the network is extremely dense, the clustering coefficient of 0.5 may be almost impossibly small. The question boils down to: *what is the reasonable context?*

The question of context is all about what the *expected* distribution of this quantity is within the context. The expectation can come from known mechanisms of network dynamics, a set of real networks, or an ensemble of "null"[6] models.

Confirmatory analysis happens when the researcher has a concrete hypothesis about the network. For instance, one may want to test a hypothesis that a given social network is strongly clustered. You can always measure the clustering coefficient of the network, but how to go from a specific clustering coefficient value to labeling the network as "strongly" or "weakly" clustered? The researcher can choose a *null model* and measure the clustering coefficient of the network under the null model, which can then be compared to the actual number, and we can ask how far the actual data is from the null model.

---

[4] The clustering coefficient (Sec. 12.4) measures how many triangles (cycles of length 3) form between nodes. Densely interconnected groups often have many such triangles.

[5] Keeping in mind, of course, that another factor may be *causing* the clustering. Such a factor, which may be unobserved, may be associated with partisan membership.

[6] As in the null hypothesis in the statistical hypothesis testing framework.

Null models demonstrate what the data would be like if an important property or properties were not present but other properties remained, and they generally do so by randomizing—in some manner—away the particulars of that property while preserving the particulars of other properties. What would be the clustering coefficient of the network if nodes formed the same numbers of links but at random? What would be the clustering coefficient of the network if nodes are randomly connected with the same number of links while preserving their degrees?

The distribution of a statistic under a null model can be derived mathematically or computationally, although in practice with networks it is often computational: many randomized networks are generated from the real network data, perhaps by shuffling elements of the network in some random fashion, and this *ensemble* is then compared to the actual network. In some instances, the randomization process in effect matches an existing random graph model, and after we notice this fact, we can use it to our advantage by building that random graph directly.

11.4   Graph models as null models

Null models are built by picking the quantities and properties from the real network that should be preserved. For instance, the simplest null model is the *Erdős–Rényi (ER) random graph model*[7] where links are placed uniformly at random between pairs of nodes. An ER graph is defined with two parameters, the number of nodes and the number of links.[8] By measuring N and M for a real network, then building an ER graph at random with those same parameters, we have created a null model[9] with the same number of nodes and edges as the real network, but where all other properties are randomized. As a null model, ER graphs let us test the real network for a property of interest while controlling for the number of nodes and links.

However, the ER graph model can be shown to exhibit a homogeneous degree distribution (binomial/Poisson), unlike most real-world networks, and so it may be *too random* for many cases. In such cases we commonly turn to another model, the configuration model.

11.5   The configuration model

The ER graph is often not suitable as a null model because it does not preserve the degrees of the nodes. Meaning, we keep the number of nodes and edges in our null, but hubs are gone, and low degree nodes are gone. Instead, we have "smeared out" the fluctuations in degree from node to node.[10] In practice, it is almost always important to understand the role of hubs and how fluctuations in the degrees of nodes interplay with one another. (Is the network *assortative* (degree correlated)? *Dissortative* (degree anti-correlated)?)

---

[7] Random graph models such as ER are the focus of Ch. 22.

[8] The ER graph can also be defined using the probability for a new link instead of the number of new links. These definitions are mostly equivalent for our purposes.

[9] More properly, an instance of a null model.

[10] More precisely, we have turned the degree distribution into a Poisson distribution.

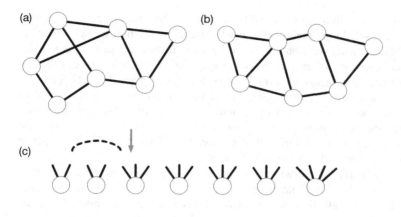

Figure 11.1 Using the configuration model to make a degree-preserving null model. (**a**) The original and (**b**) a null network. (**c**) Visualizing the construction process as randomly dropping edges between the stubs of the original nodes, in this case the first and third node. Each node begins with as many stubs as its degree in the original network. (**d**) A stublist, a convenient data structure for adding edges by randomly sampling pairs of items from the list. Here nodes are numbered by their order left-to-right in (c).

We need a way to control for more network structure in our null model than ER graphs provide. The next simplest null model would be *random graphs with arbitrary degree distributions* or the *configuration model*, which preserves the number of nodes and edges, as well as the degree sequence of the nodes. The degree sequence is a list of length N that gives the degree of each node: $[k_1, k_2, \ldots, k_N]$. Fixing the degree sequence acts as a strong constraint and this model explains many properties of the real-world networks. The configuration model is the maximally random graph defined for a given degree sequence. We feed in the real network's degree sequence and build a configuration graph null as described below, then we can compare the real data to the configuration model to see if the degrees of nodes can explain the network properties we are studying.

Building a configuration model instance

We can visualize the process of building a configuration model for a given degree sequence by starting with N nodes and zero edges and giving each node a "stub" for each edge it has in the real network (Fig. 11.1). We then insert edges into this graph by picking pairs of stubs (on different nodes) and wiring them together with an edge. This process of inserting edges at random between dangling stubs repeats until all the edges are inserted and all the stubs are connected, at which point we have built a random instance of a configuration model matching the original network's degree sequence.

However, there are some tricky details with the above description. If you pick two stubs at random, it is possible that both will be attached to the same node, in which case you would create a self-loop. Technically this fits the configuration model definition but it is probably not what we want to do, at least when the real network is simple. Multi-edges can be created in exactly the same way: two stubs chosen from different nodes can be connected with an edge even if there already exists an edge between those nodes from previous edge insertions. Steps need to be taken as edges are added to prevent these situations[11] from occurring. Usually, some random edge exchanges (see below) are performed to eliminate the self-loops and multi-edges. You can also just delete these extra edges if you're OK with not perfectly matching the degree sequence (which is less of an issue with very large networks).

A simple data structure to use when building the configuration model is a *stublist*. The stublist is a list of the nodes that contains k copies of a node with degree k. Edges are inserted by popping (selecting and removing) items uniformly at random from this list two-at-a-time. This makes it simple to keep track of the remaining stubs for any node without any extra bookkeeping, at the expense of using more memory to store the stublist. (Using a stublist does not eliminate any of the issues with self-loops or multi-edges, if these concern you.)

11.6   Computational null models (Monte Carlo)

Often, the network that you are dealing with does not fit the assumptions of these simple models because of weights, directions, or special constraints. Or, the quantities or properties that you care about cannot be analytically computed. In such cases, we can employ computational (or nonparametric) null models. The goal of these null models is to sample an expected distribution of the quantities of your interest by randomizing the existing network while controlling for chosen quantities. In other words, this can be thought as a type of (Markov Chain) Monte Carlo method.

Like the configuration model discussed above, we usually wish these computational null models to preserve at least the degree distribution of the network. We could devise many algorithms that randomize a network, for example by randomly deleting edges and then inserting them back at random. These methods tend not to preserve the degree distribution, however. Thus, many null models are based on a local *edge exchange process*, also known as *edge rewiring* or *edge swapping*, that preserves the degree or other feature.

Algorithm 11.1 describes a basic degree-preserving edge exchange randomization. A single exchange is illustrated in Fig. 11.2: you pick two edges that can be swapped while preserving the quantities of interest. Then keep swapping such edges long enough so that the network is randomized except those properties that you want to preserve. The number t in Alg. 11.1 counts the number of exchanges to perform.[12]

---

[11] A problem we happily can avoid is dealing with non-graphical degree sequences. The degree sequence is a list of integers but not all lists of integers can be made into a graph. For example, the list [1, 2] cannot be a degree sequence for a simple graph. (Why?) We avoid non-graphical sequences for our purposes here simply because we are first deriving the degree sequence from a real network, ensuring it will be graphical.

[12] Note that this formulation of the algorithm will run forever if fewer than t allowable exchanges are possible. See Exercises.

Algorithm 11.1 The basic edge exchange algorithm. This pseudocode uses functions from Ch. 8 and `get_random_edge`, a function that returns the two nodes of an edge chosen uniformly at random. This algorithm modifies G in place, so you may want to first copy G if you want to compare the original and randomized networks.

```
 1: procedure EDGE_EXCHANGE(G, t)              ▷ Perform t edge exchanges on network G
 2:     s ← 0
 3:     while s < t do
 4:         (i, j) ← get_random_edge(G)
 5:         (u, v) ← get_random_edge(G)
 6:         if i = u or i = v or j = u or j = v then          ▷ Edges not independent
 7:             continue
 8:         if has_edge(G, i, v) or has_edge(G, j, u) then   ▷ Collides with existing edge
 9:             continue
10:         remove_edge(G, i, j)                                 ▷ Perform exchange
11:         remove_edge(G, u, v)
12:         add_edge(G, i, v)
13:         add_edge(G, j, u)
14:         s ← s + 1                                     ▷ Count successfull exchange
15:     return G                                                ▷ Now randomized
```

Now, suppose we repeat the entire edge exchange process n times, each time starting over from the original network. The resulting network will be slightly different in each repetition due to the random orders in which edge pairs are chosen, and we can use this variability to measure the distribution of any network quantities we are interested in under the null assumption that the network is randomly structured (except for what is preserved, such as the degree, by the edge exchange process). Let's do an example:

11.6.1 Focal Point: Human Reference Interactome's degree correlations

Assortativity is a measure of how much connected nodes tend to be related (see also Sec. 9.3). *Degree assortativity* focuses on correlations in degree. Do nodes with many neighbors tend to connect to other nodes with many neighbors (degree *assortative*) or do nodes with many neighbors tend to connect to nodes with few neighbors (degree *dissortative*)? Assortativity r for degree can be computed by simply taking for each edge (i, j), the degrees k_i, k_j of the nodes connected by that edge and measuring the Pearson correlation coefficient between k_i and k_j. (We discuss this in depth in Ch. 12.) When $r > 0$, the network is assortative; when $r < 0$, it is dissortative; and when $r = 0$, the network is uncorrelated. Determining the correlations within a real network gives us a better understanding of how the network may have formed and whether simple random network models (Ch. 22) can be used to study the data.

Let's consider the degree correlations within the Human Reference Interactome (HuRI). For this we will focus on the giant component (Sec. 1.4) of the network and remove self-loops. Measuring the degree assortativity of HuRI gives $r_{\text{HuRI}} = -0.1179$,

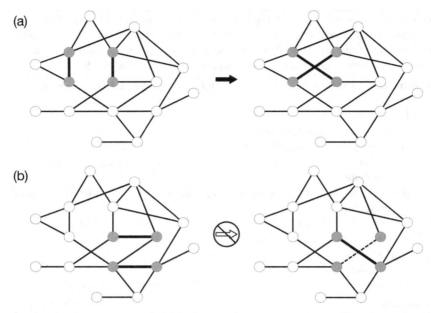

Figure 11.2 Illustrating the edge exchange process. (**a**) Suppose the two highlighted edges are chosen at random. We perform an edge exchange (or swap) by deleting those edges and inserting new edges linking the opposite pairs of nodes. This exchange preserves the degrees of all the nodes in question but, over many allowable exchanges where edges are chosen at random, the larger structure of the original network is lost. (**b**) An example of an unallowable edge exchange. In this case, the two highlighted edges should not be swapped because one of the destination edges is already present in the network (dashed line). Such an exchange would not preserve the degrees of all nodes or even the total number of edges. We reject this edge pair and sample another.

meaning that the network is degree dissortative: high-degree proteins tend to connect to low-degree proteins and vice versa. The value of r seems small, however. Perhaps it is not so different from $r = 0$? We need more context to tell. We could ask what r would be for an equally dense random network, in this case an equivalent ER graph with $n = n_{HuRI}$ nodes and $m = m_{HuRI}$. Doing so should on average give $r = 0$ but the variance across many randomized ER graphs may be wide, and we can see if the real HuRI r is compatible with the random ER r values.

However, this test is not appropriate, as the ER graph, while it controls network density, does not control the degree distribution. In fact, its degree distribution looks drastically different from that of HuRI's. And degree is the central property behind r. We want to ask, does a network with the same degree distribution as HuRI but without degree correlations give rise to values of r that are compatible with HuRI's? We can address this question using the null models introduced earlier, which generate random, uncorrelated null networks while controlling for the degrees of all nodes.

We begin by performing the edge exchange algorithm[13] multiple times, each time

---

[13] With one modification: we ensure the network remains connected while exchanges occur; see Exercises.

starting from a copy of the original HuRI network. We then measure r for each random realization. Because each edge exchange is performed independently from the same starting network, and assortativity is bounded, the central limit theorem tells us the distribution of r over many exchange realizations will converge to a normal distribution. We can therefore use a simple z-score (Sec. 4.3) to ask if the observed assortativity—call it r_{real}—is compatible with the distribution of randomized network assortativities. Computing

$$z = \frac{r_{real} - E[r_{rand}]}{\sigma_{r_{rand}}} \tag{11.1}$$

lets us answer this by asking how many standard deviations $\sigma_{r_{rand}}$ away the real value is from the mean random value $E[r_{rand}]$. We illustrate the distribution of r_{rand} compared to $r_{real} = r_{HuRI}$ in Fig. 11.3. On average, we see $r_{rand} \approx -0.04584$, still negative but much higher than the real network. Indeed, from these results, we see that HuRI's assortativity has a score of $z = -24.910$, highly significant. We conclude that HuRI's degree distribution alone does not explain its assortativity.

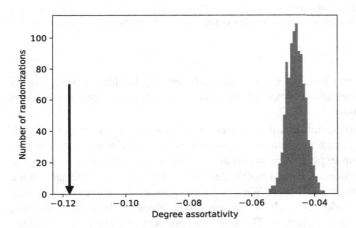

Figure 11.3 The distribution of assortativity under edge exchange compared to the observed value (arrow) for the HuRI network.

A procedural question arises. How many times should we exchange edges? What is t in Alg. 11.1? We recommend being exhaustive to get a good null. Perform multiple exchanges per edge on average: $t > M$. Usually, $t = 2M$ or $4M$ is good, although for smaller networks, where the exchange runs more quickly, it may be worth going even higher, to $t = 10M$, just to be safe. We explore the effects of t on r in Fig. 11.4. Our mean estimate of r settles down with $t \geq 2M$. And indeed, we used $t = 4M$ in Fig. 11.3.

11.6.2 Traps in the null: structural cutoffs

We glossed over something odd in the previous example. The randomized HuRI networks are also degree dissortative. How can this be? Doesn't the edge exchange al-

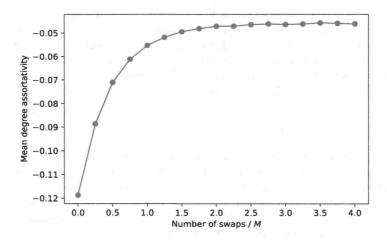

Figure 11.4   The change in assortativity for HuRI as edges are randomly exchanged. The random value settles down when the number of edge swaps exceeds $2M$. We have repeated the edge exchange algorithm 100 times over independent copies of HuRI and plot the r averaged over those trials.

gorithm explicitly destroy all degree correlations? Likewise, shouldn't a configuration model, built by choosing stubs completely at random, possess no degree correlations and therefore always have $r \approx 0$?

Yes and no. An important detail of the configuration model as we described it above is that it actually creates a *multigraph with self-loops*. As stubs are wired together (Fig. 11.1), there is no explicit step preventing the same pair of nodes from being connected more than once, creating a multi-edge, or for two stubs from the same node being connected, creating a self-loop. We can fix these issues by either checking that unique, unconnected nodes were chosen as we insert edges, or by simply deleting extra edges and self-loops afterwards. The latter will not perfectly preserve the degree sequence while the former can make it slow to generate a network. But more insidiously, trying to prevent these edges from forming may be impossible, depending on the degree sequence, and you will not be able to finish building the model. In that case, performing edge exchanges may be a way to "back out" the steps you've followed building the network, allowing you to backtrack out of the trap you've fallen into.

These traps in the configuration model also point to more fundamental aspects of the network topology. *Structural cutoffs* occur when the degree sequence involves a number of very high degree nodes, the so-called *hubs*. Because hubs have many stubs, they will be frequently chosen when sampling from the stublist as you build a realization of the configuration model. If their degree is high enough, or there are enough of them, you will be effectively guaranteed to create self-loops or multi-edges. When this happens, it means that the configuration model, which is in fact always degree uncorrelated, can only be uncorrelated when those multi-edges or self-loops exist. If we delete those edges, or exchange them away, we are actually forcing the network away from its uncorrelated state—we are preventing the existence of those edges needed to "zero out" the degree correlations. When nodes exist with degree greater than a particular value, the structural cutoff value, a (simple) network must be degree dissortative even if edges form independent of degree. This is exactly what we saw with our randomized null for HuRI giving an assortativity of $r < 0$.

A good rule-of-thumb[14] for the structural cutoff is

$$k_s \approx \sqrt{2M}. \tag{11.2}$$

If $\max_i k_i > k_s$, you should expect structural dissortativity to play a role in the network. Indeed, for HuRI, our example network, we have $k_s = 322.54$ and $\max_i k_i = 498$ and structural dissortativity, while not sufficient to explain all dissortativity, is present in the network.

As a consequence, when comparing a real and randomized network, we should ask, if the network is dissortative, is it dissortative only structurally, due to nodes beyond the cutoff, or, as we see with HURI, is the network's structure insufficient to completely explain the degree dissortativity?

 The edge exchange algorithm is a good way to confirm structural dissortativity when we test the real assortativity r. Randomize the network sufficiently, compute r_{rand} and repeat many times to find the distribution of r_{rand}. While checking the real assortativity via Eq. (11.1), check if $r = 0$ is also compatible with the null model. If it is not, the network exhibits structural dissortativity.

11.7 Case study: the degree distribution

Hypothesis testing is one of the pillars of statistical inference. Networks are not especially amenable to typical tests, as tests generally focus on summary statistics or other numeric parameters. That said, many situations can arise where a network scientist can address a question of interest using hypothesis tests on *properties* of a network.

As a case study, we'll focus in this section on the ***degree distribution***, one of the most fundamental properties of a network (see also Ch. 12). Just as the configuration model serves as a "graphical" null model for a network that preserves a real network's degree

---

[14] To see where this comes from, ask how many edges $E_{kk'}$ connect nodes of degree k and nodes of degree k' and does this number exceed $m_{kk'}$, the maximum possible number of edges between said nodes if parallel edges are not allowed. This maximum is given by $m_{kk'} = \min(kN_k, k'N_{k'}, N_kN_{k'})$, where N_k is the number of nodes with degree k (with $\sum_k N_k = N$). (To derive $m_{kk'}$, try drawing out some stubs for a value of k and k', noticing that kN_k is the total number of stubs of degree k nodes, and counting how many non-parallel edges can be inserted.) Right away, if $N_k < k'$, we will already see that multiple edges must be prevented, and likewise for $N_{k'} < k$, which means we can focus on just $m_{kk'} = N_kN_{k'}$ as a bound.

Now, for an uncorrelated network, when does $E_{kk'}/m_{kk'} = 1$? Notice that $\sum_{k'} E_{kk'} = kN_k$ and $\sum_{k,k'} E_{kk'} = 2M$. This means that $P_{kk'} = E_{kk'}/2M$ and $P_k = \sum_{k'} P_{kk'} = kN_k/(2M)$. For an uncorrelated network, we have $P_{kk'} = P_k P_{k'}$. Our ratio is then

$$\frac{E_{kk'}}{m_{kk'}} = \frac{2MP_{kk'}}{N_kN_{k'}} = \frac{kk'N_kN_{k'}}{2MN_kN_{k'}} = \frac{kk'}{2M}.$$

Assuming this ratio becomes 1 when $k = k' = k_s$, we have $k_s^2/(2M) = 1$ or $k_s = \sqrt{2M}$.

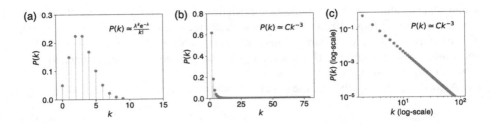

Figure 11.5   Example degree distributions. (a) A Poisson-like degree distribution. (b) A power-law-like degree distribution in linear scales. (c) Same distribution as in (b), but in log–log scale.

distribution, other statistical models can be used to study the degree distribution itself. Treating the degrees of nodes as observations of an integer-valued random variable, statistical models of *count data* can then be employed.

First we will discuss some basic exploratory techniques to examine a degree distribution, then we will discuss inferring a functional form for the distribution. We follow this with a real-world case study using the Malawi Sociometer Network.

11.7.1   Measuring a degree distribution

So how should we measure, plot, and estimate the degree distribution of a network? The first and most obvious way to plot the degree distribution is to simply plot the probability $P(k)$ as a function of the degree k in linear scale, with bars, points, etc. (see Fig. 11.5a). Here, each point is the value of $P(k) = N_k/N$ where N_k is the number of nodes with degree k as calculated from the data.

However, this plot quickly becomes problematic when the network is large and the degree distribution is highly skewed. Many real-world networks exhibit heterogeneous, heavy-tailed degree distributions, including the power-law distribution and log-normal distribution [15] In that case, with a linear scale, we will not be able to see much information about the distribution (Fig. 11.5b). Thus, it is usually more useful to consider alternative plots.

Plotting for heavy tails   Faced with the possibility of a skewed degree distribution, plot the distribution in both linear and logarithmic scales, as shown in Fig. 11.5. The plot, especially in a log–log scale, can tell us whether the distribution exhibits a heavy tail and, if it does, whether it looks linear in the log–log scale. If the degree distribution does not exhibit a heavy tail, then we can directly go to the inference steps we discuss below. On the other hand, if the degree distribution exhibits a heavy tail and spans multiple orders of magnitude, even the plotting can be non-trivial.

First, because the degree can span many orders of magnitude, for many values of k we have little data, $N_k = 0$ or close to it. Therefore, a direct plot of the raw degree distribution produces a possibly noisy plot like the one shown in Fig. 11.6a. We can

---

[15] Curious for more on these distributions? Check out Secs. 12.3 and 22.6.

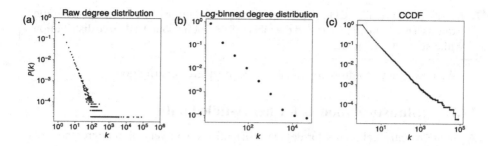

Figure 11.6   Three common ways to plot the degree distribution for the case of a power-law degree distribution.

address this with *logarithmic bins*—bins that grow in width at the same rate as the log-scale—and plot a histogram. Because apparent distance in the logarithmic scale is determined by the ratio of the bin edges (or a multiplicative factor), if you set the first bin to be $[1, r)$, the second bin will become $[r, r^2)$, the third bin will become $[r^2, r^3)$, and so on. For instance, if you set $r = 2$, the bins will be $[1, 2), [2, 4), [4, 8), \ldots$. Because the width of the bins are not equal, we need to normalize the number of data points in each bin by the width of the bin to obtain the *frequency density*, which is what we plot in the log-binned plots. This is shown in Fig. 11.6b.

> ⚠ Be careful when using histograms with unequal bins.
>   If one bin is twice as wide as another, all else being equal, we expect to see twice as many data points in the wider bin. Therefore, we must normalize the number of data points in each bin by the width of the bin to obtain the *frequency density*, which is what we should plot for histograms with unequal bins. In general, when plotting a histogram with bars, the number of data points in each bin should be proportional to the bar's *area*, not its *height*.

This strategy can smooth out noise, but it also throws away a lot of information— many data points are summarized into a single bin's frequency density.

A third option, and probably the best in most cases, is to plot the empirical *complementary cumulative distribution function* (CCDF) of the degree distribution (see also Ex. 4.4). First, if the degree distribution follows a power-law distribution, the CCDF should also follow a power-law distribution (with a different exponent) (Sec. 4.2.3). Furthermore, the empirical CCDF makes use of every single data point, without missing any detail of the distribution. Because of the cumulative (averaging) nature of the CCDF, it also suppresses noise. As you can see from Fig. 11.6c, the CCDF plot provides a much clearer picture of the degree distribution than the raw plot or the log-binned plot.

Use the "empirical" CCDF to plot the degree distribution if the degree distribution exhibits a heavy tail. It can reveal the most information about the degree distribution while suppressing noise.

We now turn to confirmatory analysis for the degree distribution.

11.7.2   Statistical models for heavy-tailed data

A common statistical model for degree distributions with very high-degree nodes is a power law: [16]

$$\Pr(k) = Ck^{-\gamma}, \tag{11.3}$$

for $k = 1, 2, \ldots$. Here C is a normalizing constant ensuring we have a proper probability distribution and $\gamma > 1$ is the *power-law exponent*. Because the support k is integer-valued,[17] this distribution is normalized by summing $k^{-\gamma}$ over all values of k:

$$C^{-1} = \sum_{k=1}^{\infty} k^{-\gamma} = \zeta(\gamma), \tag{11.6}$$

---

[16] Why a power law? We do a deep dive in modeling heavy-tailed networks in Sec. 22.6, including mechanisms that create power-law networks, and competing distributions such as log-normals.

[17] In fact, the power-law distribution function can be defined both in the continuous and discrete domains. Although degree is discrete, a continuous distribution is a good approximation and allows for easier analytical treatment. When defined for the continuous domain, we can calculate the constant C in closed form. Given

$$P(k) = Ck^{-\gamma} \text{ for } k \geq k_{\min},$$

where $k_{\min}$ is a cutoff parameter and the normalization constant can be calculated from the condition that

$$1 = \int_{k_{\min}}^{\infty} P(k)dk = C \int_{k_{\min}}^{\infty} k^{-\gamma}dk = \frac{C}{1-\gamma} \left[k^{1-\gamma} \right]_{k_{\min}}^{\infty}.$$

For $\gamma > 1$,

$$\frac{C}{1-\gamma} \left(0 - k_{\min}^{1-\gamma} \right) = 1 \implies C = \frac{\gamma-1}{k_{\min}^{1-\gamma}}.$$

Therefore, the power-law distribution in the continuous domain can be written as

$$P(k) = \frac{\gamma-1}{k_{\min}} \left(\frac{k}{k_{\min}} \right)^{-\gamma} \text{ for } k \geq k_{\min}. \tag{11.4}$$

In the discrete case,

$$1 = C \sum_{k=k_{\min}}^{\infty} k^{-\gamma} = C \sum_{n=0}^{\infty} (n + k_{\min})^{-\gamma} = C\zeta(\gamma, k_{\min}) \implies C = \frac{1}{\zeta(\gamma, k_{\min})},$$

where $\zeta(\gamma, k_{\min})$ is the *Hurwitz zeta function*:

$$\zeta(s, a) = \sum_{n=0}^{\infty} \frac{1}{(n+a)^s}.$$

Therefore, the power-law distribution defined in the discrete domain is

$$P(k) = \frac{k^{-\gamma}}{\zeta(\gamma, k_{\min})} \text{ for } k \geq k_{\min}. \tag{11.5}$$

which is the (celebrated) *Riemann Zeta function* $\zeta(\gamma)$ and Eq. (11.3) is also known as the *Zeta distribution*. A network with such a power-law degree distribution is called a *scale-free network*.[18]

Suppose we have a network of n nodes that we believe to be scale-free. Our data are the degrees of all the nodes in the network, $k_1, k_2, \ldots, k_n$. How can we fit Eq. (11.3) (that is, estimate γ) to this data?

A function of the form $y = Cx^a$ will follow a straight line with slope a when plotted on a log–log plot (Sec. 4.2.5). So a quick diagnostic is to count n_k, how many nodes have degree k, and plot n_k vs. k on log–log axes. If it looks like it follows a straight line—and you have enough data—there's a chance the data are (approximately) power-law.

To fit Eq. (11.3), recommended practice is to use maximum likelihood estimation (MLE; Sec. 4.3) to estimate γ. **Do not fit to a log–log plot.**[19] With MLE, we ask what is the likelihood $\mathcal{L}$ of γ given our data (the degrees of all n nodes) and (for simplicity) we assume each node degree is drawn iid from Eq. (11.3):

$$\mathcal{L}(\gamma \mid k_1, k_2, \ldots, k_n) = \prod_{i=1}^{n} \Pr(k_i) = [\zeta(\gamma)]^{-n} \left(\prod_{i=1}^{n} k_i \right)^{-\gamma}. \qquad (11.7)$$

We then seek the value of γ that maximizes $\mathcal{L}$, which occurs at the γ where $\partial \mathcal{L} / \partial \gamma = 0$. The maximum of the log-likelihood $\ell = \log \mathcal{L}$ occurs at the same γ, because log is a

---

[18] Why "scale-free"? See Sec. 12.3.

[19] That Eq. (11.3) follows a straight line with slope $-\gamma$ on a log–log plot implies that we can estimate γ by doing a linear regression on $\log(n_k)$ vs. $\log(k)$. This is not appropriate, as the log-transform breaks the assumption of linear regression that residuals are iid. Specifically, linear regression assumes that the errors are normally distributed which, for the log-transformed data, would mean:

$$\log P(k_i) \sim -\alpha \log k_i + \epsilon_i, \ \epsilon \sim \mathcal{N}(0, \sigma^2).$$

However, to satisfy this assumption, $P(k)$ should have log-normal fluctuations across the whole range of k, which is clearly not the case. For instance, say, $P(k = 2) = 0.1$ and $P(k = 100) = 0.0001$. Then, $\log_{10} P(k = 2) = -1$ and $\log_{10} P(k = 100) = -4$. If we assume a hypothetical linear model that predicts the value of $\log_{10} P(k)$ to be -2 for $k = 2$ and -3 for $k = 100$, the absolute error is the *same*: $|\epsilon_{k=2}| = |\epsilon_{k=100}| = 1$. But if we examine the error in the original (linear) space, the error is very different: $|\epsilon_{k=2}| = 0.09$ and $|\epsilon_{k=100}| = 0.0009$. In other words, if we operate in log–log scale, we overestimate the error for the high degrees and underestimate the error for the low degrees. Moreover, note that two directions of the error are not equivalent. Say, a model predicts $\log_{10} P(k = 100) = -3$ and another model predicts $\log_{10} P(k = 100) = -5$. Although the absolute errors are the same in log-scale, the error in the original (linear) space is very different: $|\epsilon_{k=100}| = 0.0009$ for the first model and $|\epsilon_{k=100}| = 0.000009$ for the second model. In other words, (OLS) linear regression would produce unreliable estimates of the degree exponent. Furthermore, it has been shown that the r^2 value of the linear regression is not a good measure of the goodness of fit of a power-law distribution and can exhibit a large value even when the power-law distribution is a poor fit to the data [109].

monotonic transformation, and ℓ is more convenient to differentiate than $\mathcal{L}$:

$$\ell = \log \mathcal{L} = -n \log \zeta(\gamma) - \gamma \sum_{i=1}^{n} \log k_i \tag{11.8}$$

$$\frac{\partial \ell}{\partial \gamma} = -n \frac{\partial}{\partial \gamma} \log \zeta(\gamma) - \sum_{i=1}^{n} \log k_i \tag{11.9}$$

$$= -n \frac{\zeta'(\gamma)}{\zeta(\gamma)} - \sum_{i=1}^{n} \log k_i = 0 \implies -\frac{\zeta'(\gamma)}{\zeta(\gamma)} = \frac{1}{n} \sum_{i=1}^{n} \log k_i, \tag{11.10}$$

and we can solve this last expression for γ numerically after computing the average of $\log k_i$ from our data.[20]

To demonstrate this approach, we built a synthetic (Ch. 22), power-law network with $n = 5000$ nodes and power-law exponent $\gamma = 5/2$. In Fig. 11.7 we plot n_k/n vs. k and overlay the MLE best fit of Eq. (11.3). Solving Eq. (11.10) gives us an estimate $\gamma_{\text{MLE}} = 2.47178$, very close to the true value.

Power laws in real network data   Figure 11.7 gives strong evidence that the method works well, but it's a synthetic example and there's some limitations to consider. For one, we have a relatively large sample, and estimating a power law is difficult for smaller networks. Likewise, a *pure* power law is quite simplistic and real networks are unlikely to be so simple. In real-world data, it is rare to see a clean power-law distribution that spans multiple orders of magnitude. Often, the power-law distribution is truncated due to sampling bias or the finite size of the system. For instance, the maximum degree can only reach N, the number of nodes in the network and thus by definition, the power-law distribution is truncated at $k_{\text{max}} < N$.

These issues point towards more realistic distributions than Eq. (11.3) and more advanced estimation methods than Eq. (11.10). One approach is to include an *exponential cutoff* in the distribution, for example:

$$\Pr(k) = Ck^{-\gamma}e^{-k/\kappa}. \tag{11.14}$$

Here the exponential term $e^{-k/\kappa}$, where κ is a second parameter, will eventually dominate the power-law term, cutting off how far out in k the "tail" of the distribution can be.

---

[20] The continuous case, while an approximation, is instructive and often just as accurate. For continuous k, the log-likelihood given our data is

$$\ell(\gamma) = \log \left[\prod_{i=1}^{N} \frac{\gamma - 1}{k_{\text{min}}} \left(\frac{k_i}{k_{\text{min}}} \right)^{-\gamma} \right] \tag{11.11}$$

$$= N \log(\gamma - 1) - N \log k_{\text{min}} - \gamma \sum_{i=1}^{N} \log \frac{k_i}{k_{\text{min}}}. \tag{11.12}$$

Solving for the γ that maximizes the log-likelihood function ($\partial \mathcal{L}/\partial \gamma = 0$) gives us the MLE of the degree exponent:

$$\hat{\gamma} = 1 + \frac{N}{\sum_{i=1}^{N} \log \frac{k_i}{k_{\text{min}}}}. \tag{11.13}$$

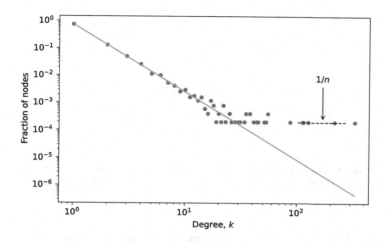

Figure 11.7 Estimating a power-law exponent for the degree distribution of a synthetic network of $N = 5000$ nodes. The best fit distribution is very close to the true distribution. Notice (1) how noisy the plot is, even with so many nodes, and (2) that we have a "data horizon" at $y = 1/n$ as the smallest value of n_k we can observe is $n_k = 1$. This is a limitation of our sample (and how we binned it) not a discrepancy of our model. (These effects were also visible in the "raw" panel of Fig. 11.6.) Indeed, looking for this horizon when examining binned data is good practice, to make sure the normalization (and even the number of data points) is as you expect.

Another approach, quite popular, and in a sense the opposite idea, is to define a minimum value of k for which Eq. (11.3) holds, and then infer γ only for $k \geq k_{min}$. While popular, this idea should be approached with caution, as we are *eliminating observations from our data*, a potentially drastic step. (If you choose to do this, be sure to check how many nodes have $k < k_{min}$; hopefully it's not too many!) In either case, such models are more complicated, including more parameters and requiring more effort to fit to data. Care is often needed when selecting between more and less parsimonious explanations of data—such are the tradeoffs we must confront when analyzing data.

Are heavy-tails always power laws? Although there exist many mechanisms that produce power-law distributions—it is one of the "naturally occurring" distributions— that does not mean that all heavy-tailed distributions that look straight in log–log scale should follow a power-law distribution. There are other "natural" distributions that look similar to power-law distributions.

One such distribution is the log-normal distribution. As explained in Ch. 4, the log-normal distribution is like a normal distribution on the logarithmic scale. Log-normal distributions can sometimes show a nearly linear tail when plotted in log–log scale. Meaning that one may struggle to distinguish a power-law distribution from a log normal, especially if the power-law is truncated (Eq. (11.14)). As you can see in Fig. 11.8, a log-normal distribution (solid line) can look almost linear in log–log scale, and quite similar to a power law with an exponential cutoff.

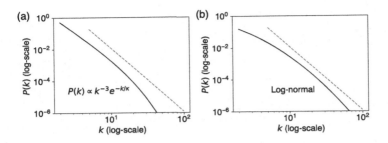

Figure 11.8 A power law with an exponential cutoff (**a**) and a log-normal (**b**) can look quite similar on a log–log plot, especially empirically at small sample sizes. The dashed line provides a pure power-law guide for the eye.

> ✅ If you are unsure which functional form to fit to data, try multiple fits using MLE, and compare them using likelihood ratio tests. Software packages exist for doing so in general and for heavy-tailed distributions in particular [13]. It is best if you have good reasons for each possible form.

11.7.3   Malawi Sociometer Network degree distribution

Let's dive into some real data. As a real-world case study of using statistical methods to explore a network's properties, let's consider the Malawi Sociometer Network. This is a dynamic network (Ch. 15), but here we simply focus on the static representation where two nodes are connected if a link between them occurs at least once during the experiment's data window.

We begin by examining the degree distribution of the network in Fig. 11.9. This network is relatively small, $n = 86$, and the largest degree is $k = 31$. From these values alone we can pretty much rule out any chance that the data will exhibit a scale-free degree distribution; even if the underlying network does exhibit hubs, our data are simply too small to see them.

Given that we should pass on the scale-free model from the preceding section, how to begin modeling this network's $\Pr(k)$? A good starting point, especially considering this network comes from a temporal process, is a Poisson distribution:[21]

$$\Pr(k) = \frac{\lambda^k e^{-\lambda}}{k!}.$$ (11.15)

This equation gives the probability for observing k events if events occur independently from one another and at a constant rate λ. By assuming events are independent from one another, the Poisson is the natural null model for data that count events, and we can consider our node degrees in just such a way—the degree of a node being the number of times that node encountered other nodes during the experiment's data window.[22]

---

[21] A strange expression, but derivable from a binomial distribution; see Sec. 4.2.5.

[22] This is in fact a bit oversimplified, as we are actually considering the number of *unique* nodes being encountered. Nevertheless, a Poisson is still worth comparison to the data as we begin our modeling efforts.

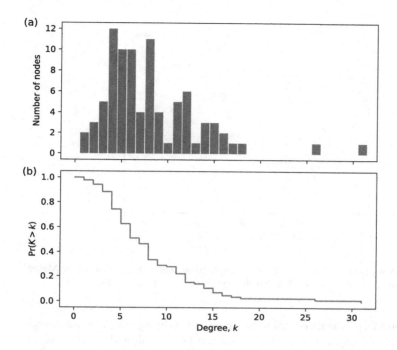

Figure 11.9 The degree distribution of the Malawi Sociometer Network. (a) A histogram, counting the number of nodes with degree k. (b) A cumulative distribution (ECDF), $\Pr(K > k)$. For smaller datasets such as this, the ECDF can provide a less noisy picture (Sec. 11.7.1).

Does a Poisson distribution provide a good fit to the Malawi Sociometer Network degree distribution? Fitting the Poisson to data is as simple as computing the mean, since the MLE of $\lambda = \sum_i k_i / n$. (Try deriving this MLE yourself!) In Fig. 11.10 we compare the MLE fit of Eq. (11.15) to the data. The fit is reasonable, but notice an obvious deviation away from the Poisson estimate beginning at $k \approx 9$. We can do better.[23]

Knowing a property of the Poisson distribution could have immediately clued us into a problem without even performing the MLE fit shown in Fig. 11.10. The data we are studying are *overdispersed*. Overdispersed data are those where the variance exceeds the mean; for our data, $\langle k \rangle = 8.07$ and $\mathrm{Var}(k) = 26.1$. A Poisson distribution has a particular property: the mean and variance are both equal to λ. A Poisson will never fit well when the variance is over three times the mean!

So we can rule out[24] a Poisson distribution. What should we do instead?

We have lots of options when it comes to modeling count data, but it's a good idea to keep it simple, as we did when starting with a Poisson, so it makes sense to go for the

---

[23] It's also good news for the experiment that the Poisson does not work very well. If it did, we would have evidence that the participants are just bouncing around at random as they meet one another. This would be odd for a real social network, and we would likely feel the need to question the veracity of our data.

[24] We may want to confirm this with a statistical test.

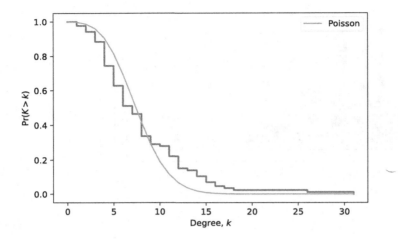

Figure 11.10 How well does a Poisson distribution explain the degree distribution of the Malawi Sociometer Network? The fit is reasonable, but there is an obvious deviation above $k \approx 9$.

next simplest option. Let's again consider the values of k as being drawn from a Poisson distribution but now instead of assuming λ is a constant for all nodes, let's imagine that λ may vary somewhat from node to node. This *heterogeneous Poisson* is meant to capture a notion of some nodes having a higher propensity for making contacts than other nodes, a more complicated situation than a pure Poisson distribution, but still one that is not too bad. It turns out (see below), if we assume the distribution of λ across nodes follows a particular form, we end up with an especially convenient distribution for k, the *negative binomial distribution*:

$$\Pr(k) = \binom{k+n-1}{n-1} p^n (1-p)^k, \tag{11.16}$$

where $n > 0$ and $0 \le p \le 1$ serve as fit parameters.[25] The mean k is given by $\lambda = n(1-p)/p$ and the variance of k is $\sigma^2 = n(1-p)/p^2$. However, using $p = \lambda/\sigma^2$ to rewrite the variance in terms of λ and n reveals why this distribution is helpful for overdispersed data:

$$\sigma^2 = n\frac{1-p}{p^2} = \frac{n}{p^2} - \frac{n}{p}$$

$$= \frac{n}{\lambda^2}\sigma^4 - \frac{n}{\lambda}\sigma^2$$

$$\left(1 + \frac{n}{\lambda}\right) = \frac{n}{\lambda^2}\sigma^2$$

$$\Rightarrow \sigma^2 = \lambda + \frac{1}{n}\lambda^2. \tag{11.17}$$

---

[25] Typically, the negative binomial distribution is interpreted as the probability that k iid Bernoulli trials are needed until n successes are observed, assuming each trial succeeds with probability p.

The reciprocal of n acts as the dispersion parameter, capturing overdispersed data where $\lambda \neq \sigma^2$. When $\alpha = n^{-1} \to 0$ we recover the (homogenous) Poisson distribution so we can check how heterogeneous our data is by fitting Eq. (11.16) and checking how far α is from zero.

ⓘ The negative binomial distribution as heterogeneous Poisson. The negative binomial distribution arises from a heterogeneous Poisson if we assume that k follows a Poisson conditioned on rate λ while λ is distributed according to a Gamma distribution

$$\Pr(\lambda) = \frac{\beta^\alpha}{\Gamma(\alpha)} \lambda^{\alpha-1} e^{-\beta\lambda}$$

parameterized by $\alpha > 0$ and $\beta > 0$, where $\Gamma(\alpha) = \int_0^\infty t^{\alpha-1} e^{-t}\, dt$ is the Gamma function. As λ is unobserved in our data (a latent parameter) we ask what is the distribution of k alone by marginalizing (integrating) out λ:

$$\Pr(k) = \int_0^\infty \underbrace{\frac{\lambda^k}{k!} e^{-\lambda}}_{\text{Poisson}} \underbrace{\frac{\beta^\alpha}{\Gamma(\alpha)} \lambda^{\alpha-1} e^{-\beta\lambda}}_{\text{Gamma}}\, d\lambda$$

$$= \frac{\beta^\alpha}{k!\,\Gamma(\alpha)} \int_0^\infty \lambda^{k+\alpha-1} e^{-(1+\beta)\lambda}\, d\lambda$$

$$= \frac{\beta^\alpha}{k!\,\Gamma(\alpha)} \Gamma(k+\alpha)(1+\beta)^{-k-\alpha}$$

$$= \binom{k+\alpha-1}{\alpha-1}(1+\beta)^{-k-\alpha}\beta^\alpha$$

$$= \binom{k+\alpha-1}{\alpha-1}\left(\frac{1}{1+\beta}\right)^k\left(\frac{\beta}{1+\beta}\right)^\alpha.$$

Finally, choosing $\alpha = n$ and $\beta/(1+\beta) = p$, we recover Eq. (11.16).

Returning to our degree distribution, we fit the negative binomial distribution to our data using maximum likelihood (Fig. 11.11), finding parameter estimates of $p = 0.3552$ and $n = 4.4455$. Compared to the Poisson distribution, we see much better agreement with the Negative Binomial. If we are happy with this model, we can use it to better understand and potentially even predict the degrees of the Malawi Sociometer Network nodes.

While agreement with our data in Fig. 11.11 is excellent visually, we are faced with a *model selection* problem. It's not surprising the two-parameter Negative Binomial fits better than the one-parameter Poisson. We should finish the analysis with a bit more quantitative support. For one, we can ask if $\alpha = 0$ is statistically compatible with our MLE estimate of $1/4.4455 \approx 0.225$, in which case we cannot rule out the Poisson ($\alpha = 0$) over the Negative Binomial ($\alpha > 0$). Better, we can do a likelihood ratio test, in this context a type of *overdispersion test* [91]. We'll leave these remaining steps as an exercise for the reader.

These examples show how exploratory analysis and hypothesis tests can be applied to network data by extracting numeric quantities from the network, then visualizing and modeling their distributions. We've focused on maximum likelihood fitting procedures,

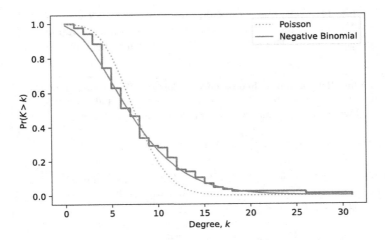

Figure 11.11 A heterogeneous Poisson (Negative Binomial) fits the degree distribution of the
Malawi Sociometer Network quite well.

but of course Bayesian inference can be used as well[26] And while we have worked with
a relatively simple (but important) quantity, the degree distribution, many other network
properties can be modeled. We consider a variety of such properties in Ch. 12.

11.7.4   Other tools from statistical inference

Hypothesis testing is not the only way confirmatory statistical inference can be applied
to network data. More generally, statistical models can be proposed for a data set, and
the parameters controlling those models can be inferred using, for instance, Bayesian
inference. One advantage of tackling the data from this perspective is that we can use
statistical models that are more directly connected to the network, as opposed to using
models of network summary statistics such as the degree distribution.

As a brief example, consider building a statistical model for the probability that
two nodes i and j are connected, that edge (i, j) exists. We expect this probability will
depend on one or more parameters which we'll represent generically with $\boldsymbol{\theta}$, giving us
the conditional probability

$$\Pr(i \leftrightarrow j \mid \boldsymbol{\theta}). \tag{11.18}$$

To model the whole network, $\Pr(G)$, convert this to a likelihood over all the edges. And
we mean *all* the edges: we need to consider the probability that every edge in G exists,
and every edge not in G does not exist:

$$\mathcal{L}(G \mid \boldsymbol{\theta}) = \prod_{(i,j) \in E} \Pr(i \leftrightarrow j \mid \boldsymbol{\theta}) \prod_{(i,j) \notin E} (1 - \Pr(i \leftrightarrow j \mid \boldsymbol{\theta})). \tag{11.19}$$

---

[26] And by moving beyond point estimates of model parameters, we can capture uncertainty in those
estimates.

Depending on how we structure $\boldsymbol{\theta}$, this can capture all manner of network models, from basic random graphs such as Erdős–Rényi graphs (Sec. 22.3) to the *stochastic block model* (SBM) (Sec. 23.2) and *exponential random graph model* (ERGM) (Sec. 23.4.1). Moreover, starting with $\mathcal{L}(G)$, we can perform a Bayesian analysis to understand the distribution(s) of our parameter(s) given the observed network:

$$\Pr(\boldsymbol{\theta} \mid G) = \frac{\mathcal{L}(G \mid \boldsymbol{\theta}) \Pr(\boldsymbol{\theta})}{\Pr(G)}. \tag{11.20}$$

This expression comes from using Bayes' theorem (Ch. 4) and requires a *prior distribution* $\Pr(\boldsymbol{\theta})$ for the model parameters. Computing the *posterior distribution* $\Pr(\boldsymbol{\theta} \mid G)$ mathematically can be challenging, typically because the normalization $Z = \Pr(G)$ is intractable; fortunately, computational approaches, specifically Markov Chain Monte Carlo (MCMC),[27] are able to generate many random samples following this distribution, and we can then examine how those samples are distributed, standard practice for Bayesian inference. With access to the posterior, we can better get a handle on what the data are telling us, and even perform tasks such as link prediction (*"that link doesn't exist in G but our model gives it very high probability"*) and uncertainty quantification (*"that link exists in G but our model gives it very low probability"*).

We explore the technical details of these kinds of statistical models in Ch. 23.

11.8 Reporting considerations

After completing a statistical analysis using one of these network null models, you will need to report your findings. Does the null model explain the structural features of the network or not? Further, even when performing exploratory analysis, you will still want to document properly any choices made when using exploratory tools and summary statistics.

The tension and challenge when reporting analyses comes from the need to be concise. Exhaustively describing every miniscule detail will lead to a paper or presentation too long and probably too boring for most readers. You want to give the minimum amount of information to inform a reader (who may be yourself in the future) of *everything* they need to sufficiently reproduce your analyses, but no more.

For statistical tests, standard reporting guidelines exist which we recommend following whenever possible. Describe and motivate the test, as appropriate, and report the number of observations that went into the test, the test statistic, and the p-value. Refrain from using the term "significant" when reporting research involving hypothesis tests, unless referring specifically to the results of a statistical test.

For the configuration model, report any preprocessing to the network (for example, retaining only the giant connected component), describe how multi-edges and self-loops were dealt with, and how many random realizations of the configuration model were created. Likewise, for computational nulls such as Alg. 11.1, report any parameters

---

[27] There are issues with scalability. If G is very large, it may be too costly to evaluate all the terms in the likelihood, particularly if $\boldsymbol{\theta}$ is complex, making MCMC so slow that it may not be feasible to use, at least for richer models. That said, never say never when it comes to Moore's Law—an intractable computation yesterday may be feasible today.

such as the number of exchanges, and any further details such as whether other network properties were checked during the exchange process (for example, were edge exchanges rejected when the network became disconnected). Report if a criterion was used to determine the number of exchanges and report how many independent realizations were created.

Of course, other applications of statistical inference should be properly reported as well. If distributions were fit or regressions performed, report goodness-of-fit statistics as appropriate. If a Bayesian model was studied, provide details on how the posterior was sampled. If multiple statistical models were employed, account for whether a more complex model gives a better fit only because it has more fit parameters (model selection measures such as Akaike Information Criterion, Bayesian Information Criterion, or Minimum Description Length are commonly used for this purpose [84]; we'll see these measures in various contexts later). In general, include confidence intervals or other measures of uncertainty.

To learn more on reporting guidelines for statistical analyses, see remarks at the end of this chapter.

11.9   Summary

This chapter covered ways to explore your network data, using visual means and basic summary statistics, and how to apply statistical models to validate aspects of the data. Exploratory data analysis (EDA) is a pillar of statistics and data mining and we can leverage existing techniques when working with networks. However, we can also use specialized techniques for network data and uncover insights that general-purpose EDA tools, which neglect the network nature of our data, may miss.

Complementing EDA, statistical models can be defined for properties of the network, such as the degree distribution, or for the network structure itself. Fitting and analyzing these models then recapitulates effectively all of statistical inference, including hypothesis testing and Bayesian inference. As part of this confirmatory data analysis (CDA), network null models are often needed to compare against the real network, and the configuration model is a prime choice, although a null model can always be built from the real network by randomizing it using the degree-preserving edge exchange process.[28]

In general, given the breadth of forms that network and network-adjacent data will take, a good grounding in inference methods, not just for network data but for *most* forms of data, is useful for any scientists working with network data.

Bibliographic remarks

Exploratory data analysis was developed most completely in the seminal book of Tukey [465]. While some of the advice of Tukey is outdated (for instance, remarks on generating graphics have been obviated by modern computers), much of Tukey's wisdom remains extremely valuable.

---

[28] That said, randomizing a network while preserving *more properties* than the degree of nodes is often very challenging, and general-purpose algorithms are not currently known.

The Erdös-Rényi random graph model was introduced by Erdős and Rényi [148, 149] and by Gilbert [182]. This was truly seminal work, as it was the start of random graph theory, being the first time probabilities were introduced into graphs. The configuration model, also called the *Molloy–Reed model*, was introduced and studied by Molloy and Reed [313, 314] A useful deep-dive into the technical details of the configuration model is given by Fosdick et al. [167].

Degree assortativity and more general *mixing patterns* were introduced by Newman [332]. Structural cutoffs, a key property of networks that should be accounted for whenever considering degree assortativity, was developed by Boguñá et al. [59]; we summarize their calculation in Sec. 11.6.2.

Much has been written about power laws and estimating power-law distributions from data. For a general review, see Newman [343]. The maximum likelihood estimators for a power-law distribution, first derived by Seal [421], were heavily popularized by Clauset et al. [109], which also introduced the estimation of the power-law cutoff ($\Pr(x)$ is a power law only for $x > x_{\min}$) using a Kolmogorov–Smirnov (KS) test. Alstott et al. [13] provide a useful Python package implementing these and other, related methods. While heavily used, a recent line of work [201, 373] has pointed out that biases in these methods may occur under some circumstances. In our example fitting the degree distribution of the Malawi Sociometer Network, we treated the degrees as count data. A good reference for the statistics of such data is Cameron and Trivedi [91]

The statistical models that we touched upon, the stochastic block model (SBM) and the exponential random graph model (ERGM; sometimes also called the p^* model), have long and rich histories of studies. The SBM was first introduced by Holland et al. [219]. For a classic overview of block models and SBMs in the context of social networks, see Wasserman and Faust [485]. ERGMs, arising from the work of Holland and Leinhardt [218] and Frank and Strauss [168], among others, are also common statistical models for social networks; Robins et al. [398] and Lusher et al. [284] give good introductions. We discuss these models and more in Ch. 23.

Reporting considerations are a perennial concern for statisticians. For readers interested in learning more, the PLOS reporting tutorial[29] is a good start. The tutorial includes links to field-specific (but still generally useful) guidelines such as the SAMPL guidelines for biomedical studies [262] and the MDAR checklist for life sciences [285]. Makin and Orban de Xivry [288] and Munafò et al. [323] also provide general discussion of good and not-so-good research practices.

Exercises

11.1 (**Focal network**) Plant–pollinator network. Explore the degree distributions in this network. Note that we say "distributions" *plural* as there are effectively two: one for plants and one for pollinators. Because links only exist between these two groups, there will not be links between plants or between pollinators. Nevertheless, we can consider "nearest" neighbor pollinators as those observed to pollinate a common plant. Describe some similarities and differences between

---

[29] https://plos.org/resource/how-to-report-statistics/

nearest neighbor pollinators. You may wish to apply both EDA and CDA techniques.

11.2 *A never-ending exchange I.* As remarked in the text, the basic edge exchange algorithm will enter an infinite loop and never end if no allowable exchanges are present. Describe a scenario where you will never be able to find an allowable edge exchange according to Alg. 11.1.

11.3 *A never-ending exchange II.* Modify Alg. 11.1 to prevent an infinite loop by abandoning the exchange process after reaching a given number of attempts without a successful exchange.

11.4 Modify Alg. 11.1 to preserve not only the degrees of every node but also the overall connectivity of the network. How does this additional constraint make edge exchange more difficult?

11.5 Suppose your network has a binary attribute $x_i \in \{-1, 1\}$ associated with every node i. Modify Alg. 11.1 to preserve the correlation on edges of x. Meaning, any new edge (a, b) inserted in place of existing edge (i, j) must have the same attributes $(x_i, x_j) = (x_a, x_b)$. How might this additional constraint make edge exchange more difficult?

11.6 *Optimization with edge exchange.* Algorithm 11.1 can be used to randomize a network structure while optimizing a property. The classic example is assortativity. Describe and then implement a modified Alg. 11.1 that chooses rewiring options that *increase* degree correlations. (*Hint*: when selecting two edges (i, j) and (u, v) to rewire, there are (up to) two possibilities: creating edges $(i, u), (j, v)$ or edges $(i, v), (j, u)$. Examine the degrees of the nodes when looking at either possibility.)

Chapter 12

Understanding network structure and organization

In this chapter, we focus on measures that quantify a network's structure and characterize how it is organized. They include both general-purpose measures and those specialized to particular circumstances, which allow us to better "get a handle" on the network data, and network science over the years has generated a dizzying array of useful measures. They are used for both exploratory and confirmatory analyses, which we have just discussed in the previous chapter. In concert, these two chapters form a central core around which network data can be analyzed and understood. With the measures of this chapter, we can understand the patterns in our networks; with the statistical methods of Ch. 11, we can put those patterns on a firm foundation.

12.1 Micro-to-meso-to-macroscale

An enduring thread of network research is that of *scales*. Networks are defined at the level of individual nodes (how do two nodes become connected?) but they are organized into larger and larger structures (what are these three clusters that constitutes the whole network?) as we move up the scale. Depending on the scale that we focus on, we can see completely different aspects of the network.

To understand the structure of a network means understanding it across these scales. Methods and measures are often tailored to a specific scale, but it is important to understand that networks exhibit important structural patterns across—and between—all scales, including intermediate (or "meso") scales (e.g., motifs (Sec. 12.6), communities (Sec. 12.7), and hierarchy (Sec. 12.8)). Given data, understanding the presence—or absence—of such patterns can reveal the principles behind the organizational pattern of the network.

In the context of networks, *microscopic* or *microscale* refers to the structure of the network at its smallest unit, typically a node and its neighbors. *Macroscopic* or *macroscale* means the opposite, the overall global structure of the network, usually captured by statistics and distributions. In between these two extremes, *mesoscopic* or *mesoscale* structure captures often the most interesting and important organization of networks.

And, of course, they are not independent; structures at different scales interact with each other and produce one another. For instance, the microstructure—the local neighborhood of a node—may be dictated by the mesoscale structure (e.g., communities) and the mechanisms behind the formation of those mesoscale structures; on the other hand, the microscale mechanisms and structure may lead to the network's mesoscale structure.

12.2   Egocentric networks

The smallest aspect of a network is that of an individual node and its immediate neighbors, the nodes with which it is connected. Despite its limited extent in the network, this *egocentric network*—the subgraph taken from the original network by retaining the ego, the ego's neighbors, and any links between them—is a fundamental unit of analysis with rich information. A node within a system cannot immediately access the global network structure; their only "awareness" of the network is through their neighbors and they are the only ones who can directly affect or be affected by the node. This fundamental interaction structure is captured as the egocentric network. Researchers sometimes intentionally collect data on the egocentric network of nodes, even if larger network data is available, either for efficiency or because the data can be better assessed for validity.

What are the most basic questions about the structure of egocentric networks? Two fundamental notions are the size of the neighborhood and the connectivity of the neighborhood. Take a node at random from the network, call it the *ego*. How many neighbors does it have? The answer, of course, follows the degree distribution: $P(k)$, $k = 0, 1, \ldots, N - 1$. It tells us the probability a randomly chosen node (our ego) has k neighbors.

We can also ask how well connected the neighborhood is, which is same as asking: how many of the ego's neighbors also know each other? This comes from the *clustering coefficient* that we will cover shortly below.

Together, the degree and clustering coefficient summarize the basic structure of the egocentric network. We can look at their distributions across the network, and we can look at how they relate to one another. For instance, are high-degree nodes, nodes with large neighborhoods, also highly clustered? If each possible triangle is equally likely to be present in the network, regardless of the ego, then we expect the number of triangles to grow linearly with k^2. We can use this to guide our exploration: make a scatter plot with a point for each node where the horizontal axis is k^2 and the vertical axis is the number of triangles. Does it exhibit a straight line across all the points or does it deviate? Looking at this plot, do some points stand out as unusual, perhaps they have unusually high or low clustering given their degree? Those unusual points likely merit further investigation.

While the degree and clustering coefficient summarize the ego by telling us how large and how interconnected its neighborhood is, we can go beyond them. For instance, suppose you remove the ego from its egocentric network, meaning you only consider nodes who are neighbors of the ego and the links, if any, between those neighbors. How many connected components does this "egoless egocentric network" contain? This quantity, which captures not only how dense with triangles the egocentric network is but how those triangles are distributed among the neighbors, hints at how many distinct groups or communities the ego belongs to, which is often a critical aspect of the position of the ego in the network. It may capture, for instance, whether the ego acts a broker between many different groups that, without the ego, would not otherwise be connected.

12.3 Degrees and degree distributions

Building off egocentric networks, where the size is determined by the ego's degree, we can look at the distribution of degree across the network. We've encountered the degree distribution $P(k)$ before (Sec. 11.7.2). It tells us the probability that a random node has k neighbors. Specifically, given a network with N nodes, we count the number of nodes with a given degree k (denoted as N_k) and then calculate the probability function:

$$P(k) = \frac{N_k}{N},\qquad(12.1)$$

which is called the *degree distribution*. Note that this is a *discrete* distribution defined on integer values of k.[1]

The degree distribution has a profound impact on the network's structure and dynamics. It captures how homogeneous the network is, how well information or diseases can spread through the network, and how robust the network is to random failure or intentional attack. Many networks can be well understand by the degree distribution, especially if they are not too "correlated" (which we can also measure, Sec. 12.5).

Measuring the degree of a node is straightforward (unless there are errors, Chs. 10 and 24) but going from the observed degrees to the degree distribution brings some subtly. It's not always as simple as counting up the number of nodes with degree k and making a histogram. We discussed these issues with some examples in Ch. 11, as examining the degree distribution of a network is one of the most fundamental steps to understanding a network using EDA and CDA techniques. The degree distribution is also the most common feature to control in a null model, and this reason is why the configuration model (Secs. 11.5 and 22.4) is so commonly used and studied.

12.3.1 Degree heterogeneity

One of the most profound features of networks is the presence of *hubs*, nodes with exceptionally high degree. When the degree is very skewed (roughly speaking, the mean degree is much higher than the median), the resulting highly connected hubs will lead to many different properties. For example, hubs make a network highly robust to damage: it will take many lost nodes or edges[2] to disconnect such a hub-heavy network [9, 111, 112].

---

[1] Continuous approximations are often helpful (Sec. 11.7).

[2] Many elements lost at *random*. Heterogeneous networks are known to be highly vulnerable to targeted attacks, where high-connectivity elements are more likely to fail.

A classic functional form that captures this heterogeneity is a *power-law distribution* (Ch. 4),[3]

$$P(k) = Ck^{-\alpha}, \tag{12.2}$$

When considering a power-law distribution, note that $k^{-\alpha} \to \infty$ as we approach $k \to 0$. Therefore, to make the distribution function well-defined, we need to have a minimum value of k, $k_{\min} \geq 1$ from which the power-law function holds. A power-law network is often called *scale-free*.

ⓘ Why are power-law distributions called "scale-free"?

A normal distribution $\mathcal{N}(\mu, \sigma^2)$,

$$P(k) = \frac{1}{\sqrt{2\pi\sigma^2}} \exp\left(-\frac{(k-\mu)^2}{2\sigma^2}\right), \tag{12.3}$$

has a clear *scale*. Samples drawn from it will tend to be nicely and evenly distributed around the "center" of the distribution, the mean μ, and any fluctuations will be finite, due to the variance σ^2—most samples will not be too far off the mean. The same holds for the similar Poisson distribution,

$$P(k) = \frac{\lambda^k}{k!} e^{-\lambda}, \tag{12.4}$$

which likewise has a clear "scale" defined by λ and finite fluctuations from its variance (also λ).

What about a power law? Let's consider the mean and variance, or equivalently the first and second moments, $\langle k \rangle$ and $\langle k^2 \rangle$ (Sec. 4.3). For simplicity, we'll focus on the continuous power-law distribution,

$$\langle k \rangle = \int_{k_{\min}}^{\infty} kP(k)dk = C\int_{k_{\min}}^{\infty} k^{1-\alpha}dk = \frac{C}{2-\alpha}\left[k^{2-\alpha}\right]_{k_{\min}}^{\infty},$$
$$\langle k^2 \rangle = \int_{k_{\min}}^{\infty} k^2P(k)dk = C\int_{k_{\min}}^{\infty} k^{2-\alpha}dk = \frac{C}{3-\alpha}\left[k^{3-\alpha}\right]_{k_{\min}}^{\infty}. \tag{12.5}$$

From these formulas and the divergence of the harmonic integral, $\int_{k_{\min}}^{\infty} k^{-1}dk \to \infty$ (similarly, $\sum_{k=k_{\min}}^{\infty} k^{-1} \to \infty$ for the discrete case), we can see that the mean $\langle k \rangle$ will diverge if $\alpha < 2$ and the variance $\langle k^2 \rangle - \langle k \rangle^2$ will diverge if $\alpha < 3$. If we encounter a power-law distribution with exponent $\alpha < 3$ (which frequently occurs in real-world networks), the variance, and sometimes even the mean, will diverge, completely unlike a normal or Poisson. With an infinite variance, some samples may be arbitrarily large! In other words, it becomes impossible to define a *characteristic scale* of the distribution.

---

[3] The log-normal distribution also captures heterogeneity and points to a different network mechanism; Ch. 22.

Scale invariance Another way to think about scale-freeness is the property of *scale invariance*. Let's ask: what is the ratio between the number of nodes with degree $2k$ and k? And how does this ratio changes across different values of k? If the distribution follows a power law, the ratio is

$$\frac{P(2k)}{P(k)} = \frac{C2^{-\alpha}k^{-\alpha}}{Ck^{-\alpha}} = 2^{-\alpha}. \tag{12.6}$$

It is a constant that does not depend on k—even if we replace k with $k' = mk$, we get the same ratio! This property—scale invariance—does not hold for other distributions. For instance, in the Poisson distribution, the ratio

$$\frac{P(2k)}{P(k)} = \frac{\frac{1}{(2k)!}\lambda^{2k}e^{-\lambda}}{\frac{1}{k!}\lambda^{k}e^{-\lambda}} = \frac{k!}{(2k)!}\lambda^{k} \tag{12.7}$$

does depend on k.

Here's an intuition for what scale invariance does. Imagine a world where people's heights are distributed according to a power law, compared to our "normal" world. If we can observe everyone else's height, and nothing else, can we determine how tall or short we are? When heights follow a distribution with a scale, we can do this, ranking our relative size, but for a power law, without a well-defined scale, it will not be possible. However short or tall we are, what we would see—the relative distribution of the sizes of other people—surprisingly, is the same!

We take a deep dive into models and mechanisms for scale-free networks in Sec. 22.6.

12.4 Clustering and transitivity

Clustering (triangles) is a fundamental measurement in understanding how a network is organized and how it functions. For one, very large, sparse networks will tend to have few if any triangles, at least if links are formed at random. Yet large, sparse, *real-world* networks almost always exhibit far more triangles than we would expect assuming connections are random. This property highlights not only how far from random real networks tend to be but also hints at the underlying mechanisms for how links often appear: *triadic closure*.

There are several ways to measure triadic closure in a network. Here we discuss three: local clustering coefficient, global or average clustering coefficient, and transitivity.

Let T_i be the number of triangles that pass through a node i or, equivalently, the number of edges between neighbors of i. The *clustering coefficient* of node i is given by

$$C_i = \frac{T_i}{\binom{k_i}{2}} = \frac{2T_i}{k_i (k_i - 1)}. \tag{12.8}$$

The term $\binom{k_i}{2}$ captures the maximum value of T_i, where every pair of i's k_i neighbors are connected. So C_i measures the fraction of possible triangles involving node i, with $C_i = 0$ indicating no triangles, or no clustering, and $C_i = 1$ indicating complete clustering.[4]

The clustering coefficient provides a local view of clustering at the level of individual nodes. This is valuable information: a node that exhibits a large clustering coefficient is *embedded* in a dense network of neighbors. Such nodes would play a different role in the network from those with low clustering coefficients, which are likely to connect different parts of the network. The clustering coefficient summarizes this information quantitatively.

 Consider examining the distribution of C_i across the nodes in the network. Identifying typical and atypical values of C_i can pinpoint unusual nodes or organizing patterns. A bimodal distribution of C_i, for example, may imply a core–periphery (Sec. 12.12.2) organization.

Given that C_i is local to node i, how can we aggregate the information across the network? One approach sometimes used is to simply average C_i over the nodes in the network: $\sum_i C_i / N$. This value captures the expected clustering coefficient of a randomly selected node in the network. This is a useful statistic but has the following caveats: Low-degree nodes often have higher clustering coefficient simply because it is much easier to have larger clustering coefficient for nodes with fewer neighbors; the denominator is much smaller (scales with the degree squared). This, along with the fact that most networks tend to be dominated by the nodes with small degree, opens the door for those nodes to dominate the average.

 While commonly done, it is important to be aware of the caveat of using the average of the local clustering coefficient (Eq. (12.8)) as a global measure of clustering.

A preferred alternative to averaging C_i over nodes is to define an explicitly global quantity, the most common quantity being the *transitivity* T:[5] the proportion of (undirected) two-paths (also known as triads or connected triples) that are closed to form triangles:

$$T = \frac{\text{Number of triangles}}{\text{Number of connected triplets}} = 3\frac{\text{Number of triangles}}{\text{Number of two-paths}}. \tag{12.9}$$

---

[4] The egocentric network is a clique.
[5] Here T is not to be confused with T_i.

(The factor of three accounts for the fact that a single triangle contains three two-paths within it.)

Both clustering and transitivity provide lenses to view the network's propensity for triadic closure—whether nodes with common neighbors tend to be (or tend to become) connected themselves. Triadic closure is often driven by *homophily* (like with like), the idea that connections tend to form between nodes that are similar. These notions of similarities and whether links in the network follow from such similarity, brings us to our next set of structural measures, *mixing patterns*.

 Clustering measures how often triangles appear in the network and span from local to global measures.

Clustering (local) For a node i, describes the proportion of triangles that i belongs to (Eq. (12.8)).

Transitivity (global) For a network, counts the proportion of triads (connected triples) that are closed to form triangles (Eq. (12.9)).

12.5 Mixing patterns and correlations

What kinds of node pairs do edges connect? Are there certain properties of the node pairs that make them more likely to be connected? These questions are addressed with measures of *mixing patterns*, how nodes "mix" (link) with one another. Usually we do not see the formation of new links (exceptions being for temporal networks), so these measures of mixing patterns are observational: what associations, if any, exist between nodes that share links compared to nodes that do not?

The most straightforward way to examine such associations in a real network is to take each edge in the network, measure a property of each node connected by that edge, and consider the association in those properties using a statistic such as a *correlation coefficient*. For a numeric property, one can also imagine this using a scatter plot: each point in the plot is an edge in the network, the horizontal coordinate of the point is the measured property of one node in the edge and the vertical coordinate is the property of the other node. The association between the two nodes then measures the relationship between the properties.

Perhaps the most commonly studied form of mixing pattern is based on the most fundamental property of nodes—their degree: do high-degree nodes tend to connect to other high-degree nodes or to low-degree nodes, or is there no association? Usually, this is a measure of *degree assortativity*. (Although *assortativity* technically refers to any such mixing pattern, it is common to simply use "assortativity" as a placeholder referring specifically to degree assortativity computed using Pearson correlation coefficient.) We previously discussed degree assortativity in Ch. 11.

To measure the degree assortativity of a network, simply examine each edge one at a time to make vectors of degrees $\mathbf{k}_l$ and $\mathbf{k}_r$ for the "left" and "right" nodes in each edge.[6] Then, compute the *Pearson correlation coefficient* between them:

$$r(\mathbf{k}_l, \mathbf{k}_r) = \frac{\sum_i \left(k_{li} - \overline{\mathbf{k}}_l\right)\left(k_{ri} - \overline{\mathbf{k}}_r\right)}{\sqrt{\sum_i \left(k_{li} - \overline{\mathbf{k}}_l\right)^2}\sqrt{\sum_i \left(k_{ri} - \overline{\mathbf{k}}_r\right)^2}} \tag{12.10}$$

where the sums run over the lengths of the vectors ($i = 1, \ldots, |\mathbf{k}|$) and $\overline{\mathbf{k}}$ is the mean of the elements of the vector.

Speaking of taking means, we should take some care here. The mean of these vectors is not just the average node degree:

---

i Some subtleties arise when summary statistics are computed over edges. For example, the average degree when computed over nodes is

$$\langle k_i \rangle_{\text{nodes}} = \frac{1}{N} \sum_{i=1}^{N} k_i = \frac{2M}{N}, \tag{12.11}$$

as the sum of the degrees of all nodes is twice the number of edges. But the average degree computed over *edges* is actually

$$\langle k_i \rangle_{\text{edges}} = \frac{1}{2M} \sum_{i=1}^{N} k_i^2. \tag{12.12}$$

This difference happens because a node i with degree k_i contributes once to the node average, regardless of the value of k_i, but contributes k_i times to the edge average, as each of its k_i edges will be a term in the edge average. Counting k_i each of k_i times gives the k_i^2 term in the sum.

What's especially nice with assortativity is that we do not need to limit ourselves to degree. As we examine the edges of the network, instead of building vectors of degree ($\mathbf{k}_l$ and $\mathbf{k}_r$), we can use any quantity x_i related to node i. We then build two vectors $\mathbf{x}_l$ and $\mathbf{x}_r$, as before, and compute $r(\mathbf{x}_l, \mathbf{x}_r)$ using Eq. (12.10).[7] Assortativity based on node attributes, which we saw in Sec. 9.3, is especially worth examining, as it connects the network structure to the attribute.

  Pearson correlation is not the only form of association. Consider using other measures for mixing patterns, such as *Spearman correlation*, which measures how monotonically related are the quantities.

In general, social networks tend to be degree assortative (positively correlated $r > 0$) while technological and biological networks tend to be degree dissortative (negatively

---

[6] For an undirected network you can add the degrees of nodes to *both* vectors to ensure symmetry. The vectors will then have length $2M$. For a directed network, you will want to examine the correlation for the four combinations of in- and out-degree, a more complicated situation.

[7] For non-numeric or categorical x we may want to adapt a different measure of association, but this is usually straightforward. Doing so also relates the assortativity for a categorical variable to *modularity* (Sec. 12.7) [342].

correlated $r < 0$) [332]. Indeed, we saw in a case study on HuRI (Sec. 11.6), that it was significantly degree dissortative.

Structural cutoffs When studying assortativity, keep in mind that networks can induce correlations in potentially unexpected ways (Sec. 11.6.2), due often to finite-size effects:

> ⚠ Keep in mind *structural cutoffs* when assessing measures of assortativity.
>
> A network with links that are truly neutral, where any pair of nodes can be linked, can exhibit degree assortativity depending on the network's degree distribution. High-degree hub nodes, if present, can have so many available connections that they must link to one another even if links are neutral. Two hubs may have such a strong prevalence for linking together that, in order to be neutral, *multiple links* must form between them. Yet, unless we consider the (usually rare) case of multigraphs, this is not allowed structurally. Without enough edges between the hubs, the network will be degree dissortative, not because of how links form in the network but due to structural constraints [59].

A good practice to assess the presence of a structural cutoff for degree assortativity is to compare the measured r to that obtained from the configuration model with the same degree distribution. When doing so, be sure to use realizations of the configuration model that are simple graphs, not a multigraph or a graph with self-loops. See Ch. 11 for more details on using the configuration model as a null baseline.

12.6 Motifs and graphlets

Another organizing pattern worth examining in a network is the distribution of frequently occurring subgraphs, which are usually called *motifs* or *graphlets*. "Motifs" usually refers to subgraphs of specific sizes (usually three or four nodes), while "graphlets" refer to subgraphs of any size, but we often use the terms interchangeably.[8] We have already encountered a triangle motif, a triplet of nodes with all pairs connected, when examining egocentric networks and clustering. The prevalence of triangles throughout a network gives a lot of information about its organization and structure, but many other motifs can occur in a network.

The number of possible motifs grows exponentially with the motif size. For three nodes, there are two distinct (i.e., nonisomorphic) connected, undirected graphs. These are the "two-path" and the triangle:

---

[8] Graphlets are more commonly used in the computer science community whereas motifs are more common in the network science and biology.

> **ⓘ** *Graph isomorphism* captures whether two graphs are the *same*, having the same structure or topology, irrespective of how we order or label the nodes. If we can take one graph and relabel all the nodes in some way to match exactly the other graph, and after doing so the edges are also the same, then the two graphs are *isomorphic*. Otherwise, they are *nonisomorphic.*[a]
>
> ---
>
> [a] More precisely, graphs $G_1 = (V_1, E_1)$ and $G_2 = (V_2, E_2)$ are isomorphic if and only if there exists a bijection $f : V_1 \to V_2$ between the node sets of G_1 and G_2 such that for any $(i, j) \in E_1$, we have $(f(i), f(j)) \in E_2$.

Notice that each triangle contains three two-path motifs, a point we discussed in Sec. 12.4 when defining transitivity (Eq. (12.9))—motifs overlap even when they have the same number of nodes. The universe of motifs within even a moderately sized network is vast.

Motifs are especially interesting in directed networks. Directed edges greatly expand the number of possible motifs on a set of nodes as now each edge can point in one direction or the other, or both. For example, what was a triangle motif in an undirected network could, among other possibilities, be a *feedback loop* or a *feedforward loop*. These possibilities have distinct consequences when it comes to real-world systems such as metabolic networks, where such loops play a role in regulating metabolic processes.

Here is an illustration of the 16 distinct three-node motifs that can occur in a directed network (arrows denote directionality, double-ended arrows denote reciprocated edges: $i \leftrightarrow j = i \to j$ and $j \to i$):

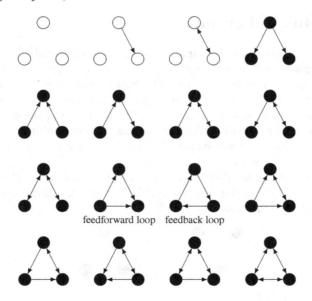

Open circles denote disconnected motifs. Often, these are not considered when conducting motif analysis. The empty motif on three nodes, for example, is usually not very interesting.

Motif analysis usually hinges on a baseline to compare against. We aren't simply looking for particular motifs, but any motifs that appear a significant number of times. To find these requires a null model to compare against: what would the motif distribution be for a network with the same degrees but no organization of how links form? That is the configuration model, which we discuss in Ch. 11. For each motif of interest, we can count how many times we observe the motif in the real network, and measure the mean and variance of those counts over many realizations of our null model. Then we can rank motifs by a z-score, telling us how many standard deviations above (or below) the expected count the motif is in the real network. These highly over-expressed (or under-expressed) motifs will often be most interesting to study.

One of the challenges with motif analysis is knowing which motifs are most interesting. Enumerating all the possibilities and then ranking them against a null with a z-score will give some guidance, but this approach tends to be limited to small-size motifs, usually three, four, or maybe five nodes. At five or more nodes, we usually cannot *exhaustively* find all the motifs in the network, the number of subgraphs to examine is simply too big.[9] Instead, we must rely on randomly sampling the larger motifs, which can work but also introduces uncertainty due to our lack of perfect information.

12.7 Communities

At the next scale above motifs, we have *communities* (or *modules*). In general, communities are taken to be groups of nodes that are densely connected to each other while being sparsely connected to nodes in other groups (Fig. 12.1; see also Fig. 1.3). There exist a number of specific definitions that make this intuitive notion exact.

Community detection, one of the most studied areas in network science, is the problem of finding these groups in a given network. Inferring community structure can reveal useful and important information. For example, in a PPI network, the community a protein belongs to can tell us about the functional roles of the protein, informing us about protein complexes and pathways that the protein may participate in. If we observe many proteins in the same community all fulfilling a particular role, that is supporting evidence that other proteins in the community, whose roles are not well understood, may also fulfill that role.

> **ⓘ** *Communities*, also known as modules, clusters, or groups, are sets of nodes that are densely connected among themselves while being sparsely connected to other sets of nodes.

Usually, we consider communities to be a *partition of the nodes* of the network into disjoint groups (we discuss breaking this assumption later). In other words, each node in the network is assigned to exactly one community and no communities are empty. The number of communities in a partition may vary from 1 to N, both trivial partitions where every node is in the same community to every node is in a separate community, respectively. More interesting partitions occur between these extremes.

---

[9] Some algorithms do exist for speeding up motif discovery [488, 213], but these cannot completely overcome the combinatorial explosion we face.

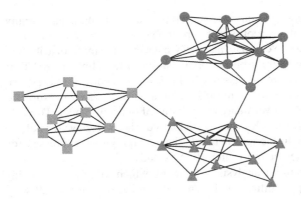

Figure 12.1 A network with three communities. There are many *intra*-community edges and few (only four) *inter*-community edges.

Although any number of data structures can be used to store the community partition (e.g., a set for each community), a mathematically convenient choice is to use a group membership vector. Assign a number to each community and create a $N \times 1$ vector **g** where g_i is the ID of the community that node i belongs to. It won't really matter what the number g_i is. We only need to distinguish the different communities, so we only care whether nodes share a community ($g_i = g_j$) or not ($g_i \neq g_j$).

> Be careful when considering community structure in *directed networks*. Just as directionality complicates connectedness (an undirected network is either connected or not, but a directed network can be unconnected, weakly connected, or strongly connected), it greatly complicates community structure.

To distinguish "good" from "bad" community partitions requires a concrete way to measure the quality of the partition. There are many ways to do this. We can ask how well the data agree with a statistical model of community structure. Or, we can start from a fixed definition of what we want a community to be and then measure how well the partition satisfies that definition. For the former, we discuss the *stochastic block model*, a statistical approach to community detection, in Ch. 23. Here, we will focus on the latter approach, which is often an easy and useful way to get started with community detection.

Measuring community quality: modularity

Think about a good quality measure for community partition. This measure should capture the intuition that there are many links within the communities but few links between them. At first this sounds straightforward—just count the number of links where both nodes are in the same community compared to where they are in different communities. However, this has a trivial solution: place the entire network in one community. Now 100% of links are internal to the community and 0% are external. It's a perfect community but it's also completely trivial. We need a more meaningful definition to measure community quality.

Overcoming this problem, the most commonly used expression [10] for measuring the quality of a set of communities is *modularity*. Modularity asks whether the number of edges within communities is greater than we would expect if edges were distributed at random. If we have more edges within communities than expected, then we can say that our communities indeed form a good partition of the network. The aforementioned trivial example where all nodes are in a single community would have zero modularity because, having controlled for the total number of edges in the entire network, the number within this giant community is exactly what we would expect at random, because the community *is* the entire network. We can also think about "anti-communities" where the density is actually lower than if edges were distributed at random.

The actual number of edges can be computed by looping over each edge (i, j) in the network and asking if the edge falls within a community: does $g_i = g_j$? We can also express this as a sum over all nodes using the adjacency matrix $\mathbf{A}$ (Eq. (8.1)):

$$\sum_{i=1}^{N} \sum_{j=1}^{N} A_{ij} \, \delta(g_i, g_j), \tag{12.13}$$

where $A_{ij} = 1$ if i and j are connected, zero otherwise, and we have put the $g_i = g_j$ "question" into each term of the sum with a delta function: $\delta(x, y) = 1$ if $x = y$, and zero otherwise. If all nodes were in the same community, Eq. (12.13) would achieve its maximum value of $2M$ (it's twice the number of edges because we double count each node pair as (i, j) and (j, i)); we often divide by $2M$ to look at the fraction of edges within communities.

The second ingredient to modularity is the null model—what is the expected number of edges that fall within communities if edges are distributed at random? Among possible null models, here we again pick the *configuration model*—a null model that preserves the degree of all nodes in the network (see also Ch. 11). So we ask, if edges were distributed at random while respecting the degree of every node, what would be the expected number of intra-community edges?

Under the configuration model, how many edges do we expect to land between node i with degree k_i and node j with degree k_j?[11] Suppose i is chosen to receive a link. The probability that j in particular is picked to link with i is given by the fraction of stubs that connect to j: $k_j/2M$ (Ch. 11). But j has multiple opportunities to connect to i, one for each of i's links, so the expected number of i's links that land on j is given by summing each of those probabilities k_i times: $k_i k_j/2M$. Then, summing across all

---

[10] But not the first and certainly not the last. Many notions of community quality extend from earlier, related studies of *graph bisection* (Ch. 25), a classic problem of finding how to split a network into two approximately equally sized components by cutting as few links as possible.

[11] At first, this might seem like a silly question—there should be either no edge or 1 edge—but remember we are talking about an expectation. The configuration model defines an ensemble of random graphs. [12] We may encounter 10 networks out of 100 where an edge was present and 90 without the edge, giving an expected number of 0.1 edges. This is what we want to compare against the real network where, yes, either 0 or 1 edge will be present between i and j.

[12] Technically, it is more complicated. The configuration model actually defines a multigraph and even in one network there may be more than 1 edge between a given pair of nodes.

pairs of nodes that share a community, we get:

$$\sum_{i=1}^{N} \sum_{j=1}^{N} \frac{k_i k_j}{2M} \delta(g_i, g_j). \tag{12.14}$$

Lastly, *modularity* Q is defined by taking the difference of Eq. (12.13) and Eq. (12.14), the actual minus the expected number of intra-community edges, and dividing by $2M$:

$$Q = \frac{1}{2M} \sum_{i,j} \left(A_{ij} - \frac{k_i k_j}{2M} \right) \delta(g_i, g_j). \tag{12.15}$$

Normalizing by $2M$ gives Q a range of -1 to 1, making interpretation easy. Large values of $Q > 0$ indicate community structure, small values $Q < 0$ indicate anti-communities, and negligible values of $Q \approx 0$ indicate an absence of community structure, at least according to the communities encoded by $\mathbf{g}$.[13]

Finding communities

Modularity by itself is a quality measure, it's a function of the network and the communities encoded by $\mathbf{g}$. It measures how strong a given community partition is, but doesn't find it. Community detection is the process of finding those communities, partitioning the nodes of the network into groups. We can think of this as a clustering problem,[14] assigning similar nodes to the same group, where we use the network structure to perform this assignment.

Ideally, we would find our communities by examining each partition of the nodes, measuring its Q, and selecting the partition with the largest Q. Unfortunately, this is effectively impossible.

> ⚠ We can never hope to look at all possible communities for a given network.
>
> The number of ways to partition N nodes into groups is given by the *Bell numbers* [401]. These numbers grow incredibly quickly. For $N = 3$, there are 5 different partitions. For $N = 5$, 52 partitions. For $N = 100$, which is not an especially large network, there are over 10^{115} partitions—many more partitions than atoms in the *observable universe*. The sheer scale prevents us from finding the best communities: unless the network is very small, exhaustive, brute-force enumeration can never examine every possible partition.

Given the challenge of finding an exact solution, and the appealing, practical nature of the problem, many heuristic community detection approaches exist.[15] The most common algorithms work to optimize modularity. In other words, they use Q as an objective function and seek out $\mathbf{g}$ using some heuristic method to make Q as large as

---

[13] We also discuss the *modularity matrix* in Ch. 25.

[14] Here we mean "clustering" as in finding clusters of points such as with k-means clustering (Ch. 25), not clustering as in the clustering coefficient. An unfortunate collision of jargon.

[15] There is even a joke that the definition of a network scientist is someone who has come up with a community detection algorithm.

possible. We can't hope to summarize all the methods that have been devised—likely every optimization algorithm out there has been applied to community detection—but the most popular such method is probably the *Louvain method* [57], a fast, intuitive and effective method. Briefly, this method works by first assigning each node to its own community, then merging communities by taking a node i and computing what Q would be if i belonged to any of its neighbors' communities, then moving i to the community that increases Q the most. This repeats until no move increases Q. Then, a new network is built where each node is a community in the original network, and that network is analyzed exactly as before. This recursive step, combined with efficiently computing how Q changes when only one entry in **g** changes (see also Ch. 27), allows the method to scale up to very large networks.

However, a problem with this algorithm, identified by Traag et al. [461], is that in the process of moving a node i out of one community and into another it is possible that the remaining nodes in the community are no longer themselves connected, if it was i that served to bridge those other nodes. Surprisingly, this can happen even though moving i increases the overall value of Q. Traag et al. [461] introduce the *Leiden method* as a faster alternative that guarantees connected communities when moving nodes.

 Some community detection methods are fast computationally, while others are slow. If you want to use a method with complexity $O(N^3)$, for instance, on a large network, it will never finish running. Be aware of the method's computational complexity, especially when your network is large (see also Ch. 27).

Problems with modularity

Modularity was introduced as a community quality measure to avoid the trivial partition of the entire network as one community. For this purpose, it works very well, but it has some drawbacks, specifically due to its reliance on the configuration model as its null model.[16] It's worth calling out these problems given modularity's popularity.

One problem is that a network can have a large value of Q without having meaningful community structure. Random graphs can exhibit high-Q partitions due simply to stochastic fluctuations [196, 392]. Perhaps more troubling, even deterministic networks such as trees and lattices can have high-modularity partitions, not necessarily because they are modular but because they deviate from the expected structure of the configuration model [24]. These concerns underscore the need in general to be circumspect when comparing networks and null models.

Another way the null model leads to problems with modularity is due to its global nature. The configuration model assumes nodes anywhere in the network are likely to be connected, with that likelihood depending only on node degree. Yet communities are local structures, small modular groups of nodes, and it may be too extreme to assume nodes in very far apart communities have any probability to be connected.

Suppose you have found the communities in a network, such as those illustrated in Fig. 12.1. Now imagine a new component suddenly appeared in this network, perhaps

---

[16] In principle, any null model can be used in place of the configuration model in Eq. (12.15). In practice, however, only the configuration model has been used by researchers with any regularity.

doubling the number of nodes, but those new nodes share no connections to the original nodes. Would this change our communities? We expect those new nodes to have no effect because they are not connected in any way to the existing communities. But this is not always what happens with modularity. By growing the network, we have increased M, the total number of edges, which does change the value of Q for our partition. Does changing Q also change which partition has the highest Q? Unfortunately, yes. It turns out that Q has a bias in favor of communities of a given size, irrespective of the structure of the network. Or, more precisely, it is biased against communities below a specific size. This is called the *resolution limit*.

---

i The *resolution limit* is usually illustrated with the example of two completely dense subgraphs connected by a single link within a larger network of M edges. Intuitively, such subgraphs naturally form two communities and it would be a mistake to put them into a single community, so let's compare Q for one partition where each subgraph is a community (call them community 1 and 2) and a second partition where we have merged the communities into a single group.

First, write modularity as a sum over communities instead of a sum over node pairs (Ex. 12.2):

$$Q = \frac{1}{2M} \sum_{i,j} \left(A_{ij} - \frac{k_i k_j}{2M} \right) \delta(g_i, g_j) = \sum_g \left[\frac{m_g}{M} - \left(\frac{K_g}{2M} \right)^2 \right], \qquad (12.16)$$

where m_g is the total number of edges in community g and $K_g = \sum_i k_i \delta(g_i, g)$ is the total degree of nodes in g.

How does Q change when we merge communities 1 and 2? All other groups are unchanged so we need only consider terms in the sum related to groups 1 and 2 after they are merged compared to before. When merged, the new group will have $m_1 + m_2 + 1$ internal edges (the +1 is for the single edge between the groups) and a total degree of $K_1 + K_2$, leading to a net change in Q of:

$$\Delta Q = \underbrace{\left(\frac{m_1 + m_2 + 1}{M} - \left(\frac{K_1 + K_2}{2M} \right)^2 \right)}_{\text{when 1 and 2 are merged}} - \underbrace{\left(\frac{m_1}{M} + \frac{m_2}{M} - \left(\frac{K_1}{2M} \right)^2 - \left(\frac{K_2}{2M} \right)^2 \right)}_{\text{when 1 and 2 are separate}}$$

$$\qquad (12.17)$$

$$= \frac{1}{M} \left(1 - \frac{K_1 K_2}{2M} \right). \qquad (12.18)$$

Modularity will increase by merging the two groups when $1 - K_1 K_2 / (2M) > 0$ or $K_1 K_2 < 2M$.

Even though our intuition about communities tells us that two dense groups with only one edge between them should be separate communities, if we seek to maximize Q we will invariably merge those groups, failing to detect them, if they are sufficiently small relative to the total number of edges in the network.

The resolution limit has practical consequences for larger networks. Communities with total degree $\approx K < \sqrt{2M}$ will tend to be missed when maximizing modularity. If we assume the best case where those small groups are completely dense, then the number of nodes in those groups is $\approx \sqrt{K}$, and we'll tend to miss communities smaller than $(2M)^{1/4}$ nodes. When $M = 1000$, for example, we'll lose communities smaller than ≈ 6 nodes. These are small groups, but meaningful in a social network. Nowadays, social network data are quite large, with online providers having social networks of hundreds of millions of social ties. If $M = 250$ million, we'll be unable to see communities smaller than 150 nodes, which is almost certainly larger than most social groups [135]. Despite the effort spent making algorithms for modularity that scale to very large networks, modularity maximization is often best avoided for such networks due to the resolution limit.

> ⚠️ Be aware of the resolution limit when using modularity-based community detection methods on large networks.
> When the network is large enough, modularity will be maximized by merging small communities together, making it impossible to discover those communities.

Note that the resolution limit can also be exploited to "fine-tune" modularity to seek out communities of a particular size. A parameter γ is introduced [391, 322] in Q to weigh the relative contribution of the null model:

$$Q = \frac{1}{2M} \sum_{i,j} \left(A_{ij} - \gamma \frac{k_i k_j}{2M} \right) \delta(g_i, g_j). \tag{12.19}$$

When $\gamma = 1$ we recover the original modularity; $\gamma < 1$ will seek out larger communities while $\gamma > 1$ will favor smaller communities. Sweeping through values of γ can also reveal a *hierarchy* of nested communities. We discuss hierarchical organization in Sec. 12.8.

Fuzzy and overlapping communities

One of the major assumptions thus far has been that communities form a partition of the nodes. Each node belongs to exactly one community. Why should this be so? One can imagine a social network, say, where a node belongs to a group of peers and a group of family members. Those two communities thus overlap at that node.[17] And beyond social networks, we can imagine a similar structure occurring in other domains. In a PPI network such as HuRI, for instance, proteins often fulfill multiple roles within the cell and we should anticipate those roles manifesting in the network as densely interconnected groups of proteins that overlap at multi-function proteins.

Over the years, a variety of methods for finding fuzzy and overlapping communities have arisen. One of the most intuitive and successful is called *Clique Percolation* [354]. Briefly, Clique Percolation works by defining a clique size k, identifying cliques (completely connected subgraphs) of k nodes, and then looking for cliques that are adjacent to one another: two k-cliques are adjacent if they share $k - 1$ nodes. We can imagine clique adjacency as taking a k-node template and "rolling" it over the network from

---

[17] The communities no longer form a partition of the nodes but instead form a *cover* of the nodes.

one clique to an adjacent clique. A community is defined as the maximum portion of the network that can be covered by rolling that template, the union of all the adjacent k-cliques. Defining communities in such a manner allows for overlap because k-cliques can intersect without being adjacent if they share few nodes. Not only does this method allow for overlap, it also allows for *local* inference of a community: one can roll the clique from a given starting point and identify the community surrounding that initial point without needing to examine the entire network.

> ⚠ Finding overlapping communities is more challenging than a partition— an objective function that is not constrained to place nodes into disjoint groups is susceptible to a problem we call "**stacking**." If a very good community is found according to our objective function, and we allow overlap, then we can improve our objective by including a copy of that high-quality community. We can maximize it even more by including more copies, stacking identical communities on top of one another. We need to prevent redundant communities from appearing. With a partition, this is not a problem. The net consequence of this is typically an objective function that allows overlap but *punishes* too much of it, leading to low-overlap or "fuzzy" communities.

Pervasive overlap

So far we have been using "fuzzy" and "overlapping" somewhat interchangeably. We believe a distinction should be made based on the prevalence of overlap. *Fuzzy communities* are where most nodes belong to one community but a few belong to two or more. Overlap, or what we call *pervasive overlap* is when **most or all** nodes belong to two or more communities.

The distinction has important consequences. We usually think of communities as a coarse-graining of the network, even to the point of creating a "super-network" where nodes represent communities of the original network, as the Louvain method does. This won't necessarily be the case with pervasive overlap. If most communities are small, say around five or six nodes, and most nodes belong to, say, four or five communities, there will be almost as many communities as there are nodes! Such a situation is actually quite plausible in a social network, for example, where most or all people belong to a few small groups, with the number of groups per person being not too far away from the number of people per group.

But pervasive overlap even calls into question the foundational assumption of communities as groups densely connected among themselves and sparsely connected to other groups, as we illustrate below.

Chess and the paradox of pervasive overlap   The *rook graph* (Fig. 12.2) contains a node for each square of the chess board and two nodes are connected when the rook, which can travel any distance along the same row or column of the board, can move between those squares in a single turn. This graph is a simple illustration of what we call the paradox of pervasive overlap: each row and column is completely connected (forming a clique) and each row and column is thus a perfect community. Every node

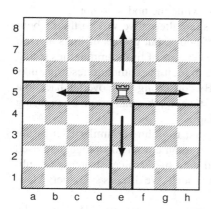 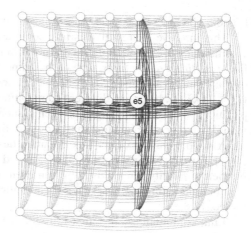

Figure 12.2 The rook graph and the paradox of pervasive overlap. Every row and column of nodes in the rook graph is a completely dense subgraph, making it the perfect community. Yet every node in a community has as many links exiting the community as it does within the community, breaking the definition of communities as being both densely intraconnected and sparsely interconnected.

therefore belongs to two communities, one for its row and one for its column—pervasive overlap. But this means that each node in a given community has as many links exiting that community as it does within its community, breaking the definition of communities as being both densely intraconnected and sparsely interconnected. When overlap is pervasive, both connections can be dense.

What this example illustrates is the sea change that occurs when we move beyond fuzzy communities to pervasive overlap, where the number of community memberships per node meets or exceeds two. In such a circumstance, we need to confront our intuitive notions of communities as decomposing networks down into smaller, nearly disjoint groups. We can have many groups in a network, and those groups can be densely interconnected while still being meaningful.

Link communities—finding overlap with a partition

Consider another fundamental assumption of communities: communities are groups of nodes. Whether communities must be disjoint or can overlap means we are assuming whether nodes belong to only one group or more than one group. Assuming nodes belong to one group may make sense in some contexts, but not all: in social networks, for instance, it is unlikely that people participate in single social contexts. However, consider the links instead. While it seems unlikely for a node to partake in a single context, a link is more likely to be driven by a single reason. Two people have both have friends and family, multiple contexts, but they may know each other because they are friends. The link, we assume,[18] has a single context while the nodes do not.

This intuitive idea motivates link communities, an alternative way to think about communities. Here communities are not groups of nodes, but groups of links and we find *a partition of the links*. One of the immediate benefits of link communities is that overlap appears automatically. A node i can belong to multiple communities if its links $(i, j), (i, k), \ldots$ belong to different communities. By converting node overlap back to a partition, we avoid many of the problems, such as stacking, that can occur when trying to build an objective function that allows for overlap.

A simple way to detect link communities is first to define a similarity measure $s((ij), (uv))$ between pairs of links (i, j) and (u, v), then cluster links together based on their similarity. This clustering can even build up a hierarchical structure of sub-communities, communities, super-communities, and such. We discuss such hierarchies in Sec. 12.8.

Link communities were first introduced by Evans and Lambiotte [153] and Ahn et al. [5].

Local community detection

Sometimes when dealing with a very large network you won't need to (or even be capable of) identifying its full community structure. It may be that you have a particular node of interest and you want to find, not all communities, but only the community structure surrounding that node. A local community detection method can be used for this purpose.

A simple local method for finding the community of nodes containing a starting node i is as follows [26]. Define an ℓ-shell as all the nodes j that are ℓ steps away from i along shortest paths. (The 0-shell is i itself.) Such shells can be found locally using breadth-first search to spread outward from i. Define the out-degree $K_i^{(\ell)}$ of shell ℓ as the total number of links from nodes in shell ℓ to nodes in shell $\ell + 1$. (We use the subscript i to denote that the shells depend on the initial node i.) Given the traditional definition of a community as densely intraconnected while being sparsely interconnected, we anticipate that the shell out-degree will drop suddenly when reaching the "border" of i's community. We capture this by introducing a parameter α and continuing to spread outward from i until $K^{(\ell)}/K^{(\ell-1)} < \alpha$ (or we reach the entire connected component containing i). While this depends on a free parameter α, which may be a disadvantage because it needs to be tuned, this simple method works efficiently and is completely local, requiring little more to implement that breadth-first search.

A concern with this method is that it assumes nodes on the border of i's community are equidistant from i. Given how dense a community typically is, this will usually happen, but if one remains concerned, a simple tweak is to move nodes into i's community one-at-a-time instead of shell-by-shell [23]. To do this, let C be the community containing i and B be the "border" of C, nodes that are neighbors to nodes in C but in C themselves. We grow C one node at a time starting from $C = \{i\}$. As C changes,

---

[18] Of course, this is also an assumption: that links only fulfil a single role or occupy a single context. It's certainly possible for a link to arise due to multiple factors, two coworkers who also go to school together, for example. But we argue that assuming a single context for a link is a weaker assumption, and a more common scenario, than assuming a single context for a node.

we update B as well. To grow C, we choose the node v in B that contains the most neighbors in C and the fewest neighbors outside C. Specifically, select the v in B with the smallest value of

$$\frac{1}{k_v}\left(k_v^{\text{out}}(C) - k_v^{\text{in}}(C)\right),$$ (12.20)

where $k_v^{\text{out}}(C)$ is the number of neighbors of v not in C, $k_v^{\text{in}}(C)$ is the number of neighbors of v in C, and $k_v^{\text{in}}(C) + k_v^{\text{out}}(C) = k_v$. This method is also local as it requires knowing only the neighbors of nodes in C and B.

Local community detection was first introduced by Bagrow and Bollt [26].

12.8 Hierarchy and cross-scale structure

A hierarchical organization may exist in your network, relating the small- and large-scale structure of the network. Communities are one way hierarchy may form in a network, where smaller communities are nested within larger communities. We can visualize this hierarchy as a tree, called a *dendrogram* (Fig. 12.3), that tracks the lineage of nodes (or links, in the case of link communities) through the set of nested communities. We can also look at networks across these scales by defining "super-networks" where nodes in the super-network represent communities in the original network. Often such networks are weighted, with edge weights between super-nodes counting the number of links between communities in the original network.

A dendrogram can be computed by merging groups of nodes (or links) together, or by dividing them. The difference is whether the dendrogram is computed from the bottom up or from the top down. These approaches are called agglomerative clustering and divisive clustering, respectively.[19]

One of the most influential methods for dividing a network into a hierarchy of nested communities is due to Girvan and Newman [183].

ⓘ The *Girvan–Newman algorithm* divides up a network into its hierarchy of communities.

An elegant method for determining a network's hierarchy is to use shortest paths and then rank edges based on how many shortest paths use them. If many shortest paths all pass through one particular edge, that edge is important, acting as a bottleneck. If we remove that edge, or a few such edges, the network may even split apart.

This observation motivates the Girvan–Newman algorithm [183]. First, edges are ranked based on importance, then removed in order of decreasing importance until eventually the network starts to break apart, revealing levels of community structure (hierarchy). Unfortunately, there is an important issue to confront: after removing the most important edge, the ranking of the remaining edges may change. Shortest paths can potentially change quite drastically when important edges are taken away. (Think of all the traffic that would be rerouted if a large highway were

---

[19] Such clustering methods are not limited to networks, although we focus on that use case here. Any clustering of points can be considered.

closed.) Girvan–Newman addresses this by re-ranking edges after removal. Edges are ranked, the most important edge is removed, the remaining edges are ranked again, the most important remaining edge is removed, and so forth until no edges are left. As edges are removed, the network will gradually break apart, revealing connected components that we take to be communities and sub-communities along the hierarchy.

(An interesting historical aside: Girvan–Newman continues dividing a network up until nothing is left, but to find a single partition of communities instead of a hierarchy, we want to determine what point to stop dividing, and use those connected components as the communities. Modularity Q was originally introduced to find the best division point of GN. Only later were methods introduced for optimizing Q directly.)

Ranking edges using the number of shortest paths they carry is called *edge betweenness*, a measure we discuss in detail in Sec. 12.9. Edge betweenness is computed by looking at the shortest path from every node to every other node, which is an $O(N^2)$ calculation. Girvan–Newman (in the worst case) does this calculation M times, making it effectively very expensive, $O(MN^2) = O(N^3)$ for most networks. We encourage using Girvan–Newman for smaller networks, as it works exceptionally well, but its cubic complexity is prohibitive even for moderately sized networks.

Using the Girvan–Newman algorithm, we analyze the hierarchy of a small network in Fig. 12.3. This network contains eight small communities that group into three larger communities, with two of the three themselves also grouping together. We first draw the network (Fig. 12.3a) using edge width to visualize the betweenness of edges. (For GN, this is only the first computation of edge betweenness.) It's clear that the highest betweenness edges are the bottlenecks or bridges that connect different communities. We then proceed to delete edges according to GN while tracking the connected components of the network. As the network disconnects, we progressively reveal the community structure across scales. We track which nodes belong to which cluster in a dendrogram tree in Fig. 12.3b. To draw the tree, nodes are ordered left-to-right such that the larger group is to the left. The height of the merger point is proportional to the number of nodes being merged.[20] These coordinates don't change the structure of the dendrogram but make it easier to understand visually.

A second way exists to envision hierarchy in a network, also using a tree structure. Here the nodes of the network occupy different positions across levels in a hierarchical tree, in much the same way as managers, vice presidents, and chief officers span upwards in the reporting structure of an organization. This hierarchy differs from that of a dendrogram where all elements of the network occupy the bottom level, and the mergers of those elements into groups of different sizes are represented at higher positions of the dendrogram tree. While a dendrogram is computed from an analysis of the network, a reporting tree must often be developed using ancillary or side information.

---

[20] More generally, many dendrograms use the vertical dimension to encode additional information. For example, if groups merge based on a numeric criterion such as merging when a similarity metric between the two groups exceeds a given threshold, the vertical position of the merger may be placed on a scale representing that similarity threshold.

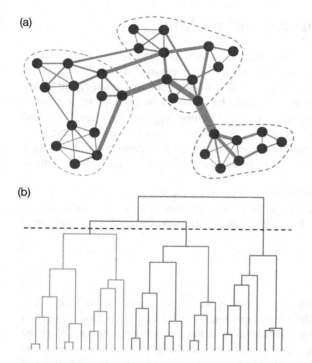

Figure 12.3 A hierarchy of communities. (**a**) A small example network, containing eight small communities of three-to-five nodes grouped into three larger communities (circled). The width of edges is drawn proportional to their edge betweenness, the number of shortest paths that use that edge. High betweenness edges act as bottlenecks or bridges between the communities. (**b**) The dendrogram of communities found by the Girvan–Newman (GN) method. The dashed line indicates the cut corresponding to the circled communities in the top figure. The bottom of the tree represents the nodes in the original network (can you match the nodes in the original network?), while the merger points as we move upward describe how communities merge as we move upward in the hierarchy, according to GN.

In fact, large (human) organizations may have both tacit and explicit reporting structures, with the former often being the key to understanding the organization's operation and (dys)function.

Lastly, yet another way to consider hierarchical organization is through recursion. We can imagine taking a network and replacing each node in the network with a copy of the original network. Iterating this process can build a larger, more deeply recurrent network. Recursive network models are especially interesting theoretically, as we can exploit recurrent relationships within mathematical expression to find novel solutions to otherwise intractable quantities. However, such recursiveness is generally less important when it comes to data-driven network analysis—although it can be used to motivate and study new analysis methods.

The interplay between hierarchy and other organizing patterns in networks such as communities remains an active and interesting area of study.

12.9 Centrality measures and ranking

Another way we approach networks is by ranking the nodes and edges, usually based on some notion of importance, trying to capture how "central" the node or edge is within the network. In other words, for each node i we compute a number, c_i, capturing how important the node is, then we can rank the N nodes in order of most to least important. Likewise, for edges we can compute a number c_{ij} for each edge (i, j), and rank those edges. A wide assortment of such measures of *centrality* are available to us when studying a network's structure, some quite simple, others more complicated (and potentially computationally costly). Let's discuss some of the more popular measures.

 Use centralities to identify and rank important nodes and links.

Most measures determine a number c_i for each node i or c_{ij} for each edge (i, j) in the network, with the larger the number, the more important the node or edge.

Degree centrality

Perhaps the simplest form of centrality for nodes is actually something we've encountered before: *degree centrality*. Here we simply take $c_i = k_i$. A node is more important when it has more connections. High-degree hubs are the most important nodes. Ranking nodes by their degree is quite simple of course, and we shouldn't expect it to reveal too much insight over more complex centrality measures. Nevertheless, many notions of importance correlate with degree, and degree is an inexpensive, local computation,[21] so degree centrality in practice is often an effective place to begin.

Spectral centrality measures

Various centrality measures can be defined based on eigenvectors of network-related matrices (Ch. 25). These include eigenvector centrality, Katz centrality, and PageRank.

Eigenvector centrality   Using the adjacency matrix $\mathbf{A}$ to represent the network allows us to define centrality using an inheritance argument: node i is central if the neighbors of i are central. While this definition sounds circular, we can actually make this computable. Suppose c_i is proportional to the total of c for i's neighbors:

$$c_i = \mu \sum_{j \in N_i} c_j = \mu \sum_{i=1}^{N} A_{ij} c_j, \tag{12.21}$$

where μ is some constant of proportionality. This equation can be written for all i at once, revealing the key to eigenvector centrality: $\mathbf{Ac} = \mu^{-1}\mathbf{c}$, where $\mathbf{c} = [c_i]$. The centrality we want is an eigenvector of $\mathbf{A}$ with eigenvalue μ^{-1}. Of course, $\mathbf{A}$ can have multiple eigenvalues and eigenvectors, but if we want to rank nodes by their elements

---

[21] By local we mean that each node's degree can be computed without looking at all the other nodes in the network, making degree centrality inexpensive to compute.

in the eigenvector, an eigenvector with non-negative elements will be most helpful. This can be guaranteed for connected networks when we look at the largest eigenvalue of $\mathbf{A}$: by the Perron–Frobenius theorem, the eigenvector of the largest eigenvalue of a matrix with non-negative elements (of which $\mathbf{A}$ is an example) will have non-negative elements. Thus we can rank nodes i by the value of the ith element of the leading eigenvector of the adjacency matrix.

> **!** *Numerics of eigenvectors.* Note when working with eigenvectors numerically that they are defined relatively, meaning that if $\mathbf{v}$ is an eigenvector, $a\mathbf{v}$ is as well, for a scalar $a \neq 0$. Essentially, it is the *same* eigenvector, because it points in the same direction (is parallel) as $\mathbf{v}$. This means that the ranking from a vector $\mathbf{v}$ for a given node i is only relative, in that the relative ranking of node i vs. node j, $c_i/c_j = v_i/v_j$ is the same for any vector parallel to $\mathbf{v}$. Keep this in mind when comparing different ranking vectors.

Katz centrality   Eigenvector centrality has a few problems. One is that it doesn't handle directed networks very well. A solution is to include a little bit of additional centrality β for nodes independent of the centralities of their neighbors, $c_i = \mu \sum_j A_{ij}c_j + \beta$, so now i receives some centrality even if it has no neighbors. (The value of β is not so important as its effects can be included in μ; Ch. 25.) We can now in principle solve this for a new, different centrality measure known as *Katz centrality* [237], which can be used for undirected as well as directed networks.

PageRank   *PageRank*, the celebrated idea behind Google Search [75], is motivated by incorporating node degree into the inheritance argument of Katz centrality: $c_i = \alpha \sum_j A_{ij} \frac{c_j}{k_j^{\text{out}}} + \beta$. The idea being that nodes with many neighbors are "spread out" compared to nodes with few neighbors: If I have a high-degree neighbor with many ties, the centrality they contribute to me should count less when I update my centrality than another neighbor with fewer other ties, because the high-degree neighbor is spreading their centrality to many more nodes than the low-degree neighbor. (We use out-degree here because, if the network is directed, centralities update outward from a neighbor j.) PageRank comes from choosing $\alpha = 1 - \beta$.[22]

> PageRank is closely related to the HITS (or Hubs and Authorities) algorithm [244].

Distance centrality measures

We would expect distances between nodes to play a (ahem) central role in measures of centrality. Distances, usually quantified by the lengths of *shortest* paths,[23] form the basis of distance-based centralities.

---

[22] While PageRank continues to be celebrated as the idea behind Google Search, it is likely no longer a key signal for Google's (proprietary) search ranking method, as it can be manipulated [507].

[23] For unweighted networks, shortest paths, sometimes called *geodesics*, can be found using graph algorithms such as breadth-first search or depth-first search (Sec. 4.1). For weighted networks, at least those with only non-negative edge weights, we usually define minimum cost paths, where the cost is the sum of the edge weights of the edges on the path. These paths can be found using Dijkstra's algorithm or the Bellman–Ford

Let ℓ_{ij} be the length of the shortest path between nodes i and j. Assuming the network is connected, there are $N - 1$ shortest paths from a given node i to other nodes in the network. Closeness centrality is defined as the reciprocal of the average of those $N - 1$ shortest path lengths:

$$c_i = \frac{N - 1}{\sum_j \ell_{ij}}. \tag{12.22}$$

The idea of taking the reciprocal is just that it reverses "distance" and "closeness." Many variants on closeness centrality exist, and it needs to be adjusted for disconnected networks (for which some of the $\ell_{ij} \to \infty$).

Another way to gain insight about the most important parts of a network is by examining not the lengths of shortest paths but how those shortest paths are distributed throughout it. Shortest paths tell us how to (optimally) navigate the topology of the network and a link serving as an important conduit between parts of the network will participate in many shortest paths. If only a single link exists to connect, say, two parts together, then every shortest path that connects a node in the first part to a node in the second part must go through that link. Links that carry many paths become very central.

Edge betweenness[24] makes this notion concrete. First, consider the shortest paths between every pair of nodes.[25] Each path from source node s to target node t is a sequence of edges: $[(s, v_1), (v_1, v_2), \ldots, (c_k, t)]$ where v_i are the intermediary nodes visited as we move along that path. We compute edge betweenness by examining each path and keeping a count for every edge of every time we observe that edge participating in a path. Specifically, the betweenness of edge (i, j) is the fraction of shortest paths containing (i, j):

$$c_{ij} = \sum_{s,t \in V} \frac{\sigma(s, t \mid i, j)}{\sigma(s, t)}, \tag{12.23}$$

where the summation runs over every distinct pair of nodes s and t, $s \neq t$, $\sigma(s, t)$ is the number of shortest paths from s to t and $\sigma(s, t \mid i, j)$, $i, j \in E$, is the number of shortest paths from s to t that pass over an edge i, j. We divide by $\sigma(s, t)$, the number of shortest paths between s and t, to account for ties: multiple shortest paths may exist between s and t.

A similar centrality can be defined for nodes by counting occurrences of node v_i in the shortest paths instead of occurrences of edges (v_i, v_{i+1}). Let $\sigma(s, t \mid v)$ be the number of shortest paths from s to t that pass through v. The *node betweenness* of v is

$$c_v = \sum_{s,t \in V} \frac{\sigma(s, t \mid v)}{\sigma(s, t)}. \tag{12.24}$$

Range betweenness is another interesting measure of edge centrality. Here we ask, for a given edge (i, j), what would be the distance between i and j if that edge did not exist? We can compute this by removing the edge, and finding the shortest path

---

or Floyd–Warshall algorithms. The latter two can also handle negative edge weights.

[24] We encountered edge betweenness in the Girvan–Newman algorithm in Sec. 12.8.

[25] It's common to encounter ties, where two or more shortest paths of equal length exist between the same two nodes but running over different edges. It is best to keep a count for these ties: let $\sigma(s, t)$ be the number of shortest paths between nodes s and t.

between i and j using, for example, breadth-first search. If the distance changes quite dramatically between i and j, then we can reasonably conclude that the edge is important: no "shortcut" alternatives exist and that edge is in many ways a bottleneck within the network. Conversely, if the distance changes only slightly, then we can conclude the edge is not overly important as another "backup path" is close by.

In fact, range betweenness anticipates more complex, non-local measures of a network's structure. If we delete edge (i, j) from the network, then all shortest paths carried by that edge will be gone. Alternate paths will be followed instead, in principle, changing betweenness values across the network.

Normalization Equations (12.23) and (12.24) may need to be rescaled. The numbers of shortest paths will depend on the total number of nodes in the network:[26] for a connected network of N nodes, there are at least $N(N-1)/2$ shortest paths. If we wish to compare the betweenness centralities of nodes or edges in two different networks, we may want to rescale c by a measure of size, to ensure that any differences are due to how shortest paths are distributed in the different networks and not merely due to the sizes of the different networks.

Directionality As with some spectral centralities, directed networks introduce complexity when it comes to distances. Shortest paths follow edge directions and so we encounter the same problems with strongly connected components as before. For example, $\sigma(s, t) > 0$ while $\sigma(t, s) = 0$. Here we again have in principle double the number of shortest paths, as the shortest path from s to t is no longer necessarily also the shortest path from t to s.

12.10 Distances and connectedness

Distances are important in networks. Unlike in a spatial system, typically we consider the distance ℓ_{ij} between nodes i and j to be equal to the minimum number of "hops" one must make to transition from i to j along edges in the network. We have encountered such *shortest paths* before: shortest paths are used in closeness centrality and betweenness measures. Shortest paths and shortest path lengths can be found using *breadth-first search* (BFS) or *depth-first search* (DFS) for unweighted graphs, and search algorithms for weighted networks include Dijkstra's algorithm and more.[27] Many fundamental

---

[26] If the network is disconnected, the number of shortest paths with finite length depends on the distribution of nodes per connected component.

[27] When edges have weights, the weights can mean two opposite things: larger weights mean longer paths, or larger weights mean shorter paths. If larger weights mean longer paths (e.g., time to traverse the edge), the shortest path is defined as the path with the smallest sum of the weights of the edges, and vice versa. In weighted networks, finding shortest paths cannot be done by the simple breadth-first search algorithm, but requires more sophisticated algorithms such as Dijkstra's algorithm. The reason is that a path with many hops can be "shorter" than a path with fewer hops but with larger weights. Therefore, we have to keep track of the smallest sum of the weights of the edges traversed so far to each node as well as the "parent" from which we reached the node. Although we will not explain the algorithm in detail, it is one of the most well-known algorithms and can be easily found in the textbooks, online resources, and software packages.

properties of a network relate to shortest paths, including information flow and spreading dynamics, as the shortest paths provide the conduits through which such dynamics occur.

Several distance-based summary statistics are common for networks. The network's *average shortest path length* (APSL), given by

$$\langle \ell \rangle = \frac{1}{\binom{N}{2}} \sum_{i \neq j} \ell_{ij}, \tag{12.25}$$

where the sum runs over all $\binom{N}{2}$ pairs $i \neq j$, is one measure of the typical "spread" of the network. ASPL is closely related to closeness centrality (Eq. (12.22)), $\langle \ell \rangle = \frac{1}{N} \sum_i \frac{1}{c_i}$, essentially by construction. Another distance measure is the *eccentricity* of a node i, the length of the longest shortest path from i to some other node in the network:

$$\epsilon(i) = \max_{j \in V} \ell_{ij}. \tag{12.26}$$

Eccentricity measures how "unusual" the node is based on its remoteness in the network's topology, like an anti-centrality. Eccentricities can be found by first computing the shortest paths between every pair of nodes. Two network-wide summary statistics come from the distribution of node eccentricities, the *radius* and the *diameter*:

$$r = \min_i \epsilon(i), \tag{12.27}$$

$$d = \max_i \epsilon(i). \tag{12.28}$$

The diameter, the maximum distance between any nodes, in particular, has attracted a great deal of attention, thanks to the "small world" or "six degrees of separation" idea [309]: very large networks can have especially small diameters, thanks to the existence of efficient routes between nodes.

Connectedness

What happens when the network is disconnected? We can no longer get from any node to any other node—paths only exist between pairs of nodes in the same connected component. The diameter of the network does not exist (or we can take it as $d \to \infty$), and many other quantities such as ASPL may need to be redefined to only consider paths that exist (path lengths $\ell < \infty$).

It's quite common for a network to be disconnected, and you can even miss this fact. We recommend always checking if the network is connected [28] and, if not, measuring n_{cc}, the number of connected components, N/n_{cc}, the average component size, and N_{gcc}/N, how many nodes are in the largest (or, giant) connected component. Often, the network is disconnected, but most nodes are connected, N_{gcc} is close to N. If this isn't the case, be sure to recognize it, and consider looking at other statistics such as the variance of component sizes, or examine the distribution of component sizes.

---

[28] To check, run BFS once and see if you reach every node in the network. If you don't, the network is disconnected.

Connectedness is complicated for directed networks. Here a network can be dis-connected, weakly connected, or strongly connected. The distinction between weak and strong comes from whether paths exist in both directions: in an undirected network if a path exists from i to j it must also exist from j to i. This is not the case for directed networks and a strongly connected network is one where both directions exist for all i and j while a weakly connected network is one where a path exists between some i and j only in one of the two directions.

12.11 Size and density

At the largest scale, it's worth emphasizing two simple statistics for understanding a network's global structure, the number of nodes N and the number of edges M. These quantities are so simple they often go unmentioned. Yet, we encounter these quantities everywhere and they are worthy summary statistics we can use with our data, particularly when the network becomes complicated. For example, if a network is changing over time (Ch. 15), we may wish to look at N and M for different time periods. Or, as we work to extract the network (Ch. 7), we can consider how our choices change N and M.

Two natural ways to combine N and M are often useful. The first is the *average degree*:

$$\langle k \rangle = \frac{1}{N} \sum_{i=1}^{N} k_i = \frac{2M}{N}. \tag{12.29}$$

The second is the *density* of the network, how many possible edges actually exist:

$$\rho = \frac{M}{\binom{N}{2}} = \frac{2M}{N(N-1)}. \tag{12.30}$$

Density can also be expressed as $\rho = \langle k \rangle / (N-1)$; it compares the typical number of neighbors $\langle k \rangle$ to the maximum possible $\max k = N - 1$.

Don't underestimate the usefulness of these simple quantities.

12.12 Other organizing patterns

So far, we have a discussed many of the most common organizing patterns and how to find them in a network. But many other patterns are possible, and here we briefly describe some of the most important. We also discuss ways of creating your own statistics to measure organizing patterns of interest, using the measures in this chapter as a starting point.

12.12.1 Bipartiteness

A network is bipartite when its nodes can be divided into two sets with no overlap, and where edges that originate in one set only terminate in the other set; nodes in the same set are not connected. Often, bipartiteness is by construction. For instance, a network of scientists and scientific articles can be constructed where nodes fall into two sets (representing scientists with one set and articles with another) and a link exists when a scientist authored or coauthored an article. By construction, this network must be bipartite: links only exist between scientists and articles.

While it may be that a network is bipartite accidentally or due to chance, meaning there exists a partition of the nodes that happens to satisfy bipartiteness, this is generally rare, especially for large networks. This means that the general focus is not on discovering whether the network is bipartite, but instead on designing measures of the network that account for bipartiteness.

For example, the standard clustering coefficient is not meaningful here, as no triangles (three-cycles) can occur by definition in a bipartite network. Instead, a clustering coefficient can be built using *four*-cycles [397]:

$$CC_{\text{bipartite}} = \frac{4C_4}{L_3}, \tag{12.31}$$

where C_4 is the number of four-cycles (cycles involving four distinct nodes) in the network and L_3 is the number of three-paths (paths of three nodes). Every four-cycle contains four three-paths, hence the 4 in Eq. (12.31), in exactly the same way that every triangle contains three two-paths, so this bipartite clustering coefficient follows analogously from the traditional clustering coefficient.

12.12.2   Core–periphery structure

Nodes in many networks can be divided into two groups, a core and a periphery. The core is the set of densely connected, central nodes while the periphery is nodes that are sparsely connected both to themselves and to the core. The presence of such *core–periphery* (C–P) structure has implications in a variety of contexts. In world economics, for instance, a dense core can arise among trading partners as they scale up manufacturing of increasingly complex goods, while those in the periphery are left focusing on simple trade, such as agriculture [329]. In transportation or other forms of exchange, it may be most efficient to organize flow so that a large region is served efficiently by sparsely connecting it to a small, central core of highly connected nodes (representing airports or cities, say).

Detecting and measuring core–periphery structure   The classic approach to determining C–P structure, due to Borgatti and Everett [66], seeks to identify binary labels x_i for each node i where $x_i = 1$ if i is in the core and $x_i = 0$ if i in the periphery. For a maximally dense, perfectly C–P network, peripheral nodes will only connect to core nodes and core nodes will connect to all nodes. The entries of the adjacency matrix $\mathbf{A}^{\text{CP}}$ of this network can be expressed in terms of the $\mathbf{x}$ labels: $A_{ij}^{\text{CP}} = x_i + x_j - x_i x_j$. The C–P quality, Q^{CP} (not to be confused with modularity) can then be measured using

$$Q^{\text{CP}}(\mathbf{x}) = \sum_{i,j} A_{ij} A_{ij}^{\text{CP}}(\mathbf{x}). \tag{12.32}$$

Finding the $\mathbf{x}$ that maximizes Q^{CP} then identifies which nodes are in the core and which are in the periphery. A drawback of this approach is the strictness of the binary decision variable $\mathbf{x}$. This makes optimization harder and it may be overly restrictive—there may not be a sharp cutoff clearly distinguishing core nodes from periphery nodes. Thus, Borgatti and Everett also relaxed the problem by introducing a non-negative "coreness" score $\mathbf{c}$ for nodes and modeled the ideal $A_{ij}^{CP} = c_i c_j$.[29]

Many other measures of C–P structure have since been introduced. One set of important generalizations is to move from a single C–P division to multiple cores [400], allowing interplay between community structure (Sec. 12.7) with a periphery [248]. In all cases, expect to compare your C–P structure to that of an appropriate null model, often the configuration model. If the configuration model exhibits an equally strong C–P presence as your real data, you should question whether the structure is significant.

Do sampling effects cause C–P?   Imagine there are two groups of nodes, one group that is well observed or well tested in the dataset, and another that is poorly observed or tested. For example, the data generating process "watches" nodes in one group, and sees all of their links. Some of those links point to other nodes that are also observed well, but others point to nodes that are not watched. Those latter nodes are only seen from the perspective of the watched nodes and little is known about them. In particular, if two nodes in this second group are linked, the data generating process will not be able to capture that link. What will the effect be? The well-observed group of nodes will look like a core and the poorly-observed group will look like a periphery.

> ⚠ Core–periphery structure can arise due to intrinsic network properties or from biased data measurement: core nodes are well measured and peripheral nodes are poorly measured. Keep the possibility of sampling effects in mind.

While such a scenario may sound artificial, it can happen in practice. Suppose you are working with a social network derived from the billing records of a mobile phone provider. Those billing records cover the provider's subscribers only. You see non-subscribers, but only when they call or text with a subscriber. Activities between non-subscribers are absent from these observational data. It is plausible that you may observe core–periphery structure predicated on the subscriber/non-subscriber divide. Or, at least, it is plausible enough that you should investigate it as a possible explanation for any core–periphery pattern you do observe.

As with many structural measures, if you identify a core–periphery structure in a network, it is likely to teach you about either how it functions or how it was measured.

12.12.3   Nestedness

Nestedness in some ways is an extension or generalization of core–periphery structure, adding a notion of hierarchy (nesting) to the C–P. Typically, it is studied in bipartite networks, especially in the area of ecology, where nested networks have been observed quite often.[30]

---

[29] Modeling $\mathbf{A}^{CP}$ as a product has nice benefits, connecting the optimization of $\mathbf{c}$ to the network's eigenvector centrality [66].

[30] Ecologists are interested in *mutualistic networks*, such as plant–pollinator networks. A plant–pollinator network is a bipartite network where nodes fall into two groups based on whether they are a plant or pollinator. Links exist when observations were made of a particular pollinator interacting with (pollinating) a plant.

A network is said to be nested when a given node's neighbors tend to be a subset of the neighbors of another, higher-degree node; low-degree nodes are "nested" within the neighborhoods of high-degree nodes. This nesting pattern won't occur for all nodes, so *nestedness* is a measure of how prevalent the pattern is.

> ⚠ C–P structure is related to, and often confused with layered structure, such as nestedness.

What does a propensity for nestedness tell us about a network? Ecologists have debated its consequences for some time, with much of the argument centered on how it improves the robustness of the network against perturbations. For instance, a network of wildflowers and bees that pollinate them that is highly nested is said to be more robust against climate change or the appearance of an invasive species than if it were not so nested. While this research often focuses on mathematical models of population dynamics that are specific to ecology, we can expect nestedness to help with dynamical stability more generally. For example, instead of flowers and bees, we may have a network of source code files and programmers who edit those files. Such a collaboration network may reflect a healthier or more efficient organization when it is nested, and the quality of code may be higher when there is turnover in the programmers as employees join and leave the organization, than if less nestedness were observed.

> ⓘ Nestedness is especially of interest for bipartite networks. Although many ways of quantifying nestedness work for other networks (unipartite networks), measures are often designed first for the bipartite case.

Measures   Ideally, we want a number to quantity the degree of nestedness in a network we're studying: "nestedness" = 0 for a totally un-nested network; "nestedness" = 1 for a perfectly nested network. However, it's not so simple, and a large variety of possible nestedness metrics have arisen over the years. We discuss a few of the more popular approaches.

The first set of measures we consider are referred to as *gap-counting measures*. The name comes from visualizing the adjacency matrix of a network. If the network is heavily nested, we can order the rows and columns (equivalent to renumbering the nodes) so that the first rows and columns contain the most ones (those nodes being the highest degree) while later rows and columns are both progressively sparser and their nonzero entries fall underneath or to the right of preceding nodes' nonzero entries. Gaps in the ordering indicate a lack of nestedness, and gap-counting measures work by counting how many such gaps appear. The first, the number of unexpected absences N_0, counts how many times node i is *not* connected to a node j with larger degree than i's lowest-degree neighbor. Let $k_i^{\min} = \min_{j \in N_i} k_j$ be the smallest degree of a neighbor of i, and

$$N_0 = \sum_i \sum_j (1 - A_{ij}) \mathbb{1}_{k_j > k_i^{\min}}, \qquad (12.33)$$

where the *indicator function* $\mathbb{1}_x = 1$ if x holds, otherwise it is 0, tracks whether the node j being examined is sufficiently high degree. Correspondingly, we can define the number of unexpected presences N_1 from the reverse perspective, counting how many times i is a neighbor of a node j with smaller degree than i's highest-degree neighbor. Let $k_i^{\max} = \max_{j \in N_i} k_j$ and

$$\mathcal{N}_1 = \sum_i \sum_j A_{ij} \mathbb{1}_{k_i^{\max} > k_j}. \tag{12.34}$$

Another class of nestedness metrics are *overlap measures*. These measures ask how much the neighborhoods of lower-degree nodes overlap with the neighborhoods of higher-degree nodes—how often are the neighbors of lower-degree nodes also neighbors of higher-degree nodes? We discuss one of the most popular, NODF (*Nested Overlap and Decreasing Fill*). We consider a bipartite network of N_A nodes of type A and N_B nodes of type B, with $N = N_A + N_B$. In a perfectly nested network, a pair of nodes i and j, both of the same type and with $k_i > k_j$, will have k_j neighbors in common, and fewer than k_j common neighbors in an imperfectly nested network. We can use this to measure nestedness by measuring how close the actual number of overlapping neighbors $O_{ij} = \sum_u A_{iu} A_{ju}$ [31] is compared to the maximum k_j, across all node pairs of the same type:

$$\mathcal{N}^A = \sum_{i,j} \frac{O_{ij}}{k_j} \mathbb{1}_{k_i > k_j}, \tag{12.35}$$

where the sum runs over all pairs of nodes i, j of type A. We define $\mathcal{N}^B$ likewise and the nestedness is assessed with the total NODF:

$$NODF = \frac{\mathcal{N}^A + \mathcal{N}^B}{\binom{N_A}{2} + \binom{N_B}{2}}. \tag{12.36}$$

The denominator acts as a normalization for the maximum possible values where every pair of nodes of a given type has maximum overlap. While not without flaws, NODF is likely the most popular measure of nestedness.

Other categories of nestedness measures are *distance-based measures* and *spectral measures*. Distance-based measures reorder the rows and columns of the adjacency matrix $\mathbf{A}$ to make it as nested as possible, identify the nonzero entries which cause the network to deviate from perfect nestedness, then compute the distance from their matrix location to the position they would fall in for the perfectly nested matrix. The idea being that the more entries that break the pattern of nestedness, and the more extreme (more distant) those entries are after ordering $\mathbf{A}$, the less nested the network will be. The most common distance-based measure is the nestedness temperature, for which several algorithms can be used to compute it [292]. Spectral measures, on the other hand, use the eigenvalues of the network's adjacency matrix, specifically, the spectral radius $\rho(\mathbf{A}) = \max_i |\lambda_i|$, the largest magnitude eigenvalue of $\mathbf{A}$. Why? Among all connected bipartite networks of a given size (nodes and edges), the network with the largest spectral radius will be perfectly nested, a theorem due to Bhattacharya et al. [55]. This means the larger the spectral radius, the more nested the network is. Usually, we want to assess the significance of the spectral radius of the real network by comparing it

---

[31] The sum runs of over all nodes u of the type opposite that of i and j as, for a bipartite network, only links between nodes of different types are possible.

against random null models (how likely does an appropriately chosen null model exhibit a spectral radius at least as large in magnitude as the real network's spectral radius).

Mariani et al. [292] give an in-depth, accessible review of nestedness in networks.

12.12.4 Rich clubs

An idea which has proven quite useful in network neuroscience, and is closely associated with nestedness and positive degree correlations, is the *rich club*. A rich club in a network is one where high-degree nodes preferentially connect among themselves, acting "clubby" in the sense that they are biasing their connections towards other rich (i.e., high-degree) nodes.

The rich club is measured with the *rich-club coefficient* (RCC):

$$\varphi(k) = \frac{M_{>k}}{\binom{N_{>k}}{2}} = \frac{2M_{>k}}{N_{>k}(N_{>k} - 1)}. \tag{12.37}$$

Here $N_{>k}$ is the number of nodes within the network with degree greater than k and $M_{>k}$ is the number of edges that exist between those nodes (where both ends of the edge land on a node with degree greater than k). The denominator $\binom{N_{>k}}{2}$ measures the maximum number of edges which can land among nodes with degree greater than k and so φ captures how dense the subnetwork is among those "rich" nodes. Despite the name, the RCC in Eq. (12.37) is not a coefficient but a function of k.

Suppose we measure $\varphi(k)$ for a network and find it to be an increasing function of k, meaning that the network gets denser as we narrow our focus on the subgraph of higher and higher degree nodes. Does this mean there is a rich club? We should worry about whether φ's behavior is just an artifact of the network's degree distribution— after all, when looking at high-degree nodes, they will have many neighbors and many opportunities to connect and increase their density.

To resolve this and better understand the effects of φ we need, as discussed in Ch. 11, an appropriate baseline or null model to compare against. In this case, the natural choice is the configuration model (Sec. 11.5) which assumes the degree sequence of the network is fixed but nodes otherwise follow no pattern in how they connect to one another. We then ask what φ looks like for such a network. This null is particularly relevant because it controls for the size of the rich club while asking what the density would be when no rich-club bias is present. To compare the actual network to this null, we can rescale the RCC:

$$\rho(k) = \frac{\varphi(k)}{\varphi_{\text{ran}}(k)}, \tag{12.38}$$

where $\varphi(k)$ is given by Eq. (12.37) and φ_{ran} is the same but for the random null model. [32]

12.13 Choosing and designing measures

Given the breadth of organizing patterns in a network, it can be overwhelming to know where to begin. What measures matter and what should we prioritize in our analyses?

---

[32] To ensure representative statistics, it is best to generate many realizations of the null model, compute φ for each and report the average ρ.

To address this question in depth requires considering the entire life cycle of your study (Ch. 5).

It also requires considering your goals. If you need to communicate your results to stakeholders, account for their technical backgrounds and choose measures you can communicate clearly and accurately. If you are building an automated data analysis pipeline, choose measures that work automatically and don't require manual inspection. In all cases, ensure the data are high quality (Ch. 10) and well understood (Ch. 11) as best as possible.

Some measures will be informed by how your network data were gathered and processed, as well as what data are available to augment your network. A directed network unlocks many interesting directions to study such as the extent of reciprocity. Motif analysis (Sec. 12.6), while relevant in all networks, is especially useful in directed networks. Bipartite networks and weighted networks open the door to even more measures. And beyond the network, with a rich collection of metadata, for example, you can begin to relate the network structure to the values of the metadata, using tools such as assortativity (Sec. 12.5) or centrality (Sec. 12.9). How does community structure relate to the metadata? Communities (Sec. 12.7) can themselves expand your metadata, for instance, by using the membership vector g as a new column in your node attribute matrix.

Often, the best measure to use will be custom to your application. The measures we've discussed in this chapter are only a starting point, a huge variety of possibilities are open to you. Consider recombination if you are interested in designing new measures. For example, assortativity can be measured with different node attributes and we can use network statistics beyond the common node degree. One possibility: assortativity with egocentric component size (Sec. 12.2). New centralities can also be introduced by weighting them with other network properties; for example, range betweenness weighted by a node centrality. Through recombination, you can build towards new measures using the base of understanding provided by the pieces we have and, when chosen relevant to your task, novel insights can be attained.

12.14 Summary

We've taken a tour of statistics and measures used to quantify and understand the structure of a network. This area has been central to much of network science, and a huge variety of material is available to us, spanning across all scales of the network (Sec. 12.1)—this chapter is really just the beginning.

The statistics and measures introduced go hand-in-hand with the exploratory and statistical analyses discussed in Ch. 11. After all, we need quantities to measure in the network to drive and ground our exploration. These measures also empower subsequent downstream tasks, such as summarizing and comparing networks, which we discuss in Ch. 14, and network machine learning models, which we discus in Ch. 16. Further, when using these quantities, careful use of null models is often critical to assessing whether the structures found by a measure are meaningful. Likewise, a mismatch between a measure and a null model is often at the root of problems with methods for studying networks, as we saw in the case of modularity, the predominant quality measure for network communities.

Studies of networks with these measures often reveal issues with sampling and biased measurement (Ch. 10). For example, core–periphery structure (Sec. 12.12.2) can be an intrinsic feature of a network or it can arise due to a bias where some nodes are well measured (broadly speaking) and others poorly measured. A core–periphery algorithm can help us understand the presence of this structure but not necessarily its nature, what caused it to appear.

ⓘ Network statistics Here is a handy reference for some of the more important network statistics described in this chapter.

(Local) clustering coefficient	$C_i = \dfrac{T_i}{\binom{k_i}{2}} = \dfrac{2T_i}{k_i\,(k_i-1)}$	(p. 170)
Transitivity	$T = 3\dfrac{\text{Number of triangles}}{\text{Number of two-paths}}$	(p. 170)
Degree assortativity	$r(\mathbf{k}_l, \mathbf{k}_r) = \dfrac{\sum_i \left(k_{li} - \overline{\mathbf{k}}_l\right)\left(k_{ri} - \overline{\mathbf{k}}_r\right)}{\sqrt{\sum_i \left(k_{li} - \overline{\mathbf{k}}_l\right)^2}\sqrt{\sum_i \left(k_{ri} - \overline{\mathbf{k}}_r\right)^2}}$	(p. 172)
General assortativity	$r(\mathbf{x}_l, \mathbf{x}_r)$ using node quantities x_i	(p. 172)
Modularity	$Q = \dfrac{1}{2M} \sum_{i,j} \left(A_{ij} - \dfrac{k_i k_j}{2M}\right) \delta(g_i, g_j)$	(p. 178)
Degree centrality	$c_i = k_i$	(p. 188)
Closeness centrality	$c_i = \dfrac{N-1}{\sum_j \ell_{ij}}$	(p. 190)
Edge betweenness centrality	$c_{ij} = \displaystyle\sum_{s,t \in V} \dfrac{\sigma(s,t\,|\,i,j)}{\sigma(s,t)}$	(p. 190)
Node betweenness centrality	$c_v = \displaystyle\sum_{s,t \in V} \dfrac{\sigma(s,t\,|\,v)}{\sigma(s,t)}$	(p. 190)
Range betweenness	ℓ_{ij} if edge (i,j) was absent	(p. 190)
Average shortest path length (ASPL)	$\langle \ell \rangle = \dfrac{1}{\binom{N}{2}} \displaystyle\sum_{i \neq j} \ell_{ij}$	(p. 192)
Node eccentricity	$\epsilon(i) = \displaystyle\max_{j \in V} \ell_{ij}$	(p. 192)
Radius and Diameter	$r = \displaystyle\min_i \epsilon(i), \quad d = \max_i \epsilon(i)$	(p. 192)
Connectedness statistics	$n_{\mathrm{cc}}, \quad N/n_{\mathrm{cc}}, \quad N_{\mathrm{gcc}}/N$	(p. 192)

Numbers of nodes, edges	$N, \quad M$	(p. 193)
Average degree	$\langle k \rangle = \dfrac{1}{N} \sum_{i=1}^{N} k_i = \dfrac{2M}{N}$	(p. 193)
Density	$\rho = \dfrac{M}{\binom{N}{2}} = \dfrac{2M}{N(N-1)}$	(p. 193)

Bibliographic remarks

Egocentric networks are of perennial interest, given how their local views of the network provide tools for studying networks where global data are unavailable. For a good overview of egocentric networks in the context of social sciences, see Perry et al. [370]. Ugander et al. [467] in particular, use the "egoless" egocentric network to study how something like a disease or an idea spreads over a social network, especially how the number of connected components in the egocentric network relate to spreading.

The clustering coefficient was introduced in the famous "small world" paper of Watts and Strogatz [486], one of the works that kick-started the modern era of network science in the late 1990s. Although sociologists had long considered transitivity and triadic closure measures well before this paper [485], the simplicity of Watts and Strogatz's definition has made it quite popular since then. Assortativity and mixing patterns, the connections between correlation coefficients and edges, was developed by Callaway et al. [89] and Newman [331, 332]. Motifs were popularized by Milo et al. [311], especially in the context of biological networks. For a recent survey, including of the algorithms that help address the challenge of finding larger motifs, see Ribeiro et al. [395]

An enormous amount of research has gone into community structure and discovery. Discovering communities, or community detection, is also known as graph partitioning and occasionally (and confusingly) graph clustering. For some recent introductions and surveys, see Porter et al. [376], Fortunato [164], Newman [335], Schaub et al. [416], and Fortunato and Newman [166]. The history of community detection traces back to the graph partitioning algorithm of Kernighan and Lin [239], although the problem is closely connected to that of network flows, which has an even longer history [121, 162, 163]. Early developments of modularity include Girvan and Newman [183], Newman and Girvan [338], and Newman [334]. The modularity resolution limit was highlighted by Fortunato and Barthélemy [165]. Mucha et al. [322] show how tuning the modularity objective function can generalize it to different types of networks. Local community detection, trying to find one community without looking at the entire network, was introduced by Bagrow and Bollt [26]. Overlapping or fuzzy communities were popularized by Palla et al. [354], who also introduce the clique percolation method. Link communities, which eliminate the complexity of node overlap by finding a partition of the links, were introduced by Ahn et al. [5] and Evans and Lambiotte [153].

Core–periphery structure in networks was formalized by Borgatti and Everett [66]. For an overview and detection method of C–P, which also highlights connections to community structure using block models, see Rombach et al. [400].

Nestedness, first formalized by Patterson and Atmar [363] has become popular especially in ecology, where mutualistic (bipartite plant–animal) networks are increasingly studied, but nestedness is broadly interesting in many other contexts. Ulrich et al. [468] devise the "taxonomy" dividing nestedness measures into gap-counting, overlap, and distance measures. The gap-counting nestedness measures $\mathcal{N}_0$ and $\mathcal{N}_1$ were developed by Patterson and Atmar [363] and Cutler [118], respectively. NODF, one of the more popular overlap measures, was devised by Almeida-Neto et al. [11]. Nestedness temperature, often considered the most popular measure, was introduced by Atmar and Patterson [17]. For an excellent recent introduction and review on nestedness, including methods to calculate it and different areas where it may be used, see Mariani et al. [292].

The rich club was introduced by Zhou and Mondragon [509] to describe how levels of the Internet's autonomous systems are organized. It was originally presented as a function of the rank of the node when ranked by degree, but later work presents it as we did here, considering φ as a function of degree itself [113]. That work, Colizza et al. [113], also emphasized the need to compare φ and φ_{ran}. The rich club has been particularly adopted by the network neuroscience community; for example, see van den Heuvel and Sporns [470] for discussions.

Exercises

12.1 Degree mixing with self-loops. Suppose you are working with a network and you discover it has self-loops, meaning edges i, i are present. How does this affect calculations of the network's degree assortativity (Eq. (12.10))?

12.2 Equation (12.16) changes modularity from a double-sum over nodes to a single sum over communities. Derive the sum-over-communities expression beginning with the double-sum-over-nodes expression.

12.3 (**Focal network**) Using your own code, report the degree assortativity, transitivity, and average shortest path length of two focal networks.

12.4 (**Focal network**) Compare the transitivity and the average clustering coefficient for each focal network. What do we learn from the differences? Which focal network(s) if any should be excluded from these comparisons and why?

12.5 (**Focal network**) Does HuRI exhibit a rich club? Why or why not?

12.6 (**Focal network**) Partition the nodes of the Zachary Karate Club into two groups based on the two hubs or highest-degree nodes, which were the club president and instructor. Specifically, place each hub into a separate group and then place each node into the group of the closest hub (closest based on shortest path length). What is the modularity of this partition?

Chapter 13

Visualizing networks

Visualization is a powerful tool. As shown by Anscombe's quartet and similar exercises [299], visualization can reveal a wide range of patterns not easily inferred from statistics alone. Therefore, when working with network data, it is almost always a good idea to visualize the network and just take a look. For instance, suppose the network consists of 10 disconnected components. Drawing the network using a simple visualization can quickly reveal the fact that the network consists of these disconnected components. Of course, this can be learned by computing the network's connected clusters. However, if you don't visualize it, or think to perform this check, it may take a while until you realize that the network is disconnected and there are 10 components. You may even perform a variety of analyses with the wrong assumption that the network is connected. It is plausible that one may miss this important fact until all analysis is complete, if the network has not been visualized.

Although network visualization can be done in many ways, the vast majority of network visualizations take the form of 2D "ball-and-stick" (or "node-link") diagrams, where each node is represented as a symbol (e.g., a circle, a ball, etc.) and they are connected by lines that represent edges. It is intuitive and familiar (you've seen many in this book already) and thus often ideal—especially for small networks—at visualizing a network. Thus, our primary focus will be on the node-link diagram, although we will briefly talk about alternatives at the end.

In network visualization, size matters a lot. When the size of your network is small, many methods will simply work well and there are not many factors that should be considered. However, when the network is large (say more than 10,000 nodes), it becomes challenging—both computationally and aesthetically—to create a useful and informative visualizations. It is also not just the size—density also plays a role. High density means more edges (lines) that can obscure the visualization. A network where each node is connected to only 1 or 2 neighbors will be much easier to visualize than a network where each node is connected to 1,000 neighbors. Network visualizations of the latter, often humorously referred to as *"hairballs"* or *"ridiculograms,"* are a real sticking point when trying to visualize a network.

203

Large or dense networks are difficult to visualize, both to compute the visualization and to read or interpret it.

Thus, one can say that the major challenge in network visualization occurs when the network becomes large and dense. That's when we need to apply some methods to simplify the network or take more sophisticated approaches. In addition, or more importantly, we need to clarify what types of information we want to get out of the visualization. Is it about seeing the large-scale structure of the network? Is it about distinguishing different "regions" or "areas" of the network? Perhaps it is not about a single static network but about demonstrating a change between networks or within a network over time (Ch. 14)? Depending on the main objectives, drastically different visualization methods may be used. Do you just want to know the component structure? Maybe then you don't need to worry too much about creating a hairball as long as you clearly see disconnected components. Do you want to see how network communities are organized? Then it may be useful to perform community detection and use the results to simplify the visualization, for instance by considering each community as a "supernode" and considering the connections between those supernodes.

In this chapter, we will delve into these questions and discuss practical ways to visualize networks from the perspectives of network analysis, visual perception, design principles, and so on. We will begin with a discussion of the basics of network visualization, from graph layout algorithms to drawing network elements. We then discuss ways to customize the visualization, tuning it for different purposes such as for exploring the data or communicating discoveries. Researchers at times must confront networks that cannot be easily visualized—some may need significant preparation and at times it may be best to approach visualization from a non-traditional avenue. We discuss advanced visualizations techniques and non-traditional approaches that can help, as can network preprocessing steps undertaken to improve or clarify a visualization. Lastly, we cover some tools available for generating network visualizations.

13.1   Standard network visualization

We have encountered many typical network visualizations already (Figs. 1.3, 1.4, and 10.1, for example). Networks are usually visualized using a "ball-and-stick" approach (a.k.a. "node-link diagram"). Nodes are represented graphically with circles, squares, or other shapes, and edges are drawn as lines or curves between those shapes. To visualize a network then requires: (i) laying out the nodes in a visually informative (and pleasing) manner, and (ii) determining how nodes and edges are drawn, including whether nodes and edges with different properties or attributes are drawn differently from one another.

13.1.1   Layout algorithms

The graph layout problem is that of determining the graphical positions of a network's nodes. Layouts can be determined by hand (check out the "sociograms" in Fig. 1.3) but this rapidly becomes tedious and impractical for all but the smallest networks so computer algorithms are used instead. Graph layout algorithms have been the subject of research in computer science for decades. Taking a graph, especially a dense graph, and projecting it into a 2D or sometimes 3D plane for visualization is a challenging computational problem (and is not always possible when some constraints must be satisfied; Sec. 13.1.3). Fundamentally, this projection problem is one of determining the spatial coordinates of the nodes, generally with an eye towards placing nodes in a pleasing and informative way. It is of course impossible to review all graph layout algorithms, thus we will discuss a few of the most common ones, most of which share the common idea that connected nodes should be placed close together.

> **i** A *layout algorithm* is one that determines how nodes are placed within the visualization. Ideally nodes should be arranged so they don't overlap and that edges cross each other as little as possible.

Some layout algorithms are simple and largely ignore the rationale of placing connected nodes close together. *Circular layouts*, for instance, arrange nodes along the perimeter of a circle of a given radius. The layout algorithm then determines the ordering of the nodes along the circle. Sometimes, nodes are arranged in descending order by a network property such as their degree. Other times, nodes are arranged along the circle to keep them close to their neighbors. Edges are then drawn as lines or arcs connecting the nodes.

Many other layout algorithms take a physics-inspired approach to place connected nodes close together. One simple solution is to consider each edge as a spring (or any medium of force that pull nodes together), so that connected nodes are attracted to each other while avoiding being too close. Imagining this attractive force and using a physical simulation to obtain a layout is the key idea behind the "force-directed" layout algorithm.

Usually, there are other ingredients that improve the algorithm's output. The first is repulsive forces that keep nodes from overlapping. Although this spring force prevents connected nodes from overlapping, it does not prevent unrelated nodes from overlapping. Therefore, the layout algorithm usually also considers the repulsion force between nodes by treating each node as a point charge that repels others. Second, many algorithms implement "gravity" toward the center of the visualization to keep the nodes from floating away. Finally, the algorithm can also be configured to consider the strength (weight) of the edges and allow fine-tuning of the strengths of these forces. Eventually, so the idea goes, if these forces are well tuned—neither too strong nor too weak—the graph will settle down into a pleasing and informative layout.

Although force-directed layout algorithms are probably the most common, there are useful alternatives. In particular, let us talk about approaches that leverage useful vector representations of nodes. The first is to determine node coordinates using spectral information. A matrix representation $\mathbf{M}$ of the network structure is derived, typically the graph Laplacian (Ch. 25) and the elements of its eigenvectors are used as cartesian coordinates to place nodes. In other words, if $\mathbf{u}$ and $\mathbf{v}$ are two eigenvectors of $\mathbf{M}$, then each node i is placed at coordinate (u_i, v_i). Typically, the eigenvectors used correspond to the two (or three, for 3D layouts) largest or smallest eigenvalues, depending on which matrix representation is employed. Spectral graph layouts share many similarities with spectral data clustering [478].

Depending on the matrix used, spectral layouts also possess underlying physical interpretations. For example, the graph Laplacian corresponds to a discrete diffusion operator, and so its eigenvectors form the basis of the graph diffusion equation's solution space. This means that information about the time it takes a particle to randomly diffuse across different parts of the graph is captured by the eigenvectors and some of that "spatial" information can be displayed visually in plots of those eigenvectors. We discuss the spectral theory of networks in greater detail in Ch. 25.

Another more recent alternative is to use (neural) "graph embedding" methods (see Ch. 26). As we will discuss in the later chapters, graph embedding methods aim to find a low-dimensional vector-representation of the graph (usually the nodes). Once we have this representation, a dimensionality reduction method can be applied to obtain a even lower-dimensional (2D or 3D) representation. This approach has become particularly appealing thanks to the development of sophisticated methods for graph embedding and dimensionality reduction (e.g., t-SNE or UMAP). This approach can be faster than the force-directed layout, which is essentially a large physical simulation.

Finally, there are many layout algorithms that are specialized to specific types of networks. For instance, when the network is a directed tree or DAG (directed acyclic graph) or close to a DAG, the "dot" algorithm (part of GraphViz) provides useful "hierarchical" layouts [145].

13.1.2   Drawing the network

Given the spatial coordinates of nodes, found from a layout algorithm, drawing the network becomes straightforward. Suppose node i is assigned 2D position (x_i, y_i) (or similarly in 3D) in the plotting coordinate system of your visualization. Then, i can be drawn using a graphical command, such as $CIRCLE(x_i, y_i, r_i, c_i)$, where r_i is a radius for the node and c_i is its color. Likewise, edge (i, j) between nodes i and j can be drawn with a basic line command: $LINE(x_i, y_i; x_j, y_j; w_{ij})$, where w_{ij} is the width of the line representing the edge.[1,2]

From this, we realize two things. First, the fundamental challenge is the graph layout: any basic plotting library can be used to draw a network once a layout is determined. Call the library's scatter plot functionality to lay down the nodes and draw edges with line plots. Of course, specialized tools (Sec. 13.6) can make this more convenient, and provide additional features such as interactive layouts, but fundamentally the challenging step is not the drawing but computing the graph layout. Second, we see that the graphical attributes used to draw nodes and edges do not have to be the same for each node. For instance, we can choose radius r_i to visualize a feature such as i's importance, and then this radius can vary for different nodes. Using graphical attributes to reinforce network properties in your visualization is very powerful, as we discuss below.

---

[1] Typically the edges are either drawn before the nodes or drawn at a lower "z" order so that the edges do not occlude the nodes that they connect.

[2] Edges can also be visualized with arcs or splines; see below.

13.1.3 When good network visualizations go bad

Because we must project the network down to 2D (or 3D, on occasion), there can be a significant loss of information. Fundamentally, only planar graphs—those that can be drawn in a 2D plane without any edges crossing one another—can be guaranteed to visualize without information loss in two dimensions. For other cases, if the network is too large, too dense, or both, it becomes almost impossible to create informative visualization in two dimensions.[3]

In our experience, network visualizations tend to lose utility when the number of nodes or the number of edges per node becomes too large. If the network is very sparse, the visualization will not be too dense, but if there are tens of millions of nodes, it still be hard to read despite being sparse. A scatter plot (visualizing a network without any edges is essentially drawing a scatter plot) suffers in much the same way: too many points and you cannot distinguish them. Likewise, if the number of nodes is reasonable (say, under a few thousand at most) but the number of edges is very high, the final visualization will be dense and hard to parse. That said, more advanced visualizations, including techniques we discuss below, can overcome some of these limitations. Your own judgment is an important final arbiter: are you able to convey the information you need to convey with the visualization or not?

What to do with a hairball? When confronted with a network that defies easy visualization, what options are available to you? Generally you need to either transform or subset the network in some way or use a non-network or non-traditional network visualization. Properly transforming or subsetting can reduce the complexity of the network such that a readable visualization may now be feasible. As mentioned in Ch. 10, removing dangling nodes, k-core filtering, or applying network backbone algorithms may reduce the network enough to make a useful visualization feasible. Another way is *coarse-graining* your visualization by applying community detection (Ch. 12). Consider communities as super nodes and visualize only the community-level network. Another option is to focus on the layout of the nodes and not draw edges, or treat edges in ways (e.g., making them more transparent) so that they don't obscure the overall visualization. An appropriately chosen visualization alternative can still achieve your specific goals. We describe some non-traditional visualization techniques in Sec. 13.3 and discuss manipulating network data for visualizations in Sec. 13.4.

 Fixes for networks that are hard to visualize include thinning the network, aggregating nodes into super-nodes, or choosing a non-traditional visualization.

---

[3] Occasionally, the network is so simple, like a lattice or a collection of almost entirely disconnected nodes, that a visualization will be hardly useful. It will be unlikely to earn back the space it consumes and time put into making it. However, even then, a quick visualization to explore the network may help you understand it better.

13.2   Customizing your visualization

Visualization is an iterative process. The first draft visualization, direct from a graph layout algorithm or application, may be not as informative as it can be, and can be a bit boring. It's useful for quick check-ins during exploratory work (Ch. 11) such as examining how well connected the network is as different filtering thresholds are employed. But it doesn't convey much information, at least, not as much as it could.

What are some useful actions you can take to improve both the information density and readability of a standard network visualization? You can incorporate various node and edge attributes (Ch. 9) as graphical elements. You can add diagramming elements such as callouts and other annotations on top of the network. While graph layout algorithms often produce great visualizations, sometimes a small bit of hand tweaking to node placement can greatly improve the layout (although this should never been done to mislead the viewer). You can also extend the "ball-and-stick" visuals by drawing edges as curves or splines instead of straight lines.

By iteratively improving a basic visualization along these lines, a network visualization can become rich, informative, and memorable (Fig. 13.1).

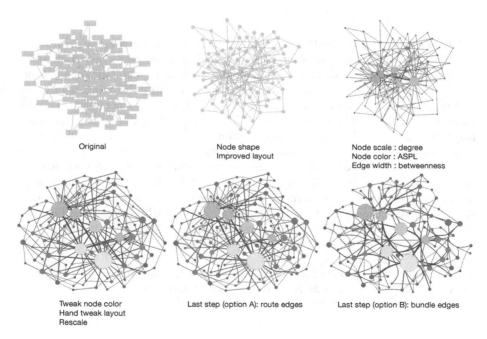

Figure 13.1   Iteratively refining a basic network visualization. An informative and memorable network visualization can be made by taking a basic network visualization, adding on the node and edge attributes as graphical elements, and, sometimes, improving the layout with small manual adjustments.

13.2.1   Reinforcing your information hierarchy

One of the simplest and yet most powerful ways to use a network visualization is
as a means to convey graphically the information about the network elements. This
information takes the form of node and edge attributes (Ch. 9). The graphical attributes
of the nodes and edges in the visualization can be mapped onto some or all of the
node and edges attributes, providing a means to convey that information directly to the
viewer.

 Use the nodes and edges in your visualization as graphical elements to "hang"
additional information from. Node size, node color, edge thickness, and more are
all available to represent further network properties or attributes.

Node attributes   Generally the nodes in the network are drawn as shapes. Visually,
besides its location within the visualization, these shapes are defined by the type of
shape (square, circle, pentagon, and so forth), the color of the shape, the shape's size,
a line defining the boundary of the same (and that line's color, thickness, and so on).
These graphical attributes can then be related to the node attributes in various ways.
For instance, a node may fall into a particular type or group (Ch. 12). Different shapes
can be used to represent different groups. Color may also be used for node attributes.
In a social network of politicians, for example, the political affiliations of the nodes
could be represented by color. Color can also be used to convey an ordinal or numeric
quantity by using a color scale. For instance, political ideology could range in strength
from right-wing through centrist to left-wing, and a color gradient fading from, say, red
to blue could be applied to the nodes' fill colors to show this information.

Edge attributes   In concert with node attributes, edges, usually drawn as lines or
curves (Sec. 13.2.3), also provide opportunities for visualizing edge attributes. Lines
are defined by their thickness and color, both of which can be associated with numeric
quantities (or categories) by defining a suitable mapping. For the widths of edges, a
basic function $w = f(x)$ can map the numeric edge attribute x to the corresponding line
width w. A linear map, $w = ax + b$, uses constants a and b to tune the minimum and
maximum thickness of lines. Nonlinear maps can also be used, such as $w = ax^b + c$ or
$w = a \log(x) + b$. These maps may overly exaggerate the distribution of x, however, so
care should be taken to ensure the data are represented accurately.

   Color can also be used to map edge attributes. A "colormap" should be defined, for
instance a gradient fading from blue to red as a numeric x goes from small to large.
A categorical x can also be illustrating with a mapping of unique colors to distinct
categories. Categorical colors can represent the types of edges, or the types of nodes
(e.g., by using the colors of the connected nodes). Another aspect is the transparency of
the color. By adjusting transparency, one can potentially mitigate edges obscuring the
nodes, especially in dense graphs. The transparency can be also tied to edge attributes.
But care should be taken when relying on color. When edge lines are too thin, color
may be hard to perceive. Some color combinations are not accessible to individuals
with color blindness, and may be lost when media do not support color, such as the

Table 13.1   Common node and edge graphical attributes and example mappings with network attributes.

Visual attribute	Variable types	Example network attributes
Node position	$\mathbb{R}^2$, $\mathbb{R}^3$	Geographic or spatial position
Node label	Categorical	Name or ID of node, group name
Node color	Categorical, continuous	Type of node, node centrality
Node size [4,5]	Continuous	Node degree, node centrality
Node shape	Categorical	Node type, group
Edge color	Continuous	Edge betweenness, edge weight
Edge thickness	Continuous	Edge weight, edge betweenness
Edge style (solid, dashed, dot-dashed)	Categorical (not recommended)	
Edge adornment	Categorical	Arrowheads for directed networks

printed page. Colormaps, if not properly designed, can introduce artifacts not present in the underlying data. Therefore, color is best used sparingly and in concert with another graphical attribute such as line thickness (Sec. 13.2.1).

Table 13.1 summarizes some common mappings between graphical and network attributes.

Choosing attributes   A graph visualization almost always benefits from adding attributes even if attributes are not present in the data. For instance, there may be no node attributes but you can always use the size of nodes to represent a network property such as their degree. This will lead to a visualization that emphasizes the degree distribution of the network, making any highly connected hubs jump out of the visualization. Or maybe instead you wish to emphasize structural bottlenecks, in which case drawing edges with high betweenness as very thick lines will immediately draw your viewer's attention. Consider also reinforcing these points by combining multiple visual attributes. Make hubs both larger and more red, emphasizing degree with size and color. Make central edges both thicker and brighter in color. Once you have chosen the message you wish to convey, you can then choose the appropriate graphical attributes to emphasize that message.

 Reinforce visual attributes with redundancy. If you use both size and color to represent the same thing, it makes the point even clearer.

---

[4] It is best to map the area to avoid misleading the viewer about the trends of a numeric quantity.

[5] The size of any node text labels can also be scaled to convey information, generally the same information as the node size itself.

13.2.2 Affordances for the viewer

Seeing a network visualization for the first time can be overwhelming to the viewer so you should take care to make their entry into your graphic as easy as possible. Options to ease the viewer's introduction will depend on the medium the visualization is shown in. In a talk or video, for instance, you can begin by showing a small piece of the network, taking a moment to describe the meaning of nodes and edges, then zoom out to show the full network, either through a sequence of slides or an animation. On the other hand, in a static medium, such as a figure in a paper, you may need to provide guidance through other means.[6] In papers, a descriptive figure caption is crucial.

> ⚠ Take care to avoid relying on visual features that are not accessible. Color blindness, for example, should be considered. Likewise, avoid overwhelming the reader with too many representations of too many attributes.

Legends   Legends provide convenient affordances for a viewer of your network visualization. With a legend, you have the opportunity to emphasize the meaning of nodes and links. You can also provide quantitative scales to interpret the attributes you added to your visualization. A color scale to represent node centrality, for instance, can be indicated with an appropriately labeled colorbar. Likewise, if the thickness of lines was used to represent a numeric edge attribute, then the legend can include a collection of lines of increasing thickness labeled with interpretable numeric values. Indeed, as you expand the number of graphical attributes in the visualization, the legend becomes more important for reinforcing your information hierarchy (Sec. 13.2.1).

Network attributes that are reinforced with multiple visual properties, such as using both size and color, should also be combined in the corresponding legend entry. Remember that the legend entries should also be easy to read and understand. We illustrate some example legends in Fig. 13.2.

Callouts and graphical emphasizers   Remember that the goal of network visualization, like any visualization, is to communicate a message via illustration. A plot can make a point clear, as can a network diagram. But don't consider yourself limited to only the elements of the plot itself or the network diagram itself. Consider adding additional text and graphics to highlight or emphasize certain aspects of the visualization. For a network diagram, consider drawing a box around the portion of the network most relevant to your point. Add a text label with some arrows calling out particular nodes or edges. Often for large networks there isn't room to label every node, but labeling a select number can be an effective way to emphasize important nodes. You can even devise a scheme to choose which nodes to label, for instance labeling the top-k most central nodes.

---

[6] Although it may still be possible using multiple figures to first show a cartoon of the network then display the full visualization, perhaps as multiple panels of a single figure.

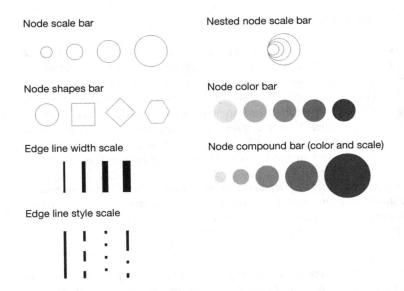

Figure 13.2 Some examples of node and link attribute legends. Figure 1.4, back in Ch. 1, shows an example of a network visualization legend.

13.2.3 Advanced visualization methods

A basic "ball-and-stick" network diagram is a great vehicle for graphical customizations and the node shapes and edge lines can be easily "overloaded" to represent node and edge attributes. Here we discuss some more advanced forms of network visualization and customization that take this even further.

Edge routing and bundling The standard network diagram uses line segments between nodes to draw edges. Straight line segments are easy to follow and often perfectly sufficient, especially when the network layout can minimize most edge crossings. But sometimes a better, more readable layout is possible by using curved lines for edges. Curved edges increase the complexity of drawing the network, as new computations will be needed to place the edges down compared to straight lines which are simply placed at the centers of the adjacency nodes once the node layout is computed.

Edge-routing methods allow edges to curve so they can move around nodes in such a way that edges cross less often. Edge routes can be computed through a variety of algorithms, often treating the problem as an optimization seeking to minimize the length of edge curves and the number of edge crossings. Edge bundling is another approach where certain groups of edges are drawn together into bundles to reduce the visual clutter of the graph while preserving its overall character. Although there are multiple methods, the most common edge-bundling method is force-directed one. To compute edge bundles, it models edges as a chain of point charges and springs, but unlike the repulsive charges used to model nodes in a force-directed layout, charges from nearby edges attract one another. These attractive forces pull together or "bundle" the edges, reducing clutter in the visualization. A key to visually appealing edge bundling is to pick which edges should interact with one another. Edge pairs that are far apart, or nearly perpendicular to one another, or where one edge is much longer than the other usually lead to unpleasant visuals if bundled together. Generally, a collection of heuristic expressions is used to determine pairwise which edges to bundle [223].

Visualizing multi-networks   Some networks consist of multiple networks. These include multilayer and multiplex networks, and networks-of-networks. One approach to visualizing multilayer networks is with a 3D stacking of 2D visualizations, sometimes called a *prism plot*. Each network layer is drawn as a separate graphic, sometimes with a foreshortened perspective added or implied to emphasize a flat planar appearance. These graphs are then arranged top-to-bottom or left-to-right to capture the notion of a stack of networks. When drawing the layers it is important to arrange them based on a meaningful ordering of the network layers. For example, an infrastructure network may have layers of increasing complexity and you can emphasize this in your visualization by stacking the low-level, concrete layers underneath the more involved abstract layers. Usually in these drawings the nodes are placed at the same relative coordinates within a given layer, so the viewer can trace changes in the network structure between the different layers more easily and without having to hunt for nodes between the layers. Both layer placement and node placement within the layers are vehicles for you to emphasize the important facets of your data and to make the viewer's job easier when deciphering your message.

Multiplex networks can also be drawn as stacked graphs, with each layer ℓ corresponding to the edges (and incident nodes) in that layer. However, it is often more common to visualize edge types using other edge graphical attributes (Sec. 13.2.1) such as edge color.

Temporal networks (Ch. 15) are sometimes visualized as multilayer networks, where each layer captures the network during a corresponding time period.

Networks-of-networks, where each node in a super-network is itself a network, are typically visualized by drawing super-nodes themselves as network visualizations. We discuss this idea below.

Of course, the fundamental challenge when visualizing multi-networks is the size of the networks. Visualizing one network is difficult if it's too big. A multilayer network consisting of 5 or 10 layers might contain 5 or 10 times more information, which may be impossible to visualize. So unless the overall network is not too big, implying that each layer of the network is not too big, it may be better to avoid visualizing the full network.

Nodes as graphical elements   While not common, on occasion, a great technique for enhancing a network visualization is to use more complex graphical elements for the nodes, instead of the basic squares, circles, and other shapes. One approach is to use a small pie chart for each node. Suppose the network was course-grained so each node in the visualization represents a community or group of nodes in the original network (Sec. 13.4). Pie charts placed at the locations of the visualization's super-nodes could then illustrate properties of the underlying group, for instance the proportions of different political party affiliations of the group members. While pie charts are generally a poor choice for visualizing a categorical distribution [110], they are very compact, so they might actually work well for nodes in a network diagram.

Besides pie charts, other graphics may be suitable for network nodes. In fact, small network visualizations can themselves be used as nodes. This may be helpful for visualizing a hierarchy or network-of-networks dataset or even illustrating relationships between small graph motifs (Ch. 12). Of course, drawing many networks can lead to a busy, hard-to-parse visualization, so this approach may be quite limited, but it can work well when there are not too many nodes in either the visualization or within the node themselves. Small motifs for instance, can sometimes be drawn quite well, as they typically only consist of a few nodes.

Animation   Animated network visualizations are powerful for showing the dynamics of a network. These dynamics can take the form of changes to the nodes or edges, such as a birth–death process or the gains and losses of edges, or they can visualize changes to any network attributes. For example, the set of edges may be fixed in time but edge weights associated with edges may vary over time. An animation could visualize this by drawing the network in each animation frame but varying the edge line widths, which represent edge weights, across frames.

Of course, animating a network is more computationally complex than preparing a single static visualization. Each frame of the animation must be drawn. You can either fix the positions of the nodes in advance and simply show the changes as each frame, or animate the whole layout process and network dynamics at the same time. The latter can be implemented as running the physical (force-directed) simulation and recording the layout at each frame while changing the network itself in real time.

For more on dynamic networks, see Ch. 15

13.3   Alternatives to "ball-and-stick" diagrams

On occasion, a network visualization task may be better accomplished using a non-traditional visualization.

One situation is when capturing dynamics in the network. While it is possible to animate the network, redrawing it over time to show how it changes, this can be done only for media that support animation, such as video. Yet in a static visualization, dynamics can still be visualized by other means. For instance, *stream diagrams*, sometimes also called *phase plots*, can illustrate the temporal dimension of the network explicitly. Here time becomes a plot axis and the network is reduced to a mostly one-dimensional representation orthogonal to the time axis. Edges are then drawn connecting points along the "network axis" for each point in time. Stream diagrams are great for illustrating specific patterns but have scalability problems. For one, a large network is generally not amenable to a pseudo-one-dimension visualization. Second, for a reasonably sized diagram, only a limited number of time periods can be drawn along the time axis. One solution is to use the stream diagram to capture the community structure of the network and how it evolves. Here one can even visualize fissions and fusions of groups with links along the time axis.

Networks as matrices   Every network structure can be mapped to a matrix representation such as the adjacency matrix $\mathbf{A}$ or an $N \times N$ similarity matrix $\mathbf{S}$. We can visualize the matrix itself, for example as an $N \times N$ grid of black and white squares,[7] which may be useful in some circumstances but has problems. Most networks are rather sparse, meaning $\mathbf{A}$ and often also $\mathbf{S}$ will be mostly empty. But beyond that, drawing the matrix as a grid requires developing an ordering of the nodes so that rows and columns can be laid out in a meaningful way. Not all networks possess a natural ordering and the same network under different orderings can produce visually very different matrix drawings. One approach to computing a consistent ordering is through hierarchical clustering. Here nodes are grouped into clusters, those clusters are grouped into super-clusters, and so forth. A tree called a dendrogram (Sec. 12.8) can be drawn to visualize this hierarchy, and ordering the leaves of the tree to prevent crossing branches can be used to arrange the rows and columns of the original matrix. Often the dendrogram and the matrix are drawn together in a visualization.

Network portraits   One particular matrix, a network "portrait," is helpful for illustrating networks as it overcomes the ordering problem. The portrait is defined based on shortest paths (Ch. 12) between all pairs of nodes. Consider a node i, called a starting node. Following along shortest paths, count how many other nodes are one step away from i (its neighbors), how many nodes are two steps away, three steps, and so forth. Now repeat these counts from each node, treating every node as the starting node in turn. The network portrait $\mathbf{B} = [B_{\ell k}]$ is defined by aggregating these counts:

$$B_{\ell k} = \text{number of starting nodes that have } k \text{ other nodes } \ell \text{ steps away.} \qquad (13.1)$$

The portrait is packed with useful summaries of network structure. The number of nodes is encoded in $B_{01} := N$, as nodes are zero steps away from themselves. The degree distribution is encoded in B_{1k}. The network diameter is captured by the number of nonzero rows of $\mathbf{B}$. Both dimensionality and regularity of the underlying network are captured by the growth and spread of nonzero values across the rows of $\mathbf{B}$. Measures for comparing networks G_i and G_j can be defined as functions of their respective portraits, $\mathbf{B}_i$ and $\mathbf{B}_j$ (Ch. 14). And most importantly for visualization purposes, the matrix is defined entirely based on the network structure and does not require an ordering of rows or columns unlike typical matrix representations.[8] This uniqueness lets network portraits serve as a stable visual representation—heatmaps of $\mathbf{B}$ are an interpretable alternative to traditional network visualizations. Even when a traditional ball-and-stick visualization is used, a portrait can be included to provide a quantitative supplemental visualization. See Fig. 13.3 for some examples of small networks and their portraits.

Multilayer alternatives   In some circumstances, it may be best to eschew the network structure itself and focus the visualization on other properties. Suppose you wish to visualize a large multilayer network with several thousand nodes per layer and 10 or more layers. It will be too much information to illustrate all those nodes and provide

---

[7] You can see some examples of this in Ch. 25.

[8] It is known as a *graph invariant*.

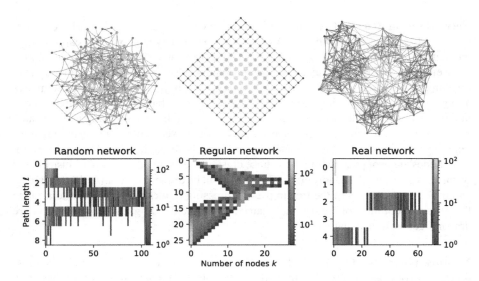

Figure 13.3   Example networks and their portraits. The random network is an Erdős-Rényi graph while the real network is the NCAA Division I football network [360]. Matrix brightness denotes the values of the portrait matrix **B** (Eq. (13.1); white indicates $B_{\ell k} = 0$). Figure from [27].

visual space for a viewer to recognize and compare different layers. But perhaps you wish to focus on some specific attributes of the network. Let's say each node in a layer belongs to a category, there are 5 or 10 categories overall, and you wish to understand how the distribution of categories varies across the network layers. You can reduce each layer to a stacked bar plot that captures the category distribution over nodes in that layer, but let's take it a step further. Let's wrap those bars around one another, transform the set of stacked bar charts into concentric rings capturing the network layers from bottom to top. Now, if there are n_ℓ nodes in layer ℓ let's draw ring ℓ as a collection of n_ℓ arcs, each colored according to the node's category. Lastly, let's arrange the arcs using a physics-based layout algorithm (Sec. 13.1.1) but with node positions constrained to move only along their arcs. This layout step will arrange the category colors such that proximity captures network structure, and can reveal patterns in both the distribution of node categories across different layers and what interplay may exist between category and connectivity.

13.4   Processing data for visualization

Network data must often be processed before it can be visualized. Typically this happens when the network itself is too dense for a 2D projection to be visually informative, but often judicious preprocessing can draw out the salient information and enable a readable final visual. Here we discuss options for filtering or transforming the network to improve its visualizability.

One can generally group data manipulations into four "verbs": filter, transform, aggregate, and sort [490]. The final verb, sorting operations, means rearranging the data in different ways. Sorting here makes sense to describe different aspects of the visualization itself, such as rearranging nodes into a more pleasing layout, but is generally not applicable to preprocessing for visualization. When specifically preparing network data prior to visualization, we can group various operations based on the remaining verbs:

Filter Remove spurious or weak links; remove singleton or low-degree nodes; window the data—remove nodes and links outside a given time period; focus on the giant connected component; focus on nodes or links meeting certain criteria such as a particular attribute value.

Transform Project bipartite network; split multilayer network; remove or ignore link directionality (if warranted); look at the graph dual.

Aggregate Merge groups of nodes into a single node (edge bundling is a kind of visual aggregation); collapse or partially collapse multilayer or temporal network.

We discuss some example scenarios.

Transform: bipartite or not Many networks follow a bipartite formulation (Ch. 1), such as a movie–actor or gene–disease network. Often it is best to visualize the network in this original form, using different shapes for the two classes of nodes. However, sometimes an effective visualization can be made by projecting the network down onto one of the two sides, usually by applying the one-mode projection. In other words, this would generate, for example, an actor–actor network where two actors are linked if they costar in a movie, or a movie–movie network where two movies are linked if they both featured at least one actor in common. When to project? Generally, if the network is not too dense following projection, and the nature of unipartite edges is more easily understood, then visualizing the projection can work well. On the other hand, if the projected network becomes very dense—which often happens—it may be best to avoid projection or find alternative ways to reduce the density of the projected network. Note that projecting a bipartite network creates a dense, clique-based network. For instance, if we project an actor–movie network onto an actor–actor network, each movie creates a clique of actors who are all connected together. Thus, it may need to be further reduced by filtering out edges. Projection will also reduce the number of nodes, and this is something to keep in mind; fewer nodes may lead to a smaller, more readable network even if the new network is more dense.

Filter: edges Suppose you have a weighted network. You try drawing every edge, but even with lines of varying thicknesses, this makes the network too dense to read. In such instances, you may use the edge weights to eliminate edges from your visualization. One approach is to define a hard threshold w_{thr} and visualize only edges (i, j) that satisfy $w_{ij} > w_{thr}$. This requires exploring different values of the threshold, of course, and defining a single threshold may not be meaningful.[9] We discuss more advanced

---

[9] One approach to defining a single threshold is to look at percentiles of the quantity you wish to threshold. For example, you may decide to retain only edges with weights at or above the 90th percentile of the weight distribution.

approaches to filtering edges in Ch. 10 that are also suitable here for preparing data for visualization.

Filter: nodes   Like edges, nodes can be filtered to reduce the complexity of the visualization. In principle, any network property or node attribute can be used to define a filtering criterion, but in practice some are more useful than others. Suppose nodes fall into two groups, A and B. Perhaps only A is worth visualizing, so you draw a network with all the B nodes (and their links) removed. Or perhaps instead you filter nodes by degree, and remove all nodes with only a single neighbor (degree $k = 1$); see the k-core decomposition in Ch. 10. Yet another approach is to filter nodes based on whether they exist in the largest connected (giant) component; drawing only the giant component is often helpful if there exist many isolates that do not need to be visualized.

Aggregate: super-nodes   Some networks are very large and it's not meaningful to draw every individual node. Instead, we can aggregate nodes into groups, then make a group–group network and visualize this new network. Now we have a drawing of "super-nodes" representing the groups of the original network. The most common approach to grouping nodes is community detection (Ch. 12). Suppose a community detection method has been used and each node i in G is assigned to a group c_i. The group–group network is then defined by taking all the nodes in G that share a group, $c_i = c_j, i \neq j$, and merging them into a single node.

Edges within a group can be ignored in the visualization while edges between nodes in different groups merged into a single edge.[10] Now we have a much smaller network to visualize. Interestingly, this "supernode" approach is at the heart of some of the most widely-used community detection methods and strategies (Ch. 12).

13.5   Emphasizing your network question in your visualization

Your visualization serves a purpose and it is worth keeping this purpose in mind as you prepare the visualization. Exploring the data and getting a rough macro-level understanding of the network by drawing it is one such purpose. Or at least attempting to, if it turns out to be too large or dense. In contrast, after you've got a good handle on the data, a visualization becomes a means for communication, and you can craft the features of the visual to emphasize the points you wish to make.

To explore a network through visualization requires keeping in mind the structural properties that can be apparent in a drawing of the network and seeing if those properties reveal themselves. One example we discussed above was whether the network is disconnected or not. When drawing the network, the connected component distribution

---

[10] Information about the original network can be retained as node and edge attributes. For example, each group (supernode) can have an associated attribute which is the number of nodes from the original network that were assigned to that group. Likewise, each group–group edge can have a weight counting how many links in the original graph were between nodes in those two groups. Either or both attribute can be used in the visualization: larger groups can be drawn as larger nodes and more heavily interconnected groups can have thicker edges.

becomes clear almost immediately: do multiple components exist and, if so, what are the sizes of the connected components? That is, is there one component containing the majority of nodes (a giant component; Ch. 12) with few (or many) very small components left over? Or is there a more even distribution of component sizes, with most components being roughly the same size? Yes, the component size distribution can be investigated from the network structure itself. But a drawing can let you quickly explore that property along with many other properties.

Properties other than connectedness can also appear in quick exploratory network visualizations. High-degree hub nodes often pop out of the drawing due to how many links they have, assuming the network is not so dense as to mask their appearance. Are there many hubs or few? And do the hubs tend to connect to other high-degree nodes or not? Whether hubs tend to link to other hubs is actually sometimes difficult to see from a visualization, as hubs tend to be uncommon and their links can be hard to trace in the presence of many other links, unless hub-links are specifically emphasized in their appearance. In contrast, it is more often easy to tell if hubs tend to connect to low-degree nodes, as they will typically (but not always) be surrounded by a fringe of low-degree or even singleton (degree one) nodes.

How else might one customize a network visualization to capture a specific question? Perhaps we are interested in associations between network properties and structure. For instance, suppose we are examining a political affiliation network, where associated with each node is a numeric left-wing/right-wing polarity score. Do hubs tend to be more right-wing or more left-wing? If we visualize the network using node size (area) proportional to degree and node color mapped from polarity score, we can at a glance tell if the hubs, large in size compared to other nodes, also tend to be similar colors. Following along these lines, suppose we hypothesize that the network is modular, with dense groups that share more links with members of their in-group than their out-group. Do we see these groups as clusters in the network drawing? If so, do nodes in clusters tend to be similar in polarity? This clustering and shared polarity can be visually easy to spot by looking for clusters of the same or similar colors.

13.6 Visualization tools

The task of generating network visualizations is common enough that a variety of software tools have been created over the years to assist us. Some tools are simple, bare-bones programs that generate a graphic only, while others provide a large graphical user interface for us to manipulate the visualization in various ways. Some visualization tools are actually larger network analysis platforms or informatics toolboxes that just happen to also provide useful visualization functions.

As with most software, the landscape of specific visualization tools is constantly in flux. Graphviz[11] is a venerable collection of command-line graph drawing tools and both NetworkX[12] and igraph[13] network libraries provide graph drawing functions. For graphical tools, those with a "click-and-drag" visual interface, at the time of writing

---

[11] https://graphviz.org

[12] https://networkx.org

[13] https://igraph.org/

(March 2023), we have had good luck with Gephi[14] and Cytoscape.[15] Cytoscape, while actually a bioinformatics application, provides high-quality network visualizations and is relatively easy to use. Many layout algorithms can be used and its "Viz Mapper" functionality provides a lot of choices to reinforce your information hierarchy, although it could stand to provide better legend affordances (Sec. 13.2.2).

Beyond specific applications, here are some useful criteria when comparing different visualization tools. The best tools should provide a variety of layout options (Sec. 13.1.1), allow mapping of network attributes to graphical attributes (Sec. 13.2.1), offer a variety of graphical export formats for saving visualization files, and ideally should allow us to manipulate the placement of nodes within the visualization, in case we need to (Sec. 13.2). Another criteria to judge a tool is performance relative to your data. If your network is large and the tool is slow, you may not be able to draw or manipulate the network. Lastly, old software rots over time, and so well-supported and actively maintained tools, ideally open source tools, are always the best to incorporate into your workflow.

13.7   Summary

This chapter toured the use of visualization techniques for network data. The common "ball-and-stick" drawing of a network is straightforward, but the key to a visualization is laying out the nodes properly and choosing the mapping between network and visual properties. Both require iteration to fine-tune the visualization.

Often networks are too dense to be drawn well, in which case you may want to filter or aggregate the network. In general, follow an iterative, back-and-forth workflow when visualizing a network. Try different layout methods and filtering steps to see what best shows the structure of the network while keeping your original questions and goals in mind.

Just as networks are a powerful organizer for connecting data attributes (Ch. 9), visualization can show those attributions with the right mapping. Consider using node size, color, or any other properties to visualize additional node or link attributes. Even network attributes such as centralities can be visualized in this way, compounding the expressiveness of the visualization by reinforcing the message.

Finally, remember that visualization is not always the endpoint of a network analysis. It can also be a useful step in the middle of an exploratory data analysis pipeline, in much the same way that traditional (statistical) visualization powers non-network data exploration. As you're working to understand a network dataset, perhaps by exploring different preprocessing steps, consider making visualizations to see the network you are getting. Even as drafts that are soon discarded, such visuals are useful for both debugging and gaining new insights into the data you are studying.

---

[14] https://gephi.org
[15] https://cytoscape.org

Bibliographic remarks

Graphs have been drawn within scientific illustrations for hundreds of years, although modern graph drawings did not appear at the origin of graph theory in the works of Euler [255]. Moreno referred to these drawings as "sociograms" [317].

Graph layout algorithms trace back to Tutte [466]. Practical interest arose at Bell Labs as communications engineers, grappling with the increasing complexity of the public telephone system, needed tools to help map out the wiring diagram (literally, in this case). This early work became the "graphviz" library, still being developed today. As with UNIX and the C programming language, science has greatly benefitted from the software developed at Bell Labs that could not be commercialized due to agreement with the United States Government.

Spectral layout methods trace back to Hall [199], while the use of physical heuristics for graph layouts originates in the work of Eades [139], then was extended by Fruchterman and Reingold into the now-classic "force-directed" algorithm. Eades considered only spring forces for edges and thought of nodes as small metal rings upon which edges-as-springs were attached. Fruchterman and Reingold [171] extended this by introducing the use of both repulsive and attractive forces. The Fruchterman–Reingold algorithm is a staple of graph drawing and remains a useful starting point for computing graph layouts.

Marai et al. [291] is a good resource for biologists looking for specific tips for network visualizations in their field.

Exercises

13.1 (**Focal network**) Make a visualization of the plant–pollinator network, using a tool described in Sec. 13.6 (or another one you've found on your own). Use different node symbols to distinguish plants from pollinators. Try several graph layout algorithms, briefly describing the results of each and report which layout seemed to best visualize the network.

13.2 (**Focal network**) Take the Malawi Sociometer Network, write a short computer program that deletes any singleton (degree 1) nodes, repeating this process if new single nodes are created until none remain, then:

    (a) Draw this network using a circular layout. Draw nodes as circles large enough to be seen in your visualization but not so large they obscure other elements of the network.

    (b) Redraw the network using a physics-based or force-directed layout. Use the size and color of nodes to represent their degree.

13.3 Draw a network of interest using a force-directed layout, but use the size of nodes to represent their degree or another centrality measure (e.g., betweenness centrality; see Ch. 12).

13.4 (**Advanced**) Take a network of interest and apply a community detection algorithm to find a new network where nodes represent communities in the original network and weighted edges represent the number of edges between nodes in different communities. Draw this community-level network using a physics-based or force-directed layout. Use size and color of nodes to represent community size and use thickness to represent edge weight.

Chapter 14

Summarizing and comparing networks

Realistic networks are rich in information. Often too rich for all that information to be easily conveyed. Summarizing the network then becomes useful, often necessary, for communication and understanding but, being wary, of course, that a summary necessarily loses information about the network. Further, networks often do not exist in isolation. Multiple networks may arise from a given dataset or multiple datasets may each give rise to different views of the same network (Ch. 7). In such cases and more, researchers need tools and techniques to compare and contrast those networks.

Summaries can fulfill this purpose: by reducing networks to summaries, we make it easier to compare one network to another. Comparing networks becomes critical when contrasting the results of a study (networks before and after treatment), when networks change over time (network in week two versus network in week one), and when the upstream task is explored (network from data processed by method one vs. by method two). It's very common to have multiple networks, or multiple versions of a network, making comparison a common task.

14.1 Summarizing networks

Given a network, how to generate a suitable summary? Answering this nuanced, complex question is our goal here. For any network of sufficient complexity, a summary must necessarily involve a tradeoff. Not all the features of the network can be contained in the summary; we must pick and choose what to leave in and what to leave out.[1] Choosing a summary first means determining the proper medium of the summary. Should it be a table of summary statistics? Should it be another (smaller) network? Perhaps a visualization of the network (Ch. 13) is a valuable summary?

Let's tackle these each in turn.

---

[1] Apologies to R. Seger.

14.1.1   Summarizing networks with statistics

The focus of Ch. 12 was on understanding the structure of a network using a variety of measures and methods. Many network statistics exist to describe the network structure. For instance, the average degree, the transitivity, the average path length, and more; even simple statistics such as the number of nodes and links. The numeric values of these statistics allow us to quantify the size, connectivity, and density of the network. A network with a high average degree (relative to the number of nodes) will be quite dense. Likewise, a network that exhibits many triangles will show a high transitivity. These quantities alone, the number of nodes and edges, the average degree, and the transitivity, while just scratching the surface, already provide a basic summary of a network:

	Nodes	Edges	Average degree	Transitivity
Zachary's Karate Club	34	78	4.588	0.2557
Plant–pollinator (bipartite)	104	81	1.558	0
Developer collaboration	679	3628	10.69	0.4475
Flavor network	376	917	4.878	0.4719
HuRI	8,272	52,548	12.71	0.05585
Malawi sociometer (weighted[2])	86	347	8.07	0.3699

By choosing a suitable set of statistics, and using the values of those statistics as the basis of your summary, you are in effect projecting networks down into a space of reduced dimension—a "summary space." Is doing so effective? It depends in part on the goals of your summary and the choices of your summary statistics.

Some diagnostics can guide you if you wish to interrogate your summary statistics. If you make a small change to the network, such as rewiring an edge (Ch. 11), how does the position of the network in your summary space change? Does it have no effect, a small effect, or does it drastically change the location of the network? Depending on the type and magnitude of the change to the network, you may expect only a small change. If so, and you observe a large change, then you may want to revisit your choice of summary. Perturbing the network in this way is a small-scale example of some of the null models used in Ch. 11, and another way to interrogate your summary is to check the same statistics on those null models.

Lastly, another question to ask is whether your set of statistics is appropriate. Do the values capture the information you need? Is something missing? Often there are connections between certain statistics. For example, the average degree is a function of the number of nodes and links in the network, and relying on those three alone captures some, but not all, aspects of the network: your summary may be information poor. Perhaps all you need to compare is the size of the network and its density, but you will often want (or need) to incorporate further information into the summary. Consider adding information on the clustering coefficient, or average shortest path length, or degree correlations. These statistics all capture further information about

---

[2] We discuss the transformation of this temporal network to a static, weighted network in Ch. 15.

the network's structure and organization. The most effective summary is not always maximally succinct, but instead maximally readable and informative.

 Consider using *network cards*, a tabular network summary and reporting tool we have proposed[3] [22]. Network cards are concise, readable, and broadly applicable to all types of networks.

14.1.2 Summarizing networks with visualizations

A visualization can be a powerful means to convey information about a given network. Many of the statistics discussed in the preceding section (and Ch. 12) are amenable to graphical plots. In addition, network visualizations are useful summaries. We dedicated Ch. 13 to the details of network visualization, including techniques and advice on how to generate useful, readable visualizations.

Chapter 13 also discusses the limitations of network visualizations and those limitations hold for summaries as well. Fundamentally, basic visualizations fail when networks become very large and very dense, forcing us to try to tease out meaning from "hairballs" or "ridiculograms."

With that said, if your network appears to be suitable for a network visualization, it is likely that the visualization will also be a useful summary of the network. At a glance, a reader familiar with network drawings will be able to glean useful information about the size and density of the network.

14.1.3 Summarizing networks with other networks

Deriving a new network from an existing network can also be an effective form of summarization. Very large networks are difficult to visualize. Their size makes it slow to draw the layout and the resulting visual may be too dense to read clearly. A smaller summary network can alleviate these problems.

One approach is to create a "super network" by exploiting the structural properties of the original network. Community structure is a natural starting point (Ch. 12). First, infer the communities using an algorithm.[4] Finding the communities means you have a mapping from each node to the community to which it belongs.[5] Use this mapping to define the super network by creating a new network where each node represents a community in the original network and these "super nodes" are linked when the nodes in the original network are linked.[6,7]

Beyond using community structure to derive a smaller summary by coarse-graining the network, other approaches are possible. An "effective network" can be defined

---

[3] https://github.com/network-cards

[4] This makes the result depend on the algorithm being used. Hopefully, the algorithm is appropriate, but this is something the researcher should determine.

[5] Or communities, if nodes can belong to multiple communities.

[6] This can also define a weighted network where the weights are the number of links between the two communities or another summary of their connections.

[7] A similar definition can be used for *overlapping* communities: two super nodes are connected when the original communities overlap by sharing nodes.

by identifying connections between nodes that capture some notion of similarity and building a network topology from that. Often dynamical information is used to find this network. For instance, a network with time series associated with nodes can lead to an effective network extracted from correlations between nodes' time series. If many time series exhibit weak correlations, the resulting network will be dense but not necessarily meaningful, so filtering the network by removing weak links can also be considered a summary using another network. However, filtered networks sit on a subset of the original nodes and therefore we don't feel this cleanly fits the idea of using a different network as a summary. See Ch. 10 for more on thinning overly dense networks.

14.1.4   Other summaries

Combinations   Don't underestimate the power of combining summaries. A small table of network statistics included within or alongside a network visualization can lead to a reinforcement effect. Bring the readers attention to the most salient details by highlighting a key detail in both the network visualization and the summary table, such as by using node size to capture node degree in the visualization and reporting the mean, median, and variance of the degree distribution in the summary table. Many such options are possible.

Model parameters   When models are fit to the network, parameters are usually used to govern the fit. Usually these are statistical parameters, such as the matrix of edge probabilities for a stochastic block model (Ch. 23) or a vector of regression coefficients for an exponential random graph model (Ch. 23), inferred by a fitting procedure. Among other uses, these parameters can serve as summaries of the network being fitted, in effect, acting as additional statistics (Sec. 14.1.1) to describe the network. An interpretable model can be especially informative as a summary, but regardless, it is important to quantify the goodness-of-fit of the model, as appropriate; a poorly fitting model will generally not be suitable as a summary.

Metadata   For networks with associated attributes, metadata describing nodes and links (Ch. 9), summaries of those metadata and using those metadata are both worth exploring. In other words, we can treat the metadata as data in its own right, and summarize it using any manner of descriptive statistics. We can also relate those metadata to the network itself, for example, reporting an assortativity using a particular node attribute (Sec. 9.3).

Qualitative summaries   Lastly, don't discount the usefulness of *qualitative summaries*. Our focus is strictly on quantitative measures, but a simple human-readable written description of the network is often helpful. We can think of this as a captioning process narrated by the researcher and guiding the reader through the key facets of the network. One of the main criticisms of qualitative methods, of course, is the potential for bias and subjectivity, so good practices should be followed whenever possible [462, 86]. In particular, it should always be clear to the reader that the summary is qualitative in nature.

14.2 Comparing networks

Comparing two or more networks is a problem closely associated with summaries. By reducing the complexity of the network down into a summary, we can define a comparison or similarity measure between two networks by comparing their summaries, often an easier task. Here we discuss why and how to compare networks.

> ⓘ Defining and measuring *graph similarity* is an integral part of network comparison.

14.2.1 Why compare networks?

On occasion, a researcher working with network data may have a single network to deal with, but this is the exception and not the rule. Of course, multiple datasets, say from different experiments, will lead to multiple networks. But even if you are working with a single dataset, in general, you can expect to work with multiple networks.

The most obvious reason for encountering multiple networks in a single dataset comes from the upstream task (Ch. 7). Working from your original data, you defined processing steps that brought you from that data to the network now before you. Were those steps correct? What does the network look like if you change part of your upstream processing? Studying these questions will lead you to alternative networks, created by varying your upstream processing. And you will now be faced with multiple networks. Comparing them can give you insight into the effects of your network processing steps.

Another reason for encountering multiple networks comes when you want to understand different parts of a single network. In a multiplex or multilayer network, you may want to compare one layer against another. For example, how does the neural network derived from an organism's electrical junctions differ from the network formed by its chemical synapses [101]? Or how does a social network extracted from mobile phone data compare to one gathered from email records? You may also wish to compare different components of a disconnected network: how do the largest and second largest components differ?

Dynamic networks are a further situation where comparison can come into play. Suppose you have data describing how a network changes over time. How does the network look at the beginning of your data window compared to the end? For instance, what's happening in the social network of students over their summer break compared to during the school year? By taking "snapshots" of the network over different time periods, we reduce the complexity of the evolving network to a collection of static networks, and comparing those networks can help us understand how a network changes over time.

(Dynamic networks are of such importance we dedicate Ch. 15 specifically to them.)

14.2.2 Comparing pairs of networks

Given two networks G_1 and G_2, a comparison can be made by several means. Comparisons may be qualitative but usually we consider quantitative comparisons, meaning we seek a numeric measure of similarity (or, equivalently, distance) between the two networks.

When nodes are consistent between G_1 and G_2, such as when $V_1 = V_2$, we can compare their edge sets directly. This can be as simple as counting the number or fraction of edges present in either network that are present in both networks (e.g., *Jaccard index*). A more challenging problem is when we do not know that nodes are the same between the two networks, but we still have options. This problem has been studied extensively in the field of *graph kernels*, which refer to a class of functions that can compute the similarity of two graphs. Here, we do not discuss graph kernels in detail but instead introduce a few simpler approaches.

Using statistics or features Earlier, we discussed summarizing networks by computing a set of statistical quantities or features. This summary can also power a
comparison measure. For network G_i, let $\mathbf{v}_i$ be a vector of statistics. Then, introduce a vector *similarity function* $s(\mathbf{v}_i, \mathbf{v}_j)$ that compares pairs of vectors. Vectors
are given a higher value of similarity by this function the more similar, or closer,
that they are. If every statistic is numeric, this could be a cosine similarity or other
common measure of similarity (or distance), but you may also find it helpful to
define a bespoke similarity function. For instance, if one of your statistics is a
binary quantity, such as a true or false attribute, you may want to build that into
your similarity function. In general, a lot of flexibility is introduced with all the
possible similarity functions; we recommend taking care in your choices and try
to pick a function that is meaningful for your application as well as simple and
interpretable, to the extent possible.

One particular feature that can power comparison is the distribution of motifs
(Sec. 12.6). First, select a set of motifs that are relevant to your use case. Then
compute the number of occurrences of each motif in each network. This gives
you two vectors of counts, one for each network, which you can then compare
against using a similarity function. (Often you want to normalize the vectors so
they sum to one or have unit "length.") Comparison with motif counts can be
quite effective [381, 427].

Using network portraits The distance or shortest path length distribution is an important summary of a network's topology, capturing a wealth of information on
the organization of the network at all scales. This distribution powers network
portraits (Ch. 13), a useful tool for visualization. But portraits can also be used
for network comparison: a similarity between two networks can be defined using a similarity between their corresponding portraits. Given that a portrait is
a graph invariant (isomorphic graphs have the same portrait) [27], comparison
using portraits is quite natural—no matter how we "label" or "order" the nodes
in the network, we get the same portrait similarity. Bagrow and Bollt [27] define
and explore a distance measure for comparing portraits.

Using embedding Similar to comparing with features, any network embedding[8] can
power a comparison (in fact, summarizing a network with a set of statistics
is *essentially* a kind of embedding). More generally, an embedding maps each

---

[8] Here we mean an embedding of the entire network, as opposed to an embedding of all the nodes, say, within a network.

network to a position in the *embedding space*. Then comparison follows by defining a similarity measure or distance between the positions of two points[9] in the embedding space. We discuss embedding methods further in Ch. 16 and Ch. 26.

This is not an exhaustive list of comparison measures, but covers most practical approaches. One missing measure is *graph edit distance*. This quantity counts how many edits (edge insertions, deletions, etc.) must be made to one network to turn it into the other network.[10] Edit distance was quite popular, but has somewhat fallen out of favor, at least for the larger networks we typically study, as computing it is far too expensive for such networks.

From any of these approaches, one can define a similarity measure $s(G_1, G_2)$ to quantify just how similar are the two networks. Usually s is non-negative, $s \geq 0$, with larger values for more similar networks, but some similarities are normalized such that $s = 1$ for identical (or at least maximally similar) networks and $s \rightarrow 0$ as network pairs become maximally different.

14.3 Clustering networks

One particular application powered by comparison measures is network clustering. Specifically, given a large collection of networks, clustering means to group related networks together. By determining the relationships between the different networks, we can learn more about the properties of those networks—what similarities or differences they have, what do "typical" networks in the collection look like, whether some networks are highly unusual, and so forth. And a network comparison measure determines whether or by how much two networks are related.

 Here network clustering should not be confused with clustering within a network, measured with transitivity or the clustering coefficient. An unfortunate collision of terminology.

Suppose you have n networks $G_1, G_2, \ldots, G_n$. For each pair of networks G_i, G_j, $i \neq j$, choose a comparison measure s and compute $s_{ij} := s(G_i, G_j)$. We assume that more similar network pairs have a higher value of s_{ij}. Otherwise, if the comparison measure is interpreted as a distance or dissimilarity measure, we can rescale or transform it in some way, for instance by $s_{ij} = 1/d_{ij}$ or $s_{ij} = d_{max} - d_{ij}$, so that larger values indicate more similar networks. We also assume that s is a symmetric function ($s_{ij} = s_{ji}$) and that $s_{ij} \geq 0$.

Given our comparison measure which we now interpret as a similarity measure, we can construct from the n networks an $n \times n$ similarity matrix **S**—this matrix is then sent to clustering methods. The similarity matrix is the key here. While any of a variety of

---

[9] Here we're talking about pairwise comparisons, but other statistics could be useful. Suppose you have a n networks, looking at the average of the positions (the *centroid*) of their n embedding points can tell us how atypical a network is.

[10] Specifically, the minimum number of edits to make the two graphs *isomorphic*.

clustering methods can be used,[11] as we discuss below, it is the choices that go into the similarity matrix that play the strongest role. As you work, consider both the effects of the clustering algorithm you are using and the features that go into the similarities.

Briefly, we discuss two possible classes of clustering methods to use on the similarity matrix.

14.3.1 Hierarchical clustering

Hierarchical clustering goes beyond finding groups to finding subgroups and super-groups, and builds a hierarchical tree (Sec. 12.8) relating, in our application, the n networks. Hierarchical clustering methods fall into two categories, agglomerative or divisive, based on whether you begin with n different clusters each containing a single point (network) that you iteratively merge together to build a tree, or you begin with one cluster containing n points that you iteratively divide apart to build a tree.

Here we describe an agglomerative method. Initially, each network is in its own cluster and our similarity matrix describes the similarities between any pair of clusters. Take the two most similar clusters (the maximum off-diagonal entry in our similarity matrix) and merge those clusters together. Now we have $n - 1$ clusters, with one cluster containing two networks. We can then repeat this step, merging the next two most similar networks, and the next two after that. But this describes merging two individual networks into a cluster of two. What happens when we want to merge two clusters that themselves contain multiple networks? In principle this may happen while we still have unmerged individual networks. In fact, it might be quite common. Imagine the two most similar networks are actually tied, or nearly tied with a third network, forming a triple of very similar networks. We may want to merge those three networks together into a cluster before we begin building clusters with the other networks. This is where the hierarchical nature of the clustering becomes so useful, giving us new information on the relatedness of the set of networks.

To use the similarity matrix to merge together clusters, and not just individual networks, we define a *linkage criterion*. First, represent each cluster as a set (an unclustered network can be considered a set of size 1). Then, for any two clusters U and V, the linkage criterion we use to determine whether to merge them is a function of the similarities of the networks $u \in U$ and $v \in V$. For example, *single-linkage clustering* uses

$$\max_{u \in U, v \in V} s_{uv}. \tag{14.1}$$

Here we look for the largest similarity among each pair of networks with one network in either cluster. At each step of the algorithm, we determine which pair of clusters has the largest criterion and we merge those clusters. Repeating these mergers then builds the hierarchical tree of networks. Thus the linkage criterion (single linkage is one of many) acts as a function mapping us from pointwise similarity to cluster-wise similarity. Single-linkage clustering is the prototypical agglomerative hierarchical clustering method.

---

[11] The prototypical clustering algorithm k-means (Sec. 25.7) is not directly applicable, however. Here we have the matrix of similarities between n points, but k-means requires the points themselves, their coordinates in a space. That said, one method we discuss will create such points which we then cluster using, for example, k-means.

Note that many common clustering methods, in particular k-means clustering, require more information than the similarity matrix. The k-means method, for instance, works by placing k new points, known as "centroids," down alongside the n points to be clustered, and then iteratively moving the centroids until they are located at the centers of k clusters. Here the actual coordinates of the n points being clustered are needed, because you are computing distances not between other points (which is effectively given by transforming the similarity matrix) but between points and the centroids, which are not present in the similarity matrix.

Such a limitation can be overcome by creating a meaningful space in which to position your networks. This is the approach used by spectral clustering, which "embeds" the networks based on eigenvectors and then applies a clustering algorithm such as k-means. Keep in mind that such an embedding may be unneeded, if you can go back from your similarity matrix to the underlying similarity measure. For example, if you defined a similarity based on a set of network summary statistics, you can keep those statistics as the "space" in which the centroids are placed and moved about. All this is to say, as you compare networks, don't throw the baby out with the bathwater.

14.3.2 Spectral clustering

Working from the $n \times n$ similarity matrix $\mathbf{S}$, spectral clustering uses its matrix properties to extract information about the clustering of points. Generally, another $n \times n$ matrix is computed from $\mathbf{S}$, and the k most "important" eigenvectors of this matrix are assembled into an $n \times k$ matrix. This new matrix represents the n networks in a k-dimensional space, and finally we can now apply a clustering method to the points in this space. Even more, the similarity matrix is often modified such that it become what is in effect another network! Nodes in this network, however, represent entire networks from the original n networks. Such an interpretation is often fruitful.

Many questions come to mind. What matrix is computed? Which eigenvectors to use? Which clustering method to use? All are worth asking. The standard approach is to threshold the similarity matrix to make an adjacency matrix $\mathbf{A}$ for a new network:

$$A_{ij} = \begin{cases} 1 & \text{if } S_{ij} > \sigma, \\ 0 & \text{otherwise,} \end{cases} \tag{14.2}$$

where σ represents some similarity threshold. The graph Laplacian $\mathbf{L} = \mathbf{D} - \mathbf{A}$, where $\mathbf{D}$ is the degree matrix, a diagonal matrix with diagonal entries $D_{ii} = \sum_j A_{ij}$ is commonly used, as the eigenvalues and eigenvectors of $\mathbf{L}$ capture lots of information about the new network we've made (Ch. 25). Taking the first k eigenvectors of $\mathbf{L}$ and making a $n \times k$ matrix $\mathbf{U}$ whose columns are eigenvectors of $\mathbf{L}$, we then apply k-means clustering to points that are the rows of $\mathbf{U}$. Several alternatives and options exist, but this is the core idea behind spectral clustering.

We discuss spectral clustering (and k-means) in greater detail in Ch. 25 (Sec. 25.7). For more information on the broader problem of clustering, please see our bibliographic remarks at the end of this chapter.

14.4   Summary

Networks are complex and we rely on tools to understand them. Summarizing such complexity, with rare exceptions, must lose information about the network. Our goal when searching for the best summary is, in effect, to lose only inessential information, where our problem or purpose controls what is and what is not essential. It is worth always keeping in mind what information is essential and what is not, to guide us towards worthwhile summaries.

To complicate matters even more, networks rarely exist in isolation and many research problems have multiple networks to study. Comparing different networks is often a key step to address a research problem. Network summaries connect directly with network comparison, as many comparison methods work by first summarizing the networks then comparing the summaries.

Bibliographic remarks

The roots of network summaries and network comparison extend back well into questions from graph theory, including graph edit distance and graph matching. Gao et al. [172] provide a useful, recent survey on graph edit distance while Conte et al. [116] review the landscape of graph matching.

Liu et al. [278] give a wide-ranging survey of network summarization. While a wealth of algorithms are available, more work is needed, including standards and best practices, when using network summaries to report on data [22].

Network comparison and graph similarity are likewise broadly studied areas. Readers may wish to consult Hartle et al. [203] for a recent survey and Soundarajan et al. [440] for useful guidance on comparing networks. Ghosh et al. [181] is a recent survey of graph kernels.

For those interested in learning more on clustering, a general and foundational problem in statistics and machine learning, we recommend Xu and Wunsch [499], von Luxburg [478] on spectral clustering in particular, and Kaufman and Rousseeuw [238].

Exercises

14.1 (**Focal network**) Take a focal network of interest and summarize it using (i) a brief, written description and (ii) 3–5 network statistics (including some beyond those used in this chapter). For the statistical summary, choose statistics that capture distinct aspects of the network's structure. Provide a brief justification for each statistic.

14.2 Define (with justification) an appropriate similarity measure using the statistics from Ex. 14.1 According to this measure, which two focal networks are most similar? Least similar? (Given the focal networks, is this question even meaningful?)

14.3 (**Focal network**) The plant–pollinator network is bipartite. A basic summary using statistics may need to be adapted for such a bipartite network. Describe a few ways to specialize a summary for a network that is bipartite. What is the most important difference you should capture in the set of statistics compared to summarizing a unipartite network?

14.4 Comparing networks using summary statistics of network features may require normalizing those features: if one feature covers a much larger range than another, it may dominate in the final distance or similarity calculation. Consider the summary table shown in Sec. 14.1.1.

   (a) Which of those statistics may end up overwhelming the others and why?

   (b) Provide a normalization scheme to correct for this.

Chapter 15

Dynamics and dynamic networks

Some networks, many in fact, vary with time. They may grow in size, gaining nodes and links. Or they may shrink, losing links and becoming sparser over time. Sitting behind many networks are drivers that change the structure, predictably or not, leading to dynamic networks that exhibit all manner of changes. The focus of this chapter is describing and quantifying such dynamic networks, recognizing the challenges that dynamics bring and finding ways to address those challenges. Dynamic network data brings along practical issues as well, and we discuss working with date and time data and file formats.

15.1 Dynamic networks and dynamics *on* networks

Let's distinguish two sources of dynamics. The first, which plays a role in network analysis but is not our focus here (we discuss it shortly), considers dynamical processes that run on the network topology. Imagine, for example, assigning a time-dependent variable $x_i(t)$ to each node i in the network. This variable tracks a dynamical process across the network where the neighbors of i play a role in how x_i changes. That is, $x_i(t + 1)$ is a function not only of $x_i(t)$ but also, among other things, $x_j(t)$, for each node j which is a neighbor of i. The network structure affects the dynamics of x by dint of the neighborhoods. Examples of such dynamics mediated by a network structure include the Kuramoto model [2], voter model [438], and many other models that fall in the class of interacting particle systems [274, 275] and, more generally, dynamical systems [375].

The second source of dynamics is that of the network structure itself. Here the graph representing the network changes with time: $G(t) = (V(t), E(t))$, meaning the node and edge sets are both time-dependent or *indexed* in time.[1] Nodes can appear

---

[1] Note that time can be discrete or continuous. For the discrete case, time increments in steps, with each timestep $t = 1, 2, \ldots$. This makes it easy to consider changes to the network. Continuous time, where t is now real-valued instead of integer-valued, is more challenging, as now we need to consider mathematically an uncountable index set.

or disappear, as can edges. Evolving network structures and organizing patterns can then manifest: triangles and other motifs can emerge and disintegrate, communities can appear, growing or shrinking over time. The overall density of the network may vary, perhaps periodically, or not.

---
i A dynamic network is also called a *temporal network*.

Many questions arise due to the network's dynamics:

1. Does the network change slowly over time, or drastically? Can we predict what the network will look like in the future, or is its evolution too chaotic?

2. What kinds of changes occur? Are changes due to changes in the node set or the edge set? Both?

3. Do many nodes appear, leading to a growing network? Do nodes disappear, leading to a shrinking network? Are nodes exclusively added over time, exclusively removed, or do both occur together?

4. How do edges changes? Do edges appear or disappear? Both? Do new edges tend to form between existing nodes? Perhaps the node set is static in time and only the edges between those nodes change?

5. How best to quantify or represent the network dynamics? Is it meaningful to simplify the dynamic network with a static network?

6. Can we use information from the dynamics of the network structure to better understand the network?

15.2 Representations

The first challenge with a dynamic network structure is how to represent it. The two most common and complementary approaches are edge-event sequences (*event representation*) and network snapshots (*snapshot representation*). A third, the *signals representation*, appears when time series are available for nodes but the network structure is absent and needs to be inferred.

15.2.1 Event representation

Here we treat the network as a sequence of edge events, where each edge event is a tuple $(u, v, t, \Delta t)$. (We assume $\Delta t > 0$.) This tuple tells us that edge (u, v) appeared at time t and disappeared at time $t + \Delta t$ (which may be the end of our data window if the edge never disappeared). The full dynamics of the network comes from a sequence of the form

$$\text{events} = \{(u_i, v_i, t_i, \Delta t_i), i = 1, 2, \ldots\}. \tag{15.1}$$

Of course, this is quite general. We may be studying data where duration is fixed across all events, for example, in which case we can omit Δt. A useful bookkeeping device is to decompose the full set of events (Eq. (15.1)) by edge or by node:

$$\text{events}_{uv} = [(t_i, \Delta t_i) \mid (u, v, t_i, \Delta t_i) \in \text{events}],$$
$$\text{events}_u = [(t_i, \Delta t_i) \mid (u, v_i, t_i, \Delta t_i) \in \text{events}].$$
$$(15.2)$$

(We use $[\cdot]$ to denote these as sequences, not sets, ordered in time.)

An equivalent way to consider the event representation is to define a network G containing every edge that appears in at least one event, then define an edge attribute that lists all the time periods $(t, \Delta t)$ when that edge occurred. The length of this attribute can vary from edge to edge, as some edges may appear in more events than others.

15.2.2 Snapshot representation

Another way to treat the dynamics of the network is to divide time into specific periods, sometimes called *windows* (see also Sec. 10.5.4), and aggregate all the edge events in a given period into a single network. This gives us a sequence of networks $G(T_1), G(T_2), \ldots$ for periods $T_1, T_2, \ldots$. We call these *snapshots* or *snapshot networks*. This representation can be derived from an event sequence by taking all the nodes and edges that occur during[2] a time period,

$$V(T) = \bigcup_{t_i \in T} \{u_i, v_i\},$$
$$(15.3)$$
$$E(T) = \{(u_i, v_i) \mid t_i \in T\},$$

and constructing a corresponding network $G(T) = (V(T), E(T))$. (Here $E(T)$ uses the set notation we describe in Ch. 4.)

A word of caution. To properly include nodes and links in the snapshot networks, you need to ensure that the period when a given node or link is present in the network intersects with the time period of the snapshot:

 Account for event duration when building snapshots. An event i that begins at time t_i and lasts until time $t_i + \Delta t_i$ may be present in more than one snapshot depending on the duration Δt_i. This can be lost if only using t_i from Eq. (15.1) to define the snapshot graphs.

While this sounds obvious, care may be needed. For example, it is naive to assume that a node or link is present in the window if the earliest time a node or link appears is before the time window and the latest time a node or link appears is after the time window—a node or link may be present for multiple periods before and after the window but not during it.

The sequence $G(T_1), G(T_2), \ldots$ invites us to consider a sequence of corresponding adjacency matrices $\mathbf{A}(T_1), \mathbf{A}(T_2), \ldots$ where $A_{ij}(t) = 1$ if i and j are neighbors at time

---

[2] We use the notation $t_i \in T$ to denote this, but a more proper notation would consider the intersection of the interval $[t_i, t_i + \Delta t_i]$ with the interval T.

t and zero otherwise, or even an adjacency "tensor" A_{ijt} where a third index has been added to capture time. And we could also consider the sequence as a single multilayer network, where each layer represents one period. While fundamentally the same, such different viewpoints on the network can help with different problems.

Of course, dividing the network into snapshots raises an important question: how to define the periods $T_1, T_2, \ldots$? Sometimes it may be obvious. In a social network, it may make sense to have one snapshot per day, or perhaps one snapshot per week. In a financial network, perhaps business quarters would justify four snapshots per year. In other data it may not be so clear, perhaps there are no obvious cycles to base your snapshots on.[3] One choice in this situation would be to base snapshots on the data: for example, define a fixed number of events n and then choose periods such that each period contains n events. Fundamentally, what has occurred is that the snapshot representation returns us to the *upstream task* (Ch. 7), and with a dynamic network we should take the same care with our snapshots as we did when extracting a static network.

15.2.3   Signals representation

Yet another way to consider a dynamic network is from a multivariate time series, or a set of N time series, one per node. This sounds like the dynamics on networks discussed above where each node i has an associated time-varying quantity $x_i(t)$, but it differs in that we do not know the network—we only know the time series. The most common example is in neuroscience, where functional imaging data gives us activity levels for different regions of the brain but nothing on how those regions are connected. From the time series, we can devise an $N \times N$ similarity matrix, using a measure such as correlation between every pair of time series. We can extract a static network from this matrix, as we discussed in Sec. 10.5, but we can also in principle extract a time-dependent network.

A snapshot representation can be extracted fairly easily: divide the time series into snapshot periods, compute a similarity matrix for each period, and extract a static network from each similarity matrix. When doing so we may want to use overlapping snapshots, creating a sliding-window effect, that may be more robust. Many other approaches exist, utilizing ideas from statistical inference and signal processing. See Dong et al. [133] and Masuda and Lambiotte [297] for overviews.

15.2.4   From dynamic to static

Often a dynamic network is too much to bear, and you may find yourself wishing to have a static network. Maybe you want to apply a particular data processing algorithm which only works for a static network. Or you'd rather begin with something simple and easy to work with before investigating the dynamics. Or you may simply be limited in capacity, with a computer that cannot handle the full dynamic network (this is

---

[3] Even in more clear situations, you may need to consider less obvious possibilities. For example, a social network that describes a workplace, such as an office building or factory, may need to be divided into snapshots based on work shifts, not work days. Likewise, you may want to make two snapshots for each week, separating weekdays and weekends. This choice leads to snapshots of varying durations, something to keep in mind.

increasingly less of an issue as computers improve in memory capacity and computing power, but networks can get big). Keep in mind that information is necessarily lost when neglecting the dynamics, and you will need to judge whether a static representation is either appropriate or sufficient for your research—it may be neither.

Probably the simplest way to generate a static network from a dynamic one is to simply take the *cumulative network* by accumulating all nodes and links.[4] This cumulative network is defined such that every node and edge that ever appeared in the dynamic network is present in the static network. This accumulation may be fine for simple networks that change slightly, but it has significant potential downsides too. For instance, paths may exist in the static network that are impossible in reality, as the edges of those paths were never present simultaneously.

Significant information can be lost in this accumulation as well. An edge that appeared and disappeared many times and an edge that appeared only once are on equal footing in the cumulative network. One obvious way to address this is with *weights*: assign to each link in the static network a weight counting how many times the link appeared (how many events) over the course of the data or, if appropriate, the total duration of time ($\sum \Delta t$) the link was present. When extracting this static network, it may need further processing. For example, many rare edges may occur. Consider using a thinning technique such as backbone extraction (Ch. 10). Many such techniques exist tailored specifically to dynamic networks.

15.3   Quantifying dynamic networks

Many network statistics discussed in previous chapters can be applied to dynamic networks. But many also need to be modified because they are either ill-suited or entirely ill-defined for dynamic networks. On top of that, we also have information on event times, which we may wish to quantify. Let's discuss some examples:

Degree We can count node degree several ways. First, we can take node i's degree to be the number of unique nodes $j \neq i$ that have at least one event with i: $k_i = |\{j \mid i, j \in \text{events}\}|$. But we may want to define a weighted sum to count for how often j is a neighbor of i. This can distinguish a node with many fleeting neighbors from one with many long-term neighbors.

We may also want to consider the time-dependent degree $k_i(t)$ by counting the number of neighbors of i during some time period denoted t. If we aggregate the events into network snapshots, we can take the degree of the node in each snapshot, essentially defining the degree as the number of edges connecting the node that activate at least once during a snapshot period. One issue here is if we are looking at many snapshots, or otherwise aggregating over small time periods, k_i may be small or even zero most of the time. A solution is to define a time-weighted average degree, like a rolling mean of degree, across snapshots.

---

[4] This is equivalent to the snapshot representation with a single snapshot covering all events; see the upper left of Fig. 10.1 for an example.

Clustering Clustering considers triangles, and now we need to consider not just whether nodes a, b, c form a triangle but whether they do so at the same point in time, a concept known as *temporal coherence*. To define the clustering coefficient for a node, we need to consider every time period where a two-path exists with that node in the center, and then ask if and for how long during that period do the three nodes form a cycle. Let T_{uv} be the set of all time periods where edge u, v exists, which can be computed from Eq. (15.1). Then the temporal clustering coefficient [263] of a node u is

$$C(u) = \frac{\displaystyle\sum_{v_1 < v_2; v_1, v_2 \neq u} \left| T_{uv_1} \cap T_{uv_2} \cap T_{v_1 v_2} \right|}{\displaystyle\sum_{v_1 < v_2; v_1, v_2 \neq u} \left| T_{uv_1} \cap T_{uv_2} \right|}, \tag{15.4}$$

where the sums run over every pair of nodes v_1, v_2 such that neither node is u and not separately counting both v_1, v_2 and v_2, v_1. From this we can define a measure like transitivity or an average of C over the nodes.[5]

Paths Temporal versions of shortest paths and related quantities such as betweenness can be defined by accounting for the time-ordered existences of edges as we move along a particular path. One subtlety is that paths in the cumulative (one-snapshot) network always over-count the number of temporally valid paths. In the cumulative network, we assume all the edges on a path exist simultaneously, but this may not be and often isn't true. A path between nodes i and j is temporally valid over a period T if we can start at node i during $t \in T$ and follow edges from i to j forward in time such that those edges exist at the times we need to move on them. We call such paths *temporal paths* and we say i and j are *temporally connected* or that j is *reachable* from i if a temporal path exists from i to j. (Note that, even for undirected networks, temporally valid paths are not necessarily symmetric, we may be able to reach j from i during T but not vice versa.) We can define the temporal path length as the number of edges traversed or, sometimes, the total duration $t_j - t_i$, where t_j is the earliest time we could arrive at node j if we leave node i at time t_j. Shortest path lengths are usually defined based on such "travel times"; several definitions exist [297].

Finding temporal paths allows us to define temporal connected components. The temporal component of i is the largest set of reachable nodes j. This gets complicated quickly, as reachability is not only not symmetric but also not transitive: Just because we can reach node v_2 from v_1 and we can reach v_3 from v_2 does not necessarily mean we can reach v_3 from v_1. It may be that the path between v_2 and v_3 occurs *before* the path between v_1 and v_2. By the time we reach v_2 from v_1, the path onward to v_3 is gone. These effects make connected components more complicated in temporal networks—for example, a node can belong to multiple connected components at the same time![6] And finding temporal components

---

[5] For an average clustering, we recommend taking an average of $C(u)$ weighted by how often u is present in the network, which can again be calculated from Eq. (15.1). Doing so ensures the contribution of u to the average depends on how often u is present in the network.

[6] See how difficult temporal networks are? We can understand the appeal of ignoring time!

becomes computationally challenging. Unlike in a static network where BFS can find a connected component in $O(M)$ time, finding all temporal components is known to be an NP-complete problem [297].

Centralities Temporal paths allow us to adapt betweenness centrality (Eqs. (12.23) and (12.24)) and closeness centrality (Eq. (12.22)). Another measure, *temporal efficiency*, similar to closeness centrality, is

$$\text{eff}(i) = \frac{1}{N-1} \sum_{j \neq i} \frac{1}{\ell(i, j)}, \tag{15.5}$$

where $\ell(i, j)$ is, for example, the shortest (or fastest) path length observed at any time. Efficiency is sometimes used in place of closeness centrality as it is less biased by disconnected nodes (which appear more often in temporal than static networks) [297].

Many other temporal centralities can be used, including time-dependent eigenvector centralities. One interesting approach is to build a "supra"-matrix that aggregates all the $N \times N$ matrices (such as $\mathbf{A}(t)$) from the snapshots into one larger matrix. The aggregation process can also be interpreted as a multilayer network and spectral properties of the supra-matrix can be used for centralities and ranking. See Taylor et al. [454] and Masuda and Lambiotte [297] for more details.

Communities Communities are especially interesting in dynamic networks. Just as the events lead to evolving sets of nodes and edges, a dynamic network can have communities that grow, shrink, appear or disappear. Different communities can merge into one, and one community can split in two or more. A temporal community detection method should not only find communities but connect together related communities based on their changes, even building an "ancestry structure" relating "parent" and "child" communities.

Methods to find dynamic communities involve adjusting modularity (Eq. (12.15)) to account for time or by using a multilayer representation aggregating the snapshots. We can also treat the snapshots as separate networks, find the communities in each, and use a "matching" algorithm to find which communities in snapshot T_s are closest to which communities in snapshot T_{s-1}. Other approaches include extending matrix factorization to tensor factorization, and applying it to the adjacency tensor with elements A_{ijt}, or extending the stochastic block model to temporal networks and performing inference of block membership. For a detailed overview of such methods, see Masuda and Lambiotte [297].

15.3.1 Change measures and change detection

With network statistics, even simple ones like degree, now being time-dependent, we may want to explore such measures as time series. For example, what does $k_i(t)$ look like over time? In effect, this question points us to a time series exploratory data analysis (Ch. 11), where we look at the networks and their structural measures (Ch. 12) over time.

Further, suppose the network is relatively static until some particular time τ when it suddenly changes. Can we detect this *changepoint* from our data? If the network suddenly became very dense, this would be evident in a sudden increase in $k(t)$, but what about other structural changes? For those we may wish to use a network comparison measure (Ch. 14) and compare adjacent snapshots.

Changepoint detection—and the related question of anomaly detection—is often formulated as a statistical inference problem. The strategy is to build a statistical model whose parameters θ depend on a "switchpoint": $\theta(t) = \theta_1$ if $t \le \tau$; θ_2 if $t > \tau$. We can then perform a hypothesis test on whether $\theta_1 \ne \theta_2$ or we can build a Bayesian model and look at the posterior distributions for θ_1 and θ_2 (the posterior of τ even tells us when the change occurs). Masuda and Lambiotte [297] discuss network changepoint detection in more detail.

15.3.2   Time-dependent attributes—dynamics on networks

For the most part, we are focusing on networks that are themselves dynamic, the nodes and edges change. But we can also consider dynamics on top of a network. Consider a dynamical process $x_i(t)$ associated each node i that evolves in time. Often this is something network scientists consider as a mathematical model; $x(t)$ could come from a simulated epidemic running over the network, for example. But we could also have data for such a process. How best to consider it?

For a static network, we recommend considering $x(t)$ as a node attribute (Ch. 9) that varies with time. We can treat each node's $x_i(t)$ separately, and quantify it using any time series or other appropriate statistics. We can also compare between nodes $x_i(t)$ and $x_j(t)$. We can make this comparison at specific times or across time using any manner of summary statistics and comparison measures, such as the correlation coefficient between x_i and x_j. If we focus on nodes that are adjacent, we arrive at an assortativity measure (Sec. 12.5). We can either measure the correlation between x_i and x_j at each time t, giving a time-dependent assortativity across edges, or we can devise a time-independent similarity between x_i and x_j using all times and then incorporate that similarity into an assortativity.

An especially interesting, but also challenging, situation is when we have both dynamics on networks and dynamics of the network. Now we need to understand how $x(t)$ evolves in time on a network that is itself dynamic, no mean feat.

15.4   Null models

Chapter 11 describes ways to understand a static network structure by comparing it to null models, often randomizing the original network while preserving its degree using monte carlo (the edge exchange method) or by building a new network from scratch using the degrees of the original network (the configuration model). Such methods are applicable to dynamic networks as well, but we can do even better by accounting for the temporality of the network in our null.

If we suspect time plays an important role in our network, a temporal null model, chosen appropriately, can dampen or remove the effects of time through randomization.

Considering the event representation, we can randomize events across links or random-ize the order events occur on links. And while doing so we may wish to preserve various quantities, such as the number of events on a given link or the time between events.

Some possible null models:

Random times A simple randomization is to distribute the same number of events but uniformly at random over the full time span of the network, call it the interval $[0, t_{max}]$ or, if we wish to preserve the times when edge u, v exists, the interval $[\min(\text{events}_{uv}), \max(\text{events}_{uv})]$. This preserves the number of events, and the time span of the events but otherwise breaks all other statistics. For one, the distribution of *interevent times*, or waiting time between events, is not the same.[7]

Interevent shuffle Here we explicitly control for the times between events by shuffling the interevent times. If we think of the time span of the edge, $[\min(\text{events}_{uv}),$ $\max(\text{events}_{uv})]$, as being divided up by the events into time segments that add up to the full time span, we simply shuffle those segments. Now the interevent times are fixed, but when those interevent times begin and end is random. Note that doing this requires fixing the first and last interevent time, which preserves the first two and last two events. If there are four or fewer events, you won't be able to carry out this shuffle.

Across-edge shuffle Here we preserve when events occur but randomize which edge they belong to. One way to do this is to build two lists from the events, a list of edges and a list of times, randomly permute the list of times, then put the list of edges and times back together to create a random events list:

$$\text{events}_{null} = \left\{ (u_i, v_i, t_{\sigma(i)}, \Delta t_{\sigma(i)}), i = 1, 2, \ldots \right\}, \tag{15.6}$$

where $\sigma(i)$ in the subscript of t represents a random permutation across all event indexes $i = 1, 2, \ldots$.

Edge exchange The temporal analog of edge exchange, here we exchange the endpoints of edge pairs in the sequence of events, and randomly distribute the original edges' events between the new edges.

Complicating matters further, we can also apply multiple randomizations one after the other, testing even more the role of temporality by destroying it further in our null. In fact, a broader mathematical framework classifying these and other temporal nulls as different levels of a graph *ensemble* (Ch. 22) was recently proposed [175].

What about snapshots? While we've discussed null models in terms of the event representation, where it's generally easier to envision the nulls, the same procedures can be applied to the snapshots by first randomizing the events, aggregating them into the snapshots $G_{null}(T_1), G_{null}(T_2), \ldots$ and then proceeding with further analysis. Depending

---

[7] In fact, this randomization turns the event sequence into a Poisson process, where events occur at random and without memory, meaning the probability for a new event to happen does not change given when previous events happened.

on what the null preserved, we can even compare the real and random snapshots on a per-period basis ($G(T)$ vs. $G_{\text{null}}(T)$) using some techniques outlined in Ch. 14. That said, if you only have the snapshot networks and not the event times, you can still randomize the $G(T)$ individually but information such as interevent times will not be available, so not every null model can be used.

15.5   Visualization

15.5.1   Dynamic and static network visualization

Visualization can show how a dynamic network changes. One option is to create an animation (Sec. 13.2.3), drawing the network and then changing the drawing as the network changes, but animated visualizations are limited to certain contexts such as webpages or slideshows. A static network visualization can still be used on dynamic networks in a few ways. One is to build a larger network that contains one or more snapshots and visualize those. One can draw several network snapshots next to one another, with common node positions across time—essentially, treating time as the layers in a multilayer network and drawing a prism plot (Sec. 13.2.3). This may work well for a small network but a viewer may struggle to scan across snapshots if the network is too large.

Another way to incorporate temporal information into a static visualization is by mapping time-related quantities to graphical attributes. For example, node color can be used to represent age, the time when the node first appeared in the network, with, say, blue nodes being very young, and red nodes being old. Link age can also be mapped likewise. We can even combine these ideas with the previous approach by visualizing multiple snapshots using graphics for various temporal attributes. Some creativity can go a long way.

 Consider using node and link attributes based on temporal quantities in a static visualization of a dynamic network.

As you can imagine, however, it's easy to overdo it. Static visualizations can struggle when too much is happening. The network could be very large, or change very quickly, or have lots of different changes occurring simultaneously—it can be hard to capture when nodes or edges are removed as well as added. As with most visualizations, we have to fight against overloading the graphics and rendering them unreadable.

One possibility to overcome this is to tailor a *changepoint* visualization. Pick two time periods[8] and look at the difference in the networks. How many nodes exist in the first period but not the second? How many nodes exist in the second period but not the first? How many were in both? Then draw the network as the union of the network structure in the two periods, showing all nodes, including those that were born and those that died. Pick some visual attributes to map the node status (born, died, survived, etc.). From the visuals[9] we can then see how much change may have occurred and

---

[8] It may be possible to extend this beyond two periods but comparison becomes difficult and it may be too hard for the viewer to interpret the final product. Care is needed when balancing complexity and readability.

[9] And visuals can and often should be complemented with quantitative analysis.

if that change followed a pattern. Perhaps neighbors of newly created nodes tend to die, indicating a spreading process or at least some network association between the addition and removal of nodes. Perhaps nodes that survive the longest tend to be high degree, or low. By focusing not necessarily on the full dynamics, but on one point, we can narrow the complexity of the visualization and tailor it for what is (hopefully) the most important moment in the network's dynamics.

As with the snapshot representation itself, the key to this changepoint visualization is determining the two time periods. As a researcher you may have a specific hypothesis in mind that can clue you into which periods to look at. But otherwise you may wish to do some exploration first. For example, vary the duration of the two periods and slide them along your data window. As you do, examine a network property such as the density of the network in the two periods or how much the networks differ. Changepoint detection can also help you find the best snapshots to use. Network differences can be measured by looking at similarities in nodes and edges; see also Ch. 14 on comparing networks. See Ch. 10 for techniques specific to windowing data.

15.5.2 Other visualizations

Beyond creating node and edge attributes related to time and generating a static network visualization, non-network visualizations are also often helpful. Any kind of time series visualization may be appropriate, especially if working with the signals representation. Most network statistics are now time-dependent, so even basic plots of $\langle k \rangle$, number of nodes or edges, and more as functions of time may be helpful. Visualizing the distribution of Δt may be useful, or even the number of events over time. Some more involved visualizations can help certain analyses. For instance, if you have found temporal communities, then an "alluvial" diagram (also called a Sankey diagram), showing the communities as flows, can highlight their sizes over time.

Space-time matrices   We have found one uncommon non-network visualization to be quite helpful, which we call, somewhat humorously, the *space-time matrix*. For a dynamic network consisting of $T_{\max}$ snapshots, $G(1), G(2), \ldots, G(T_{\max})$, let M and N be the numbers of edges and nodes that appear in at least one snapshot:

$$M = \left| \cup_{t=1}^{T_{\max}} E(t) \right|, \quad N = \left| \cup_{t=1}^{T_{\max}} V(t) \right|. \tag{15.7}$$

The space-time edge matrix $\mathbf{E}$ is an $M \times T_{\max}$ matrix with one row per edge and one column per snapshot. Likewise, the space-time node matrix $\mathbf{N}$ is the $N \times T_{\max}$ matrix with one row per node and one column per snapshot. The entries E_{eT} and N_{vT} of these matrices equal the number of events involving edge e or node v, respectively, during snapshot period T:

$$E_{eT} = |\{\text{events}_{uv} \mid t_i \in T; (u, v) = e\}|,$$
$$N_{vT} = |\{\text{events}_v \mid t_i \in T\}|. \tag{15.8}$$

In other words, the matrices tell us how much activity is happening across the elements of the network over different time periods.

We find plotting these matrices, especially **E**, to be fascinating. Let's look at them for the Malawi Sociometer Network, a temporal focal network.

This undirected network came from contact observations between participants wearing sociometers, small sensors that detect and record when they are in proximity with other sensors. Some data filtering was performed by Ozella et al. [353] after data collection and we are left with observations of edge events at $\Delta t = 20$s over a 14-day period. These data follow the form of Eq. (15.1) without Δt readings. For the most part, until now, when working with this network, we have reduced it to a static, weighted network by creating a snapshot representation with a single period. But now, let's explore its temporality.

We construct daily snapshots and proceed to build **E** and **N**. Looping over the edge events per snapshot, we count the number of events involving edge u, v and node u, populating the entries of **E** and **N**. We order the rows of these matrices by the first appearance times of the corresponding edges and nodes: $\min (\text{events}_{uv})$ for edge u, v and $\min (\text{events}_u)$ for node u so the first edge that appears is the first row of **E**, the second edge that appears is the second row, and so forth (and likewise for **N**). We similarly order the columns of **E** and **N** by snapshot period (day). Figure 15.1 shows the final matrices.

Information about the network immediately jumps out at us in Fig. 15.1. We see that about one-third of edges, all appearing on day 1, often frequently reappear throughout the remaining snapshots. The remaining two-thirds of edges, meanwhile, which tend to

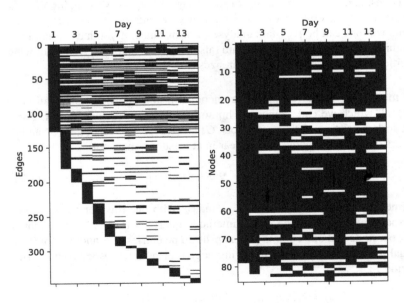

Figure 15.1 "Space-time" matrices for the Malawi Sociometer Network. Note that we have binarized the matrices into zero and nonzero entries, for ease of printing. A bit more information is available using a color scale for the matrix element values, but even the binarized view is very informative.

appear after day 1, tend not to reoccur on later days. In other words, **E** clearly shows a kind of temporally "unrolled" core–periphery structure. Fascinating!

Figure 15.1 also shows **N**. The view we see is less exciting, but still important. Most nodes appear immediately on day 1 and persist, appearing in most subsequent snapshots. This tells us that the fleeting edges in **E** generally lie between preexisting nodes; if **N** showed a pattern like that in **E**, the network would be growing over time.

Of course, this view doesn't capture all possible temporal information, how could it? For one, among the two-thirds of fleeting (peripheral) edges, it is not obvious which nodes they connect to just from the matrix. This could be investigated. In general, we should expect many other patterns to appear in these data. Visualizations and quantitative analyses using some of the techniques described in this and other chapters, can extract such information.[10]

15.6 Further considerations

Some practical details specific to dynamic networks and their data warrant further discussion.

15.6.1 Storing dynamic network data

Many data formats can store a dynamic network (see also Ch. 8). We recommend a simple format, a temporal edge list. Each line of a temporal edge list takes the form

```
node <coldelim> node <coldelim> timestamp <rowdelim>
```

where `<coldelim>`, usually a comma or tab character, separates columns, `<rowdelim>`, usually a newline, separates rows, and `timestamp` indicates when the link occurred.

When links exist over durations we recommend using a start timestamp and either a stop timestamp to indicate the duration, or the duration itself. (Links that appear and disappear multiple times would have multiple rows in the file, with an associated start, stop duration for each appearance.) This is equivalent to recording Eq. (15.1).

Notice in this representation that augmenting the edge list with time information maps nicely to the edge attributes[11] discussed in Ch. 9. Essentially, time is another source of edge attributes with which we can describe the network.

15.6.2 Times and timestamps

Working with network data requires working with underlying data, and that often includes times and timestamps. Time data are no fun to deal with. How can you tell in a computer code that "12 Dec 2012," "December 12th, 2012," and "12/12/12" are all the same date? How can you deal with time zones? Daylight saving time?

Timestamps, written recordings of dates and times, usually take one of two forms. The first, ***epoch-based***, defines an "epoch" reference time and a unit of duration, then each time becomes a count of how many time units since the epoch. The UNIX "Epoch"

---

[10] Interested? We apply a statistical model, the *edge observer model*, to these data in Ch. 23.

[11] Or, more precisely, multi-edge attributes.

time is one example, measuring the number of seconds that have elapsed since 00:00:00 UTC, 1 January 1970.[12] If you see timestamps in your data that are recorded only as numbers, you probably have an epoch-based timestamp.[13] The second form of timestamp, *language-based*, follows natural language records of time, such as the "December 12th, 2012" or "12/12/12" examples we discussed.

An epoch-based timestamp format is easy to deal with computationally once you know the epoch time and the duration unit. But these two facts are not readily contained in a list of numbers. Does "34095" mean 34 thousand seconds or 34 thousand minutes? Hopefully, from context you can rule out some possibilities, but ideally the data you are working with properly document these two facts.

A language-based timestamp, on the other hand, is more challenging to work with computationally, as you need to parse the text of the timestamp into a data structure used by your code. All modern programming languages come with support for parsing timestamps and performing operations on dates and times. The challenge really comes when the language-based timestamps are messy, not following a single written form. If every timestamp is written in the form "12/12/12" you may be fine (assuming you know it's always day first or month first). But when different timestamp formats show up, which often happens with data produced manually by people or data coming from different sources, you may need to produce code that handles multiple formats. Some libraries exist for inferring timestamp formats in an automatic fashion, but it is best not to rely on them if you can.

If you are producing data with timestamps, we recommend taking a hard stance and always use the international standard for times and dates, called *ISO 8601*. This standard records dates as YYYYMMDD or YYYY-MM-DD using four-digit years and zero-padded two-digit months and days (example: 2012-12-01), and times as Thhmmss or Thh:mm:ss using a "T" to denote time and a 24-hour clock (example: T16:20:00). Dates and times are combined using <date><time> (example: 2012-1201T16:20:00). Time zones can be included as time offsets from universal time (UTC) (example: 2012-12-01T16:20:00-05:00). While verbose compared to some formats, this standard is generally the most supported and least ambiguous. It also has the attractive property that alphabetically sorting timestamps in this format will also sort them chronologically.

15.7 Summary

Dynamic or temporal networks have special circumstances both in how you prepare the data and in what you do with the generated network. Note also the distinction between a network whose structure is changing (dynamics *of* networks) and a network with a static structure where a dynamic process is running on top of that structure (dynamics *on* networks). It can be acceptable to neglect temporality, avoiding much complexity and nasty issues, but sometimes we have no choice but to confront the mess of time if we want to truly understand our network.

---

[12] Wonderfully, UNIX time excludes leap seconds. What could possibly go wrong?

[13] Assuming they really are timestamps and aren't confused with elapsed times or durations.

Steps to deal with temporal networks include quantifying the dynamics or finding static representations. The former remains a developing field with many approaches but no broadly accepted standards; the latter is often necessary when you wish to apply a network analysis method that does not account for temporality.

Many dynamic networks come from time series data and some time series may require processing natural language date and time information. Use libraries to assist with this processing as much as possible; when generating your own datestamps and timestamps, favor the ISO 8601 standard.

Bibliographic remarks

Thorough treatments of dynamic networks, particularly theoretical aspects, are given by Holme and Saramäki [221, 222] and Masuda and Lambiotte [297].

For those interested in learning more about time series analysis in general, we recommend Kirchgässner et al. [243], Shumway [430], and, for time series forecasting, Hyndman and Athanasopoulos [227].

Exercises

15.1 Given two snapshot networks, $G(T_s)$ and $G(T_{s+1})$ define a simple comparison measure to quantify how similar they are. (*Hint*: unlike most of the comparison measures discussed in Ch. 14, here we can safely assume that nodes are consistent between the networks.) Briefly describe the intuition behind your comparison, what it measures and why.

15.2 (**Focal network**) Devise daily snapshots of the Malawi Sociometer Network network.

  (a) Which snapshot is densest?

  (b) Which snapshot changed the most compared to the preceding snapshot?

15.3 (**Focal network**) Compute the *total appearance time* for each node in the Malawi Sociometer Network, defined as the elapsed time between the earliest and latest appearance of that node.

  (a) How many nodes are "long-lived" and how many are "short-lived"?

  (b) What is the mean and median appearance time for nodes?

  (c) What does the distribution of total appearance time look like? Make a histogram or ECDF and interpret.

  (d) Total appearance time may not be the best measure for tracking a node's lifetime. What information is lost when using total appearance time and what may be a better measure?

15.4 (**Focal network**) Repeat Ex. 15.3 but for the total appearance time of *edges* instead of nodes. Compare node and edge quantities.

15.5 (**Focal network**) *Link prediction I*: Divide the Malawi Sociometer Network into two snapshots, G_1 and G_2, where G_1 covers the first half of the network's time span and G_2 covers the second. Use the following baseline link prediction rule: Predict that a link exists in the second snapshot if it existed in the first snapshot; otherwise, predict it does not exist.

    (a) How many predicted links were actually present in G_2? How many predicted links were absent? How many links not predicted to be in G_2 were actually present?

    (b) Interpret the "predictability" of the network based on how well or how badly the baseline link prediction rule works.

15.6 (**Focal network**) *Link prediction II*: Divide the Malawi Sociometer Network into G_1 and G_2 as in Ex. 15.5.

    (a) Use the first snapshot to compute a link prediction scoring function (Sec. 10.6.2). Apply the scoring function to each pair of nodes.

    (b) Use the second snapshot to validate (how?) the scoring function.

    (c) Repeat your analysis but with two snapshots, where the first snapshot covers every day but the last day, and the second snapshot is the last day. How do your results change compared to the 50/50 split?

Provide a brief write-up describing the details of your work on this exercise.

Chapter 16

Machine learning

We live in the era of machine learning. Machine learning, also called statistical learning, is the science and technology of building predictive models that generalize from data. The advancement of statistical learning and deep learning, supported by a deluge of data and computing power, has created enormous progress: now computers can identify objects in a photo more accurately than humans; computers can diagnose some diseases and predict health outcomes as accurately as doctors; computers can draw photo-realistic paintings from simple written prompts; computers can write documents and answer questions in natural language. Machine learning has not only disrupted many industries, it's revolutionizing how science is done. Problems that seemed to defy solution, such as protein structure prediction, have been effectively solved by a collaboration between scientists and machines.

Because of its ubiquity, network data are also heavily leveraged by machine learning. Thus, scientists working with network data (and all forms of data) can benefit from the tools and techniques of machine learning, as we explore in this chapter. There are several types of machine learning methods: supervised learning and unsupervised learning are the primary categories and then there are self-supervised learning, reinforcement learning, and more. Because the space of machine learning is vast and the book's focus is not machine learning, we will focus only on a subset of machine learning problems and techniques that are most relevant to network data.

We begin with an overview of machine learning problems common to network analysis before proceeding into the background and practice of machine learning.

16.1 Common network machine learning tasks

Although numerous machine learning tasks exist with network data, there are several fundamental tasks that cover a lot of our use cases. Traditionally, these tasks were tackled by designing network properties (Ch. 12) and using those properties as "features" for traditional, general-purpose machine learning models like logistic regression. However, with the advancement of neural networks, most state-of-the-art methods use graph embedding or graph neural networks, methods designed specifically for network data and which often determine features automatically.

251

16.1.1 Machine learning with *and near* network data

Typical machine learning problems (such as classifying spam emails) and the general methods employed to address them (Sec. 16.2) are not particularly connected to network data. Nevertheless, general-purpose methods are helpful to the network researcher in a variety of ways.

For one, non-network data may need to be analyzed during the course of a study. A computational humanities researcher, for instance, may wish to apply classifiers to written documents describing the nodes of her network. Or a systems biologist may want to use dimensionality reduction on gene expression data as part of the upstream task of extracting the network. These cases work with data *near* the network, but in the traditional guise of supervised or unsupervised learning. Second, many network prediction tasks we describe below such as link prediction can actually be addressed using non-network-specific methods or a combination of network-specific and general-purpose methods. Sometimes, it may be better to use methods tailored to network applications, but not necessarily always. Many state-of-the-art network methods require large amounts of data (or large numbers of networks) and if your networks don't yield enough information, you may actually have better results with general-purpose methods.

If predictive models will be part of your research, it is best to have a good grounding in how they work and how they fail. And be sure to have good data hygiene (Sec. 16.6).

16.1.2 Node classification

One fundamental prediction task for network data is "node classification," where we need to predict a given node's class or label. For instance, we can imagine an online communication network of people where each person may belong to one of (or none of) multiple political parties. The task is to predict the party affiliation (label) of a person based on information about them (node features) as well as that of people around them (network features).

16.1.3 Link prediction

Another common task is "link prediction," as discussed in Sec. 10.6.2. Link prediction refers to the task of predicting missing links from a network data. Link prediction is ubiquitous when dealing with network data because the data is rarely complete and perfect (Ch. 10). One of the most direct examples would be the suggestions that social media platforms make. On Twitter, Facebook, and other social media services, you can see the feature where the service suggests some people to follow or send a friending request. These suggestions are made through link prediction.

At the simplest level, it makes use of the network structure around each pair of people. Let's say Alice and Bob are not friends on Facebook, but share 20 friends (they are all friends with both Alice and Bob). In such a case, it is reasonable to assume that Alice and Bob may know each other or at least know about each other through their shared friends. Unconnected users with many common friends are more likely to be connected (or *become* connected) than users with few or no common friends.

Link prediction is, however, not confined to social media services. Many problems related to network data can be formulated as link prediction problems. Biological networks are rarely "complete" and a large fraction of links between biological elements are missing.[1] For instance, protein–protein interaction networks are constructed with various experimental and data-mining methods and the overlap of links probed by different protocols is not that high, indicating that most links are never seen. Predicting likely missing links is therefore a very important problem that can guide biologists to discover new interactions.

Another example is the protein folding problem, where we need to predict the 3D structure of proteins from their sequences. To do this, we should be able to predict the long-range interactions between amino acid pairs that are located far apart in the sequence. One of the breakthroughs of DeepMind's AlphaFold is exactly this—being able to predict the links between distant amino acid pairs.

16.1.4 Community detection (clustering)

Another fundamental machine learning task is community detection (Sec. 12.7). It is a form of the clustering problem, a canonical unsupervised learning task. The goal of community detection is to find groups of nodes that are densely connected to each other while being sparsely connected to the rest of the network. A common class of community detection algorithms define a quality function (e.g., modularity) and then optimize it. Another approach, covered in Ch. 23, uses statistical inference: a statistical model is first defined that captures the relationship between the observed network structure and the communities and then the best parameters of the model (or their distribution) are estimated from the data.

The community detection problem is closely related to both the node classification and link prediction problems. A node classification task may be formulated as a community detection problem where node labels represent the communities. A link prediction task may be formulated as a community detection problem because it is usually assumed that the nodes in the same community should have a higher probability of being connected to each other than another pair that belong to different communities.

16.1.5 Graph representation learning (embedding)

Finally, there is a "meta" task, learning representations (Sec. 16.3) for a network. This is usually called "graph representation learning" or simply "graph embedding." In this task, rather than making specific predictions, we want to obtain a generally useful representation of nodes, edges, or even the whole network. Learning good representations is an essential part of machine learning and often good representations can automatically addresses downstream tasks. For instance, excellent performance can often be achieved by simply applying good graph embedding methods to obtain node embeddings and then combine those embeddings with simple machine learning methods (e.g., logistic regression). We discuss graph embedding in Sec. 16.7.

---

[1] We discuss techniques to estimate how many links are missing in Ch. 24.

In addition to the network itself, we often have attributes, additional data describing the nodes, links, or both (Ch. 9). These data can themselves be used as features for machine learning but recently, new methods have been devised that are able to learn the representation of nodes by *combining* attributes with structural information from the network. The idea is that we can train a neural network to effectively combine and aggregate the node attributes through their local network neighborhood to make use of both structural (graph-level) and local (node-level) information. These methods are usually called "graph neural networks." We will examine this approach in more detail in Ch. 26.

16.2   Supervised learning

Tom Mitchell, one of the pioneers of machine learning, broadly defines machine learning as [312]:

> A computer program is said to learn from experience E with respect to some class of tasks T and performance measure P, if its performance at tasks in T, as measured by P, improves with experience E.

The most straightforward type of machine learning that perfectly fits this definition is *supervised learning*. In supervised learning problems, such as labeling an email as spam or not, we have examples of what we wish to predict, called *training data*, that can guide our "learner" or model as it is built. The training data should have both the "hints" and the "answers"; the "answers" are called *labels* (or "ground truth") and they are what we wish to predict (e.g., whether a given email is a spam or not) and the "hints" are (usually) called *features*, the measurements that we can use to predict the label.

The pair of known hints and answers "supervise" the training of the model to improve its predictive abilities. Some examples of supervised learning problems and their training data include:

- predicting whether an email is spam or not after being given a large body of prelabeled spam and non-spam messages,

- classifying the type of animal present in a photograph after being given a large body of prelabeled animal images,

- forecasting the future value of a stock given the trading histories of all stocks in a market.

The general supervised model can be described as follows. Let $\mathbf{X}$ be an $n \times p$ data matrix where each row, sometimes denoted $\mathbf{x}$, represents a datapoint (an observation) and each column represents a feature (or variable). Let $\mathbf{Y}$ be an $n \times 1$ column vector called the target. The target $\mathbf{Y}$ contains our labels and is available in our training data, but not in general, and this is what we wish to predict using the information in $\mathbf{X}$, assuming that $\mathbf{X}$ can inform us enough to predict $\mathbf{Y}$ with some accuracy. In general, the link between the data matrix and the target is given by

$$\mathbf{Y} = f(\mathbf{X}) + \epsilon. \tag{16.1}$$

Here $f(\mathbf{X})$ is an unknown, row-wise function of $\mathbf{X}$ and ϵ is a zero-mean, random noise term independent of $\mathbf{X}$. Because ϵ has a mean of zero,[2] we can use the mean of $f(\mathbf{X})$ as a replacement for $\mathbf{Y}$: when the target is not available we predict it using information from the features:

$$\hat{y} = f(\mathbf{x}), \tag{16.2}$$

where $\hat{y}$ is the prediction for a datapoint and $\mathbf{x}$ is a vector of features describing that datapoint. The fundamental challenge, however, is that f is unknown. Supervised learning proceeds by finding a functional "proxy" for f, denoted $\hat{f}$ (read "f-hat"):

$$f(\mathbf{X}) \approx \hat{f}(\mathbf{X}; \theta). \tag{16.3}$$

Here $\hat{f}$ is a general purpose function that includes a vector of fit parameters or coefficients θ that we can alter to better approximate f. "Learning" is the process of modifying these parameters to improve the approximation using training data. The proxy function $\hat{f}$, or simply the *model*, can be any of a wide range of models, from linear regression equations to deep neural networks.

16.2.1 Regression and classification

There are two general types of supervised machine learning problems: *regression* and *classification*, determined by whether the target $\mathbf{Y}$ contains continuous/numeric (regression) or discrete/categorical entries (classification). The most basic method to perform a regression task would be *linear regression*, and the most basic method to perform a classification task would be *logistic regression*.

From Eq. (16.1), linear regression can be written as

$$\mathbf{Y} = \mathbf{X}\beta + \epsilon, \tag{16.4}$$

where $\mathbf{X}$ is a $n \times (p+1)$ matrix with each row as features with a constant term (1) added to each row (hence $p+1$ columns), β is $(p+1) \times 1$ column vector called *parameters* or *coefficients*, and ϵ is a zero-mean random *error* term or *noise* that should be independent of $\mathbf{X}$. The extra constant column is added to represent the intercept coefficient β_0 of the linear model. In other words,

$$\mathbf{X} = \begin{pmatrix} 1 & x_{11} & \cdots & x_{1p} \\ 1 & x_{21} & \cdots & x_{2p} \\ \vdots & \vdots & \ddots & \vdots \\ 1 & x_{n1} & \cdots & x_{np} \end{pmatrix}, \beta = \begin{pmatrix} \beta_0 \\ \beta_1 \\ \vdots \\ \beta_p \end{pmatrix}, \text{ and } \epsilon = \begin{pmatrix} \epsilon_1 \\ \epsilon_2 \\ \vdots \\ \epsilon_n \end{pmatrix}. \tag{16.5}$$

The β can be learned from the training data by minimizing errors. The error term is usually squared and summed and, if so, this is called the *ordinary least squares* (OLS) model:

$$\|\epsilon\|^2 = \|\mathbf{Y} - \mathbf{X}\beta\|^2 = \sum_{i=1}^{n} \left(y_i - \sum_{j=0}^{p} X_{ij}\beta_j \right)^2, \tag{16.6}$$

---

[2] This assumption is still general: if ϵ did not have a zero mean, we would subtract off its mean and fold it as a constant into f.

where $\| \cdot \|$ is the Euclidean or ℓ_2 norm. Linear regression is useful for predictions but is often outperformed by more advanced, nonlinear methods such as neural networks. Yet linear methods have a very distinct advantage: interpretability. Once the fit parameters β are learned, we can easily reason about the relationship between an observation $\mathbf{x}$ and y. Specifically, if one feature x_i changes by a unit amount and all other features x_j $(j \neq i)$ are held fixed, then our model predicts that y will change by an amount β_i. In comparison, for other methods, such as neural networks, we cannot as easily isolate the effects of particular features on the final predictions.

Logistic regression—although it is literally called "regression"—is probably the most widely used binary classification model.[3] Logistic regression is still a regression model in the sense that it tries to predict the *probability* for a data point to have a given $(0, 1)$-label, rather than the label itself, and that its fit parameters can be interpreted similar to a linear regression. Because probabilities must be between zero and one, we generally should not run a linear regression model as it won't be limited to $y \in [0, 1]$. Instead, we need a function that obeys those bounds. A choice of function that turns out to be quite convenient is the *sigmoid function*:

$$\frac{1}{1 + e^{-x}}. \tag{16.7}$$

This function goes to zero as $x \to -\infty$ and goes to 1 as $x \to \infty$, obeying the bounds we need and smoothly transitioning between the two at some point. By applying linear (affine) transformations to x (i.e., $x \to \beta_1 x + \beta_0$), we can slide the transition point back and forth and we can make the transition steeper or more gradual. These "degrees-of-freedom" allow us to fit the model to training labels (0's and 1's at different values of x), although a fitting method other than OLS must be used [224].

Why the choice of sigmoid (Eq. (16.7)) instead of another function? A clever trick is to convert the probability into the *odds*, which is defined as the ratio between the probability (p) and the inverse probability $(1 - p)$:

$$\frac{p}{1 - p}. \tag{16.8}$$

Applying a logarithm gives the *log-odds* (also called the "logit"):

$$\log \frac{p}{1 - p}. \tag{16.9}$$

The log-odds is useful because, unlike p, it can vary across the whole range of real numbers: as $p \to 0$, the log-odds approaches $-\infty$ and as $p \to 1$, the log-odds approaches ∞. And if we assume that p follows Eq. (16.7) we get:

$$\log \frac{\frac{1}{1+e^{-x}}}{1 - \frac{1}{1+e^{-x}}} = \log e^x = x. \tag{16.10}$$

---

[3] Logistic regression is designed around predicting one of two possible labels for each datapoint, which we can just refer to as 0 or 1. However, it can naturally be extended to more than two labels, using what is called *multinomial regression*. For simplicity here, we focus on the binary case.

In other words, the sigmoid is the *inverse* of the log-odds. This means that, even though our target variable is a probability, we can formulate a *linear* regression problem[4] for its log-odds:

$$\log\left(\frac{p}{1-p}\right) = \beta_0 + \beta_1 x_1 + \beta_2 x_2 + \cdots + \beta_m x_m + \epsilon. \tag{16.11}$$

This can be also written as

$$p = \frac{1}{1+e^{-z}}, \tag{16.12}$$

where $z = \beta_0 + \beta_1 x_1 + \beta_2 x_2 + \cdots + \beta_m x_m + \epsilon$ and z now represents the linear transformations we can use to move the sigmoid around to fit the training data. This sigmoid function *linking* the right-hand side of Eq. (16.11) to the probability is also called the *logistic function* and that is why this method is called *logistic regression*. In fact, the logistic regression model is but one example of *generalized linear models* that use various *link functions* to model different types of target variables.[5]

Once we can model the probability, we can perform a binary classification task: when the predicted probability $\hat{p}$ is larger than $1/2$, we can assign $\hat{y} = 1$; otherwise, assign $\hat{y} = 0$. And, while this idea is couched in binary classification, this idea can be generalized to multi-class classification problems as well, through multinomial (also known as softmax) regression.

16.2.2 Neural networks

Neural networks have long been studied as part of artificial intelligence and machine learning. While promising in principle, it was not until recently that their practical significance was achieved, with the wild revolution of machine learning in the 2010s powered by neural networks. Neural networks are a general-purpose class of machine learning methods that are suitable to supervised learning, unsupervised learning, and anything in between.

Let's describe the basics of the prototypical neural network, the fully connected feedforward network. This network can be used for both regression and classification problems. Many variations exist, such as convolutional neural networks which are ideal for image data, but the principles are roughly the same. (We discuss neural networks for unsupervised problems below.) The neural network consists of a set of nodes called units or neurons that are arranged in some number of successive layers:

---

[4] Abusing notation slightly, here we use $x_1, x_2, \ldots$ to represent an observation's value for each variable, and the p we are estimating corresponds to that observation. Being more explicit, we would write p_i and $x_{i1}, x_{i2}, \ldots$ to note we are describing an observation i.

[5] Furthermore, we can imagine taking many different logistic regression models and nesting them together recursively. Models that do this are called neural networks.

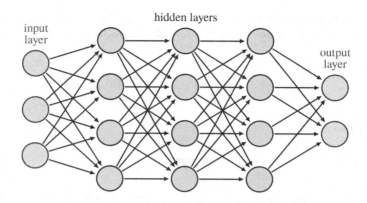

The particular arrangement of the units, the number of layers and the numbers of units per layer, is called the *architecture* of the network. In our fully connected case, each node in one layer is connected to every node in the next layer (this is what we mean by "fully connected"). Nodes in the first layer, called input nodes, receive the original data being analyzed. Nodes in the last layer, called output nodes, compute the final output of the network, used for predictions. As data move through the nodes between the input and output, it is transformed using an "integrate-and-fire" idea drawn in analogy to biological neurons. Specifically, the output z of a given unit is a weighted sum of its inputs $\mathbf{x}$, that was passed through a nonlinear *activation function* σ: $z = \sigma(\mathbf{w}^\mathsf{T}\mathbf{x})$, where $\mathbf{w}$ is a vector representing the weights on each input to the unit and $\mathbf{w}^\mathsf{T}\mathbf{x} = w_1 x_1 + w_2 x_2 + \cdots$.[6] Often, σ is the sigmoid function, but a variety of other options are possible.[7] The outputs at one layer then form the inputs to the next layer and the process repeats until the final layer whose outputs form the final prediction of the neural network. Our data dictate the shape of the output layer. For example, if we wish the network to be used for a binary classification, we would have two output nodes, one assigned to each possible label, and when data are fed through the network, we take as our predicted label whichever of the two outputs is larger. Just as our data shapes the output layer it also shapes the input layer. If we are performing a regression with 10 features, for instance, we would have 10 units in the input layer, one for each feature. If our data was text, we may have an input unit for each unique word in the language. And if we were working with images we might have an input unit for each pixel in the image.

As mentioned, a weight w is assigned to each link between units. These serve as the fit parameters and are learned via training. They are (typically) initialized at random, then training data are passed through the network, the final output is compared to the training label and the weights are adjusted to "nudge" the network's output towards the desired label. The process repeats many times until the weights are tuned and the network (hopefully) is well trained. The weights are usually updated with an algorithm called *backpropagation* [408]. For a fully connected network, we can combine all the weights between nodes in adjacent layers into a weight matrix, and compute all the weighted sums simultaneously using matrix multiplication. A neural network is essentially a large

---

[6] Usually, each unit also has a *bias* term, which acts similar to the intercept coefficient in a linear regression model.

[7] They must be nonlinear. Why?

combination of matrix multiplications passed through nonlinear functions.

A major drawback of neural networks is their black box nature: unlike simpler methods such as linear and logistic regression, it is often challenging to interpret or explain how and why a neural network predicted what it did. This is the interpretability–flexibility tradeoff: neural networks work by being highly flexible and able to capture complex, nonlinear patterns in data, but in doing so they lose interpretability over simpler, more "rigid" methods.

16.3 Unsupervised, self-supervised, and representation learning

Unsupervised learning problems lack the training data used by supervised methods to learn fit parameters. There is an **X** but there is no **Y**. Then how can any learning happen? One answer to this conundrum is that it is possible to detect, even without any labels, some interesting patterns in the data. One classic example is *clustering*: finding groups of data points that are very similar to one another while being different from other clusters. Many clustering methods exist, the classic being k-means clustering. Another example is *dimensionality reduction*: using an algorithm to recognize simpler, lower-dimensional representation of your high-dimensional data. Numerous dimensionality reduction methods have been developed, including principal component analysis (PCA), various matrix factorization techniques, t-SNE (t-distributed stochastic neighbor embedding) [471], and UMAP (uniform manifold approximation and projection) [302].

Unsupervised methods can also be combined with supervised learning. In *semi-supervised learning*, we have a small set of labeled training examples and a much larger set of unlabeled data. If the dataset contains useful features, then the data points with similar ground truth labels should be clustered together. Therefore, the clustering (unsupervised learning) can, in principle, be used to efficiently expand the small number of labels to the entire dataset.

Another answer to the conundrum is that it is often possible to create a supervised task from the unlabeled data by leveraging patterns in the data itself. For instance, if we have a large corpus of natural language sentences, we can think of a prediction task where we use all the previous words to predict the next word in each sentence. Likewise, if we wish to train a computer vision model and we have many photos but no supervising labels, we can create a training task by, for instance, taking each image and removing a small patch of that image, then ask the model to predict the correct missing patch from a set of small patches. In these cases, we are using the data itself—the patterns in the data—to produce labels for training. This is called *self-supervised learning*.

Self-supervised learning is deeply connected to *representation learning*. Representation learning focuses on obtaining useful *representations* (features) of the data elements. Many self-supervised tasks work by generating representations: a large language model for example becomes good at predicting words because it has generated computationally meaningful representations for those words, representations that capture the semantics and relationships between the words. The goal of representation learning is not just to make predictive models focused on the self-supervised task, but instead to obtain the most useful representations of an image, a word, a sentence, or whatever we are interested.

But how can this be useful? It turns out that learning good representations is one of the most important secrets to building powerful models in machine learning.

16.3.1   Neural networks for unsupervised problems

As discussed above, neural networks appear in both supervised and unsupervised problems. Neural networks need to be trained, generally using backpropagation and stochastic gradient descent. But these algorithms require training labels (you need errors if you want something to backpropagate), so how can they be used for unsupervised problems.

Many innovations come from transforming unsupervised problems into supervised problems. As an example, consider one type of neural network, the *autoencoder*:

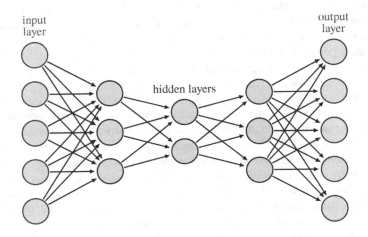

This network takes a d-dimensional vector of data as input ($d = 5$ in the illustration) and uses *that same vector for the output*. No other labels are necessary. (Notice the input and output layers in the illustration are the same size.) Between the two is a bottleneck of hidden layers, and the network must learn to compress the input so that it can successfully reconstruct it after passing through the bottleneck. In other words, the neural network is learning to do (nonlinear) dimensionality reduction. Further, the learned weights become useful *representations* themselves.

An autoencoder is one example of using neural networks for unsupervised problems.

16.3.2   Importance of representation

Although we do not fully understand how and why many machine learning models work, there is already lots of practical knowledge that has been established. One of the most important lessons from machine learning research is that identifying good *features* for your data—or learning good *representations* from your data—is what matters most in building powerful machine learning models.

If critical information is missing from the features, no fancy model can salvage it; on the other hand, it is easier to find a model that effectively extracts the information in the features as long as the features contains the signal. In other words, a simple model with useful features (information) tends to perform better than a state-of-the-art model with useless features. Therefore, *feature engineering*—the process of identifying and crafting useful features—has been historically recognized as a critical step in building a successful machine learning system in practice.

In fact, the huge progress of deep learning in recent years is largely thanks to the fact that deep neural networks excel at *learning* good features and useful representations. The lesson was that letting the model discover and learn features that are useful for the task at hand works much better than trying to hand-craft features, as long as the model that we are using is powerful enough and we have enough data—both conditions are satisfied with deep neural networks and massive datasets.

In the case of computer vision models, for instance, traditional feature extraction methods tended to be confined to low-level features such as sharp "edges" in the image and high-level features (human faces and their components such as eyes and noses) are not easily identified. By contrast, deep neural networks tie the process of learning both low-level and high-level features to the final prediction task, thereby extracting the highly useful and impossible-to-hand-craft features that help the task most. One of the meanings of the word "deep" in "deep learning" is this idea of connecting the feature discovery process itself to the prediction task.

And therein lies the power of representation learning. Representation learning allows powerful models (especially neural networks) to learn extremely rich representations of complex data such as images, videos, words, sentences, and so on. These representations can then be leveraged to solve a wide range of prediction tasks. A good example is *large language models* (LLMs). LLMs are "pre-trained" using text with a self-supervised task that asks the model to predict the next word in a sentence, fill the blank, fix the sentence, and the like. Through this pre-training and representation learning, LLMs have been shown to learn the high-level patterns of natural language extremely well—well enough to achieve human-level performance in some natural language understanding tasks (e.g., question-answering tasks). Once you have these LLMs that "understand" the patterns of natural language very well, then we can ask it to solve any other problems that involve natural language. For instance, the BERT (Bidirectional Encoder Representations from Transformers) model [129] has been used to solve a wide range of natural language understanding tasks, including question-answering, text classification, and so on, by *fine-tuning* the pre-trained LLM for a specific task using new training data. The incredible part is that the larger the pre-trained model becomes and the more pre-training it went through, the better it performs at specific tasks with small training data. In other words, scaling up the model's size and pre-training improves the quality of the representation and the quality of representation can then improve any tasks that use the representation. Furthermore, another type of model (decoder models such as GPT [383]) has been found that does not even need to be fine-tuned to a specific task! Instead, it is possible to simply *ask* the model as if it is a human and the model will *answer* the question [79]. This capacity of *few-shot* or *zero-shot* learning is opening up a new era of machine learning research and applications.[8]

---

[8] Along with a variety of thorny ethical questions.

Representations and networks   The same principle applies when building machine learning models with network data. If you are building a traditional model that relies on hand-crafted features, then it is absolutely critical to identify and engineer useful features that you can obtain from the network. But a lot of recent research on machine learning with network data is focused on ways to set up neural networks or other models so that we can learn good representations of the network—i.e., "graph embeddings"—then use such learned features to perform prediction tasks such as node label prediction and link prediction. We discuss ways of using machine learning with network data shortly.

16.4   Overfitting, bias–variance tradeoff, and regularization

As we have just discussed, the features included in the model limit the extent of information the model can use to learn the patterns in the data. Once we throw away a feature, the information uniquely contained in that feature cannot be recovered. Then a natural question is: is it always better to include more features?

The answer is: not necessarily. The more features we include, the more we are prone to *overfitting*. A great example is the case of a simple regression problem. Let's say we are predicting the price of houses using their square-footage x alone. The simplest linear regression model would be the following:

$$\hat{f}(x;\beta) = \beta_0 + \beta_1 x. \tag{16.13}$$

But we can also create a slightly more complex linear[9] regression model:

$$\hat{f}(x;\beta) = \beta_0 + \beta_1 x + \beta_2 x^2 + \beta_3 x^3 + \beta_4 x^4 + \beta_5 x^5 + \cdots + \beta_d x^d. \tag{16.14}$$

This is a *polynomial regression* model as we're fitting a polynomial of degree d. The complexity of the model, and the number of parameters it contains, is given by d. What would happen with this model?

As we include more parameters by raising d the model will be able to fit the training data better and better.[10] But will this be a *better model*? Although this will reduce the error in our training data, the model overfits and begins to capture noise. It will not generalize well. If we throw new data points at it, the model would fail, especially with more and more parameters. We see this occur with some synthetic data in Fig. 16.1a–c. Here we created some ground truth data (Fig. 16.1a), and it is reasonably well-fitted when d is not too big (Fig. 16.1b). That model, with $d = 3$, should predict new data from the same underlying function fairly well. But larger d, especially in Fig. 16.1c where we chose $d = n$, the number of data points, while it fits the data we have perfectly,[11] it

---

[9] By the way, "linear" regression can absolutely use higher-order (nonlinear) features. What makes it linear is that the model is a linear combination of the features, a weighted sum where the fit parameters are the weights.

[10] Remember that, as John von Neumann put it, "with four parameters I can fit an elephant, with five I can make him wiggle his trunk."

[11] With a parameter for every data point, we have enough degrees-of-freedom to perfectly capture or *memorize* the data.

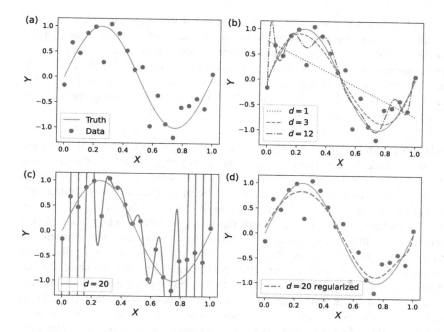

Figure 16.1 An example of overfitting and regularization using polynomial regression. (**a**) A small example of $n = 20$ data points generated by adding a small amount of noise to the ground truth function. (**b**) Different polynomial fits $\hat{Y} = \beta_0 + \beta_1 X + \beta_2 X^2 + \cdots + \beta_d X^d$ of degree $d = 1$ (linear fit), $d = 3$, and $d = 12$. We see the curve start to deviate away from the ground truth at higher d, meaning that the fit is beginning to capture noise. (**c**) At $d = n = 20$ we have as many fit parameters as data points, and our curve perfectly fits the data points, interpolating them. But between the points it varies wildly, indicating that it will perform poorly at predicting any new observations. (**d**) A fit with the same number of parameters as (c), but now regularized. Despite having so many parameters, the quality of the fit is greatly improved. This model will likely predict new observations much better than the fit in (c).

varies wildly between those points—any new observations, drawn from the underlying example function, will be badly predicted by such a fit.

This phenomenon is called the *bias–variance tradeoff*. The *bias* refers to the error of the model. The worse the model predicts, the larger the bias is. The *variance* refers to the generalizability of the model. The worse the model generalizes, the larger the variance is. The bias–variance tradeoff is a fundamental property of statistical predictive models. The bias and the variance are connected: as you decrease the bias of a model, you inevitably increase the variance of the model. Conversely, this means if you increase the bias of the model, you decrease the variance, and it may make *better* predictions.

The traditional, statistical approach to this overfitting problem is to build the model carefully and then perform thorough diagnostics. This is strongly motivated by the need to accurately estimate the parameters of the model. For instance, a social scientist would start from certain theories and hypotheses, which dictates what kinds of features are measured and included in the model. Then the scientist would go through a series of tests to make sure that the model does not have problems of overfitting, or that the data violates any of the model assumptions such as multicollinearities.[12] By contrast, machine learning focuses on the prediction task and the model is built to perform the prediction task as well as possible. It is good to be able to estimate the relevant parameters, but it is not the primary goal. Therefore, there is a strong incentive to include as many features as possible.

Regularization   An important solution to the challenge of overfitting with complex, many-parameter models is *regularization* (or *shrinkage*). Regularization is a technique that penalizes the model based on its complexity. For instance, the model that swings widely (Fig. 16.1c) would be considered complex because it has so many "important" parameters that should be accurately estimated to fit the data. We can make a simpler model by eliminating parameters, but we can also take a complex model and simplify it by enforcing constraints on the values of the parameters, preventing them from taking on arbitrary values. This "squeezes" the complex model down to something effectively simple, whichs turns out to be a very, very effective strategy for making accurate models.

   Regularization techniques come in many forms, but for regression models like we've discussed so far, the general idea is to make parameter values that are large in magnitude more "expensive" than smaller values. For instance, two of the most common regularization methods for regression models are *ridge* and *LASSO* regression. The ridge regression penalizes the complexity of the model by considering a model's "cost" as:

$$\|\beta\|^2 = \sum_{i=1}^{p} \beta_i^2. \tag{16.15}$$

In other words, we use the Euclidean norm or ℓ_2 norm of the parameter vector β. We want this norm to not be too big; if it were that would indicate a model that is wildly varying the values of its fit coefficients, likely overfitting the model.

   We incorporate this cost into our model by changing its optimization. Instead of just minimizing the sum of squared errors $\|\epsilon\|^2$ (Eq. (16.6)) by freely picking β, now we try to minimize $\|\epsilon\|^2$ but subject to a limit on how big the parameters can be:

$$\text{Find } \beta \text{ to min } \|\epsilon\|^2 \text{ subject to } \|\beta\|^2 < \alpha. \tag{16.16}$$

Here α is a new quantity we pick to represent our "budget" or how much we allow our parameters to vary. When $\alpha \to \infty$ we have no limit and we revert back to the original OLS fit. As we lower α, we place stricter limits on the fit parameters, *shrinking* them to smaller and smaller values. The new term α is an example of a *hyperparameter*; unlike β it is not estimated from fits to the data but must somehow be determined before fitting occurs.

   Usually with optimization problems such as Eq. (16.16), we can write them in a second form, called the *dual problem*. Instead of minimizing one term subject to a limit

---

[12] For a linear regression, the fit becomes undetermined if one feature is a linear combination of other features. This is called *multicollinearity*. For data where one feature is close to a linear combination of other features, a fit is possible but it will be unstable: small changes to the data points will drastically change the fit parameter values. This makes the model particularly untrustworthy.

on a second term, we can simply minimize a weighted combination of both terms. For Eq. (16.16), the dual formulation is:

$$\text{Find } \beta \text{ to min } \|\epsilon\|^2 + \lambda\|\beta\|, \tag{16.17}$$

where λ is related to α. (Notice we recover OLS fits when $\lambda = 0$.) By eliminating the constraint, we sometimes have an easier, although equivalent, optimization problem to solve. This is the usual form that ridge regression is presented. To demonstrate how powerful regularization is, Fig. 16.1d shows the same high-parameter polynomial model from Fig. 16.1c but with a ridge penalty included (we use $\lambda = 10^{-3}$ in the figure). The fit quality is greatly improved over the un-regularized model.

Another regularization technique similar to ridge is called LASSO. Its penalty is:

$$|\beta| = \sum_{i=1}^{p} |\beta_i|. \tag{16.18}$$

In other words, LASSO uses an ℓ_1 norm while ridge uses ℓ_2. This may seem like a minor difference, but they have fundamentally different implications for the types of the models that they encourage. A simple way to think about these two regularization methods is that you want to use LASSO if you want a *sparse* model—where you throw away more features from the model by setting some $\beta_i = 0$—and use ridge if that is not your priority (ridge may make β_i smaller but not exactly zero).

We've focused on regularizing OLS models, but regularization is not limited to linear models. Neural network models often use various regularization techniques during training. This makes sense because deep neural networks regularly employ a huge number of parameters (weights in the neural network), sometimes even much greater than the number of data points. Although neural networks seem much less prone to overfitting issues, it is often beneficial, even required, to use regularization techniques.

Double descent and neural networks Regularization produces a very important phenomenon called *double descent*, which is exhibited by highly expressive models such as deep neural networks. It is called double descent because, as we increase the number of parameters, the *test error* of the model first decreases and then increases (as expected by the bias–variance tradeoff), but then decreases *again* as the number of parameters increases further! This is highly counterintuitive and seems to violate the principle of the bias–variance tradeoff. But this indeed happens, not only in neural networks, but also in simple regression models. The key ingredient that makes this possible is regularization. When regularization is in place, the model can sometimes generalize better when over-parametrized! Specifically, when the flexibility of the model is greater than the number of datapoints, there exist numerous possible solutions that minimize the error function—they can "memorize" the training data. Yet, among these many possible solutions, only the simpler and more generalizable ones tend to be selected because of the strong constraints imposed by the regularization. Despite memorizing the training data, including its noise, they still generalize well. Although the power of deep neural networks is still in ways a mystery, the double descent phenomenon does not necessarily contradict the bias–variance tradeoff.

So how to regularize a neural network model? It is possible to apply similar ideas as we did for regression models—penalizing large weights in the network. But there exist many creative solutions as well. One famous technique is called *dropout*. Dropout is a technique that randomly drops some of the neurons (like *half* of the neurons!) in the neural network during the training process. These dropped neurons are invisible for some portion of the training data, then returned and other neurons, also randomly chosen, are dropped. The idea is that we want to force the neural network to be able to perform the task even if an unknown but substantial fraction of neurons are missing. This forces the network to spread the "signals" across the network rather than "remembering" the answers.

Another interesting aspect of neural networks is that the training process of the neural network itself is known to be *self-regularizing*. The most common way to train neural networks is using *stochastic gradient descent* (SGD).[13] Neural networks are usually trained by feeding in the training data, seeing how well the output matches the training data labels, then slightly adjusting the network weights, sending the training data through again, adjusting again, over and over to improve the network. The SGD algorithm uses the gradient of the model—the direction and the amount that we need to adjust the parameters of the model, according to backpropagation—from a *random subset* of the training data, different for each iteration, rather than the full dataset. In other words, the SGD algorithm makes training more stochastic than gradient descent using the whole dataset, as the model never knows exactly what data it will see. Although this method was originally developed so that we can speed up training, there is a somewhat unexpected side-effect—the training process is self-regularizing, is less likely to become trapped in local optima, and overall is more robust because it is messier and noisier.

16.5   Model selection

An important concept closely tied to the bias–variance tradeoff, overfitting, and regularization is *model selection*. Model selection is the process of selecting a specific model from a set of competing models. For example, should we use a neural network or a linear regression? Should we use a linear regression with ridge regularization or should we use one with LASSO? If we decide to use polynomial regression, what order of polynomial is best? Whether choosing between different models, different fitting methods for the same model, or even different hyperparameter values, model selection methods help us make our choices.

The primary question in model selection is: *how can we know which model is better?* This question is intimately linked to the bias–variance tradeoff. A better model should not only have low bias, but also low variance. In other words, a good model should not only *fit the training data well* but also *generalize well*. But how can we know its generalizability without testing it on new data?

One way to think about model selection is to assume that *generalizability* or the variance of the model can be approximated by the *complexity* of the model, which can be quantified by the number of parameters in the model or something similar. Then

---

[13] With backpropagation being used to compute the gradient in error that is "descended."

we can define measures that capture the tradeoff. For instance, the *Akaike Information Criterion* (AIC) is defined as follows:

$$\text{AIC} = 2k - 2\ln(\hat{L}), \tag{16.19}$$

where k is the number of estimated parameters and $\hat{L}$ is the maximized likelihood of the model.[14] The more parameters we have (high variance), the larger AIC becomes; the larger the likelihood of the model (low bias), the smaller AIC becomes. When faced with multiple models or multiple options (such as hyperparameters) within a model, we can compute each choice's AIC and select the model by picking the smallest.

Another, probably more common, approach to model selection is simulating a scenario of model generalization by utilizing the data we have. Specifically, we can split the training data available to us into a smaller training dataset and a dataset for model selection (often called a "validation set"). We then train the model using only the newly reduced training set and calculate the bias from it, then apply the model to the held-out validation set to estimate how well it generalizes. By ensuring the validation data is previously unseen to the model, we ensure it is not overfitting to it, or memorizing it. If the model is memorizing the training data, then we will see it perform badly on the newly revealed validation data. So we then select the model that performs and generalizes well. In practice, "cross-validation," a related technique, is often employed to efficiently use the dataset on hand. We will discuss these methods, known as *resampling techniques*, in more detail in the next section.

16.6 Data hygiene and evaluation

When we apply machine learning to data, one of the most critical things is to practice good *data hygiene*. Usually, in most other things, if you screw up, you would score worse. However, in machine learning, your performance may look *better* when you screw up![15] This is of course because of the bias–variance tradeoff—it is easy to overfit while it is difficult to obtain a *generalizable* model. To ensure the model performs well not only with the dataset that we currently have but also with any future data that we will have, we need to think carefully about how to train our models and how to measure their performance.

What screw-up are we talking about here? The key is to prevent *information leaking* from the validation dataset into the training process. If we have somehow accidentally included information from our validation set within the training data, then our attempts to validate the model will be polluted because the model already sees something about the validation examples. The more information leaks into our training, the more prone to overfitting our models will be.

Here's a simple and quite insidious way leakage can occur. Suppose we have a feature x_i and we want to rescale its values to have zero mean and unit variance. (This is often necessary for models and data where the typical magnitude of values in one

---

[14] The likelihood $L(X)$ of data X assuming a statistical model is the probability of X given the model's parameter(s) θ: $L(X) = \Pr(X \mid \theta)$. The maximized likelihood $\hat{L}$ is the likelihood that comes from the "best" parameters $\hat{\theta}$, the parameters that maximize L.

[15] At least until you measure the performance with another dataset.

feature is very different from those of another feature.) This rescaling can be done with a z-score, here for the jth value of x_i:

$$z_{ij} = \frac{x_{ij} - \langle x_i \rangle}{\sigma_{x_i}}, \tag{16.20}$$

where $\langle x_i \rangle$ and σ_{x_i} are the sample mean and sample standard deviation, respectively, of the values in x_i. All well and good, we can now use z_i instead of x_i. This is standard practice, so what's the problem? Well, suppose we compute z_i and then split the data into training and testing. The mean and standard deviation in Eq. (16.20) were computed over *all* the data points, and thus the model we fit will have information it shouldn't, lurking in the sample statistics. The correct approach is to apply Eq. (16.20) *after* we split the data. Doing so captures the fact that we can't rescale data using observations we don't see. And if you think this will have only a small effect, think again: it can cause quite a false performance boost if you have lots of features and all are being transformed.

Training–test split   First of all, for the sake of simplicity, let us focus on supervised learning. To be able to predict, machine learning models should generalize from the training data. But how can we measure their generalizability? As mentioned above, we can *estimate* it using proxies such as the number of parameters or by using the data splitting strategy to simulate the generalization scenario.

The latter approach is more commonly employed because it is a more direct test of generalizability. In essence, we need to find another dataset against which we can *test* the performance of our model. This new dataset can come from the future (if you can wait) or from a different system. But a much easier way is to simply split our existing dataset in two: *training data* and *test data*. (Usually we perform the split randomly. For each observation, with probability p it goes into the training data; otherwise with probability $1 - p$ it goes into the test data.) We can now train using the training data and then test using the test data to evaluate the generalizability of the model. Simple enough, right?

Although training–testing splits allows us to *test* the generalizability of the model, it is post-hoc. It only tells us how well the already-trained model generalizes, but cannot help us much in selecting the best model. To be clear, it is still possible to train multiple models on our training dataset and test them on the test set to choose the best model from the pack. However, this process—using the test data to select the best model—is not ideal because whenever we use the test data, the information only in the test dataset that allows us to estimate the generalizability of our models will *leak* to us. If we keep testing numerous models on the test set, eventually it will overfit, just as if we used the full data, not the split data, to train our models. Therefore, ideally, we want to avoid using *any information* from the test set in our training and model selection process, and this is the key to good data hygiene.

Training–validation/evaluation–test split   When we need to perform model selection, a simple way to avoid information leakage from the test dataset into our training data is simply to split the training set *again* to create another test set (we usually call it "validation set"). This validation set can then be used to evaluate and select models and hyperparameters. Although we can do even more "Russian doll"-like splits, the more splits we have, the less data will we be able to use to train the model.

Cross-validation Another very popular method for model selection and validation is *cross-validation*. With a single training–testing split we would only have a single measure of how well the model performs and it would be difficult to tell if that value is typical or not. Cross validation, on the other hand, uses multiple data splits to estimate the typical out-of-sample performance of a model. For instance, a 10-fold cross validation will split our data into 10 equal-sized sets. Once we have 10 sets of data, 9 of them can be used together as a training set and one will be used as a test set. After we estimate the performance on that test set, we put it back into the training set and take out another set to serve as the test set. We retrain the model from scratch on this new training set and re-evaluate on the new test set. We can do this for each of the 10 possible splits and this will produce 10 different results, which can then be averaged to assess the performance of our model.

Although k-fold cross validation[16] is most commonly employed, there are many other techniques. For instance, "leave-one-out" cross-validation uses a single data point as the testing set while the whole dataset except this single data point is used to train the model. This is then repeated n times, each time using one of the n observations as the test set.

Evaluation metrics Our procedure for splitting the training data to infer how well a model would generalize gives us unseen data for testing, but we still need to measure the trained model's performance on the testing data. How to evaluate performance? Many such metrics exist. For regression tasks where our model is predicting a numeric quantity, we want to compare a test value y_i to $\hat{y}_i = \hat{f}(x_i)$, the prediction our model $\hat{f}$, trained on the training data, makes on the corresponding input x_i. Common metrics are *mean squared error* (MSE),

$$\text{MSE}(y, \hat{y}) = \frac{1}{n_{\text{test}}} \sum_{i=1}^{n_{\text{test}}} (y_i - \hat{y}_i)^2, \tag{16.21}$$

where n_{test} is the number of data points held in the test set, or the *coefficient of determination* (R^2)[17]

$$R^2(y, \hat{y}) = 1 - \frac{\sum_{i=1}^{n_{\text{test}}} (y_i - \hat{y}_i)^2}{\sum_{i=1}^{n_{\text{test}}} (y_i - \bar{y})^2}, \tag{16.22}$$

where $\bar{y} = \frac{1}{n} \sum_{i=1}^{n_{\text{test}}} y_i$, but many other metrics exist.

---

[16] $k = 10$ is the most common number of folds.

[17] A common "gotcha" in practice is to discover a problem with a model by finding bizarre values of R^2. We expect, because it is a squared quantity, for R^2 to always be positive, but this is not the case. If the model is making predictions that are so bad, it may be doing worse than a model that is just predicting the average value of y, in which case the second term in Eq. (16.22) will be greater than one. If you encounter negative R^2 scores, reconsider your modeling and evaluation methods.

For classification tasks where our model is predicting a label such that test value y_i is now a categorical quantity, we use metrics that compare how often the predicted category $\hat{y}_i$ matches the test label. Here are some metrics for binary classification tasks, with only two possible categories, but many multi-class metrics are commonly used. The most natural binary metric is probably the *accuracy*

$$\text{Accuracy}(y, \hat{y}) = \frac{1}{n_{\text{test}}} \sum_{i=1}^{n_{\text{test}}} \mathbb{1}_{y_i = \hat{y}_i}, \tag{16.23}$$

where the *indicator function* $\mathbb{1}_x = 1$ if x holds, otherwise it is 0. That said, accuracy has some problems. For instance, on imbalanced data sets where most y_i are in one category, performance will appear inflated. It also doesn't distinguish where errors are happening; are you making false positives or false negatives[18]? Thus, it is common to employ more specific metrics, *precision* and *recall*:

$$\text{Precision} = \frac{TP}{TP + FP}, \tag{16.24}$$

$$\text{Recall} = \frac{TP}{TP + FN}, \tag{16.25}$$

where *TP*, *FP*, and *FN* are the numbers of true positive, false positive, and false negative evaluations, respectively. The intuition: Precision measures, whenever the model predicts a positive, how often it is correct. Recall measures, whenever the data are positive, how often the model is correct.

Working with two metrics is generally fine, but sometimes we need a single numeric quantity to evaluate the classifier. Usually, we then combine Precision and Recall using an F-score:

$$F_\beta = \left(1 + \beta^2\right) \frac{\text{Precision} \times \text{Recall}}{\beta^2 \text{Precision} + \text{Recall}}. \tag{16.26}$$

This is the harmonic mean of Precision and Recall, and β is a weight that tells us how much more important one is over the other; the F1-score, where $\beta = 1$ weighs them equally.

While we've described some of the common evaluation metrics, many others are used for both classification and regression. Please see our remarks below if you are interested in learning more.

16.7   Graph embedding

Graph embedding has emerged as a powerful tool for various machine learning and analysis tasks with network data. Graph embedding converts a network into a *vector representation*: a node embedding would learn a continuous and dense vector representation for each node, an edge embedding would do the same for edges and a whole graph embedding would do the same for each network. Once we find the vector representations, they can be readily plugged into machine learning models and pipelines to

---

[18] Assume that the two categories are $y = 0$ and $y = 1$. A false positive is when you predicted $\hat{y} = 1$ for a test point where $y = 0$. A false negative is when you predicted $\hat{y} = 0$ but $y = 1$.

help with downstream predictions. The vector representations can be used to visualize the network data as well.

Let's start from the adjacency matrix (Eq. (8.1)). In the adjacency matrix, each row and column vector captures the neighborhood of a node and actually is an N-dimensional vector representation of the node. However, the adjacency matrix is not a good representation because it is neither *dense* nor *continuous*; it contains only zeros and ones and is sparse. Consider two random nodes from a large, sparse network. The probability that they have any overlap in their neighbors is very low. This means that the corresponding *vector representations* of these two random nodes will probably be orthogonal to each other, without any overlapping components. In other words, most node pairs will appear completely unrelated to each other in the vector space. If we took, say, the dot product[19] between the vectors to compute their similarity, it would almost always be zero. Even if a pair of nodes has neighbors in common, because of the non-continuous nature of the adjacency matrix, we will not have a fine-grained, useful measurement of similarity between them.

A simple way to overcome the issue of sparsity is to define a more appropriate form of similarity between nodes. For instance, instead of measuring the similarity between two nodes by the number of common neighbors they have, we could measure how close they are located to one another in the network. We can even use random walk trajectories to measure the similarity between nodes, for instance by calculating the transition probability between nodes, how likely or how often a random walker moving over the network moves from one node to the other. Now the representation we have—still N-dimensional—is continuous and dense. Is this the best representation we can have? One important drawback of this representation is that it is not compact—each node is represented by a vector of size N. It turns out we benefit from vector representations that are dense and sit in dimensions $r < N$.

16.7.1 Compact representations

One of the most popular approaches to find compact representations has been *matrix factorization* methods. A square[20] matrix $\mathbf{S}^{N \times N}$ can be factorized, into multiple matrices, for instance, with *singular value decomposition* (SVD): $\mathbf{S} = \mathbf{U}\boldsymbol{\Sigma}\mathbf{V}^T$, where $\mathbf{U}$ and $\mathbf{V}$ are $N \times N$ unitary matrices and $\boldsymbol{\Sigma}$ is a $N \times N$ diagonal matrix containing *singular values* on the diagonal. When we *truncate* $\boldsymbol{\Sigma}$ be retaining only the r largest singular values, we compute a *low-rank* approximation of $\mathbf{S}$. This approximation helps with prediction by eliminating noise in $\mathbf{S}$, improving generalization, especially when most singular values are small and we can choose r such that $r \ll N$. It also allows us to find the most meaningful, *compact* representations for our data, because we can treat the rows of $\mathbf{U}$ as vector representations but keeping only the r most relevant columns, giving us dense, compact r-dimensional vector representations.

When $\mathbf{S}$ is non-negative, for example, when it is an adjacency matrix, we may instead wish to use *non-negative matrix factorization* (NMF):

$$\mathbf{S} \simeq \mathbf{WH}, \tag{16.27}$$

[19] Or, cosine similarity.

[20] SVD is not limited to square matrices although we only treat that case here.

where $\mathbf{W}$ and $\mathbf{H}$ are $N \times r$ and $r \times N$ matrices usually with $r \ll N$. The benefits of this approach is that it captures the matrix being non-negative and gives representations that are dense, continuous, but also compact—each node is represented by a vector of size $r \ll N$.

Another approach to find such representations is graph embedding algorithms. For a network with N nodes, the objective of graph embedding is, just like matrix factorization, to learn an $N \times r$ matrix, where $r \ll N$ is the dimension of the embedding and each row will be the vector representation of the corresponding node. There are numerous approaches, but we can categorize them into several classes of models. The first class of methods—simply called graph embedding—does not use any additional information other than the network structure. Usually, structural proximity between nodes is captured through random walks or other graph traversal methods and then translated into a vector representation. The most straightforward approach in this class may be "DeepWalk" and "node2vec," where the random walk trajectories, the sequences of nodes visited at random as a random walker bounces around the network, are considered "sentences" and fed into the standard word2vec model.[21] Second, there are methods that make use of node and edge attributes, in addition to the structural information. These methods are usually referred to as "graph neural networks" and we will examine that approach briefly below and in more detail in Ch. 26.

Although it was not obvious at first, it was discovered that neural network-based representation learning, including graph embedding, is an *implicit* matrix factorization [269] (Ch. 26). Instead of going through the process of explicitly constructing the similarity matrix $\mathbf{S}$ and then factorizing it into low-rank matrices, neural graph embedding methods directly obtain the factorization (representation) by using neural networks to learn the representation from some self-supervised prediction tasks. This means that we do not have to create the large, dense matrix $\mathbf{S}$. Therefore, it is inherently space-efficient and scalable. Furthermore, thanks to the usage of stochastic gradient descent or similar methods in training, the training process is also computationally efficient and robust [270].

16.7.2   Applying neural networks for graph embedding

The most canonical graph embedding methods use random walks to generate sequences of *nodes* and treat them as *"sentences,"* from which the "word" (node) vectors are learned. Just like word2vec or similar models use natural language sentences to learn word embedding, this class of methods use the same neural architecture to learn node embeddings. DeepWalk and node2vec are two well-known, basic variations. Random walks are fundamental to understanding a network's structure and how it affects dynamics. We discuss random walks further in Ch. 25.

When applying these methods, as these methods use random walks, we need to pay close attention to structural properties of the networks that affect the random walks. For

---

[21] Word2vec [307] is a very influential natural language processing technique. It works by taking a large corpus of text as a sequence of words, and training a neural network to predict a word given the words that surround it, its context. This self-supervising, context–word prediction task leverages the "distributional semantics" hypothesis of linguistics [202, 159] and gives rise to "word vectors," representations that are semantically meaningful.

instance, one must ask the following questions and decide whether to process the network further to make it more suitable for embedding methods. Is the network *connected*? If not, random walks will not connect the components and be completely disjoint across the components. As a result, the learned embedding will not be able to capture any useful information regarding the relative location of nodes in different components. Is the network *directed* or undirected? If the network is directed, then we may want to examine the prevalence of dead-ends and dangling nodes, areas where a random walk will become trapped if it enters because it has no way to exit, as such areas can create artifacts in the embedding. Dead-ends in the network may not be reached often by the random walks and therefore, the corresponding nodes may not be represented well in the embedding. Is the network *weighted*? Can the weights be interpreted in the random walk's perspective? If the network is weighted and the weights are meaningful, then we may want to make sure to use that information because random walks can be heavily affected by the weights. Is the network *bipartite*? If the network is bipartite, then the bipartite structure can produce artifacts in the embedding.

Other embedding methods also have their own requirements and assumptions. Even if they do not rely on random walk processes, it is always critical to understand the assumptions and requirements of the methods and make sure that the network is suitable for the methods. The types and structural properties of the network are important to consider.

16.7.3 Graph neural networks

Graph neural networks (GNNs) make use of both node-level attributes and the network structure by aggregating the node attributes from the neighboring nodes. Therefore, they tend to be used when rich node attributes are available. Because they make use of more information, GNNs also tend to be more powerful than simpler graph embedding methods. However, that is not always the case due to the complexity of the models and the amount of available data. When the network is small or the node attributes are not sufficiently informative, GNNs may not be able to learn how the network is organized. For instance, GAT (Graph Attention Network) needs to learn "how to pay attention" to certain nodes based on the attributes. This learning process tends to be more data-hungry and therefore it may not always work well for small network data. In such cases, simpler graph embedding methods, or even other traditional machine learning models, may be more suitable.

Some of the canonical graph neural network models are: Graph Convolutional Networks (GCN) [242], GraphSAGE [200], and GAT [475, 77]. Graph Convolutional Networks (GCN) is a generalization of convolutional neural networks developed for computer vision tasks. The idea is to use convolutional filters to aggregate node attributes from neighboring nodes. GraphSAGE allows a more flexible operation as well as ways to sample neighbors and perform inductive learning (learning on unseen nodes). Graph Attention Networks use the idea of the "attention mechanism" [474], which has been extremely successful in creating the *transformer architecture* upon which many large language models, as well as computer vision models, are now built.

We will discuss these in more detail in Ch. 26.

16.8   Challenges and practical considerations

Above, we discussed why it is extremely important to maintain good data hygiene in supervised learning tasks. One critical challenge with network machine learning is that we often have only *one* instance of a network that is not amenable to resampling or splitting such as cross-validation. The data splitting process itself can destroy salient information that is useful to perform the machine learning task. For instance, let's assume that we are working on a link prediction problem on a social media network. Can we split this network into a training, validation, and test set? If we try to devise a method of assigning nodes or links to different sets, then we may encounter links that span between sets. For instance, suppose node i is in the training set but has several neighbors in the test set. If we train on i's degree, we will be under-counting its actual degree due to the missing test neighbors.

Likewise, if we try to construct predictive features from the network topology, it can be challenging to make them independent. If we wish to predict y as a function of two features x_1 and x_2, it will be easiest if x_1 and x_2 both give useful, *different*, information about the value of y. But many networks force relationships between features or those features are related by definition. For example, if one feature is node degree and another is clustering, these two quantities are themselves correlated. While not an insurmountable problem, predictive models typically perform better when useful features are unrelated.

Lastly, the network structure also affects the use of traditional machine learning methods applied to node attributes. Usually such methods assume observations are iid, but the network structure will introduce non-trivial relationships between the rows of an attribute matrix—the network guarantees that observations are *not* iid.

16.8.1   Feature engineering

While it's often under-emphasized in machine learning research, enough cannot be said about feature engineering. Feature engineering is the practice of devising what features to measure or include in your predictive model. In other words, what are the columns of our data matrix $\mathbf{X}$? A predictive model in general[22] is powered by a collection of variables, or features. We should take care to gather the features we need, even going back to the drawing board and gathering new data (Ch. 6).

What happens if you don't have the right feature? It is certainly possible that we missed something important, something so related to our prediction target that our performance would be far greater were the feature available. Unfortunately, it can be hard to conclude if this is the case. Fortunately, with domain knowledge of the data we can often reason, at least somewhat, about what we do have, and determine if there is enough signal among the available features.

We may also want to explore various manipulations of our input data matrix $\mathbf{X}$. It may be that applying various transformations such as taking the log will be more useful.

---

[22] In some problems such as computer vision, neural networks have become powerful enough to avoid the need for feature engineering. Instead, the neural network is able to develop features as needed out of the raw data. In this area, it is the neural network's form or *architecture* that needs careful "engineering."

We could generate "indicator" features, for example finding values that are outliers and then creating a new binary feature to indicate outlier/inlier for an existing feature.

It is also helpful that some methods, particularly more flexible, black box methods like neural networks, depend less on feature engineering. In computer vision, for example, feature engineering, trying to devise useful numeric summaries of images, was the focus for many years. The advent of deep learning eliminated that, as it was shown that neural networks build both simple and complex features from an image's pixel values as they propagate through the layers of the neural network. Such feature composition is helpful for network data as well, although feature engineering still plays a role when it comes to non-network attributes that are associated with the nodes or edges of the network: even if they correlate closely with the network structure, it is still possible for useful, non-redundant information to be present in an unseen feature.

Closely related to feature engineering is *feature selection* and *feature importance*. Feature selection is the task of finding if there is a subset of our features that are most helpful for the predictive model. Selection is especially helpful in data that are very high dimensional, **X** has many columns, and we aren't sure which features are most important. Feature importance, then, determines which features contribute the most to the model's performance. Some methods, like linear or logistic regression, automatically tell us the importance of features by giving interpretable statistics for the fit parameters. Other methods, notably tree-based methods like decision trees and random forests [425, 315], have standard measures of importance because we know, inside the model, when a feature is used and when it is not. On the other hand, black box methods like neural networks are notorious when it comes to feature importance. Often it is not possible to tell which feature(s) contributed and which did not inside a neural network.

16.8.2 Model diagnostics

Suppose you've built a predictive model. Is it working well or poorly? Diagnostics can help answer this question. We've covered specific evaluation schemes like resampling methods and evaluation metrics like MSE, R^2, and F-scores in Sec. 16.6. One issue to be mindful of is when your training data are highly imbalanced. Suppose you are classifying nodes into two categories, A and B, but 90% of nodes in your example networks fall into category A. In this case, a model that always assigns a node to A will actually work very well. Indeed, it should be correct 90% of the time. But such a model is probably not going to be very useful or insightful to us. We should be mindful of data balance, take a healthy dose of skepticism, and always anticipate such explanations for good model performance.[23]

16.8.3 Dataset shift

A serious challenge facing predictive models is dataset shift, sometimes called distributional shift.[24] When the training data become outdated, a predictive model can be

---

[23] And another problem to be mindful of is leakage.

[24] We can further divide data shift into feature shift or covariate shift, where the input features change, and label shift, where the response or target change.

expected to perform poorly. It won't generalize to new data because the new data have shifted away from what it was trained on.

One of the most insidious aspects of dataset shift is that resampling methods, the traditional means of inferring the performance of a predictive model, are poorly suited to capturing its effects. Suppose you used cross-validation to measure the performance of a fitted model. What you have done is taken a single dataset and divided its observations into disjoint training and testing sets.[25] But the testing set, although unseen by the model, always comes from the same pool of observations as the training set. There can be no dataset shift within resampling.

What remedies can we turn to when faced with dataset shift? Merely detecting the shift is already an important task. If model performance in practice begins to fall, we may suspect shift is to blame. With network data, we can investigate how the current network data compares to the training network data, using, for example, techniques and ideas from Ch. 14. We can also investigate if node or edge attributes (Ch. 9) have changed, a classic question of exploratory (and perhaps confirmatory) data analysis. It may also be that missingness patterns (Ch. 10) have changed.

Some research has considered remedies for dataset shift, but more work is needed. One approach has been to apply *transfer learning* methods. Transfer learning is the general idea of taking a predictive model trained for one intended task and modifying it to work on another predictive task (think of taking an AI system trained to play chess and transferring it to play Go). A shift in the training data is analogous to a shift in the task, or so the argument goes.

16.9   Summary

Machine learning has taken the world of data science by storm, and studies of network data often rely upon machine learning tools. Researchers and data scientists should be well versed in using such tools, including both general-purpose machine learning methods and methods tailored to the complexities of network data. Predictive models can play both upstream and downstream roles, helping researchers clean and process data to build their network and to study and work with the network once it is built. Methods that transform networks into meaningful representations are especially useful for specific network prediction tasks, such as classifying nodes and predicting links. Despite all the amazing advancements in machine learning, one must not lose focus on the fundamentals, including practicing good hygiene when it comes to evaluating a predictive model's performance, notable examples being recognizing leakage and detecting dataset shift. Often, a healthy dose of skepticism will go a long way to making sure your predictive models are helpful and reliable.

Bibliographic remarks

Machine learning has a rich scientific history going back to the roots of artificial intelligence. Nilsson [348] provides a fascinating tour of AI's early history. Readers

---

[25] Cross validation does this multiple times to average over different breakdowns of training and testing observations.

wishing to learn more technical details are well served by Russell et al. [411], recently updated with a new addition, and the classic work of Bishop [56].

Many textbooks cover machine learning, some treating predictive models and learning in the context of traditional statistical methods and others taking the modern vantage point of neural networks and deep learning. James et al. [230] provide a fantastic introduction to statistical learning, which can be supplemented by the more technical work of Hastie et al. [205]. Hastie et al. [206] is a great, in-depth treatment of regularization, the practice of forcing overly parameterized models to generalize well, in the context of statistical learning. We continue to find surprises within over-parameterized models, including the recent exploration of double descent, which overthrows our classic intuition of overfitting [49, 506]. Another excellent book on statistical inference, Efron and Hastie [142], provides a wonderful, succinct treatment of neural networks and their training. Goodfellow et al. [186] is a useful overview of modern neural network and deep learning methods. And with so much of machine learning, including neural networks, predicated on linear algebra, in particular matrix decomposition and factorization, Strang [446] is an excellent dive into the intersection of learning and linear algebra.

It must also be said that there are profound ethical and safety concerns when it comes to using machine learning, particularly black box methods. From misuse in the criminal justice system, to automating disinformation in online social media, to accidents caused by experimental self-driving cars, AI systems have real-world, deadly consequences. O'Neil [352], give an exciting, and worrying, general audience overview of how big data and algorithms reinforce discrimination and cause societal harm. Readers wishing to avail themselves of even more existential dread may enjoy Bostrom [69], as we ponder the delightful question: Will AI research lead to a super-intelligence that causes humanity's extinction?

Exercises

16.1 Consider the polynomial regression example in Sec. 16.4. Reproduce Fig. 16.1d. Then, make a plot showing average fitting error as a function of polynomial order d. Include $d < n$, $d = n$, and $d > n$ (go out to at least $d = 2n$). Interpret the curve.

16.2 Neural networks always use nonlinear activation functions. Why? What does a neural network with linear functions end up doing?

16.3 Suppose you are analyzing a link prediction algorithm. Think of this predictive model as a binary classifier: given a pair of nodes, predict 0–no link, 1–link.

   (a) You find that a baseline classifier that always predict "no link" works well. Why? What metrics would show this and what does it mean?

   (b) Interpret performance: What does it tell us if the link predictor has very high *precision* but very low *recall*?

16.4 In an ideal world, training and testing data are unrelated to one another. This can be achieved in a tabular dataset, for example, when observations (rows) are independent from one another. We can then split the rows at random without fear of data leakage. But it's not so simple to split a dataset X where each row represents a node in some network G.

(a) What about G makes leakage more or less of a concern?

(b) If we were to split X into training and testing, what may be a good strategy for doing so?

16.5 **(Focal network)** Consider the Malawi Sociometer Network.[26] How predictable is this network? Divide the data into two parts based on time and make a weighted network for each part. Split the data so that equal numbers of days of data are covered in each part.

Build a predictive model (a classifier) to predict the presence or absence of an edge between a given pair of nodes. Use the first network as training data and the second network as test data.

(a) What model did you use? What features of the network are most useful for predicting edges?

(b) How well does your predictive model perform? Compare your predictions to a baseline model that predicts the test network is identical to the training network.

(c) **(Advanced)** How do your predictions depend on where in time the data were split?

---

[26] See also Exs. 15.5 and 15.6.

Interlude — Good practices for scientific computing

About this interlude

Before moving on from the applied focus of Part II to the theoretical aspects of working with network data in Part III, here we take a brief detour to discuss practical aspects of doing data-driven and computationally driven research. Our advice is applicable to network data, of course, but is more general, and we hope useful to any computational research you may conduct.

Chapters 17 and 18 tackle a major bugbear of computer work: keeping track of what you have done and what you have changed and, just as importantly, why. Record-keeping is critical for good science but computer work is different in some fundamental ways compared to the traditional laboratory setting. We discuss ways to cope.

Computational work is driven by computer code. You will read, write, and run code as you work. You want code to be reliable and understandable above all else, with computational efficiency being a further desirable property. What are good strategies for producing quality scientific code? Software engineering principles help, but coding for software and coding for science can be quite different. Turn to Ch. 19 to learn more.

In general, this interlude is structured around advice for computational work that is relatively independent of specific tools. Computer software is a high-churn environment, with packages and programs always coming and going. Any advice we give for very specific software is sure to be soon obsolete. That said, we round out the interlude by discussing specific tools to help you work in Ch. 20.

Chapter 17

Research record-keeping

As you conduct research, you must maintain a record of your work. This record can be used by yourself, for reflection and even inspiration. It can also be used by others, say members of your research group who need to reproduce your work at a later date. Good record-keeping ensures your collaborators are well informed, and your closest collaborator is yourself, in the future, so you will benefit most of all from good record-keeping practices.

Illustrating the need for organized records, Kanare [232] relays an interesting anecdote about the discovery of Uranus and the French astronomer Pierre Lemonnier. William Herschel, not Lemonnier, is credited with discovering Uranus in 1781. However, Lemonnier had made multiple observations of Uranus in 1763 and 1769. But, as described by another astronomer, Francois Arago, he never realized he was looking at an undiscovered planet because his notes were in such disarray. Arago wrote, after seeing Lemonnier's notes:

> [...] Lemonnier's records were the picture of chaos. [Alexis Bouvard] showed me, at the time, that one of the observations of the planet Uranus was written on a paper bag which once upon a time held powder to powder the hair...

Perhaps with a more organized and systematic note-taking system, that great discovery would not have passed by Lemonnier.

Here we discuss record keeping, like maintaining a lab notebook. Historically, lab notebooks were analog, pen-and-paper affairs. Now with so much work being performed on the computer and with most scientific instruments creating digital data directly, most record-keeping efforts are digital. In this chapter, we cover strategies for keeping records with a focus on computer-based work. In subsequent chapters we take a deeper dive into record-keeping specific to data followed by coverage of good practices for coding.

17.1   Establishing a research record

One of the fundamental challenges with developing a good record-keeping habit is that it takes time. You need to move slower, more deliberately, at least a little. In our rush to get through the slog of coding up an interesting hypothesis—what is the answer, I must know!—we can get sloppy. Yes, we can go back afterwards and clean up our code, take down notes about the day's work, but short-term memory is fragile, and important details can be lost. We argue that it is best to keep records contemporaneously, as events occur. But, acknowledging the challenge in practice, we also advocate for a practice of double-checking and cleaning up afterwards, for those times when we lose the path.

While it slows us down in the short-term, over time there is a huge cumulative benefit to good, accurate research records. By documenting the paths we have traveled, hypotheses we have considered, and questions we have investigated, we can avoid repeating ourselves, we can reflect on past work to inspire ourselves, we can teach our students and collaborators more quickly what we have done, and we can build papers and presentations from our records without starting from scratch. These effects make us work more quickly and more reliably.

In general, we will assume the records being kept are private to yourself and possibly close collaborators. Open science is valuable and gaining traction, and there is value in keeping open records of the process, not just the results, of the research. But right now this is far from the norm, and "working in public" is rare for scientists. That said, think of the broader implications of your records as you take them: while unlikely, they may be involved in a lawsuit or made public somehow.[1] Keep in mind that formal records may be accessed in ways you do not anticipate, and plan accordingly.

17.1.1   Documenting your work and your ideas

We have found that good practice is adopting a future-first mindset. At all moments, remember that what you are creating is not for the here-and-now, but for the time stretching in front of you. Have you captured enough information so that a far-away descendent can understand—and reproduce—what you are doing?

The future comes at you fast, and that far-away descendent may be yourself. Even just a few months in the future (say, enough time for a journal paper submission to go through a round of peer review) and you will be shocked how much front-of-mind knowledge has been lost, especially if you are busy with other things.

 Write for the future, both yours and others.

At first, when you are deeply embedded in one problem, it can be difficult to write down everything you need because you may not realize how much there is to know—it's all tacit, right at your fingertips. But with experience, you will soon begin to realize what information in your work is not being recorded in the moment, and then you'll see what you need to be writing down!

---

[1] Climate change deniers going after climate scientists have made inroads by forcing the release of otherwise private emails and research records [271].

Capture and reflect on your ideas. Keeping notes of what you're thinking about won't be very helpful if you don't use those notes. (They may help a bit, the act of writing something down often helps solidify it in your memory.) One strategy to consider is some form of *spaced repetition*, where you prompt yourself with previously recorded ideas. Usually, spaced repetition is done with flashcards and helps students memorize ideas or concepts, but here we can simply use these prompts as we work to see if any new connections form. An idea we had months ago, wrote down and promptly forgot, may serendipitously be brought to the fore at random precisely when we realize its implications in a different area. Another approach is to turn out a (simple) taxonomy of ideas, adding categories or tags to your ideas, describing their current feasibility and what other collaborators may know about or be interested in that area. Sorting through and re-tagging your ideas may be unhelpful busywork, but a little bit can go a long way to prompting new research projects.

17.1.2 Defining and adopting record-keeping standards

As mentioned earlier, it's challenging to keep good records while you work. One strategy for maintaining useful, actionable records (records you can glean information from in the future) is through adopting a standard.

Think of your record-keeping standard as filling out a template or following a checklist while you work.[2] Indeed, checklists are well known as a good strategy for ensuring that a repeating procedure is followed through each time: doctors and pilots now routinely run through checklists for surgery and flying, improving their safety and consistency [176].

What's a good checklist for keeping records of your computer work? Here's a starting point:

- ☐ Who worked and where?
- ☐ When did work begin and end?
- ☐ What research question(s)[3] did you work on?
  - ☐ What were the results?
  - ☐ What questions should be (re)prioritized based on these results?
  - ☐ What new research questions were you led towards?
- ☐ Where were records taken as you worked?
- ☐ What files changed?
  - ☐ What files were changed by you?
  - ☐ What files were changed by your code?
- ☐ What code was run?
  - ☐ Outcomes from that code?
- ☐ *Final check.* Can someone else successfully repeat your research with these records?

---

[2] This idea of a template recalls a physical paper lab notebook. Often lab notebooks pages are printed with boxes for names, dates and other records for noting entries. Keep this in mind with our digital records.

[3] You may want to generalize this to include "tasks" more broadly. For instance, "Set up download script" isn't really a research question. But whenever possible, framing the work as a research question is good practice to keep you focused on, well, research.

Some further questions may come to mind as you consider the practice of your record-keeping:

What does your research group do?   If you work in a group, like a research laboratory or data science team, what is the expected convention? Are traditional lab notebooks kept by all members of the group, perhaps even using a procedure for storing notebooks in a locked cabinet? Or maybe there is a shared knowledge base or "wiki" maintained on the group's computers? Does someone track the lab notebooks or monitor the wiki?

What does your field expect?   If you work in a certain research field or sector of industry, a convention for records may already be in place. Field-specific conventions are often taught as part of training, and allow researchers to move more quickly between research groups or companies. If you are new to a field, consider asking a more senior colleague or mentor what conventions are expected.

Lastly, keep in mind that there are often legal and ethical aspects to documenting your work. An employer may have strict requirements for record-keeping to fend off patent lawsuits. For example, it may be standard practice to have a second lab member initialize pages of a notebook to indicate that they can act as witness to any discoveries.

17.2   Backups and backup practices

An implementation detail closely tied to good record-keeping is maintaining backups of your digital files. Backups are important. Digital data are rather ephemeral and all it takes is a small fault to destroy the files on a piece of hardware.

Here is a good mantra for copies of critical information: *two is one and one is none*. Assume anything that exists in only a single place doesn't exist at all. Make sure it's replicated—backed up.

Backups vs. mirrors

Having a copy or, better, multiple copies of your files replicated somewhere else is important, but is not necessarily the end of good backup practice.

Imagine you accidentally delete some files locally without recognizing it (unseen deletions are an especial risk when writing file-manipulating programs). If your remote backup is then synchronized to be an exact copy or *mirror* of your local files, then those deleted files will be removed remotely as well. Your files are now lost. Now imagine you did not delete those files but modified them in some way. When those modifications are synchronized to the remote backup, the earlier contents of the files are lost. Despite having a second copy, you have no backup.

The best solution to this quandary is to copy not just the current state of your files but also a history of the changes to those files. You can then "rewind" that history when you need to retrieve a missing file or recover the contents of an updated file.

> ⚠ A remote copy or mirror of your files is not always enough. Ideally you want to back up the full *history of changes* to your files. "Rewinding" the history will allow you to retrieve deleted files and recover the earlier contents of modified files.

Keep in mind that many cloud file storage providers, such as Dropbox, Google Drive, and others, do not keep a full history by default.[4] If you use such a service as a backup, good. But keep in mind these concerns and consider whether you would benefit from more coverage.

17.3 Summary

Keeping good records of your work is important. These records inform your future thoughts as you reflect on the work you have already done, acting as reminders and inspiration. They also provide important details for collaborators, and scientists working in large groups often have predefined standards for group members to use when keeping lab notebooks and the like. Computational work differs from traditional "bench" science and this chapter has described practices for good record-keeping habits in the more "slippery" world of computer work.

A close corollary to research record-keeping is data provenance, record-keeping for your data. We turn to this next.

Bibliographic remarks

Kanare [232] is a classic well worth studying for more insights into keeping scientific records. A more recent study of research record-keeping best practices in the context of medical research but still of broad interest is Schreier et al. [418]. Of particular note is their hierarchical breakdown of best practices for individuals, leaders of research groups, and departments or institutions.

---

[4] Some services may offer a truncated history of changes or provide histories as an additional paid service.

Chapter 18

Data provenance

Data provenance or data lineage refers to the detailed history of how data was created and manipulated, as well as the process of ensuring the validity of such data by documenting the details of its origins and transformations. For instance, we want to know where (by whom) a dataset was created and what was the process used to create it. Then, if there were any changes, such as fixing erroneous entries, we need to have a good record of such changes.

18.1   Why should we care?

We care because things can go terribly wrong if we don't. Imagine that your year-long analysis was based on a wrong, or simply outdated, version of the dataset! Or imagine your analysis was based on a wrong assumption on the meaning of a column in the dataset. Not ensuring data provenance can potentially lead to career-ending disasters.

Imagine a file generically named data.txt containing some tables of numbers but without explanation. No header names, no readme.txt you can find. Then imagine you discover files data-v1.txt, data-v2.txt, data-v2-final-new.txt. Which file should you work with? Who made the different files and when? Good research practice means we must have answers to these questions. We must both understand the provenance of the data and ensure that the provenance remains accessible moving forward.

18.2   Best practices for data provenance

The key to data provenance is record-keeping while you work. New data replaces old data. Data will be changed by your code, then the code will change. Data will be changed by your collaborators, without explanation. Having a personal practice of record-keeping is critical, and an eternal vigilance.

18.2.1   Document, document, document

The most important step is to document details about your dataset. Are you constructing
your dataset? Then document each and every step of the construction process metic-
ulously. Have you received the dataset from someone else? Then document each and
every step of the data acquisition process. Who sent you the data? When? Were there
multiple updates?

Then document what is in the dataset. Document each and every column of the
dataset. What is the meaning of each column? What is the data type of each column?
What is the range of each column? Some columns may use pre-defined *data vocabularies*
such as ICD-10 (International Classification of Diseases, 10th Revision) codes; some
columns may contain free text. Together, the detailed documentation that explains a
dataset is known as the *data dictionary*.

Then document your processing of the dataset. A good way is *never touching the
dataset by hand* and doing everything through a script. The script can then serve as a
record of the data processing steps. For instance, suppose there is a typo in the dataset
and you need to replace some words. First, create a separate data table that contains all
the replacements to be made then write and run a script to apply those replacements.
The replacements table and the script together document your changes.[1] And the script
and data can be woven into a workflow using workflow tools (Ch. 20).

Documentation is eternal: when (not if) data change, the documentation will need
to change with it.

Data identifiers and filenames   It should never be a mystery where a file came from
or what it means. Appropriate identifiers are key to good provenance. Identifiers can
be URIs (uniform resource identifiers) or DOIs (Digital Object Identifiers), but at the
most basic level we can think of them as filenames. Identifiers should be unique and
follow a consistent naming convention. Consistency prevents confusion and forgetful-
ness around the data; once you design or learn the naming pattern, you can quickly
derive the identifier. Choose identifies that give enough information so an interested
party can figure out where the data fits into the research. We recommend identifiers
that incorporate project name, author name (if appropriate) or data source name, a brief
description of the data (say two–three words) and a rough indication of the file format
(which can be as simple as a file name extension). We see that `inferred-social-
network_bagrow_node-attributes_v20221001.json`, while long, is more informative
than `node-attributes.json`. Avoid special characters and spaces in filenames, and
document any abbreviations used to prevent names becoming overly long.

> ⚠ Avoid generic file names, such as "f.txt," "result.dat," "output.csv," etc. Even for
> the briefest programming session, be fastidious with filenames.
>
>   A corollary: if you encounter a codebase or dataset containing such generic
> filenames, be wary of problems and skeptical of the data provider's commitment
> to data provenance good practices.

---

[1] Consider also recording metadata such as when the script was last run and why. This may also be captured
automatically with a logging framework.

Always use a unique filename such that you can determine where in your code the file was made even without running the code. Add "slugs" as unique parts of the name when building file names programmatically.

Data versioning As data are modified, you'll need to track different versions. If not, you may wind up with different copies of the data and be unsure which to use. Even worse, the different copies may have undergone separate revisions, causing a diverging history that you will need to reconcile. We recommend three practices to help alleviate this problem. First, use data versions; we recommend a timestamp-based approach (see also Sec. 15.6.2), not version numbers. And never indicate "version new" or "version final" as you will soon have "version new new," "version final new," "version new final 2," etc. That way lies madness. Second, when possible, log whatever the latest version of the data is, alongside older versions. Three, avoid discarding older versions: you never know if you will need to consult them in the future. The second and third practices interact: as you revise data, update the "latest version" metadata on *all* prior versions. This can be a burden if not automated, but it is good practice.

 Avoid "free-floating" version identifiers. When confronted with "data-new.txt," "data-new-final.txt," and "data-new-new.txt," you will struggle to tell which to use.

 Use timestamps as version numbers for data.

 Don't rely on filesystem metadata such as "last modified" dates. These can be notoriously ephemeral, one disk migration or operating system upgrade away from being rewritten.

Checksums A computational tool that can help with tracking data provenance is the *checksum*. Checksums are small blocks of data, computed by an algorithm, that describe our data. Crucially, if our data change by a tiny amount, the checksum will be completely different. Checksums are often used to check for errors when data are transmitted. If the checksum on the received data has changed compared to the original checksum, we know a change happened; conversely, if the checksums are equal, there is a very high probability the data are unchanged. Checksums are also very important for cryptographic security, but for data scientists, integrity checks are the most valuable use of checksums. We recommend recording checksums whenever you generate an important data file to help you distinguish the file and detect if errors in the data have appeared. All computer platforms come with built-in commands for computing checksums.

 Compute and save checksums of important data files. Keep them in a readme or other documentation file, along with the date when they were computed and other relevant details.

Raw data should be read-only   A simple step to safeguard data is making raw data read-only. Different computer systems and database processes provide file access permissions in different ways. Flagging data as read-only forces you to never change the data by hand. You can only derive new datasets from the raw data. While not foolproof, this extra layer of permission does provide a valuable safety check, particularly against accidental modifications (such as from buggy code).

18.3   Backups

As discussed in Ch. 17, backups are critical when working with digital files. While computers are astonishingly reliable these days, it is still easy to lose information.

Backing up data presents challenges and opportunities when it comes to tracking data provenance. First, backups necessarily mean multiple copies of the data will exist. You need to keep track of and ensure all copies are synchronized, updated, and otherwise tracked.

What happens when data files are dynamic? For example, an experiment may be ongoing and data continue to be generated. Or the data come from a web service and new outputs are collected, say, daily. Follow the data versioning practices we described above as part of your backup practice. Together, good identifiers and versioning can help maintain provenance when backing up data.

Finally, consider using checksums to track different versions of a backed-up file. Checksums are again helpful here, to track the integrity of your backups. Keep logs of checksums separate from the files themselves so that they are available in the event of data loss.

18.4   Summary

Data provenance is a central challenge when working with data. Computing helps but also hinders our ability to maintain records of the work we do with the data. The best science will result when we adopt strategies to carefully and consistently record and track the origin of data and any changes made along the way. While such strategies generally take time and effort to implement, making them seem tedious in the short term, over time your research will become more reliable and you and your collaborators will be grateful.

Bibliographic remarks

The overview of record-keeping best practices by Schreier et al. [418]. includes helpful information for data provenance. An influential review by Gray et al. [191] is also worth consulting, particularly the discussion of scientists choosing between files and databases. Edwards et al. [141] discuss pain points between data management and interdisciplinary collaboration.

Chapter 19

Reproducible and reliable code

Most scientists receive training in their domain of expertise but, with the possible exception of computer science, students of science do not receive much training in computer programming. While software engineering has brought forth good principles for programming, training in software engineering does not translate completely to scientific coding.

19.1 Coding for readability

Boswell and Foucher [70] introduce what they call the "fundamental theorem of readability" when it comes to code:

> Code should be written to minimize the time it would take for someone else to understand it.

We agree wholeheartedly with this sentiment and believe it should be a constant guiding practice to follow whenever coding.

Much ink has been spilled on good coding practices (see the end of this chapter for reading recommendations). We summarize our thoughts here, with an emphasis on advice for making code that is readable (and thus, useful) to yourself and others in the future.

Simple program organization

At the large-scale, when you are writing code you need to make decisions about how to organize your data and functions. Should you use a class to track network data as objects? Should you use a dataframe for network-associated metadata? What computations will you perform most often and should they be incorporated into functions?

We recommend using a simple program organization. Your project should consist of a central library of reusable code, preferably in a single file, along with a number of "task" scripts, scripts focused on performing specific jobs. Place all your library import statements and reusable functions into the central file and use a standard name for that file across all projects (we use funcs.py for Python scripts) then import from that central file at the start of any task scripts.[1] These task scripts are dedicated to single operations (retrieve and process data, exploratory analysis, statistical tests, . . .). Keep task scripts short and single-purpose.

 Use a *standard comment block* at the top of each script explaining its purpose. Make a template for this comment and use it consistently across all files.

 Mark the trailhead.

When you're trying to understand a large amount of new code it can be difficult to know where to begin reading, where is the starting point of the computation. *When writing code, always make sure the starting point is clear to any reader.* The best way to do this in complex scripts is with a "main" function (usually called main). Ideally, a reader of your main function will be able to glean the overall outline of your computation, even if they do not (yet) know exact details. Make sure the main function or other starting point is described at the top of your script file, so a reader will know where to begin.

Favor homogeneous data structures   Often, especially when working with APIs, you will find yourself working with a complex nested data structure containing all sorts of information. This makes it difficult to tell what any given record will contain, either the fields present in that record or what the fields may mean. Further, if you are looping over that structure, each record may present different data and your code inside the loop will need to check for what is present. Avoid these situations with a homogenous data structure: make each element store the same information in the same way.[2] Avoid surprises.

Don't repeat yourself—the DRY principle   If you find yourself reaching over and over for the copy-paste command of your code editor, something is probably wrong. Instead, determine what code you are repeating, pull it out into a function and then use that function wherever the duplicate code was. Duplicating code is risky. If you need to edit the code at all, you will need to track down and update every duplicate. There's a high risk you will introduce bugs.

 Avoid copy-pasting code. Don't repeat yourself.

A corollary to the DRY principle, which emphasizes its strengths, is SPOT: Single Point of Truth. When introducing any data into your code, define it exactly once, in one place, and reference it everywhere else from there. A simple example is a script that reads a file. Store the file name in a variable (file_in = "data.txt"[3]) and then use that

---

[1] Task scripts can also be computational notebooks.

[2] Of course, you will in time find yourself dealing with someone else's heterogeneous data structures. When this happens, we feel your pain. When you can control the form of the data, go for homogeneity as much as possible. Someone downstream from you will be grateful.

[3] In practice, don't use such a generic filename!

variable everywhere it is needed. Do not write "data.txt" throughout the script. If you do, and you want to change the file name, you will have to track down every instance and rewrite it. While a repeated filename is quite simple, more complicated, hard-to-find bugs can creep into a large code base when constants and data are imprecisely duplicated.

Decompose your code into simple blocks. Use whitespace (blank lines) strategically to separate the code into logical pieces. Aim for single-purpose functions and perform complex tasks by combining and recombining those functions. A single-purpose function is easier to understand and debug (see also "informational dependencies" below).

 Divide and conquer. Break programs down into small pieces that you can keep in your head.

Informative variable names

At the small scale, the readability of your code depends on how you name your variables. Informative variable names can communicate both what the variables are and what their purpose is. Well written code with good variable names can be so self-descriptive that few code comments are needed.

Some advice:

Favor concrete and specific variable names over abstract and generic ones.[4]

Example: Python dictionary. Suppose we have a dictionary in Python (also known as an associative array, hash, or map in other programming languages). Dictionaries map keys to values. Compare the following possible names for a dictionary produced by a function:

```
D = build_dict(...)
dict_data = build_dict(...)
node2name = build_dict(...)
```

If we encounter D later in this program we know essentially nothing about it. If we encounter dict_data on the other hand, we know we are dealing with a dict but that's it. What's in the dict? No idea. But if we run into node2name we have some information just from the code itself: it's a dictionary that takes a node as a key and returns the name of that node.[5,6]

Encode units   If you have data that comes with units, try to put that information into the variable name. For instance, if you are recording the age of each link in a temporal network, don't use age (age of what?) but instead linkDurationSeconds. Now we know much more about what we are dealing with.

---

[4] This also goes for filenames as well as variables.

[5] We may want to be even more descriptive and use nodeid2nodename or the like.

[6] We are big fans of the notation key2val for associative arrays and even key2key2value for nested arrays. However, don't *literally* name a variable key2value, k2v, etc., as such names are too generic in practice.

Plural names for "loopables"   A plural name is a great way to clue the reader in that a variable is a data structure (list, vector, dictionary, etc.) which can be iterated over. `user = read_data(...)` is not nearly as clear as `users = read_data(...)`. That simple change conveys a lot.

Other coding practices

Comments explain why, code explains how   Comments within code are critical to explain the code—use them wisely. Your code should be written so that a reader can tell how the program works simply from the code itself. A comment can then provide the reader with information about why the code is doing what it does. Compare this alternative (comments begin with #):

```
x = x - 1 # decrement x
```

versus

```
x = x - 1 # prevent double-counting at end of list
```

The first comment tells us nothing but the second gives the motivation behind what that code does. Later on, if a change to the code eliminates that double-counting, you will know you should remove this line. With the first comment, you may be frustrated for some time figuring out what the line does and dealing with bugs.

> ⚠ Comments can lie.
>   Be wary of out-of-date comments and documentation. As code is rewritten, especially in a rush, the comments may no longer reflect the actual code. When you write code, always update comments and documentation.

Keep informational dependencies local   While working on lines 720–750 of your script, don't make it so you need to also know detailed information about variables defined on lines 125–150. Later, when you are working on line 740, you will have completely forgotten what was so important on line 130.

"Premature optimization is the root of all evil"   We want our code to run quickly. But there are two ways to think about speed:

1. The time it takes you to write the code.

2. The time it takes the computer to run the code.

Generally, (1) is far more important. And (2) often conflicts with the fundamental theorem of readability: optimizing code to run as quickly as possible almost always makes the code more complex and difficult to understand.

  Do not optimize your code for speed if you don't need to. If you need to, do not optimize your code for speed until you know it works. When it works, and you're confident it works, archive everything you can, whether it be input and output files, runtime logs, whatever needed, so that when you dive in and crank up the optimizations, you will be able to refer to the original working-but-slow version to make sure the optimized version is also correct. A very fast, incorrect code is far worse than a slow, correct code.

Do not control aspects of your code by commenting and uncommenting blocks   As you perform some exploratory analysis or computation, you may find yourself producing a script that does a bunch of different steps and you may want to learn which ones are needed and which ones are not. You may be tempted to run and rerun your script with different sections of the code disabled using code comments. Avoid this. First, manually selecting and toggling comments risks introducing errors: you may fail to comment out a specific line and cause your program to run incorrectly (and you may not notice this). Second, it becomes hard to distinguish the different results of the script as it is run. Other than your commenting, nothing has changed, you've run the same script the same way. So you won't have good records of what has been modified, unless you manually start making other changes (every time you comment out a block you also change an output filename, for example). This is also error-prone.

Instead of this manual work, take a little bit more time to structure your code with if-statements and true-false variables. Use the if-statements and your true-false control variables to control where the code goes. Now you can loop over all the different options, add more options with more if-statements, and you can log the values of the control variables whenever needed. While it takes a bit more work up front, this good practice will make both capturing provenance (Sec. 19.2) and automating your script much easier.

19.2   Coding for record-keeping

Software engineering has codified best practices for writing high-quality, reliable code. Many of those practices carry over to scientists producing code as part of their research, but not all of them do. For example, one of the most important factors leading to robust, reliable software is *resilience to errors*: you do not want your computer operating system crashing, for instance, just because you tried to read a corrupted Microsoft Word file.

But for scientists, encountering errors and problems in data is not something that should be caught and handled in silence. Instead, you should be aware of such issues so you can determine their seriousness. Are you reading an inherently messy file or did something go wrong in an earlier step in your code that you should fix? Is the credibility of your research effort called into question by these faults?

 Fail explicitly on errors. Avoid silencing or ignoring errors. It can be annoying to deal with code that doesn't always work, but you must know when errors occur and why.

Record-keeping is central to good scientific practice. Doubly so for coding. What are the ingredients of code that is reproducible and keeps good records?

1. Assume code will need to be understood when it cannot be run.

2. Document all input and output file(names) and formats explicitly.

3. Document explicitly all external dependencies.

We discuss each in turn.

Assume code won't run　Write code that is understandable even when it cannot be run. Do not rely on someone inspecting your program's output to determine what files were created, what was in those files, and the like. Play it safe and adopt a worst-case mindset.

 Do not assume your code will always work. Someone may need to understand your code but cannot run it. Write code that is understandable even for someone who cannot run the code themselves.

Document inputs and outputs　As we discussed in Ch. 18, it should never be a mystery where a file came from. If you are writing code to produce files, it is worth following some good practices. Short filenames are helpful but they should be informative. (The same goes for table names if you are storing data in a database.) Use the filename, including file extension, to make it easier to understand the contents and format of the file. Use standard extensions—don't use .txt for a CSV file, use .csv. Finally, make sure you can find the code that made a file *from the filename*. Filenames that are built by assembling strings in your code are helpful, but if the code won't run, it may be difficult to find out what code created what files. Make it obvious how filenames were generated.

 Make filenames trackable. When creating a file with your code, make it so you can easily find where in the code your file was made just from the filename.

 Include example input and output data files, even just for yourself.

Document dependencies explicitly　Good code should always document its dependencies. If you are writing a Python script and it uses the "numpy" library, make sure this is clear. Further, make sure you document what version of that library is being used. If in the future, someone is using a much newer library than the one you wrote the code for, an incompatibility could introduce bugs.

 Document all dependencies of your code, including version numbers. Ensure these are updated. If your programming language has a convention for this information (in Python, a requirements.txt file is standard), use it.

Other practices

Use version control　Version control systems (VCS) such as git (which powers GitHub) are used by programmers to keep the history of changes to a code base. This facilitates collaboration, as tracking the changes to a code base can allow multiple programmers to "merge" their edits into the code. Even without collaboration, a VCS is helpful for reproducible code. If code is revised in place, the earlier versions of the code will be lost, but if the code is tracked with a VCS, those earlier versions can be brought back. If something has happened and results from months or years earlier can't be reproduced, you can at least compare the code changes and see if something has gone wrong. We discuss VCS in more depth in Ch. 20.

 Use version control. Even when you are not collaborating with others, version control makes backing up easier and lets you record the history of your code. History-tracking can be invaluable for tracking down bugs and changes in outputs, helping you maintain data provenance.

Use logging Most programming languages come with *logging frameworks*. These provide functionality to keep a record of what your code is doing in a separate log file. Effectively, as you write your program, you are also building a small automated lab notebook. Every time you run the code, your logging function can call out and record the run time, the inputs and outputs, checksums for files (Ch. 18) and any other information you need. Of course, you can code this record-keeping yourself, using the programming language's built-in file-writing features. But in many respects, this is a solved problem, and taking a bit of time up front to learn how others do their logging can work wonders for your productivity.

We recommend adopting some conventions for logging. Record details of the current session or script run, including date and time, the name of the machine you are working from, and the user who is performing the work. Define and identify "jobs" uniquely, for example: "PROJECT_jobNN_YYYY-MM-DD" where "PROJECT" describes the project and "NN" is an integer tracking the number of the job for that day. Record what script was run, record the names and locations of all files read or written by the script, and any errors or warnings that your code gives. Depending on file size, record checksums for all input and output files. Record the checksum of the script itself (yes, you can write code that reads itself). Finally, log the final status of the script: did it exit successfully or did it end with an error?

While it takes time to set up logging, both to learn the framework and to add it to your code while you work, it will be well worth it in the long run.

 Use logging functions in your code to automatically record runtime details such as when the code was run, what inputs and outputs were made, and more. Adopt and follow a standard convention for logging formats across projects.

 Automated logging can be a good supplement for manual record-keeping such as a lab notebook, but it is not a replacement.

19.3 Summary

Coding for science is not the same as coding for software. The key to writing good data science code is to keep it simple. The challenge is finding the right balance between simplicity and functionality of what your code does. Sometimes, highly flexible general-purpose code is the right target to aim for; often, however, it can create a burden on the reader who can't see what you actually did with the code. Favor explicit practices and ensure your code is understandable to a reader who lacks the ability to execute the code—there may come a time in the future where the code no longer works, perhaps because of a dependency change or missing input file; make sure the information—the *intent*—of your code remains available to the interested reader.

Bibliographic remarks

The software carpentry project[7] provides many useful resources to help scientists improve their code [493, 494]. Boswell and Foucher [70], while not specific to scientific coding, is a wonderful text on writing code well.

---

[7] https://software-carpentry.org

Chapter 20

Helpful tools

Many tools exist to help scientists work computationally. In addition to both general purpose and domain-specific programming languages, a wide assortment of programs exist to accomplish specific tasks. We call attention to a number of tools in this chapter, with a particular focus on good practices when using them, good practices computationally and good practices scientifically.

20.1 Computational notebooks

Computational notebooks play an increasingly important role in scientific computing and working with data, network or otherwise. All data scientists should be familiar with their use and understand good practices for ensuring they are used appropriately.

Examples of computational notebooks are Jupyter, R Markdown, and perhaps the pioneer of the format, Mathematica. Unlike a normal program or script, a computational notebook divides its contents into a sequence of blocks, often called cells. Cells can be executed individually and any outputs that would normally be printed to the user's screen are instead inserted into the notebook itself. Further, notebooks allow for different types of cells: code cells or text cells, with the latter allowing the user to record important information, thoughts, etc. These text cells are really what makes a notebook a, well, notebook; often, they support rich formatting, hyperlinks, and mathematical typesetting.

The power of notebooks comes from their ability to weave together code, the results or outputs of code, and non-code writing. Notebooks implement a kind of literate programming [247].[1] The power and flexibility comes at a cost, however.

Some inherent aspects of the design of notebooks can lead users away from good programming practice. For example, the interactivity of notebooks, that cells can be edited and rerun, especially out of the order in which they are written, makes notebooks ripe for bugs. Often, it's a good idea to rerun the entire notebook top-to-bottom to make sure all the code still works—and this can suddenly reveal subtle mistakes due to notebook cells being out of order. Worse, this problem is pervasive enough that often

---

[1] Literate programming as advocated for by Knuth [247] puts prose writing and computer programming on equal footing.

times scientists won't rerun the notebook, taking the results as a finished artifact as is, and this leads to reproducibility issues as the intervening edits that got the out-of-order notebook to its final state are not present in the notebook itself. This latter point, this loss of information, harms notebooks-as-archives, despite archiving being one of a notebook's central purposes.[2] Indeed, while many factors contribute to reproducibility issues with code (Ch. 19), a large-scale study of Jupyter notebooks on GitHub [374] recently found that only 24% of publicly available notebooks executed without errors, and only 4% produced the same results!

In general, we consider notebooks to be an important tool for data scientists, but one whose strengths and weaknesses should be well understood. Use notebooks appropriately.

20.2 Pipelines

It is useful to think of computational work as being composed of pipelines: First, data are downloaded. Then, loaded into your program. Next, a processing step is applied, perhaps a filtering criteria is used to select an appropriate subset of data. Afterwards, apply a statistical method. Finally, compute summary statistics or perhaps make some plots.

Together, the linear sequence of these steps forms a pipeline that you have built. Will you find yourself repeating those steps, for instance if the data are updated? Will you need to create a new, similar but not identical pipeline for a future calculation?

Such situations arise often enough that a variety of tools have been introduced to create, maintain, and run these computational pipelines. By far the most popular, particularly in computational biology and bioinformatics, is *Snakemake* [252]. Snakemake goes beyond basic pipelines with an entire workflow management system. Using readable Python code, a workflow can be described or documented in Snakemake, and that workflow can be carried over from your local machine to remote servers to large compute clusters.

Workflows and pipelines also work extremely well at gluing together different computing tools. Need to combine Python code with some cutting-edge Julia or legacy-but-still-potent Fortran code? No worries, interoperate these languages within your pipeline. In fact, a key idea of the UNIX operating system (Sec. 20.4) is to build most functionality out of pipelines of small, single-purpose utilities. UNIX has pipes! Now you can use specialist languages where they are most suited, and glue them together in a pipeline.

---

[2] Other, more technical issues arise when storing notebooks in a version control system (Sec. 20.5). Rerunning an identical notebook and getting identical results may still lead to changes to the contents of the notebook, changes invisible to us as users but still present in the data. These changes need to be stored in the version control system and they make it difficult for users tracking the notebooks to easily understand whether changes to a notebook are important or not.

20.3 Working with remote computers

Modern personal computers are exceptionally, impressively powerful, but scientists still find themselves using large workstations and supercomputers to analyze data and perform computations. Typically, these computers are used remotely: you sit at your personal device (say, your laptop) and connect to the computer over a local network or the Internet.[3] The connection allows you to transfer files back and forth between your local and remote computers as well as execute commands and run programs on the remote computer. Somewhat recently, "cloud computing" services have sprung up which act in a similar way, though commercial providers often provide easy-to-use website interfaces. Yet fundamentally, while these providers hide the details of the computers, you are still connecting with and using a remote computer or set of computers.

Graphical vs. text-based interaction

One way to work with a remote computer is through a tool such as VNC (Virtual Network Computing) or other variants of what is called "Remote Desktopping." Remote desktops provide a graphical means of working with the remote computer. In essence, a window is open on your local computer inside of which is the graphical interface of the remote computer—its windows, desktop, files, and such. You can click and drag inside this window, type commands, and act in many ways as if you are actually sitting in front of the remote computer.

The second way to work with the remote is through a text-only interface, called a command prompt or command line interface. Here you have the ability to issue commands to the remote computer by typing them in only. (Historically, this was the only way to work with a computer, local or remote.)

The graphical user interface (GUI) and command line interface (CLI) approaches have advantages and disadvantages:

- GUIs over a network connection can be annoying to use as any delay introduced by sending the graphics over the network makes the computer feel sluggish.

- GUIs require more computing resources on the part of the remote computer. Often, remote computers are set up without running any graphical interfaces in order to use as much memory and computation for their tasks.

- CLIs have a steeper learning curve compared to a GUI, particularly if the remote computer is using a GUI you are already familiar with. However, for scientific work, CLIs are incredibly powerful and we encourage you to invest the time to learn them.

Nowadays, any computer you sit down and work at will be running a GUI. But you can quickly access a CLI inside your GUI using a "terminal" or shell program.[4] Mac

---

[3] We forget, but very early computers worked exclusively in such a manner: an operator sat at a device and communicated with a central "mainframe" computer. The difference was the device was not a computer or laptop but usually a teletype: an electric typewriter connected to a telephone line. Early computer terminals with video screens were actually called "glass teletypes."

[4] Sometimes these programs are called *terminal emulators* because they mimic (emulate) within your computer an old-fashioned text terminal connected to a remote mainframe computer.

computers come with an application *Terminal.app*, Linux OS computers have a variety of such programs,[5] and Windows computers come with a program called Command Prompt (`cmd.exe`).[6]

 Learn to use a CLI.

If you're not familiar with using a text interface for a computer, consider learning it. Complementing graphical interfaces, command line interfaces are powerful and efficient ways to productively use a computer.

The most common tasks when working with remote computers are (1) transferring files to and from the remote computer and (2) issuing commands and running programs on the remote computer. Both tasks require establishing a connection—hopefully a secure one!—between your local computer and the remote computer and often that connection is made with the same underlying method for both types of tasks.

Secure connections

To log into a remote computer works exactly like logging into a website. You create or receive a username and password and then you provide those to access your account. However, unlike using a website, when working with a remote computer you may find yourself logging in many times during a work session, and it can become tiresome to keep reentering your login credentials. The solution is to set up a *passwordless login* using a matching pair of *keys*: a *public* key which is copied to the remote computer (and can be safely viewed by anyone) and a *private* key (which is kept to your local computer and should never be copied or shared[7,8]). Together these keys allow you to log into the remote computer quickly, invaluable when you are logging in many times during a work session.

How can you log in securely without a password? Using public–private key pairs, the local and remote computers perform some calculations behind-the-scenes to convince the remote computer you are who you say you are. When you first ask the remote to log in, the remote will create a unique "challenge question" and then use your public key, already on the remote, to encrypt that challenge question. The remote will then send the encrypted question to the local (since this is sent over the network, assume a bad actor may be able to see it). The local computer can then use your private key to decrypt the question, read it, and answer it. The answer is then sent back to the remote computer. Unless the system is insecure, only someone with the private key can answer the challenge question, so when the remote receives the correct answer, it knows the

---

[5] If you're using Linux you almost certainly already know this program.

[6] Recent versions of Windows come with *Windows Terminal* which can run multiple CLIs. Windows users may also be familiar with PuTTY (https://www.putty.org), a Windows program that enables access to another computer's CLI.

[7] You can add a password to your private key for extra security if you wish, although doing so removes the convenience of passwordless login.

[8] It's quite easy to make new public–private key pairs and best practice if you need to log into one remote from multiple local computers (like a laptop and a desktop) is to generate a separate pair on each local machine and copy both public keys to the remote. SSH (discussed shortly) will find which public key is needed automatically.

local computer holds the right private key and that it is safe to assume you can access the account. Lastly, now that the remote is convinced you have access, some additional key pairs are created and exchanged in such a way that a shared secret is available on both computers and that secret can be used by both computers to encrypt and decrypt any data passing between the two. (This step also occurs when using a password for authentication.) At this point you have a secure connection or "tunnel" between the two computers.[9]

 Save time by setting up secure, passwordless login for any remote computers you regularly use.

File transfer

A CLI usually gives you commands to move or copy files. For example, on a UNIX-style computer (Sec. 20.4):

```
cp huri_ppi01.edgelist ~/archive/
```

will copy (cp) the file huri_ppi01.edgelist into a folder in your home directory (~/) called archive. In other words, we can transfer files within our local computer. But such commands also extend almost automatically to transfer files *between* computers:

```
scp huri_ppi01.edgelist remote_supercomputer.org:~/archive/
```

This command is nearly identical, with two differences:

scp Instead of using the copy program, we use the *secure copy* program (scp). The secure refers to its ability to send data over the network using a secure connection as discussed above.

remote_supercomputer.org: The destination of the copy now begins with the address of the remote computer followed by a colon (:). We can also specify a username: user@computer:/path/to/folder. The CLI naturally incorporates network addresses into file and folder names using this syntax, and thus we don't need to change much to send data between computers.

Issuing commands

Complementing the ability to send files to a computer you can access is the ability to log into that computer from a CLI. Most commonly this is done with a program called ssh (*Secure SHell*), although some alternatives exist such as mosh.[10] To do so, in a CLI window on your local computer you run the appropriate ssh command, then the contents of that CLI will show the CLI running on the remote computer. Any commands you type will actually be seamlessly running on the remote computer. Issuing a logoff

---

[9] We have of course skipped over significant technical details, in particular how the keys are computed and how the challenge question is created and answered. This use of key pairs is a form of *asymmetrical encryption*.

[10] https://mosh.org/

command on the remote will then terminate this connection and your CLI will return to that of your local computer.

SSH powers the connection by forming a secure tunnel between the two computers, through which commands and their results are sent back-and-forth. SSH also enables other commands such as scp discussed above. Indeed, most commands that send data between computers securely rely on the same library.

SSH comes with many additional commands to make it easier to use. ssh-keygen allows you to create and store public–private key pairs using different algorithms. ssh-agent provides additional security by managing control over unencrypted keys on your local computer, allowing you to keep a password for your local key but not have to reenter the password for every remote work session. You can also use a config file to save the usernames, addresses, and other details for all the remote computers (hosts) you're working with. The config file is especially helpful for creating aliases, short abbreviations for longer login details, saving you from typing in a long computer name or other information every time you want to connect to a given computer. Very handy!

20.4   UNIX—I know this system

Fundamentally, our computational work is performed on computers and those computers are powered by operating systems (OSes). While there appears to be a plethora of OSes to choose from—Macs and PCs and Linux devices—in reality they fall into two main groups: Windows and UNIX-style.[11]

What do we mean by "UNIX-style"? At the dawn of computing and into the mid-twentieth century, it was common for each machine to have a custom-made operating system. Often the first thing you needed to do with your new computer was write your OS for it! Gradually this changed. In the late 1960s, Dennis Richie and Ken Thompson, computer scientists at Bell Labs, designed an OS for a slightly out-of-date machine they were allowed to use. This machine was quite underpowered, even for the time, and so care was needed to create a minimal, efficient OS for it. Under those constraints, Richie and Thompson designed what would eventually be UNIX.[12, 13]

Why do we care? At the time of writing, every major computer operating system except one[14] is a direct descendent of UNIX. This includes Linux, macOS by Apple, all major smartphone OSes, both iPhone and Android, and, importantly for us, nearly every major supercomputer. The very limitations that Richie and Thompson wrestled with led to an efficient, flexible, and modular operating system architecture that would

---

[11] We are skipping over lots of small cases such as real-time OSes designed to power vehicles and other potentially dangerous and expensive equipment. Such dedicated systems are of little interest to us here.

[12] The name "UNIX" is a bit of a jab at a competing OS effort called "multics" coming from MIT. Multics was intended for multiple users on a large central mainframe computer. The original computer UNIX was designed for couldn't support multiple users.

[13] In order to create UNIX, Richie and Thompson also needed to create a programming language to write it in. What they created would eventually be called C, and in many respects C is the foundation of most modern computing languages, either in design or implementation. Windows, the one major OS today not descended from UNIX, is written in C, C++, and C#, all languages that owe their existence to Richie and Thompson.

[14] That OS is Windows by Microsoft, a big exception as it is extremely popular for PCs. However, it is unheard of in other contexts, such as supercomputers, and, further, Microsoft has begun providing UNIX-style compatibility with the WSL (Windows Subsystem for Linux) initiative.

eventually form the basis for nearly all major OSes. But, besides very good design, there is another reason why UNIX propagated throughout computing. AT&T, then owner of Bell Labs, had an agreement with the United States government where it would not enter into any computing businesses in return for being allowed to maintain a monopoly on telephones and telecommunication. This meant AT&T could never sell UNIX and had to give it away.[15] People like free, especially cash-strapped computer scientists and small startup companies. Thus, over time, UNIX became the starting point for many operating systems, including "BSD," which led to macOS, and Linux, which led to Android. We use the term "UNIX-style" for these UNIX descendants, which conform more or less to the major design patterns of UNIX.[16]

 Thanks to good design and a quirk of history, UNIX has become the foundation for nearly all computer operating systems, especially those used for scientific computing.

UNIX-style OSes have become the standard for scientific computing. If you ever plan to use a supercomputer (which nowadays is really nothing more than a large number of regular computers) or a cloud computing service, you will be using, at some level, a UNIX-style operating system. Take advantage of this by understanding and embracing some of its properties in your workflows.

 Consider even adopting a UNIX-style OS for your personal work machine.

The modular design of UNIX-style OSes provides us several features to help with scientific workflows:

- Pipelines become natural. It is easy to make reusable programs that interoperate (Sec. 20.2).

- Remote access is easy. The OS's modular nature includes separating interaction and display, making it easy to have the keyboard and screen come from a different, over-the-network computer. This gives us SSH (Sec. 20.3).

- Rethinking what it means to be an OS. Containers and virtual environments extend the next generation of scientific computing systems.

Pipelines

UNIX-style OSes come with an elegant mechanism, the standard streams, for building pipelines that any program can take advantage of. The most important streams are stdin (read as: "standard in") and stdout (read as: "standard out").[17] In many ways, we can

---

[15] Given that not only UNIX but the *transistor*, the very basis of the information age, came out of Bell Labs, we are all very fortunate for this agreement.

[16] A codified standard operating system design called POSIX (ISO/IEC/IEEE 9945) is based on UNIX. For our purposes, "UNIX-style" is sufficient.

[17] We omit from this discussion stderr, the standard stream for reporting error messages. It is useful to separate error messages from output so that you can record them separately, in case you need to review what happened during a failed program run.

think of each as a file: your program can "read from" stdin in a manner exactly like reading from any other file, and likewise your program can "write to" stdout. What's so nice about using these streams instead of normal files is that they can be composed. You can write one program that sends data to stdout and another program that receives data from stdin, then you can wire those programs together into a pipeline. Further, if your program is designed to use the standard streams, you can place it within any pipeline using other programs. All the programs that come with the OS already support these streams.

Here's an example in Python using streams. If you are familiar with Python, the code for reading/writing streams is almost identical to reading/writing files:[18]

```
import sys

data_in_txt = sys.stdin.read()
result_txt = f(data_in_txt)
sys.stdout.write(result_txt)
```

This simple program mock-up receives data (as a string) from stdin, computes a result using a function f(...), and then writes that result to stdout. (The streams work with text data because they are considered files; your program can convert stdin's data as needed.) In Python, all you need to begin using standard streams is to import the system module (import sys).

Here's a more specific example of a pipeline involving a text file and two programs. Suppose you've written a program, get_nodes.py, that uses stdin and stdout. We wish to send data to our program, pass our results to another program created by a colleague, then save the second program's output to a file for storage. Here's the command we would enter into our UNIX-style computer:

```
$ cat network.txt | python get_nodes.py[19] |
      process_data > result.txt[20]
```

(The $ is not typed in; we use it to represent the start of the "prompt" the computer displays showing us where to enter commands.) Besides the Python program we created, there are three things to understand in this pipeline: cat, |, and >.

cat cat, which stands for "concatenate," is a standard UNIX program for printing a text
      file (network.txt). Printing it to what? In this case, to stdout, which becomes
      stdin for the next step of the pipeline, our Python program.

| The vertical bar or "pipe" character is, you guessed it, the symbol used to wire
    together the streams to build a pipeline. Here we used it to connect the text file
    to our get_nodes.py program, and then the output of our program to another
    command, process_data.

---

[18] As far as Python—and the OS—is concerned, the standard streams *are* files!

[19] Even better, you can set a flag inside your script called a "shebang" that will tell the computer to run it with Python automatically. Then the pipeline no longer specifically mentions Python. Why is this useful? If in the future you replace your Python script with something written in a different language, your pipeline code can remain unchanged!

[20] In practice, we recommend against generic names like process_data and result.txt.

> A shorthand for redirecting stdout to a file. This lets us save `process_data` to a file of our choosing. While a program can always specify what output it will write, by taking output out of the program and putting into the pipeline, we can see and control this output ourselves without either modifying the program or, if the program lets us specify its output, learning how to do so.

This is just a taste of what UNIX-style pipes can do. In general, it's difficult to overstate the power of UNIX's simple, modular nature.

20.5   Version control

A version control system (VCS) allows multiple editors to work on shared files. Generally the users are programmers working together on a code base, but not always—this book, for instance, was written using just such a system.

The main problem a VCS addresses is handling conflicts: when two editors make different changes to the same part of a file, an algorithm cannot automatically merge those changes to make a single file. VCS systems provide functionality to pick out or replace the conflicting edits manually.

Allowing multiple editors to work separately on their own versions of a fileset also allows multiple versions of the fileset to be maintained. This is useful for software makers: you can maintain a "release" version of a package and a "beta" version of a future release at the same time, and use the VCS to switch between the versions as you work.

For researchers, version control is important for provenance: by keeping a detailed history of changes to the code, you can track who made what changes, when a particular output was first (or last) computed by the code, and you can revert to an older version of the code if you need to replicate exactly a past result. Scientists in particular benefit greatly from this use of a VCS.

ⓘ Git (https://git-scm.com) is by far the most popular VCS in use today. Git underlies GitHub (https://github.com) and many other open-source projects.
    While not as user-friendly as others, it is the most important VCS to learn, and all programmers and data scientists should be familiar with the basics of using git.

A VCS provides a set of commands that examine and modify a *repository*. The repository, the central object of interest for the VCS, is just a folder on your computer, but besides your files, within it are special files that the VCS uses to record the full history of all the files placed under version control, or "tracked" by the VCS. (You need to tell the VCS specifically which files to track.) From these files, the VCS can efficiently build or rebuild the state of the repository at any point in its history; if you frequently tell the repository about your updates using relevant VCS commands (the repository is not updated automatically), then you will develop a fine-grained history for your project. Those special files also usually contain additional information, such as the addresses for copies of that repository on remote computers, if they have been set up. Saving these details makes it easy for you to update your local repository and then send your updates to the remote copy or, conversely, for you to quickly bring in any updates from the remote repository to your local copy. Together, these commands, while they have a steep learning curve—it helps to know a bit *how* the VCS works to know *what* to do with it—smooth the road for productive work history and work collaboration.

Version control for data?

Data scientists have a further need than programmers when it comes to version control: keeping the history of a dataset.

It may be that you are working with dynamic data, where new observations are constantly being added to a large dataset. Conversely, your data may come from a laborious process of data entry, where individuals manually code in observations. Data entry can be error-prone and may require processes whereby observations are verified and revised, if needed. In both cases, the data are changing, and a researcher should be able to review those changes as needed.

A VCS appears to be ideal for tracking dynamic data. After all, what's the difference between files that represent computer code and files that represent other data? As far as the computer is concerned, they're both ones and zeros, right? Unfortunately, there are differences, not in how the computer treats those files but in how the VCS was designed. Most VCSes use algorithms designed for small, human-readable, human-writeable files, such as source code. These algorithms don't handle very large files, or large changes to files, very well, and over time many common VCS tasks will become incredibly slow, even tasks not involving those large files.

 Version control is not well suited to large data files.

It is generally intended for files made by humans, such as source code. If you are generating large files (meaning, 100s of megabytes or more), particularly binary files, you probably want to avoid tracking their contents with a VCS. Doing so is likely to slow down the system considerably.

Some options do exist for using version control with data files. One is Git Large File Storage (LFS).[21] Git LFS is an extension you connect to git that allows git to track references to the large files and not the contents of the files themselves. Another option is Dolt,[22] a database that you can interact with as if it were a git repository. Hopefully, work in this area will continue and version control for data will become easier and more popular.

20.6 Backups

Backing up your data, writing, and other work against computer failure is critical. As discussed in Ch. 17, a good backup system accommodates both off-site backups and, more importantly, full file history. While most cloud computing services can provide off-site *copies*, they generally do not keep track of file history.

---

[21] https://git-lfs.github.com
[22] https://github.com/dolthub/dolt

For your personal computer, at the time of this writing, the two most popular file history backup services are Apple's Time Machine (for macOS computers) and Windows Backup (for Windows 10 and later). Both are free and built into their respective operating systems. Both also use a locally connected external disk[23] to store backups, although there is some support for backing up over a network. We strongly encourage users to use such a service if possible.

Version control systems such as git (Sec. 20.5) are an excellent choice for backing up the file histories of code and other "person-generated" files, so long as you set up a remote location to "push to" and you frequently add and update your work. While it would be great to use a similar system for tracking changing data, as we discussed, currently these systems are generally poor for large data files.

Research data backups are a different story. Your institution may provide facilities for data archiving and indeed there may be specific requirements for doing so. Be mindful of both legal requirements and ethical considerations.

20.7 Selecting tools for yourself

Two factors will continually drive your need to evaluate and re-evaluate new and existing computational tools. The first is obsolescence of software (and perhaps hardware). As tools fall out of favor, you may need to jump ship to alternatives. The second factor is the rise of new alternatives and innovations, plus, to an extent, changes in your needs as a researcher. Software is continually churning, with new ideas for solving problems and indeed new problems arising for us to solve.

A variety of tools exist to help you work computationally. Which do you wish to use? Do you have a problem that existing tools do not solve? Do you wish to improve your productivity by incorporating a new piece of software? We can help. Here we provide some advice on selecting a tool for a problem.

Suppose you have identified a problem you wish to solve. It may be a research question or a basic computing task. Ask yourself the following:

- Q0: Will an existing tool work on this problem? If so, how painful or painless is it as a solution?

- Q0.5: Is your problem even solved by a tool or not?

- Q1: What tools are available for this problem? If you are a beginner to the problem area, this question is easier said than addressed. Over time you will become more "plugged in" to the space of tools and options and this experience should be useful guidance.[25]

---

[23] A local disk does not provide an off-site backup. One option is to use two external disks[24] kept at different locations, for example, one in the office and one at home.

[24] Two is one and one is none.

[25] This underscores our assertion that good computing scientists need to always spend some portion of their time evaluating, re-evaluating, and changing their workflows. It seems highly efficient to identify a system that works, and never change it. But if you do need to make a change after being stuck for a long time, you will be out-of-practice at evaluating exactly these decisions and more likely to decide poorly. Take care to allot a portion of time, *but only a small portion*, to self-(re)evaluation.

Suppose now you have found a tool you wish to consider. Perhaps it is an open source project. Ask yourself:

- Q2: Does the tool solve your problem? Is it intended for exactly this problem or are you using it "off-label"?

- Q3: Is the tool popular?

- Q4: Is the tool actively maintained. Software goes stale fast!

- Q5: Is it new or well established with a long history?

- Q6: Is the tool open source or commercial? Are there restrictions on how you can use the tool?

Lastly, consider other criteria:

- Q7: Are you working with or accommodating others? Perhaps you work in a research group with specific requirements or workflows. Are you making decisions on a tool that others will have to use? Will they want to use the tool?

- Q8: Other criteria to evaluate? Cost? Compatibility?

20.8 Summary

We conclude our scientific computing interlude with a tour of some specific tools that we recommend at the time of this writing. These tools help scientists work with data in a reliable, reproducible, and hopefully efficient manner, saving time and perhaps even preventing mistakes.

Of course, the world of computing is always fast moving, and it is likely that specific tools mentioned here will fall out of favor and be replaced with new alternatives. A good, working computational scientist should always spend some (small) portion of their time evaluating current tools to see if better options exist; if you never take the time to switch up your workflow, and something comes along which forces a change such as a discontinued tool, you will be out of practice and may choose your new replacement poorly. To help, we have included some advice and a brief workflow to guide you through evaluating new tools to use.

Bibliographic remarks

For readers interested in diving deeper into the world of practical computational tools and practices, we highly recommend *Effective Computation in Physics* by Scopatz and Huff [420]. Don't let the title fool you, this isn't just for physicists; it's a tour of computing for any technically minded researcher interested in sharpening their scientific computing skills.

The most common form of cryptosystem for establishing secure connections between computers is based on the RSA algorithm, which creates challenge questions by multiplying large prime numbers. For those wanting to learn more of the details, a brief overview of public key cryptography, also known as asymmetrical cryptography, by one of its founders, is given in an article by Hellman (of the Diffie–Hellman algorithm) [210].

A fantastic article demonstrating the power of UNIX pipes (written by their creator, Doug McIlroy) is Bentley et al. [52]. The program McIlroy discusses wonderfully illustrates the UNIX philosophy.

For those interested in learning more about UNIX's creation, a recent memoir by Brian Kernighan, a major contributor to UNIX, provides an interesting account of its history [240]. More broadly, *The Idea Factory* by Gertner [180] gives an overarching story of Bell Labs, the research center where UNIX (and other major inventions) was developed.

Readers interested in learning to use a version control system should consult Scopatz and Huff [420] or the resources available at the git homepage.[26]

---

[26] https://git-scm.com

Part III

Fundamentals

Chapter 21

Networks demand network thinking: the friendship paradox

Before diving into the fundamentals in Part III, let's talk about the *friendship paradox*. It is a powerful tale that cuts to the heart of network science.

21.1   Is the friendship paradox a paradox?

The friendship paradox is a profound network phenomenon connected to so many other important network phenomena, such as majority illusion, preferential attachment, epidemic spreading, and so on. The motivating question is simple:

> *Do my friends tend to be more popular than myself? Is this just me?*

This may resonate with many! But, the real question is whether this is true for *most* people in the network.

A common intuition is that this should not be the case for *most people* because both a randomly chosen person and their friends should not be *special* and thus if some are less popular than their friends, there should be others who are more popular than their friends. In other words, the intuition is that they (a randomly chosen person or their friends) are more or less *random*. But is this true?

Let's formulate the question more concretely. Here, we will take a somewhat less rigorous but intuitive approach.[1] First, we pick a random person from a social network and ask how many friends they have. In reality, the definition of being "friends" differs across people. But here let's assume either that everyone has the same definition of friends and knows exactly how many friends they have, or that we are simply looking at the number of friends in a social networking service like Facebook. Now, we randomly choose a "friend" of this person and count their number of friends. We record these numbers and put them into two bins. The first bin contains the number of friends of a randomly sampled person. The second bin contains the number of friends *that the*

---

[1] For a more rigorous calculation, see [92].

317

Figure 21.1 The probability to pick a node with degree k is proportional to the prevalence of nodes with degree k in the network ($\propto N(k)$). If there are twice as many nodes with degree 5 as those with degree 4, it is twice as likely to sample a node with degree 5 than one with degree 4.

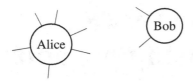

Figure 21.2 Sampling a random neighbor is like randomly grabbing one of these edge "stubs" that each node has.

friends of the randomly sampled person have. If we repeat this sampling many times, will these two sets of numbers have the same distribution or not?

Let's formalize this. We first need to think about the general probability distribution of the *number of friends* that each person has. In network science, we usually use k to represent the number of friends or neighbors in the network, which we call "degree." If we count how many people have only one friend ($N(k = 1)$), how many have two friends ($N(k = 2)$), and so on, then we can obtain the probability distribution of the degree in the network with N people:

$$p(k) = \frac{N(k)}{N}. \tag{21.1}$$

What would be the probability distribution of the first bin? The answer is $p(k)$. The probability to sample a person with k friends is $N(k)/N$ thus $p(k)$ (Fig. 21.1).

How about the second bin? This is where the question gets interesting. You may think that the distribution should be the same because it is still random. But it is not! The distinction comes from the fact that, if you have many friends, you will show up more on others' friend lists![2] In other words, let's compare Alice and Bob; Alice has six friends in the network and Bob has two friends in the network (Fig. 21.2).

This means that Alice will show up in the friend list of six other people while Bob will show up only in the list of two others. Thus, Alice will be three times more likely than Bob to be on the list of friends when we sample a random person in the network.

---

[2] Have you noticed a problem? Sit tight. We will talk about the hidden assumptions later!

Figure 21.3 A star graph with eight nodes.

21.2 Paradox under the extremes

Let's think about the most extreme cases to solidify this intuition. First, can you think of a network where the differences in the degree of the nodes are maximized? How about thinking about a network where there is a single node that is connected to *everyone else*? To maximize the heterogeneity, we can assume that everyone else has only one neighbor (the single node that is connected to everyone). This network is called a *star graph* or a *star network* (Fig. 21.3).

Let's examine how the friendship paradox plays out in this network. First of all, if there are N people in the network, the probability to pick the *star* is $\frac{1}{N}$ while that of picking others would be $\frac{N-1}{N}$. In other words, whenever we pick a random person from the network, it will be very likely a non-star. What about the next step? If we pick a non-star, there is only one neighbor to choose from—the star node. If we pick the star, then whichever neighbor we sample, it is the same—those who only have one friend. The star, because it has so many friends, shows up more frequently on the "list of friends" of other nodes, while everyone else only shows up on the list once.

In summary, except in a single case (randomly choosing the star), we will always sample someone with only one friend and that friend will have a huge number of friends. You can say that the friendship paradox holds (almost) always, and spectacularly.

Can we also think of the other extreme, where the friendship paradox does not hold at all? Because the star graph had extreme heterogeneity in the degree distribution, maybe we should think about the case where there is no heterogeneity at all. This is simple; we just create a network where everyone has the same number of friends.

Such a network is called a *regular graph*. It can form a *lattice* that has a regular pattern of connections everywhere (Fig. 21.4), or it can be a *random* regular graph. Whatever it is, the simple truth is that there is no way for these networks to exhibit the friendship paradox. Whomever we pick, their degree is the same as their neighbors (and any other nodes in the network).

From these two extremes, it seems the paradox is connected to how much variation there is in degree. When the degree distribution is extremely heterogeneous (the star), the paradox is strong; when the distribution is homogeneous (the lattice), the paradox vanishes. Now let's formalize this intuition.

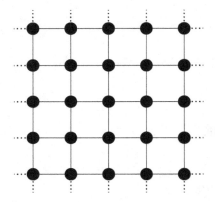

Figure 21.4   An infinite regular graph (lattice) where everyone has exactly four neighbors.

Figure 21.5   The more neighbors you have, the more likely you will be sampled as a friend ($\propto k$).

21.3   The random neighbors' degree distribution

As we reasoned above in our discussion of friend lists, the more friends you have, the more likely you will show up as a friend of someone else (Fig. 21.5). Thus, if we use $q(k)$ to represent the probability distribution of the second bin, that a random node has a random friend with degree k, then $q(k)$ should be proportional not to $p(k)$ but to $kp(k)$:

$$q(k) \propto kp(k). \tag{21.2}$$

This is not a probability distribution because if you sum over all possible k, it is not equal to one, but the average degree, call it $\langle k \rangle$, of the network:

$$\sum_k q(k) = \sum_k kp(k) = \mathbb{E}_{k \sim p(k)}[k] := \langle k \rangle. \tag{21.3}$$

Therefore, we need to normalize q to obtain a proper probability distribution:

$$q(k) = \frac{kp(k)}{\langle k \rangle}. \tag{21.4}$$

Now that we have both distributions, we can calculate the expected degree of the second bin (random friends) as well as that of the first bin (random people). What is

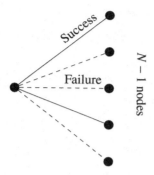

Figure 21.6 Creating the neighbors of a node in a random graph is like performing $N - 1$ iid Bernoulli trials with P.

their difference?

$$\mathbb{E}_{k\sim q(k)}[k] - \mathbb{E}_{k\sim p(k)}[k] = \sum_k kq(k) - \sum_k kp(k)$$

$$= \frac{1}{\langle k \rangle} \sum_k k^2 p(k) - \langle k \rangle$$

$$= \frac{\langle k^2 \rangle - \langle k \rangle^2}{\langle k \rangle}.$$

Wait, the numerator is the *variance* Var (k) of the degree distribution $p(k)$! What are the implications of this finding?

First, this quantity is always positive, except when the variance of the distribution is zero and everyone has the same degree. In other words, *the friendship paradox is almost unavoidable!* Whatever network you look at, as long as its degree distribution has a nonzero variance, the expected degree of *random neighbors* will be likely larger than that of *random people*. Just like many other statistical paradoxes, the friendship paradox is not a paradox.

21.4 Do random graphs show the paradox?

One interesting question we can ask here is whether a random graph exhibits the friendship paradox or not. What we mean by a "random graph" is a graph with N nodes where each pair of nodes has the equal probability P to be connected. As a result, each node will have the roughly same number of friends or neighbors:

$$\langle k \rangle = (N - 1)P \simeq NP. \tag{21.5}$$

If we focus on a single node (Fig. 21.6), we can see that the probability distribution for this node's degree follows the binomial distribution. From this node's perspective, there are about N other nodes and we test whether to connect with each of them with

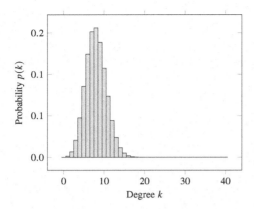

Figure 21.7   The binomial distribution with $N = 40$ and $p = 0.2$.

probability P. Therefore the degree distribution for this node will be

$$p(k) \simeq \binom{N}{k} P^k (1 - P)^{N-k}. \tag{21.6}$$

Because this node is not special and every other node is in the same situation, this is simply the degree distribution of the whole network (Fig. 21.7). And we know the variance of a binomial distribution:

$$\text{Var}\,(k) \simeq NP(1 - P), \tag{21.7}$$

which is of course not zero! In other words, even though the network is homogeneous in the sense that everyone is connected to others according to the same criterion, we could still see the friendship paradox due to the natural variance in the random process. As long as there are some nodes with larger degree than others, we expect to see—it can be tiny—the friendship paradox.

Does a large variance guarantee a strong friendship paradox?

We now understand that the friendship paradox is largely driven by the variance of the degree distribution and as long as there is nonzero variance, we expect to see the discrepancy between a random node's degree and a random neighbor's degree. Here is another useful question to think about:

> *If our network has a broad degree distribution with large variance, does this guarantee a strong friendship paradox? In other words, does the variance determines the strength of the friendship paradox?*

Again, think about this question before reading on. What do you think?

You may conclude that the equation derived above tells us that it is indeed the case that the variance of the degree distribution dictates the strength of the friendship paradox. But, there was an important, implicit assumption that we did not specify.

You see, the important assumption throughout the calculation was that when we go from a randomly sampled person to their neighbors, we can completely forget about the degree (or any other properties) of the randomly sampled person. We assumed that, regardless of the degree of the randomly chosen person, sampling a neighbor is equivalent to sampling a person based on their degree. This assumption makes sense if everything is random. However, it is not necessarily true in all possible networks.

Let's imagine a network that consists of only *cliques*[3] of many different sizes. In this example, whomever we sample, the neighbors of this person are guaranteed to have the same degree as the person we sample. But there will still be variation in degree because the different cliques will be different sizes. In other words, we can easily remove the friendship paradox completely while maintaining the variance in the degree distribution. Variance in degree alone is not enough to create the friendship paradox.

Mathematically speaking, the assumption that we implicitly made above can be understood by thinking about a conditional probability $q(k|k')$, which is the probability that a random neighbor has degree k, given that the original node has degree k'. This function describes the *degree correlation*.[4] If the network does not have any correlation between degrees of connected node pairs, $q(k|k')$ will not depend on k'. If the connections are random, $q(k|k') \propto kp(k)$. Now, our implicit assumption can be written as the following:

$$q(k) = \sum_{k'} q(k|k')p(k') \simeq \frac{1}{\langle k \rangle} \sum_{k'} kp(k)p(k') = \frac{kp(k)}{\langle k \rangle}. \qquad (21.8)$$

However, this is only true when we assume no degree correlation and completely random wiring. It breaks down in the network of cliques because

$$q(k|k') = \begin{cases} 1 & \text{if } k = k', \\ 0 & \text{otherwise.} \end{cases}$$

Therefore,

$$q(k) = \sum_{k'} q(k|k')p(k') = p(k). \qquad (21.9)$$

In other words, if there is a strong positive degree correlation—nodes are connected to other nodes with similar (same) degree—then the friendship paradox can be reduced or even vanish!

The friendship paradox emerges from a simple constraint created by networks (nodes with many neighbors should appear many times in other nodes' neighbor lists) and is therefore fairly universal. Yet, it is crucial to understand the conditions that enable the friendship paradox. It depends not only on the variance of the degree distribution but also on how exactly the network is wired, particularly the correlation between degrees of the connected pairs.

---

[3] A clique is a fully connected (sub)graph. By definition, everyone in a clique is connected to everyone else in the clique and therefore has the same degree as everyone else in the same clique.

[4] Also note that social network exhibits homophily. Due to homophily and constraints on people's time, the real social networks may exhibit a much weaker friendship paradox than expected [29]

21.5 Generalized friendship paradox

The friendship paradox can also generalize to properties other than degree because many properties of a person (or a node, in general) are correlated with their degree. For instance, let's think about a scientific collaboration network where nodes are scientists and edges represent collaborations, measured perhaps by coauthored papers. In this network, those who have many collaborators (large degree) are more likely to be those who have more papers, more citations, and more fame. In other words, degree in the collaboration network should have a strong correlation with other *measures of success* in academia. Indeed, this is what researchers have found [147]. And this is of course not limited to the scientific collaboration network. The number of "friends" or "followers" on social media is likely to have a strong correlation with various measures of success and recognition. Therefore, according to the principle of the *generalized* friendship paradox, we may be almost always surrounded by those who are better off than us in some way.[5]

In other words, you may not want to *compare yourself to your peers*. Very likely, you will see they are better off than you! Just don't think you're a loser, it's just the network (and mathematics) conspiring against you.

21.6 Summary

We delve deep into the strange world of networks by considering the friendship paradox, the apparently contradictory finding that most people have friends who are more popular than themselves. How can this be? Where are all these friends coming from? It is due to constraints induced by the network structure: pick a node at random and you are much more likely to land *next to* a high-degree node than *on* a high-degree node, because high-degree nodes have many neighbors. This is unexpected, almost profoundly so; a local (node-level) view of a network will not accurately reflect the global view. This "paradox" highlights the care we need to take when thinking about networks and network data mathematically and practically.

Bibliographic remarks

The friendship paradox was first observed by Feld [156] when studying social networks. Given the interplay between emotional health and social ties, an underestimation of your social well-being can have negative repercussions, so Zuckerman and Jost [513] explore how people estimate their own popularity and how the friendship paradox may act as a subjective factor.

The friendship paradox generalizes beyond degree to other attributes, an idea introduced by Hodas et al. [214] and Eom and Jo [147] and greatly explored since. When other node attributes (Ch. 9) correlate with degree, a generalized friendship paradox can arise. It does not always arise though, even for highly degree-correlated attributes, as recently pointed out by Evtushenko and Kleinberg [154].

---

[5] But, beware that the generalized friendship paradox may or may not hold depending on the network structure and the nature of the correlation. The generalized version is much more fragile [154, 92].

Network models are a great way to explore the friendship paradox and its generalizations. Jackson [229] uses models to explore the effects of estimating other features of a network when a person's view is biased by the friendship paradox, while Cantwell et al. [92] analyze the paradoxes across multiple network models.

Exercises

21.1 Suppose you have a network that consists of two cliques of sizes n and N with no edges between them. Recall that a clique of n nodes is a completely dense subgraph with $\binom{n}{2}$ edges.

   (a) What are $\langle k \rangle$ and $\langle k^2 \rangle$ as functions of n and N?

   (b) Suppose edges are now placed at random between the two cliques: each i, j where i is in one clique and j is in the other are connected with probability p. Does this network exhibit a friendship paradox? If so, how does it depend on p, n, and N?

21.2 Suppose a large network has a strong friendship paradox. Describe a *rewiring* strategy (Sec. 11.6) that can eliminate the friendship paradox without changing the degree sequence, or show that this is not possible.

21.3 What can happen to the friendship paradox in a network with hubs (high-degree nodes) if you delete the K highest-degree nodes?

21.4 (**Advanced**) Suppose node i with degree k_i can accurately determine k_j, the degree of any neighbor j. Using only this information, can i determine if their degree is above or below average compared to the entire network? Demonstrate how and under what conditions this is possible or show why it is not.

Chapter 22

Network models

Networks exhibit many common patterns, which we seek to describe (Ch. 12). But we also seek explanations for those patterns. What causes them? Why are they present? Are they universal across all networks or only certain kinds of networks?

One way to address these questions is with *models*. Models are one way we can search for explanations of phenomena. Propose a simple mechanism and analyze what it may tell us. For instance, in the previous chapter (Ch. 21), we considered a random network as a model to understand the friendship paradox, discovering that no special ingredients are necessary for the "paradox" to arise.

In this chapter, we explore in depth the classic "mechanistic" models of network science. Later, in Ch. 23 we turn our attention to the "statistical" models commonly used with network data.

22.1 Mechanistic and statistical models

In 1905, Einstein proposed a model for Brownian motion [144], how small particles diffuse through a medium by the random jostling of their neighbors. (This was, by the way, the first direct evidence of the existence of the atoms.) In one dimension, the model is simple: a particle sits on the origin of $\mathbb{Z}^{\pm}$ ($x_0 = 0$). At each time step, the particle moves by a "coin flip" (this mimics the random jostling). With probability p (typically $p = 1/2$) the particle moves to the right one unit ($x_{t+1} = x_t + 1$); otherwise, with probability $q = 1 - p$, the particle moves left ($x_{t+1} = x_t - 1$). The probability p acts as our model parameter and this model is known as a random walk. Over time the particle will randomly drift back-and-forth over wider and wider intervals. Even though left and right are completely symmetric (for $p = 1/2$), meaning the average position $\langle x \rangle$ of many random walks will remain at $x = 0$, the variance of x will grow. Specifically, Einstein not only showed that the probability density of walkers $P(x)$ follows a normal distribution, but also that the variance of x grows linearly with time, $\langle x^2 \rangle \sim t$.

As an alternative to this random walk model, one can appeal directly to the central limit theorem and propose a model where the particle's occupation probability follows a normal distribution centered on the origin with a variance that may depend, somehow, on time. That time dependence is then controlled by a parameter of the model. A scientist examining data from diffusion experiments may realize the variance has a linear time dependence, which can then be estimated with a regression.

While this second model captures the flavor of Einstein's results, you can immediately see the limitations. It does not *explain* where the time-dependent normal distribution comes from. It is not parsimonious because simpler explanations exist. And it fails to capture the trajectories of individual particles by immediately "skipping to the end" to propose the occupation distribution. The random walk model in contrast derives the occupation distribution from underlying principles. This is powerful.

The second model is a phenomenological or *statistical* model: its properties are imposed through probability distributions. The first is a *mechanistic* model: a simple left-right random walk mechanism was imposed, and statistical properties blossomed from it.

Yes, both types of models rely on randomness as a key ingredient, but the character of that randomness is radically different. And both types of models are of central importance when studying network (and all) data.

In this chapter, we focus our attention on primarily *mechanistic* models. Following, in Ch. 23 we consider statistical models.

22.2   Randomness in models

When dealing with network data, we usually have a *single* instance of a network. Whether it is a social network, a citation network, a protein–protein interaction network, or a road network, a particular edge is either present or absent given the definition of the network we have. However, it is often very useful to imagine the instance we have as a sample from an *ensemble* of networks that share the same properties as the network in our hand.

Randomness is a key ingredient in this imagination. In most random graph models, we imagine that the edges are wired up randomly, based on certain prescribed mechanistic rules, probabilities, and properties. In doing so, we can create many realizations of the "same" network and study their properties, as each instance may evolve differently due to randomness. From this we can understand what does a model "typically" look like. Furthermore, random graph models allow us to analytically study the properties of networks based on the *other* properties that we prescribe, making randomness, in a sense, a low-information assumption: all else being equal, when we don't want to assume a behavior, assume it is random.[1]

In this chapter, we will start with the simplest random graph model, the Erdős–Rényi model, and then move on to the configuration model where the degree sequence of nodes is prescribed, followed by a variety of other intriguing mechanistic models. We will also introduce the *generating function* approach, which allows us to analytically study such random graphs. The core ideas behind these models are also at the heart of many other network models, especially the statistical models that we will study in the following chapter.

---

[1] The same arguments for mechanistic models apply to statistical models as well.

22.3 Erdős–Rényi model

If you know nothing about a network, what's the simplest model that you can imagine? The *Erdős–Rényi model* might be it. It's probably the simplest possible model that describes a network with a certain number of nodes and edges.

22.3.1 Two basic formulations

We first imagine there are N nodes in the network. Given that we do not have any information about the nodes, the most natural assumption is that there are no special nodes and we cannot distinguish them from each other.

How about the edges? Again, the simplest assumption that we can make is that the edges are created completely at random. But there are still two basic possibilities. First, we can start with a set number of edges M and then distribute these edges randomly among all possible node pairs.[2] We call this model $G(N, M)$. There are $\binom{N}{2}$ possible node pairs; among them M are chosen to be connected. The probability that we observe an edge between two nodes is then $M/\binom{N}{2}$.

Alternatively, we can start with a set number of nodes N and then prescribe the probability p that an edge exists between any pair of two nodes.[3] In other words, for each pair of nodes, the decision to connect them or not is the result of a *Bernoulli trial* where the probability of success is p. We call this model $G(N, p)$. We can calculate the expected number of edges in the $G(N, p)$ model by multiplying all possible pairs of nodes by the probability of an edge existing between them (which is same for every pair):

$$\mathbb{E}[M] = \binom{N}{2} p = \frac{N(N-1)}{2} p. \tag{22.1}$$

We can also describe both formulations in terms of the *adjacency matrix* $\mathbf{A}$, where $A_{ij} = 1$ if nodes i and j are connected, and zero otherwise. For $G(N, p)$, we assume, for an undirected network, that we have a Bernoulli trial for each element A_{ij} above the diagonal ($j > i$) and we set $A_{ji} = A_{ij}$. There are no self-loops so $A_{ii} = 0$ for all i. For $G(N, M)$, the story is similar: we distribute M 1's uniformly at random among the elements above the diagonal, set the rest to zero, and again set $A_{ji} = A_{ij}$.

22.3.2 Degree distribution

Another important, closely related quantity is the degree of each node and the degree distribution of the network. From the perspective of a node, the process of creating $G(N, p)$ network is the same as $N - 1$ (all other nodes) Bernoulli trials where the probability of success is p. As we know (Ch. 4 and Ch. 21), this process is described by the *binomial distribution*. The probability that a node has degree k is then

$$P(k) = \binom{N-1}{k} p^k (1-p)^{N-1-k}. \tag{22.2}$$

---

[2] If you have a physics background, you may be thinking about the *microcanonical ensemble*.
[3] Again, you may be familiar with the *canonical ensemble*.

Conveniently, because we cannot distinguish nodes from each other, this probability is the same for all nodes and the degree distribution of the whole network is also the same.

As we mentioned in Ch. 4, when N is large and p is small, the binomial distribution can be approximated by the *Poisson distribution*:

$$\binom{N-1}{k} p^k (1-p)^{N-1-k} \xrightarrow{N \gg 1, p \ll 1} \frac{\lambda^k}{k!} e^{-\lambda}, \tag{22.3}$$

where $\lambda = (N-1)p$. Because of this reason, the Erdős–Rényi model is also called the *Poisson random graph model*.

However, this correspondence *does* break. As an example, let us revisit the *friendship paradox* that we discussed in Ch. 21. In that chapter, we calculated the difference between the expected degree when we sample from random neighbors $\mathbb{E}_{k \sim q(k)}[k]$ and when we sample from all nodes $\mathbb{E}_{k \sim p(k)}[k]$, where $q(k)$ is the degree distribution of the nodes we sample by following an edge. Their difference is

$$\mathbb{E}_{k \sim q(k)}[k] - \mathbb{E}_{k \sim p(k)}[k] = \sum_k k q(k) - \sum_k k p(k)$$

$$= \frac{1}{\langle k \rangle} \sum_k k^2 p(k) - \langle k \rangle$$

$$= \frac{\langle k^2 \rangle - \langle k \rangle^2}{\langle k \rangle}. \tag{22.4}$$

We can compare the binomial and Poisson distribution using this formula. The expected value of the binomial distribution that describes the degree distribution of an Erdős–Rényi network is $\langle k \rangle = (N-1)p$. The variance of the binomial distribution is $\langle k^2 \rangle - \langle k \rangle^2 = (N-1)p(1-p)$. Therefore,

$$\mathbb{E}_{k \sim q(k)}[k] - \mathbb{E}_{k \sim p(k)}[k] = \frac{(N-1)p(1-p)}{(N-1)p} = 1 - p. \tag{22.5}$$

On the other hand, because the Poisson distribution has the property that the mean and variance are the same, we have

$$\mathbb{E}_{k \sim q(k)}[k] - \mathbb{E}_{k \sim p(k)}[k] = \frac{\lambda}{\lambda} = 1. \tag{22.6}$$

If we assume that the degree distribution follows the Poisson distribution, we would expect that the strength of the friendship paradox will stay the same as the network gets denser and denser. However, as we can see from the original binomial distribution, the strength of the friendship paradox would decrease as the network gets denser. It is easy to see this difference in the extreme case: when $p \to 1$, the network approaches a complete graph where the friendship paradox should completely vanish.

This is a useful reminder that the Poisson distribution, although it is often said to be the degree distribution of an Erdős–Rényi network, is not a perfect approximation.

22.3.3 Clustering coefficient

For the Erdős–Rényi model, the clustering coefficient is easy to calculate. Recall that p is the probability that *any* pair of nodes is connected. Therefore, from the perspective of a node that is connected to two neighbors i and j, the probability that they are connected is simply p. Therefore the clustering coefficient for this node, or any other node, can be written as

$$C = p = \frac{\langle k \rangle}{N - 1}. \tag{22.7}$$

Whenever we see e_{ij} and e_{ik}, then the probability to see e_{jk} is p. So regardless of how we define the global clustering coefficient, we will get $C = p$.

22.3.4 Giant component

Imagine a number of nodes floating in space without any connections. Now let's add some random edges—increasing M or p. We will start to see connected pairs of nodes floating around rather than isolated nodes. As we keep increasing the number of edges, we will start to see larger connected clusters. If we have enough edges, *all* nodes will form a single connected component. What happens between small, isolated clusters and a single component that contains every node? If we imagine an infinite number of nodes, given that the largest component is initially a single node, then there should be a moment where the size of the largest component becomes infinite.

A *giant component* or *giant connected component* (GC) is a component that contains a nonzero fraction of all nodes in the network as $N \to \infty$, or a component that grows proportional to N as N increases.

Can you guess when the GC will appear? Let's first consider the case where the mean degree is less than 1, would there even be a GC? Probably not, because most nodes will have either a single neighbor or no neighbor at all. We probably need *at least* one neighbor per node. How about the case where the mean number of neighbors is larger than 1?

One of the most exciting properties of the Erdős–Rényi model is a phase transition. If p is too small, there will almost surely be no GC. When p crosses a critical threshold, suddenly there is a finite probability for a GC. Let's explore this briefly with a non-rigorous derivation; in Ch. 24 we present a more general calculation in the context of network measurement and missing data.

Let u be the probability that a randomly chosen node i is not a member of the GC. Suppose we pick another node j, which may or may not be a neighbor of i and may or may not be in the GC. For i to not be in the GC, either j is not connected to i (probability $1 - p$) or, if they are connected, then j is also not in the GC (probability pu). This should hold for any j, of which there are $N - 1$, so the probability u should satisfy,

$$u = (1 - p + pu)^{N-1}. \tag{22.8}$$

Define $c > 0$ such that $p = c/N$; notice that $c \approx \langle k \rangle$. Using c and taking the log, we find,

$$\begin{aligned}
\log u &= (N-1)\log\left(1 - \frac{c}{N}(1-u)\right) \\
&\approx -\frac{c}{N}(1-u)(N-1) \\
&\approx -c(1-u),
\end{aligned} \tag{22.9}$$

where we assume N is large and use $\log(1+x) \approx x$ for small x. From this, the probability $v = 1 - u$ for a node to be within the GC satisfies

$$v = 1 - e^{-cv}. \tag{22.10}$$

This equation is always solved by $v = 0$, the trivial solution. A second solution is possible, although not in a closed form, depending on c. The functions $f(v) = 1 - e^{-cv}$ and $g(v) = v$ will intersect both at $v = 0$ (the trivial solution) and, if $c > 1$, at $v > 0$. We know this because the slope of $f(v)$ at $v = 0$ is $f'(0) = c$. If $c > 1$ then $f(v)$ will be above $g(v)$ after the origin, and, since $f(1) < 1 = g(1)$, then $f(v)$ must cross $g(v)$ at some $0 < v < 1$.[4] Therefore, a giant component exists when the mean degree is at least 1.

22.3.5 Why Erdős–Rényi is often not good enough

The Erdős–Rényi model is a simple model that we can understand mathematically. As such, when there is little information about the network's structure, it can serve as a no-knowledge baseline model. However, as we've learned more about the structural properties of real-world networks, it has become clear that the Erdős–Rényi model is not expressive enough to explain the properties of real-world networks. In particular, network scientists have found that real-world networks tend to exhibit two properties that the Erdős–Rényi model does not have: *clustering* and *heterogeneity*. These properties seem to be integral to how networks are created and organized as well as how various processes unfold on networks.

First, real-world networks tend to be *sparse*; in other words, $p \ll 1$ if we consider the $G(N, p)$ model. Because the clustering coefficient of the Erdős–Rényi model is p, what is expected according to the model for most real-world networks is that the clustering coefficient is almost zero. However, real-world networks tend to have much larger clustering coefficients than what the Erdős–Rényi model predicts.

Second, real-world networks tend to have a *heterogeneous* degree distribution. According to the Erdős–Rényi model, the degree distribution is a binomial (or Poisson), a homogeneous distribution with variance similar to its mean. In other words, since real-world networks are sparse and the average degree is small, the variance in degree should also be small and thus we do not expect to see nodes with very large degree. However, many real-world networks have a *broad* degree distribution, with a substantial number of nodes reaching very large degree, sometimes close to the order of the system size N. These nodes are the hubs and many real networks evolve under mechanisms that drive hubs into existence. For instance, the air traffic network tends to have hubs because hub

---

[4] As a concrete example, if $c = 2$, then $v \approx 0.797$; nearly 80% of nodes will belong to the GC.

airports such as Atlanta and Heathrow are the most efficient way to connect numerous airports across the world within a few number of flights.

These are not just interesting observations, they have significant consequences. For example, it has been shown that if the network exhibits a power-law degree distribution, then it becomes almost impossible, especially as the size of the network increases, to stop an epidemic outbreak on the network [362]. Because of their many connections, the hubs can not only be easily reached and infected, they can also infect many other nodes quickly. The existence of these large hubs fundamentally changes the dynamics on the network.

In summary, the Erdős–Rényi network model is a convenient model to understand the basic properties of networks and can be a reasonable model in some cases. However, most real-world networks are created due to mechanisms that are far from the uniformly random edges that the Erdős–Rényi model assumes, and therefore possess properties that are not captured by the Erdős–Rényi model. Because these properties, such as strong clustering and heterogeneity, are abundant in nature and society, it is often necessary to use more sophisticated models.

22.4 Configuration model

As we have just discussed, the heterogeneous degree distribution is a common and consequential property of real-world networks that is not explained by the Erdős–Rényi model. Therefore, it makes sense to think about a model that allows us to control the degree distribution of the network. The simplest version of such a model is the *configuration model* (also known as the Molloy–Reed model) [313, 314], which we first discussed in Ch. 11.

The configuration model is defined by the number of nodes N and the *degree sequence* $\{k_i\} = \{k_1, k_2, \ldots, k_N\}$. Note that not all sequences are allowed; the sequence must satisfy the following conditions (Erdős–Gallai theorem):

1. The sum of the degrees must be even: $\sum_{i=1}^{N} k_i$ is even.

2. $\sum_{i=1}^{r} k_i \leq r(r-1) + \sum_{i=r+1}^{n} \min(r, k_i)$, for each integer $1 \leq r \leq n - 1$.

The first condition can be easily understood: whenever we add an edge between two nodes, the degrees of the two nodes both increase by 1, making the total degree increase by 2. Because each edge has two *edge stubs*, the sum of the degrees is always even and equals two times the number of edges. The second condition, roughly, ensures that there are enough low-degree nodes to connect to the high-degree nodes. If and only if a degree sequence satisfies these two conditions can we create a network with that degree sequence. Such a degree sequence is called—because the sequence can describe an actual graph—a *graphic sequence*.

As we discussed in Sec. 11.5, we can build an instance of a configuration model network by laying out the nodes without edges, only stubs representing their degrees, and randomly creating edges by wiring stubs together. Because edges are otherwise inserted at random, the configuration model is often the minimal *null model* we use for a random network with a given degree distribution—its properties let us control for the degrees of nodes only.

This makes the null probability that two nodes are connected given the degree distribution an important, frequently encountered quantity. For a network with $\sum_i k_i/2 = M$ edges, suppose we select a stub from node i. There are k_j stubs connecting node j, out of $2M - 1$ stubs. Then the probability of connecting at random to j for a single edge stub is $k_j/(2M - 1)$. Of course, node i has k_i stubs in total, any of which can land on j, so the expected number of edges between i and j is

$$\tilde{A}_{ij} = \frac{k_i k_j}{2M - 1} \approx \frac{k_i k_j}{2M}. \tag{22.11}$$

This quantity is useful in many places. For instance, *modularity* (Eq. (12.15)) uses Eq. (22.11) to ask how many edges would fall within communities if edges were randomly connected given the degree sequence.

Of course, Eq. (22.11) is an approximation as we neglect whether j has other connections already and we even included i's other stubs in the connection probability. When studying and using the configuration model, it is convenient to think about a *multigraph* with self-edges (self-loops) because this is the least restrictive model that is amenable to analytical calculation and it makes the stublist construction method efficient—while randomly connecting pairs of stubs together, we do not have to check if a parallel edge already exists or even if the two stubs end at the same node. It is still possible to obtain a network without any self-edges or multi-edges, e.g., by rewiring away such edges, but doing so can introduce biases in the network structure (Sec. 11.6.2).

22.4.1   Generating function approach

A powerful method to study the configuration model is the *generating function* approach. As you will see below, it lets us formulate combinatorial questions using algebraic expressions, although it can be puzzling at first.

In the simplest sense, a polynomial function that is defined by a series $\{a_n\}$ is a *generating function*:

$$f(x) = \sum_{n=0}^{\infty} a_n x^n. \tag{22.12}$$

This polynomial function—generating function—can be thought of as an analytical way to manipulate *counts* and *probabilities*. Just like we can solve geometry problems with equations in Cartesian coordinates, we can solve discrete mathematics problems that involve counting (probabilities are really just comparing counts of events) with generating functions.

It's also useful to think of generating functions as a *mathematical trick* to store, retrieve, and manipulate numbers. In other words, instead of trying to understand the meaning of this function—how it behaves, what it represents, etc.—we can just think of it as a trick and focus on how to manipulate it to obtain what we want. In this sense, the generating function is simply a device where we can store each number a_n that is associated with a certain outcome n by associating it with the n-th power of the polynomial. This association with a number n becomes very useful as we will see.

Once you store a number, you can also retrieve it. For instance, we can immediately get a_0 by plugging zero into the generating function: $a_0 = f(0)$. What about a_1? We just need to find a way to keep only the x^1 term in the polynomial. We can do this by plugging zero into the *derivative* of the generating function: $a_1 = f'(0)$. The derivative removes all lower terms and putting zero into the function removes all higher terms. In general, we can retrieve a_n by

$$a_n = \frac{1}{n!} f^{(n)}(0).$$ (22.13)

Multiplication as counting

Because the generating function is a polynomial, and because the *outcome* is stored in the exponent, multiplication becomes *counting*. Namely, because $x^a x^b = x^{a+b}$, we can think of multiplying two terms from generating functions as enumerating the cases with outcome $a + b$ by combining the cases with outcome a and b.

In this interpretation, a_n can be thought as the *number of ways* (or *probability*) to get a total of n. For instance, let's look at this trivial formula:

$$5x^2 \cdot 2x^5 = 10x^7.$$ (22.14)

This equation, from the perspective of generating functions, can be interpreted as:

"There are 5 ways to have 2 of something and 2 ways to have 5 of another, so the total number of ways to have 7 of these things will be 10."

This may sound silly at this point, but it can let us solve non-trivial problems. For instance, let's assume that we want to have 10 fruits in total, and we have 3 types of fruits: apples, oranges, and bananas. We want to have at least 2 apples, an even number of oranges, and at most 5 bananas. How many ways can we have 10 fruits in total under these conditions?

You can probably solve this problem by carefully enumerating all possible cases— but that's tricky. Instead, let's use generating functions. We can first write down all possible cases for at least two apples:

$$a(x) = x^2 + x^3 + x^4 + x^5 + x^6 + x^7 + x^8 + x^9 + x^{10}.$$ (22.15)

This means that there is one way to have two apples, one way to have three apples, etc. because apples are indistinguishable here. Likewise, we can write down all possible cases for an even number of oranges,

$$o(x) = x^0 + x^2 + x^4 + x^6 + x^8 + x^{10},$$ (22.16)

and at most five bananas,

$$b(x) = x^0 + x^1 + x^2 + x^3 + x^4 + x^5.$$ (22.17)

Then we multiply these three generating functions to get the total number of ways to have 10 fruits:

$$a(x) \cdot o(x) \cdot b(x). \tag{22.18}$$

Let's look at one of the terms in the product: $x^2 x^8 x^0 = x^{10}$. The interpretation of this term is: "There is one way to have 2 apples, 1 way to have 8 oranges, and 1 way to have 0 bananas, so the number of ways to have 10 fruits total with 2 apples and 8 oranges will be 1." There are other ways to have 10 fruits total, of course. For instance, we can have 6 apples, 2 oranges, and 2 bananas, written as $x^6 x^2 x^2 = x^{10}$. All of these different possible ways to have 10 fruits in total will be gathered with x^{10} in the product of the three generating functions and the coefficient will be the total number of ways to have 10 fruits (21, it turns out). Alternatively, we can use Eq. (22.13) to obtain the total number of ways to have 10 fruits. The generating function turns all the enumerating steps into multiplication!

We can consider a *probability* generating function in the exact same way. A probability generating function is a generating function that is defined by a series $\{p_n\}$ of probabilities:

$$f(x) = \sum_{n=0}^{\infty} p_n x^n. \tag{22.19}$$

Because the coefficients are probabilities,

$$f(1) = \sum_{n=0}^{\infty} p_n = 1. \tag{22.20}$$

If we consider the multiplication of two terms, say $p_m x^m \cdot p_n x^n = p_m p_n x^{m+n}$, this can be interpreted as: "the probability of having $m + n$ is $p_m p_n$, which is the probability of having m times the probability of having n." This is an obvious statement when we consider the case where we are interested in the *sum* of the values.

For instance, if we have a biased dice with probabilities $\{p_n\}$, we can think of the case where we throw this dice twice and consider the sum of the two outcomes. The probability of getting a sum of 2, for instance, would be $p_1 p_1$, which is exactly the coefficient of x^2 in the product of the two (same) generating functions. Likewise, the probability of getting a sum of 4 would be $p_1 p_3 + p_2 p_2 + p_3 p_1$, which is the coefficient of x^4 in the product of the two generating functions.

Do you see now what the generating function is doing and why we said generating functions are like an analytical way to manipulate counts and probabilities?

Representing a random graph with a generating function

Now, let's think about a generating function for a random graph with a certain degree distribution $p(k)$, or $\{p_k\}$. In other words, if we randomly sample a node from the network, the probability that the node has degree k is p_k. Given that, in random graph models, we consider the edges as randomly connected, many of the processes that we are interested in this network can be considered as a random sampling from $\{p_k\}$. In other words, the generating function that describes this network can be written as

$$G(x) = \sum_{k=0}^{\infty} p_k x^k. \tag{22.21}$$

All the properties of the generating function that we mentioned above apply here as well. We can retrieve p_k with Eq. (22.13); we can compute $G^2(x)$ to obtain a generating function that describes the sum of two randomly sampled nodes' degrees.

Moments We can also calculate more relevant quantities, such as the average degree. The average degree is the expected value of p_k,

$$\langle k \rangle = \sum_{k=0}^{\infty} k p_k. \tag{22.22}$$

This formula can be obtained from the generating function by taking the derivative of $G(x)$ with respect to x:

$$\langle k \rangle = \sum_{k=0}^{\infty} k p_k = \left[\frac{d}{dx} G(x) \right]_{x=1}. \tag{22.23}$$

How about the second moment? Taking the derivative twice would not work because we will have $k(k-1)$ terms instead of k^2 terms. This is where "considering the generating function as a trick" becomes useful. We can simply focus on how to manipulate the function to produce k^2 terms. And an easy way is to just multiply x after the first derivative and then take the derivative again:

$$\frac{d}{dx} \left(x \frac{dG(x)}{dx} \right) = \sum_{k=0}^{\infty} k^2 p_k x^{k-1}. \tag{22.24}$$

In other words, we can write

$$\langle k^2 \rangle = \left[\left(x \frac{d}{dx} \right)^2 G(x) \right]_{x=1} = \sum_{k=0}^{\infty} k^2 p_k. \tag{22.25}$$

But what would be the point? We can just calculate the second moment directly, right? Let's see some examples that will convince you that the generating function can do more than just calculating moments.

Erdős–Rényi random graph Let's think about the degree distribution of the ER random graph. From a single node's perspective, it is like a repeated coin flip (Bernoulli trial) with probability p. For each possible node, the success (having an edge) probability is p and the failure (not having an edge) probability is $1 - p$. In the language of the generating functions, we can write the single trial as

$$g(x) = (1 - p)x^0 + px^1 = 1 - p + px. \tag{22.26}$$

But we are doing the same trial $N - 1$ times. How can we write this as a generating function? Note that the quantity we are curious about, the degree k, is the *sum* of the $N - 1$ trials. In other words, we can apply the generating function's multiplication property! The probability generating function for the degree distribution of the ER random graph can be written as

$$G(x) = (g(x))^{N-1} = (1 - p + px)^{N-1}.$$ (22.27)

The coefficient associated with x^k is exactly the binomial distribution,

$$G(x) = \sum_{k=0}^{N-1} \binom{N-1}{k} p^k (1 - p)^{N-1-k} x^k.$$ (22.28)

What happens if N is large and p is small? If we let $z = (N - 1)p$, we can rewrite Eq. (22.27) as,

$$G(x) = (1 - p + px)^{N-1} = \left(1 - \frac{z}{N-1} + \frac{z}{N-1}x\right)^{N-1}$$ (22.29)

$$= \left(1 + \frac{z(x-1)}{N-1}\right)^{N-1}.$$ (22.30)

Letting $N \to \infty$ we have, because $e^x = \lim_{n \to \infty} \left(1 + \frac{x}{n}\right)^n$,

$$G(x) = \left(1 + \frac{z(x-1)}{N-1}\right)^{N-1} \simeq e^{z(x-1)}.$$ (22.31)

This function still satisfies the properties of the generating function. For instance, $G(1) = e^{z \cdot 0} = 1$ and $G'(1) = ze^{z \cdot 0} = z$. What about the degree distribution? We can retrieve the degree distribution by using Eq. (22.13):

$$p_k = \frac{1}{k!} G^{(k)}(0) = \frac{1}{k!} z^k e^{-z} = e^{-z} \frac{z^k}{k!}.$$ (22.32)

Look familiar? Yes, it is the Poisson distribution!

We have just shown, not only that the degree distribution is described by the binomial distribution, but also that, when N is large and p is small, the degree distribution of the ER random graph can be well approximated by the Poisson distribution. Do you begin to see the power of the generating function method?

Number of second neighbors and the giant component   Here is another interesting application of the generating function: can we calculate the expected number of second neighbors of a node in a network? In other words, can we estimate the number of nodes that are two hops away from a random node?

To answer this, let us first think about the situation where we follow an edge to reach a neighbor of a node. This is reminiscent of the friendship paradox (Ch. 21). When we follow an edge, we arrive at a node. For second neighbors, we care about the number of *other* edges that we can follow from this node to those other nodes (Fig. 22.1). This is

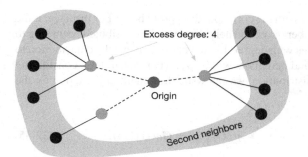

Figure 22.1 The excess degree of a node $(k - 1)$ is the number of edges that we can follow from this node to reach other nodes, assuming that we do not go back through the edge that we traversed to arrive at the node.

called the *excess degree* $k - 1$, as we do not consider the edge that we have just followed because that edge leads us backward and not to a second neighbor.

As we saw in Ch. 21, when we follow an edge, we are more likely to arrive at a node with a higher degree. The probability of arriving at a node with degree k is proportional to k; thus, the probability of arriving at a node with *excess degree* k' is proportional to $k' + 1$. In the language of generating functions, the coefficient of x^k (excess degree of k) should then be proportional to $(k + 1)p_{k+1}$. If we denote this generating function as $G_1(x)$ (to distinguish it from $G_0(x)$ that represents the degree distribution), we can consider the following form:

$$G_1(x) \stackrel{?}{=} p_1 x^0 + 2p_2 x^1 + 3p_3 x^2 + \cdots = G_0'(x). \tag{22.33}$$

Do you notice any problem with this definition?

The problem is that it does not contain the normalization of the excess degree distribution. If we substitute $x = 1$, we get $G_1(1) = p_1 + 2p_2 + 3p_3 + \cdots = \langle k \rangle$, not 1 (the normalization). Therefore, we should define $G_1(x)$ as

$$G_1(x) = \frac{1}{\langle k \rangle} \left(p_1 x^0 + 2p_2 x^1 + 3p_3 x^2 + \ldots \right) = \frac{G_0'(x)}{G_0'(1)}. \tag{22.34}$$

Equipped with $G_1(x)$, we can now calculate the expected number of second neighbors of a node in a network. The reason why we defined the generating function for the excess degree distribution is that the number of second neighbors equals the *sum* of the excess degrees of the neighbors of the node, each of which is sampled through following an edge!

Therefore, by following the sum–product property of the generating function again, we can conceptualize the generating function that captures the probability distribution of the number of second neighbors as the product of $G_1(x)$. If the node that we are considering has four neighbors, each of the four neighbors will have excess degree distributed according to $G_1(x)$. Therefore, the distribution of the number of second neighbors can be expressed by $G_1(x)^4$. But this is just one case where the focal node has degree 4. To compute the expected value, we need to consider all possible cases across the whole distribution, averaging with the corresponding probability. For instance, if the focal node has degree 1, the number of second neighbors will be distributed according to $G_1(x)^1$. And the fraction of nodes with degree 1 in the network is p_1, the contribution to the expected number of the second neighbors will be $p_1 G_1(x)^1$. Therefore, considering the whole range of the degree distribution, we can write the generating function for the number of second neighbors as:

$$p_1 G_1(x) + p_2 G_1(x)^2 + \cdots = \sum_{k=1}^{\infty} p_k G_1(x)^k = G_0(G_1(x)). \qquad (22.35)$$

Now we can even write down the generating function for the number of *third* neighbors as well: $G_0(G_1(G_1(x)))$. We can keep going!

Once we write down the generating function for the number of the first and second neighbors, we can also estimate the condition for the existence of the giant component. First, the number of the first neighbors is generated by $G_0(x)$, which is simply the degree distribution, and the expected value of $G_0(x)$ is $z_1 = G_0'(1)$. Second, the number of the second neighbors is generated by $G_0(G_1(x))$, which is the sum of the excess degrees of the neighbors, and the expected value can be computed as $z_2 = \left[\frac{d}{dx} G_0(G_1(x)) \right]_{x=1} = G_0''(1)$. Given these two expected values, when would the giant component emerge?

Let's first think about the case where $z_1 > z_2$. This means that the number of second neighbors tend to be *smaller* than the number of first neighbors, suggesting that as we travel farther away from a focal node, the additional nodes that we discover will keep decreasing. How about the opposite case, $z_1 < z_2$? This means that we will discover more nodes as we travel farther away from a focal node. Thus, we can *guess* that the giant component may emerge when $z_1 < z_2$. This intuition can indeed be confirmed by a more involved calculation with a generating function for the size of the connected components (see [340]).[5]

Overall, generating functions are powerful tools when studying network structure. Here we used them to understand the giant component of the configuration model. We'll see them again for similar purposes in Ch. 24 when we study the effects of missing data on observations of networks with community structure.

22.5   It's a small world!

Now let us turn out attention to another foundational models of network science: the Watts–Strogatz model. Before getting into the model, it will be useful to discuss the notion of "small world" first.

You've probably had the strange experience of meeting a complete stranger who happens to know someone you know. This is called the *small world phenomenon*, and it is a common experience in our daily life. How does this happen? How is this related to the properties of networks?

---

[5] We also derive the condition for a giant component without resorting to generating functions in Ch. 24, where we consider whether missing data can disconnect a sampled network.

Milgram's experiment The notion of "small world" goes back to a 1929 short story by Hungarian author Frigyes Karinthy, in which the characters believe that any two people in the world can be connected via at most five acquaintances. This idea became a concrete hypothesis and led to a series of studies. For instance, in the 1960s, Stanley Milgram, a sociologist who was also famous for his troubling obedience experiment [308], put it to a test. The "small world problem" he posed was: "given any two people in the world, person X and person Z, how many intermediate acquaintance links are needed before X and Z are connected?" With a $680 grant from the Laboratory of Social Relations at Harvard University, he designed and conducted an experiment, where he asked participants selected across the United States to send a package to a specific person living in Boston. But participants could only send the package to a person they already knew socially (along with the name of the final recipient), asking them to forward the package to someone *they* knew, and so forth, in the hopes it would eventually reach the specific recipient. By only sending packages along social ties, his experiment showed that the average number of intermediate acquaintances needed to connect two people is indeed around 6, which remarkably coincides with the number used by Karinthy.[6]

How small is small? How can we translate this idea to the language of networks? Probably the most basic way is to consider shortest path lengths, stating that networks tend to have "surprisingly small" average shortest path lengths. Another way to conceptualize the problem is explicitly considering the notion of a search process in social networks. In this case, we are interested in both the shortest paths and the search strategies that people may use (and the organization of the network that enables such search processes). Here, we will focus on the former (we discuss a "navigation" model of the latter shortly).

How small is *small*? Although there may be different definitions of smallness, usually, when the average path length scales roughly as the logarithm of the number of nodes N, $\bar{d} \sim \log N$, we call the network a *small world network*. Such logarithmic growth captures the small world notion well: N can increase by a huge factor, but $\log N$ will barely bulge.

The question now becomes, what ingredients are necessary for a small world?

---

[6] There are many important caveats of the original experiment by Stanley Milgram. First, the experiment did not measure the *shortest paths* between people, the found paths were the results of *heuristic search* processes that were not guaranteed to find the shortest paths. Second, many packages were lost in the process; due to survival bias, the packages that were lost were likely to be on longer paths than the packages that were found. In other words, the result probably underestimates the actual paths that could be found in the real world. Third, the selection of the participants was not entirely random and may have been biased towards a certain population. Fourth, there is always ambiguity regarding the definition of "acquaintance." Of course, there are more caveats, but these are the most important.

There have been other studies that broadly supported this result, most notably by Dodds et al. [131]. They used a much larger sample size and addressed some of the critical issues of the original study like the missing chains. They also found that the path lengths are around 6.

Another approach is to use online social networks to directly measure the shortest paths between people. For instance, a study estimated the average shortest path length in the Facebook social network to be around 4 and 5 [20] (Ch. 27). Note that this approach is about shortest paths rather than search, with the caveat that the estimated shortest paths can be shorter than the actual shortest paths due to spurious "friendship" links online.

Is a random network a small world? Does the basic Erdős–Rényi model, where each pair of nodes is connected with a constant probability p, exhibit a small world? It has essentially no structure, so if it were small world, we would expect small worlds to be quite common. Let's see.

First, because every node is more or less the *same* in this network, we can simplify the network further. Consider this tree network, where each non-boundary node has the same number of neighbors (3 in this particular case):

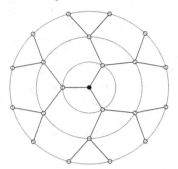

Beginning from a root node (filled), the remaining nodes are arranged in concentric rings. This network is known as the *Cayley tree* or (if $N \to \infty$) the *Bethe lattice*. We can use it as a simplified model of a random network if we assume a constant degree and the network is sparse enough that no cycles appear. Let's estimate how the average path length scales with the number of nodes. Number each "ring" (or layer) from the root node as $d = 1, 2, \ldots$. We now calculate how many nodes are within each layer. For instance, the number of nodes within the $d = 1$ layer is $1 + 3$ (or $1 + k$, where k is the degree of each node). By noticing that each subsequent layer has $k - 1$ times more nodes than the previous layer, we obtain the following formula for the number of nodes within d layers:

$$N_d = 1 + k \sum_{i=0}^{d-1} (k-1)^i = 1 + k \frac{(k-1)^d - 1}{k - 1 - 1} \simeq k^d. \tag{22.36}$$

In other words,

$$d \simeq \frac{\log N_d}{\log k} \propto \log N_d, \tag{22.37}$$

and thus, since most nodes will be in the last ring when N is large, we have a small world.

This heuristic argument demonstrates the main mechanism behind the small-world property: with each additional hop from a node to a neighbor, the number of nodes that can be reached increases *exponentially*. If the network is more or less random and large enough, the small world phenomenon is exactly what we would expect without any special assumptions!

Clustering and the Watts–Strogatz model The above example shows how exponential gain can power the small world, but it fails to capture something important about real networks: *they are not trees where every node has the same degree.* Real networks have transitivity, where edges tend to form triangles (and other loops). The Cayley tree cannot reproduce such features, and we may worry that a propensity for loops may prevent the exponential gain needed for a small world. Can we model a more realistic, small world network? Watts and Strogatz [486], in pioneering work, asked exactly this question, and proposed about the simplest possible graph model for a small world network.

The model imagines two extremes: a highly clustered network with high clustering coefficient, but with a long average path length, and a highly random, small world network with little clustering. Then it tries to find whether a middle ground—where the path length is still small but the clustering remains high—is possible. The idea is that adding a small number of *random* edges to a highly clustered network can drastically reduce the average path length without sacrificing the high degree of clustering, underlining the power of *weak ties* [190] that connects different, faraway groups of people or nodes. Clustering and small world do coexist!

The *Watts–Strogatz model* takes a one-dimensional periodic lattice, essentially a ring of nodes where each node is connected to a few nearest neighbors to its left and right, and then looks at how the diameter changes as a few of the edges in the ring were rewired to random nodes:

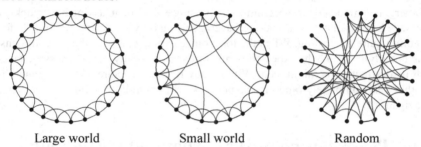

Large world   Small world   Random

If you randomize all the edges, you just have a random graph, but Watts and Strogatz found something interesting: it only takes a very few random edges, edges that could land anywhere around the ring, and suddenly you have enough "shortcuts" to drastically lower the diameter $d \sim N$ of the ring to $d \sim \log N$. Of course, a random network, unless it is very sparse, will also have a small diameter. What sets this result apart is just how few edges are needed, the middle network is still mostly a triangle-heavy ring. You can make the network much larger, and the diameter will hardly change at all, even without making the network denser.[7] It is difficult to overstate the impact of this article, especially in terms of effectively highlighting the inherent tension between *local* clustering and *global* connectivity. In many ways, it was the impetus for the modern era of network science.

Navigating a small world   The Watts–Strogatz model gives us an understanding of the network structure that lies beneath the six degrees of separation, but it can be pushed further. For one, the "ring world" of the model isn't really spatial in the sense that we

---

[7] Obviously the diameter will go down if the network becomes more dense.

humans are used to. Living on the surface of the Earth is quite different from a ring, and we should account for spatial or geographic distances too. To that end, in another pioneering work, Kleinberg [245] proposed a model that, while still quite simple, captures a bit more the notions of spatial distance. The Kleinberg model consists of an infinite 2D grid of nodes where each node is connected to its four nearest neighbors. But on top of that, each node is allowed one additional connection to a random *long-distance* node:

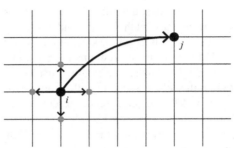

Unlike the Watts–Strogatz model, here the probability for a long-range connection from i to j depends on their distance r_{ij} on the grid, $P_{ij} \sim r_{ij}^{-\alpha}$, a power law with exponent α. Kleinberg then modeled the Milgram letter-passing experiment, asking how quickly can a message be passed between nodes on average using only local information. He found something quite interesting. When the long-range exponent $\alpha = 2$, the same dimension as the underlying grid,[8] the expected number of steps a locally-passed message takes to reach its destination is minimized. Without any further structure, the long-range links are otherwise completely random, the network will naturally maximize the efficiency of navigation![9]

22.6 Power-law networks—what's all the hubbub?

The previously discussed models, with one exception,[10] fail to address a feature common to many real networks, a broad or heavy-tailed degree distribution. Unlike Erdős–Rényi graphs, or the Watts–Strogatz ring, or the Kleinberg grid, real networks have *hubs*, a few nodes that have very high degree. Real degree distributions often do not look like a Poisson or a binomial (Secs. 11.7.2 and 12.3). To better capture real networks with models, we need models that have hubs.

Understanding the need to model hubs, Barabási and Albert [36] asked what are the minimum ingredients a model would need for hubs to appear. They arrived at a pair of mechanisms:[11] *growth* and *preferential attachment*. Either alone will not suffice, it has to be both.

---

[8] This holds in dimensions other than $D = 2$: navigation is fastest when $\alpha = D$.

[9] Kleinberg analyzed the navigation model using bounding arguments; later work brought interesting new insights into the model with exact asymptotic solutions [95, 96].

[10] That exception is the configuration model, but only because it specifies the degree sequence, and hence the degree distribution, as input. It matches a given degree distribution, but it cannot explain it.

[11] In fact, both mechanisms were discovered earlier, first by Yule [504] then by Nobel Laureate Simon [432] although not in the context of networks.

The Barabási–Albert (BA) model they proposed is a growing network. Starting from a small seed network, nodes are added, one per time step. Each new node forms links to $m \geq 1$ existing nodes. Preferential attachment (known in other contexts as *cumulative advantage*, the *Matthew effect*, or even just "*rich-get-richer*") is the second ingredient in the model; it dictates how a new node attaches to an existing node. Under preferential attachment, the probability a new node j attaches to an existing node i is proportional to k_i. Over time, higher-degree nodes will attract more new links than lower degree nodes, becoming higher-degree still. The rich indeed get richer.

Intuitively, it makes sense that preferential attachment will lead to high-degree nodes, but what exactly does the degree distribution look like? Let's analyze it mathematically using a rate equation approach [253].

We seek $P_k = N_k/N$ (Eq. (12.1)), where N_k is the number of nodes with degree k. To find it, we build an equation, called a rate equation, expressing how N_k[12] evolves over time. While the time evolution is discrete, with one new node added per step, we can make a continuum approximation and consider the change in N_k as a derivative, $\Delta N_k = dN_k/dt$. Since a new node appears at each timestep, the total number of nodes $N = t$ (neglecting the size of the seed graph), so we can equivalently use dN_k/dN.

Several changes occur at each time step. We'll account for each to build the equation for how N_k changes. First, growth: a new node is created with degree m, so N_m increases by 1. Second, N_k can *increase* if an existing node with degree $k - 1$ is attached to following preferential attachment, turning it into a node with degree k, meaning N_{k-1} decreases while N_k increases. Third, N_k can *decrease* if an existing node with degree k is attached to and becomes a node with degree $k + 1$.

What is the probability for a node with degree $k - 1$ to become a node with degree k? According to preferential attachment, it is $P(k - 1)$ with

$$P(k_i) = \frac{k_i}{\sum_j k_j} = \frac{k_i}{2M} = \frac{k_i}{2mN}, \tag{22.38}$$

where the last equation comes from the fact that every new node creates m new edges, so the total number of edges is $M \approx mN$ (neglecting the seed graph). For simplicity, let's take $m = 1$, meaning the network is a tree. This doesn't change anything fundamental, it just eliminates some additional constant terms.

Using this probability for the rates at which $k - 1$ nodes become k nodes and k nodes become $k + 1$ nodes, we can now express the change in N_k:

$$\frac{dN_k}{dN} = \delta_{k1} + \frac{1}{2N} [(k - 1)N_{k-1} - kN_k]. \tag{22.39}$$

The first term, $\delta_{k1} = 1$ if $k = 1$ and zero otherwise, captures the growth in new nodes, while the gains $[k-1] \rightarrow [k]$ and losses $[k] \rightarrow [k+1]$ capture preferential attachment.

We can begin to solve Eq. (22.39) by starting with $k = 1$, giving us $dN_1/dN = 1 - N_1/(2N)$ or $N_1 = 2N/3$ (neglecting terms that decay in time). If we continue for $k = 2$, $k = 3$, etc. we see that all solutions grow linearly in time. As we're more

---

[12] Technically, these equations describe $\langle N_k \rangle$, the expected number of nodes with degree k when averaged over an ensemble of model instances. We drop the $\langle \cdot \rangle$ for clarity.

interested in the limiting form of the degree distribution, let's eliminate time by choosing $N_k = Nn_k$, which converts Eq. (22.39) into a simpler recurrence equation,

$$n_k = \frac{1}{2}\left[(k-1)n_{k-1} - kn_k\right] = \frac{k-1}{k+2}n_{k-1}, \tag{22.40}$$

for $k \geq 2$, while for $k = 1$ we have $n_1 = 1 - n_1/2$ or $n_1 = 2/3$. Iterating on the recurrence starting from n_k until we reach n_1 gives the solution,

$$n_k = \frac{4}{k(k+1)(k+2)} = \frac{4\Gamma(k)}{\Gamma(k+3)} \sim k^{-3} \text{ (for large } k). \tag{22.41}$$

Thus, the BA model gives (for high-degree nodes) a power-law degree distribution (N_k/N) with exponent 3.

This form of degree distribution is also why these are called power-law networks. Their degree distributions follow a power law, very different from Erdős–Rényi graphs, and with many interesting properties, such as being *scale-free* (Sec. 12.3).

The work of Barabási and Albert [36] was exceptionally influential. Alongside that of Watts and Strogatz [486], it ushered in the modern era of network science. But as successful as it has been, it has attracted some controversy[13] over the years [220]. For one, it is a quite simple model, and certainly wouldn't capture every phenomenon driving the behavior of a real network. This means an *exactly* power-law degree distribution will probably not develop [78]. And if it did, it is a difficult distribution to validate statistically from a finite sample [477, 423]. Second, many of its most interesting properties rely on the asymptotic or infinite limit—technically, only an infinite network can be scale-free. Yet real network must be finite, so an exactly scale-free network cannot exist. Nevertheless, the existence of hubs and their importance are very real effects, even without an exact power-law degree distribution. To that end, we greatly prefer and recommend focusing on the heavy-tailed or broad nature of the degree distribution rather than the specific functional form, as that nature is both what matters and incontrovertibly exists in data.

22.6.1   Redirection

One implausible aspect of the BA model is that global information is needed when new edges form. Every time a new node is added and an existing node needs to be chosen for attachment, the BA model samples the new node with probability proportional to the existing node's degree. Yet it seems implausible for an edge creation, which is a local change to the network, to require examining the entire network's degree distribution— would a new student at a school need to ask how many friends *everyone else has* before they make a new social tie?

In fact, there is a lovely mechanism that *leads to* a power-law network but works entirely locally. Preferential attachment is an *effect* of the mechanism and not imposed on the model. That mechanism is connected with the friendship paradox that we discussed in Ch. 21 and called *redirection* [246, 253].

---

[13] Controversy? *Gasp!*

The *redirection model* (sometimes called the GNR (growing network with redirection) model) is similar to the BA model. A seed graph is created and new nodes i are added to the graph, one new node per timestep. The difference is in how the new nodes attach to existing nodes. At each time step, the new i first selects an existing j uniformly at random and forms (a directed) edge $i \rightarrow j$. Then, it "flips a coin": with probability $1 - r$, edge $i \rightarrow j$ remains. Otherwise, with probability r, node i looks at the neighborhood of j and *redirects its edge* to a randomly chosen neighbor of j. If this sounds too complex, we can assume that $r = 1$ and that the edge is always redirected. This is same as the case where we pick a random node, then pick a random neighbor of that node, and then connect to that neighbor. This process then continues with a new node.[14] No global information needed.

The growth mechanism in this model is the same, but a power-law network needs preferential attachment. How does that arise from redirection? The same intuition from the friendship paradox discussion (Ch. 21) applies: if a node v has many neighbors, then it is more likely that a new node will land on one of its neighbors and then redirect to v. Therefore, the random neighbors are sampled with probability $\propto k p_k$, where p_k is the degree distribution. The more neighbors v has, the more chances a new node will land on one and then redirect to v. The rich get richer.

If new nodes have multiple edges, then we can even introduce clustering. The idea is simple: we can do a random walk (keep sampling neighbors of neighbors) and keep connecting to those nodes appearing in the walk. Each time we sample a neighbor, we preferentially sample it based on their degree; by connecting to each subsequent, connected node, we introduce triangles. Now we have both preferential attachment and clustering.

Let's go back to the basic redirection model and determine the degree distribution of this model, now as a function of the parameter r. As before, we develop a rate equation for how N_k changes over time. To do so, we find the probabilities for $[k - 1] \rightarrow [k]$ and $[k] \rightarrow [k + 1]$, but now redirection may or may not happen. If redirection does not occur, then a randomly chosen node j with degree $k - 1$ is chosen with rate N_{k-1}/N. On the other hand, if redirection does occur, then the randomly chosen node j must be a *descendent* of a node v with degree $k - 1$, and every node has one ancestor so v has $k - 2$ descendants. Therefore, the rate at which j is connected by redirection is $(k - 2)N_{k-1}/N$. This is the point where the rich-get-richer mechanism emerges from redirection. A similar argument holds for the rate of $[k] \rightarrow [k + 1]$.

Putting these together, we arrive at the following rate equation:

$$\frac{dN_k}{dN} = \delta_{k1} + \frac{1-r}{N}[N_{k-1} - N_k] + \frac{r}{N}[(k-2)N_{k-1} - (k-1)N_k] \qquad (22.42)$$

$$= \delta_{k1} + \frac{r}{N}\left\{\left[k - 1 + \left(\frac{1}{r} - 2\right)\right]N_{k-1} - \left[k + \left(\frac{1}{r} - 2\right)\right]N_k\right\}. \qquad (22.43)$$

The first term in Eq. (22.42) accounts for the constant growth of new nodes, the second accounts for no redirection (probability $1 - r$), and the third accounts for redirection

---

[14] As described, the network will form a directed tree as each new node creates only a single edge. However, like with BA, the model easily extends to non-trees by allowing new nodes to connect to m existing nodes, simply by repeating the redirection mechanism once for each new edge.

(probability r). Within the last two terms are terms for the gains $[k-1] \rightarrow [k]$ and losses $[k] \rightarrow [k+1]$.

Now, compare Eq. (22.43) to Eq. (22.39). Having grouped terms, we see that the right side of Eq. (22.43) is nearly identical to Eq. (22.39): $1/(2N)$ becomes r/N, kN_k becomes $(k+w)N_k$, and $(k-1)N_{k-1}$ becomes $(k-1+w)N_{k-1}$, where $w = \frac{1}{r} - 2$. The simpler BA equation, with no offset w, led to a power-law exponent of 3, so we might expect the redirection equation to give an exponent of $3 + w = 1 + \frac{1}{r}$. Indeed it does,[15] meaning that the redirection model can produce any power-law degree distribution with exponent ≥ 2, simply by choosing r. All from a mechanism that needs only local information!

22.6.2   Copying

Lastly, there is another mechanism that can produce power-law degree distributions: the copying model. The copying model has been suggested in multiple contexts, including the web graph, citation networks, and biological networks. In the context of biological networks, it is motivated by a mechanism called *gene duplication*. When cells replicate, mistakes can occur and one type of mistake is the duplication of a gene. Having two copies of the same gene allows for more flexibility for the cell regarding how to use each gene. For example, if one copy of the gene then mutates, the cell can still use the other copy. In other words, over the course of long evolutionary processes, duplicated genes can diverge and become different genes. But note that two duplicated genes start out as the same gene, and so they have the same interaction partners; this is like copying a node to create a new node while keeping the same edges. We can also think of a similar mechanism in the citation network: when a new paper is written, perhaps largely based on an existing paper, the new paper may copy many of the existing paper's citations because they are all relevant. This is essentially like copying a node to create a new node while keeping many of the same edges.

Although there are several variations, the basic mechanism is simple: we randomly choose a node and copy it to create a new node, and then we randomly add or remove edges from the new node. This is all we need to create a power-law degree distribution. Although this may sound different from the other preferential mechanisms, it boils down to the same rich-get-richer principle. The more edges you already have, the more likely you will be chosen as one of the neighbors of the randomly copied node, therefore you will gain edges preferentially based on your degree.

---

[15] However, determining this properly means solving

$$n_k = \frac{1}{2}\{[(k-1)+w]n_{k-1} - (k+w)n_k\}.$$

The solution [253],

$$n_k = (2+w)\frac{\Gamma(3+2w)}{\Gamma(1+w)}\frac{\Gamma(k+w)}{\Gamma(k+3+2w)} \sim k^{-(3+w)},$$

does indeed give the exponent we expected.

22.7 Summary

Graph models underpin much of our understanding of network phenomena, from the small world path lengths to heterogeneous degree distributions and clustering. Mathematical tools help us understand what mechanisms or minimal ingredients may explain such phenomena, from basic heuristic treatments to combinatorial tools such as generating functions. These mechanistic models also go hand-in-hand with statistical models to help us better understand network data via inference. We now turn our attention to such statistical models.

Bibliographic remarks

While graph theory dates back to Euler, random graph theory, the merger of graph theory and probability, took a surprisingly long time to appear. Random graph theory dates back only to 1959 and the work of Erdős and Rényi [148, 149] that studied what became called the Erdős–Rényi model. Gilbert [182], a mathematician working at Bell Labs at the same time, independently contributed many of the same results as Erdős and Rényi. More properly, Erdős–Rényi graphs should be called Erdős–Rényi–Gilbert graphs.

Generating functions, a mathematical tool often used for combinatorial problems, are often very helpful when analyzing (random) graphs. Readers interested in learning more about using generating functions may wish to consult the classic *generatingfunctionology* by Wilf [492].

The modern era of network science, fueled very much by the rising popularity of the Internet and World Wide Web during the 1990s, began in 1998 with the Watts–Strogatz model [486]. Soon after, the Barabási–Albert model appeared [36] (although the works of Yule [504], Simon [432], and Price [379, 378] were clear antecedents; see Simkin and Roychowdhury [431]), and network science was in high gear. The redirection model we discussed was introduced by Kleinberg et al. [246] and Krapivsky and Redner [253]. More broadly, such growing network models are generalized by studying attachment kernels [253], which can take on different functional forms beyond the linear kernel of the BA model.

Exercises

22.1 Implement the Watts–Strogatz model. Use the number of nodes, the number of nearest neighbors, and the link randomization probability as parameters. Report the clustering coefficient and average path length (found using breadth-first search) as functions of the link randomization probability.

22.2 Compare the Watts–Strogatz model to an Erdős–Rényi graph using techniques from Ch. 14. Choose (and justify) parameters to make the comparison meaningful. As both models are random, be sure to average your comparison appropriately.

    (a) What conclusions can you draw from these comparisons?

    (b) Do any parameter values (for either model) make the two indistinguishable?

22.3 Implement the Kleinberg navigation model on a one-dimensional "ring world" like that used for Watts–Strogatz. (Be careful about measuring r appropriately.)

(a) What value(s) of power-law distance exponent α make the two equivalent?

(b) Use comparison measures (Ch. 14) to study the equivalence as a function of α.

22.4 An alternative to the Watts–Strogatz model is one where new long-range edges are added at random, not rewired from existing edges. Compare and contrast this model with the original Watts–Strogatz model.

22.5 Use the rate equations we derived for power-law networks to study the degree distribution when there is:

(a) growth but no preferential attachment,

(b) preferential attachment but no growth.

For the latter, assume the initial network consists of N nodes and zero edges.

Chapter 23

Statistical models and inference

In this chapter, we explore several important statistical models. Statistical models allow us to perform *statistical inference*—the process of selecting models and making predictions about the underlying distributions—based on the data we have. Compared to the models of Ch. 22, the focus tends to be more on statistical properties of the network rather than the microscopic mechanisms for how the network is created and evolves, although the distinction is often blurred as randomness is often intrinsic to both types of models. Statistical models can leverage powerful tools from statistics and help explore our data and the space of possibilities.

For instance, the stochastic block model assumes the network is constructed based on the block (community) membership of nodes; the probability of connection between nodes is prescribed as a set of parameters based on which blocks they belong to (which are also parameters of the model). Although no attempt is made to *explain* the community structure, this simple model allows us to write down the *likelihood function*, which estimates the likelihood of our network data given our model and parameters. The power of such an approach is that we can then infer the parameters through methods such as *maximum likelihood estimation* or full Bayesian inference. Carefully fitting a block model, for instance, can allow us to test whether any community structure is actually present in the network or create synthetic examples by sampling from the posterior distribution.

23.1 Statistical models we've seen before

Some of the simple graph models we've already encountered are in a sense statistical. For example, Erdős–Rényi graphs assume that each possible link is an iid Bernoulli trial. This makes it, in fact, the most basic statistical model of network data. As a random graph model, it's interesting to explore, showing, for instance, a percolation phase transition. But statistically, it's not as exciting, being not particularly expressive and estimating its parameters is relatively elementary (e.g., the MLE for link probability p is just the sample mean, $\hat{p} = M/\binom{N}{2}$). The configuration model is similar: while very useful as a degree-preserving null (Sec. 11.5), its inferential capacity, like the Erdős–Rényi model,

351

is limited.

Indeed, it is the expressiveness of a statistical model, and the information that we gain from fitting its parameters to data, that makes the statistical model interesting and useful with respect to data. Statistically, all an Erdős–Rényi graph can express is the overall density of the network. If we want to capture more structure, we need to invoke a more involved model.

We now turn our attention to statistical models primarily intended for inference using network data.

23.2 Stochastic block models

Block models refer to the idea and models that a network consists of groups of nodes called blocks. These blocks then dictate the connectivity of the network. *Stochastic* block models (SBM), often written "blockmodel," are the class of *statistical* models where the connectivity between nodes are probabilistically determined by the block membership of the nodes (and potentially other parameters).

23.2.1 The basic formulation

Formally, the basic stochastic block model assumes that a network consists of k blocks and every node belongs to one of these blocks. Node membership is described by a vector $\mathbf{z}$, where $z_i \in \{1, \ldots, k\}$ represents the block membership of node i. Then, in the most basic model, we assume that the connectivity between nodes is solely (and stochastically) determined by the block membership. The relationship between blocks is encoded into the block matrix $\mathbf{M}$, where M_{ij} represents the probability[1] of connection between any node in block i and any node in block j. If $z_u = 1$ and $z_v = 2$, then the probability that u and v are connected is M_{12}; if $z_u = 1$ and $z_v = 1$, then the probability that u and v are connected is M_{11}. That the connection probability for nodes depends only on what blocks they belong to is known as *stochastic equivalence*.

Notice that the SBM generalizes the Erdős–Rényi model. If there is only a single block containing all N nodes, then we have the Erdős–Rényi model with $p = M_{11}$.

The SBM is more flexible that it may first appear. We do not assume that $M_{ii} > M_{ij}$ ($i \neq j$). Although we usually conceptualize communities as assortative structures with more connections within than between, stochastic block models do not make such an assumption by default. But this flexibility comes at a cost: we have to carefully encode our assumptions and objectives into our models. As we will see soon, this may lead to some non-intuitive results.

Given the parameter set $\{k, \mathbf{z}, \mathbf{M}\}$, we can generate networks with arbitrary block structure. This capability is already useful. For instance, one can use a stochastic block model to generate an ensemble of networks with planted partitions (communities) and then use these synthetic networks to compare and evaluate community detection methods that aim to find such block structure.

---

[1] This is called the "canonical" form. In the "microcanonical" form, the block matrix prescribes the number of edges rather than the connection probability.

23.2.2 Inference

But probably a more useful application of the SBM than generating synthetic test data is *inference*. As Bayes' theorem tells us, if we have a statistical model with a computable likelihood, then we can "flip" it to infer the posterior distribution of the model parameters based on the data. In the context of block models, the data is the *adjacency matrix* $\mathbf{A}$ and the parameters are $\{k, \mathbf{z}, \mathbf{M}\}$. Identifying good parameters mean that we assign nodes into communities and estimate the connection probabilities based on the community membership. In other words, this means that based on the given network data, we can learn about the community structure, in terms of the number of communities (k) and community membership ($\mathbf{z}$).

The basic stochastic block model allows us to calculate the likelihood of a given network:

$$\mathcal{L}(G \mid k, \mathbf{z}, \mathbf{M}) = \prod_{(i,j) \in E} \Pr(i \to j \mid k, \mathbf{z}, \mathbf{M}) \prod_{(i,j) \notin E} (1 - \Pr(i \to j \mid k, \mathbf{z}, \mathbf{M}))$$

$$= \prod_{(i,j) \in E} M_{z_i, z_j} \prod_{(i,j) \notin E} (1 - M_{z_i, z_j}). \tag{23.1}$$

Note that many terms in the product are the same. This property allows us to gather terms together and simplify the formula. For two blocks r and s, let's denote the number of edges between them as e_{rs} and the number of possible edges between them as n_{rs}. Then the likelihood between these two groups can be written as the product of two terms

$$M_{rs}^{e_{rs}} (1 - M_{rs})^{n_{rs} - e_{rs}}, \tag{23.2}$$

and the full likelihood function is simply

$$\mathcal{L}(G \mid k, \mathbf{z}, \mathbf{M}) = \prod_{r,s} M_{rs}^{e_{rs}} (1 - M_{rs})^{n_{rs} - e_{rs}}. \tag{23.3}$$

Once we have the likelihood function, we can do *inference* using Bayes' theorem:

$$\Pr(k, \mathbf{z}, \mathbf{M} \mid G) = \frac{\Pr(G \mid k, \mathbf{z}, \mathbf{M}) \Pr(k, \mathbf{z}, \mathbf{M})}{\Pr(G)}. \tag{23.4}$$

Numerous methods are available to perform inference such as maximum likelihood estimation or Bayesian inference. A practical technique for the latter, Markov Chain Monte Carlo (MCMC), is standard practice. Indeed, the original inferential method for SBMs proposed by Snijders and Nowicki [437] used Gibbs sampling, a standard MCMC inference algorithm. Expectation–maximization, which we'll discuss shortly, is another approach. For a review of the SBM inference literature, see Lee and Wilkinson [266].

23.2.3 Model selection

Now we know the basic formulation of the stochastic block model. Let us ask you a question. Suppose we fit the SBM to the Zachary Karate Club, which we know has 🗲

roughly two groups. Let's not impose any constraints on the parameters and assume that our inference method can find the best possible parameters that maximize the likelihood. Will we find $k = 2$? What will $\mathbf{M}$ look like? How large will $\mathcal{L}$ be?

Will the SBM find the community structure in the network? Unfortunately, the answer is no: without constraints, there is a **trivial solution**. Consider the SBM where $k = N$, $z_i = i$, and $\mathbf{M} = \mathbf{A}$. In other words, every node is its own community and $\mathbf{M}$ simply prescribes the actual connections without any randomness. If we specify the SBM this way, the value of the likelihood function (trivially) equals 1 and no other model can be more likely. In other words, we have overfit the data with an overly complicated model.

This is why we must think about *model selection*. The SBM is expressive and can capture a wide range of block structure. By increasing the number of parameters ($k \gg 1$), it may become too expressive and begins to overfit (see also Ch. 16). To prevent overfitting, we need to think about model selection—the process of comparing different models and choosing the "best" model that achieves a good balance between *parsimony* (simplicity and generalizability) and fit accuracy. For instance, consider again the extreme case of the "perfect" model ($k = N, z_i = i, \mathbf{M} = \mathbf{A}$). Although it maximizes the likelihood of the given data, any noise or variation in the data will immediately make the likelihood go to zero. Since most data will have some noise, we must assume, it's unlikely such an overly expressive model will accurately capture the true structure.

In the case of tabular data (the usual machine learning setting), the model selection problem can be handled by resampling: randomly splitting the data (e.g., creating a validation set or doing a cross validation). We fit the model to some of the data and evaluate it with the rest of the data; forcing the model to generalize past the fitted observations help limit overfitting. For networks, however, this is much more tricky, because often we just have a single, highly interconnected set of data points. Splitting the data can easily destroy the very structure that we want to discover.

Instead of resampling, some approaches to model selection for the SBM could be to sweep across parameters (namely k) to find the best fit model for each value of k, then use an information criterion such as AIC or BIC to pick the best tradeoff between model simplicity (low k) and fit (high $\mathcal{L}$). However, keep in mind that block models are nested: a $k - 1$ block model is a special case of a k block model. Likelihood ratio tests, as pursued by Wang and Bickel [483], may be more appropriate for such cases.

Another successful approach inspired by information theory is to appeal to the *minimum description length* (MDL) principle. First, we can considering maximizing the likelihood as equivalent to minimizing the number of possible configurations (Ω) given parameters, and thus the amount of information (or entropy), $\ln \Omega$, needed to describe the data. For instance, the trivial solution with $\mathbf{M} = \mathbf{A}$ minimizes this entropy because there is only one possible configuration given the parameters. However, the MDL principle argues that we also need to consider the information necessary to describe the model itself. Because more complex models require more information to describe them (more blocks and larger $\mathbf{M}$), it balances the model's complexity against the likelihood of the data and chooses the model that requires the least amount of information to describe the data and the model. Peixoto [365] shows that appealing to MDL places a penalty on the model's likelihood, and reveals the number of detectable

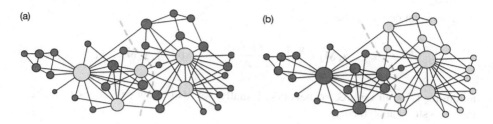

Figure 23.1 Comparing the uncorrected (**a**) and degree-corrected (**b**) stochastic block model partitions for the Zachary Karate Club [236]. The dashed line indicates the split observed by Zachary [505]. Reprinted figure with permission from Karrer, Brian and Newman, M. E. J., *Phys. Rev. E*, 83, 016107, 2011. Copyright (2011) by the American Physical Society.

blocks scales like $\sqrt{N}$, in a sense reminiscent of the resolution limit we encounter with modularity (Ch. 12).

This is an active area of research and there are, and will be, many competing approaches for model selection.

23.2.4 Degree-corrected model

Let's say you already know that there should be exactly two communities in the Zachary Karate Club network and you don't need to worry about the model selection problem. Then applying the basic SBM with $k = 2$ fixed will produce the communities that we expect, right?

Well, not so fast. Figure 23.1a shows the result, which is not exactly what we expect! There is another implicit assumption in the model that we need to address. Imagine two nodes u and v that belong to the same block. They will have exactly the same probability to be connected with all other nodes; they are stochastically equivalent, indistinguishable in the statistical sense. The implication is that every node in a block is indistinguishable in the model, but in real networks nodes are distinguished by more than their block membership; in particular, degree varies a lot among the nodes, even in the same block. Remember that the SBM does not automatically find assortative communities. Any consistent pattern can be interpreted and discovered as "block" structure by the SBM. Therefore, the basic SBM may find that grouping nodes with similar degrees is a better fit than finding "communities." This is exactly what we see in Fig. 23.1a.

A common solution to this problem is introducing an additional node-level parameter that modulates the degree of each node and consider a multigraph where multiple edges can exist between two nodes. This is called the *degree-corrected stochastic block model*.

Instead of assuming that the probability of a connection between u and v is $M_{z_u z_v}$, we assume that each element in the adjacency matrix **A** is Poisson-distributed around the mean of $\gamma_u \gamma_v M_{z_u z_v}$. As γ_i increases, the degree of node i can increase as well. Recall that the Poisson distribution is $\lambda^k e^{-\lambda}/k!$, where λ is the mean. Thus the likelihood of the graph can be written as

$$\mathcal{L}(G \mid k, \mathbf{z}, \mathbf{M}) = \prod_{i<j} \frac{(\gamma_i \gamma_j M_{z_i z_j})^{A_{ij}}}{A_{ij}!} \exp\left(-\gamma_i \gamma_j M_{z_i z_j}\right), \qquad (23.5)$$

assuming G is an undirected network. Usually, we also allow self-edges to make it easier to study analytically,

$$\mathcal{L}(G \mid k, \mathbf{z}, \mathbf{M}) = \prod_{i<j} \frac{(\gamma_i \gamma_j M_{z_i z_j})^{A_{ij}}}{A_{ij}!} \exp\left(-\gamma_i \gamma_j M_{z_i z_j}\right)$$

$$\times \prod_i \frac{(\frac{1}{2}\gamma_i^2 M_{z_i z_i})^{A_{ii}/2}}{A_{ii}/2!} \exp\left(-\frac{1}{2}\gamma_i^2 M_{z_i z_i}\right). \qquad (23.6)$$

The factor of $\frac{1}{2}$ appears in the self-edge term because creating a single self-edge "consumes" two edge stubs of the node. Compared to the uncorrected model, the degree-corrected SBM finds the partition we expected in the Zachary Karate Club (Fig. 23.1b).

23.2.5   Understanding community detection with the SBM

The stochastic block model is simple enough that it becomes analytically tractable and several interesting discoveries about it have been made, sparking an ongoing line of research [1]. Two of the most important discoveries are the *detectability limit* and *optimal recovery*.

The difficulty of inferring the hidden $\mathbf{z}$ in the SBM depends on how clearly separated the groups are. The groups should be distinguishable when $M_{ii} > M_{ij}$ for $i \neq j$, whereas if $M_{ii} = M_{ij}$ for all groups we would have a globally random (Erdős–Rényi) graph with no modular structure. In the latter case, we expect that we cannot detect the groups while in the former case we can. The detectability limit makes this precise—and shows that groups can be **impossible** to detect even when they exist (i.e., when $M_{ii} > M_{ij}$).

Consider a SBM with q groups of equal size and density. Let $c_{\text{in}} = N M_{ii}$ and $c_{\text{out}} = N M_{ij}$ (for $i \neq j$) be rescaled block probabilities that are the same for all groups: the average degree $\langle k \rangle = \frac{1}{q}[c_{\text{in}} + (q-1)c_{\text{out}}]$ for this "homogeneous" SBM. Decelle et al. [125, 126] show with an asymptotic analysis that a *phase transition* in the learnability of $\mathbf{z}$ occurs: when

$$|c_{\text{in}} - c_{\text{out}}| > q\sqrt{\langle k \rangle}, \qquad (23.7)$$

learning $\mathbf{z}$ is possible ($\mathbf{z}$ can be recovered with high probability); otherwise, detectability of $\mathbf{z}$ with any accuracy is impossible, it is believed, for *any* (polynomial) algorithm.[2]

---

[2] This was conjectured by Krzakala et al. [259], with various proofs of special cases following (see Abbe [1] for details). Also, we are glossing over some details. In general, detectability actually transitions from impossible to hard to easy, but it is argued that the hard phase, where it is possible in principle to find $\mathbf{z}$ but computationally very difficult, is narrow enough that it is unlikely for a practical inference problem to land within it.

Intuitively, Eq. (23.7) makes sense: if the difference in density within groups and between is too small relative to the overall density of the network, the groups are undetectable. But the details are still surprising. Suppose $q = 2$ and let $c_{out} = \epsilon c_{in}$. Then $c_{in} = 2 \langle k \rangle / (1 + \epsilon)$ and groups are detectable only when

$$c_{in} - c_{out} > \sqrt{2(c_{in} + c_{out})}$$

$$c_{in} > 2\frac{1 + \epsilon}{(1 - \epsilon)^2}$$

$$\langle k \rangle > \left(\frac{1 + \epsilon}{1 - \epsilon}\right)^2$$

$$\epsilon < \frac{\sqrt{\langle k \rangle} - 1}{\sqrt{\langle k \rangle} + 1}.$$

Suppose $\epsilon = 1/2$, a noticeable difference between in- and out-group links. If $\langle k \rangle = 4$, the groups will be *undetectable*, as $\epsilon > 1/3$. For $\epsilon = 1/2$, the network needs to be denser, $\langle k \rangle > 9$, for the groups to be found. Enough sparsity and meaningful groups become invisible to us.

A corollary to the detectability threshold is the development of optimal recovery methods. The same calculations by Decelle et al. [125, 126] showing the detectability transition also show that a *belief propagation* (BP; also known as message passing) algorithm is asymptotically optimal: if the z can be found, the BP algorithm will do so and is optimal in the sense that no other algorithm can have better expected accuracy. The optimality of a BP algorithm motivated the search for more scalable methods, capable of inferring the SBM for larger, sparse networks. Spectral methods are usually helpful in these circumstances due to their scalability on sparse networks. However, a gap existed where spectral methods were unable to achieve the same accuracy as the BP algorithm if the network was too sparse [326]. This gap has been closed, with some additional computational cost, by using the *non-backtracking matrix* [204]. Krzakala et al. [259] show that the spectra of this matrix is more useful for SBM inference as it relates more closely to belief propagation[3] than other matrices typically used, such as the graph Laplacian.

(We discuss the non-backtracking matrix and some spectral methods for community detection in Ch. 25.)

There are some caveats to these results. One, the derivations are asymptotic, and finite size effects will play a role (the asymptotic results are still quite accurate) [502]. Two, this does not treat the degree-corrected SBM; a heterogeneous degree distribution may actually help with detection [125]. Lastly, and perhaps most crucially, these results only hold for the SBM, which is not always the best or most appropriate model for a real network. Despite these caveats, these results still teach us useful and surprising details about this inference problem.

---

[3] The non-backtracking operator arises when linearizing BP.

23.2.6   Other variations

Numerous variants exist for block models to accommodate various network types and block structures. For instance, models exist that incorporate weighted edges (where the SBM generates a weighted adjacency matrix $[w_{ij}]$), directed edges, multi-edges, hyper-edges, and other higher-order structures. Regarding the block structure, probably the most notable models describe hierarchical structure and overlapping block structure.

Hierarchical models   A hierarchy of super-blocks, blocks, sub-blocks, and so forth can be specified by repeated use of the SBM. In other words, we can model the block matrix **M** *itself* using an SBM.[4] Treating the block matrix as a weighted graph on k nodes, a weighted SBM can generate **M**, so long as self-loops[5] are allowed (being needed for the within-group probabilities M_{gg}). In principle, we can parameterize the full hierarchical model by specifying the number of levels of SBMs and the parameters within each level, then fit to data using inference. All this is easier said than done, of course. Care must be taken when it comes to parsimony, as such a highly parameterized, nested model can easily overfit an observed network. For full details, see Peixoto [366].

Mixed membership models   Community detection methods can be divided into partitioning methods and overlapping methods. Most SBMs focus on partitions, but mixed membership stochastic block models [7] have been formulated to address the case where we wish to associate nodes with multiple blocks. Suppose each node i has an associated membership probability vector π_i, where $\pi_{i,g}$ is the probability that i belongs to group g, and $\sum_g \pi_{i,g} = 1 \ \forall i$. Using these vectors along with the $k \times k$ block probability matrix **M**, we can generate a random network as follows. For each pair of nodes s, t, draw group g_s with probability π_{s,g_s} and g_t with probability π_{t,g_t}, then connect s and t with probability M_{g_s,g_t}. In other words, $A_{st} \sim \text{Bernoulli}(M_{g_s,g_t})$. (Note that this need not be symmetric: when it comes time to generate A_{ts}, the group memberships and thus the connection probabilities in that orientation may be different.) To perform inference, we need to specify a prior distribution for π, the natural choice being the Dirichlet distribution (i.e., $\pi_i \sim \text{Dir}(\alpha)$) which ensures that the normalization condition holds. Likewise, the groups follow a categorical distribution parameterized by π (i.e., $g_i \sim \text{Cat}(\pi_i)$). The Dirichlet parameter vector α along with **M** then serve as our inferential targets, while the membership probabilities π and group pairs (g_s, g_t) act as latent variables. The presence of these latent quantities makes expectation–maximization (EM) a natural choice for performing inference. (We will use EM in Sec. 23.3; see also Sec. 24.5.) For full details, see Airoldi et al. [7].

23.3   *Witness me*: the edge observer model

Consider a network dataset derived from tests conducted on each edge. Measurements are taken and we record each time an edge is or is not *observed*. Such a data generating

---

[4] This may remind you of the Louvain method [57].

[5] Alternatively, instead of weights we can consider the block matrix as a multigraph where the number of edges between two "nodes" g_i and g_j is proportional to M_{g_i,g_j}.

process (DGP) describes many network datasets such as the HuRI and Malawi Sociometer Network focal networks. Our inferential goal is to understand an unseen or latent ground truth adjacency matrix $\mathbf{A}$ that relates to the probability that edges are observed when measurements are taken. That is, when an edge (i, j) really exists ($A_{ij} = 1$), we are likely to observe it in our DGP, although there is a chance we may not due to noise. Likewise, when (i, j) does not exist ($A_{ij} = 0$), we are likely not to observe it, although there is a chance we may observe it, again due to noise. How likely it is to make such mistakes, false negatives or false positives, will depend somehow on our measurement process's accuracy and reliability. What can we say about the latent $\mathbf{A}$ from our noisy observations?

Statistically, we model the DGP using a probability $P(\text{data} \mid \mathbf{A}, \theta)$. Here "data" acts as a placeholder for how the DGP measurements are stored (we discuss specifics below) and θ represents a set of parameters that we use to model the DGP. By manipulating this probability, we can express other probabilities; if we can calculate $P(\mathbf{A} \mid \text{data}, \theta)$, we can use it to find networks that are probable given the data, allowing us to reconstruct a network using the DGP's noisy measurements. Further, finding this probability will also reveal a way to calculate it efficiently. First, from Bayes' theorem,

$$P(\mathbf{A}, \theta \mid \text{data}) = \frac{P(\text{data} \mid \mathbf{A}, \theta)P(\mathbf{A}, \theta)}{P(\text{data})}. \tag{23.8}$$

How best to get from $P(\mathbf{A}, \theta \mid \text{data})$ to $P(\mathbf{A} \mid \text{data}, \theta)$? Marginalizing out $\mathbf{A}$ gives us the probability for the model parameters given the observations, $\sum_{\mathbf{A}} P(\mathbf{A}, \theta \mid \text{data})$. (Of course, summing over all networks is intractable in general.) The θ which maximizes this probability is our *maximum a posteriori* (MAP) estimator of θ. The log of $P(\theta \mid \text{data})$ has the same maximum and is more convenient to work with:

$$\log P(\theta \mid \text{data}) = \log \sum_{\mathbf{A}} P(\mathbf{A}, \theta \mid \text{data}) \geq \sum_{\mathbf{A}} q(\mathbf{A}) \log \frac{P(\mathbf{A}, \theta \mid \text{data})}{q(\mathbf{A})}. \tag{23.9}$$

The last step comes from Jensen's inequality[6] and holds for any probability distribution $q(\mathbf{A})$ such that $\sum_{\mathbf{A}} q(\mathbf{A}) = 1$. But if we take

$$q(\mathbf{A}) = \frac{P(\mathbf{A}, \theta \mid \text{data})}{\sum_{\mathbf{A}} P(\mathbf{A}, \theta \mid \text{data})}, \tag{23.11}$$

the inequality in Eq. (23.9) becomes an equality, and this means Eq. (23.11) maximizes the right-hand side wrt $q(\mathbf{A})$. If we maximize this expression *again* with respect to θ, we get our MAP estimate of our model parameters. This double maximization can be solved iteratively: first we maximize with respect to $q(\mathbf{A})$ (via Eq. (23.11)) then we maximize with respect to θ with $q(\mathbf{A})$ held constant. This second maximum can be found by differentiating Eq. (23.9) with $q(\mathbf{A})$ constant and solving

$$\sum_{\mathbf{A}} q(\mathbf{A}) \nabla_\theta \log P(\mathbf{A}, \theta \mid \text{data}) = 0 \tag{23.12}$$

---

[6] Jensen's inequality states that $\log \mathbb{E}[x_i] \geq \mathbb{E}[\log x_i]$ because log is concave. From this, we have

$$\log \sum_i x_i = \log \sum_i x_i \frac{q_i}{q_i} = \log \mathbb{E}_q\left[\frac{x_i}{q_i}\right] \geq \mathbb{E}_q\left[\log \frac{x_i}{q_i}\right] = \sum_i q_i \log \frac{x_i}{q_i}, \tag{23.10}$$

which is what we use in Eq. (23.9).

for θ. With θ estimated, we can then estimate $\mathbf{A}$ from our data. In fact, $q(A)$ already tells us about $\mathbf{A}$, since

$$q(\mathbf{A}) = \frac{P(\mathbf{A}, \theta \mid \text{data})}{P(\theta \mid \text{data})} = P(\mathbf{A} \mid \text{data}, \theta). \tag{23.13}$$

So $q(\mathbf{A})$ is exactly the distribution we want, the posterior probability of $\mathbf{A}$ given our data and parameters.

Putting these pieces together, we have a general-purpose method for estimating a network's structure from noisy observations of its edges. The double maximization is an example of the *expectation–maximization algorithm* [127], an elegant and often very effective technique for finding maximum likelihood parameters numerically when the likelihood function is complicated. That said, presented generically, the steps above may be a little opaque, so let's make some specific, simplifying assumptions, then apply this technique to one of our focal networks.

Independent observer model

To simplify, assume iid edge measurements, that is, each measurement is an independent Bernoulli (edge/no-edge) random variable. We assume that our measurements have true positive rate α and false positive rate β. In other words, when an edge is actually present between nodes i, j, $A_{ij} = 1$, our DGP observes it correctly with probability α and fails to observe it with probability $1 - \alpha$. Likewise, for a non-edge ($A_{ij} = 0$), our DGP correctly does not observe it with probability $1 - \beta$ and incorrectly observes it with probability β. For our data, we observe node pair i, j a total of N_{ij} times; of those observations, we observe an edge E_{ij} times.

Taken together, and further assuming the network is undirected, the likelihood of our data is

$$P(\text{data} \mid \mathbf{A}, \theta) = \prod_{i<j} \left(\alpha^{E_{ij}} (1 - \alpha)^{N_{ij} - E_{ij}} \right)^{A_{ij}} \left(\beta^{E_{ij}} (1 - \beta)^{N_{ij} - E_{ij}} \right)^{1 - A_{ij}}. \tag{23.14}$$

To get from this to the posterior, we introduce some priors. Because edges are independent, and assuming ρ is the prior probability for any given edge to exist, the full network $\mathbf{A}$ has a prior probability of $P(\mathbf{A} \mid \rho) = \prod_{i<j} \rho^{A_{ij}} (1 - \rho)^{1 - A_{ij}}$. Lastly, we'll keep things simple by assuming a uniform prior for θ, $P(\theta) = P(\alpha)P(\beta)P(\rho) = 1$, i.e., α, β, and ρ are uniformly distributed on the interval $[0, 1]$.

Applying all these assumptions to Eq. (23.8), we have

$$P(\mathbf{A}, \theta \mid \text{data}) = \frac{P(\text{data} \mid \mathbf{A}, \theta)P(\mathbf{A} \mid \theta)P(\theta)}{P(\text{data})} = \frac{P(\text{data} \mid \mathbf{A}, \theta)P(\mathbf{A} \mid \rho)}{P(\text{data})}$$

$$= \frac{1}{P(\text{data})} \prod_{i<j} \left(\rho \alpha^{E_{ij}} (1 - \alpha)^{N_{ij} - E_{ij}} \right)^{A_{ij}} \left((1 - \rho)\beta^{E_{ij}} (1 - \beta)^{N_{ij} - E_{ij}} \right)^{1 - A_{ij}}. \tag{23.15}$$

Unlike the generic probabilities we had earlier, now we have an explicit expression. Moreover, we no longer have a sum over all 2^N possible networks, but instead (after taking the log) a sum over $\binom{N}{2}$ node pairs. Much more reasonable. Continuing on, to substitute into Eq. (23.12), let's take the log and differentiate with respect to our parameters:

$$\frac{\partial}{\partial \alpha} \log P(\mathbf{A}, \theta \mid \text{data}) = \sum_{i<j} A_{ij} \frac{\partial}{\partial \alpha} \log \rho \alpha^{E_{ij}} (1 - \alpha)^{N_{ij} - E_{ij}}$$

$$= \sum_{i<j} A_{ij} \left(\frac{E_{ij}}{\alpha} - \frac{N_{ij} - E_{ij}}{1 - \alpha} \right), \tag{23.16}$$

$$\frac{\partial}{\partial \beta} \log P(\mathbf{A}, \theta \mid \text{data}) = \sum_{i<j} (1 - A_{ij}) \left(\frac{E_{ij}}{\beta} - \frac{N_{ij} - E_{ij}}{1 - \beta} \right), \tag{23.17}$$

$$\frac{\partial}{\partial \rho} \log P(\mathbf{A}, \theta \mid \text{data}) = \sum_{i<j} \left(\frac{A_{ij}}{\rho} - \frac{1 - A_{ij}}{1 - \rho} \right). \tag{23.18}$$

Substituting these into $\sum_{\mathbf{A}} q(\mathbf{A}) \nabla_\theta \log P(\mathbf{A}, \theta \mid \text{data})$ and solving for the $\hat{\theta}$ that makes this zero gives expressions for our parameter estimates. For $\hat{\alpha}$, we have

$$\sum_{\mathbf{A}} q(\mathbf{A}) \sum_{i<j} A_{ij} \left(\frac{E_{ij}}{\hat{\alpha}} - \frac{N_{ij} - E_{ij}}{1 - \hat{\alpha}} \right)$$

$$= \sum_{i<j} \sum_{\mathbf{A}} q(\mathbf{A}) A_{ij} \left(\frac{E_{ij}}{\hat{\alpha}} - \frac{N_{ij} - E_{ij}}{1 - \hat{\alpha}} \right)$$

$$= \sum_{i<j} Q_{ij} \left(\frac{E_{ij}}{\hat{\alpha}} - \frac{N_{ij} - E_{ij}}{1 - \hat{\alpha}} \right) = 0, \tag{23.19}$$

and solving for $\hat{\alpha}$ gives

$$\hat{\alpha} = \frac{\sum_{i<j} E_{ij} Q_{ij}}{\sum_{i<j} N_{ij} Q_{ij}}. \tag{23.20}$$

Along the way we introduced $Q_{ij} = \sum_{\mathbf{A}} q(\mathbf{A}) A_{ij}$. This is the posterior probability for edge i, j: $Q_{ij} = P(A_{ij} = 1 \mid \text{data}, \theta)$. This matrix is where our estimated network is found. Following the same steps to estimate our other parameters leaves

$$\hat{\beta} = \frac{\sum_{i<j} E_{ij}(1 - Q_{ij})}{\sum_{i<j} N_{ij}(1 - Q_{ij})}, \quad \hat{\rho} = \frac{1}{\binom{N}{2}} \sum_{i<j} Q_{ij}. \tag{23.21}$$

To finish building our model, we seek an expression for Q_{ij} in terms of $\hat{\alpha}, \hat{\beta}$, and $\hat{\rho}$. With our simplifications, we know that

$$P(\mathbf{A} \mid \text{data}, \theta) = q(\mathbf{A}) = \prod_{i<j} Q_{ij} (1 - Q_{ij})^{1 - A_{ij}} \tag{23.22}$$

for whatever Q_{ij} turns out to be. To find it, given our parameter estimates, let's substitute $P(\mathbf{A}, \theta \mid \text{data})$ from Eq. (23.15) into $q(\mathbf{A})$ from Eq. (23.11) and rearrange terms until we are in the form of Eq. (23.22)'s right-hand side:

$$q(\mathbf{A}) = \frac{\prod_{i<j} \left[\rho \hat{\alpha}^{E_{ij}} (1 - \hat{\alpha})^{N_{ij} - E_{ij}} \right]^{A_{ij}} \left[(1 - \rho) \hat{\beta}^{E_{ij}} (1 - \hat{\beta})^{N_{ij} - E_{ij}} \right]^{1 - A_{ij}}}{\sum_{\mathbf{A}} \prod_{i<j} \left[\rho \hat{\alpha}^{E_{ij}} (1 - \hat{\alpha})^{N_{ij} - E_{ij}} \right]^{A_{ij}} \left[(1 - \rho) \hat{\beta}^{E_{ij}} (1 - \hat{\beta})^{N_{ij} - E_{ij}} \right]^{1 - A_{ij}}}$$

$$= \frac{\prod_{i<j} \left[\rho \hat{\alpha}^{E_{ij}} (1 - \hat{\alpha})^{N_{ij} - E_{ij}} \right]^{A_{ij}} \left[(1 - \rho) \hat{\beta}^{E_{ij}} (1 - \hat{\beta})^{N_{ij} - E_{ij}} \right]^{1 - A_{ij}}}{\sum_{A_{ij} = 0,1} \left[\rho \hat{\alpha}^{E_{ij}} (1 - \hat{\alpha})^{N_{ij} - E_{ij}} \right]^{A_{ij}} \left[(1 - \rho) \hat{\beta}^{E_{ij}} (1 - \hat{\beta})^{N_{ij} - E_{ij}} \right]^{1 - A_{ij}}}$$

$$= \prod_{i<j} \frac{\left[\rho \hat{\alpha}^{E_{ij}} (1 - \hat{\alpha})^{N_{ij} - E_{ij}} \right]^{A_{ij}} \left[(1 - \rho) \hat{\beta}^{E_{ij}} (1 - \hat{\beta})^{N_{ij} - E_{ij}} \right]^{1 - A_{ij}}}{\rho \hat{\alpha}^{E_{ij}} (1 - \hat{\alpha})^{N_{ij} - E_{ij}} + (1 - \rho) \hat{\beta}^{E_{ij}} (1 - \hat{\beta})^{N_{ij} - E_{ij}}} \tag{23.23}$$

$$= \prod_{i<j} \hat{Q}_{ij}^{A_{ij}} \left(1 - \hat{Q}_{ij} \right)^{1 - A_{ij}}, \tag{23.24}$$

where we achieved the form we want in Eq. (23.24) after a bit more rearranging of Eq. (23.23) and found

$$\hat{Q}_{ij} = \frac{\rho \hat{\alpha}^{E_{ij}} (1 - \hat{\alpha})^{N_{ij} - E_{ij}}}{\rho \hat{\alpha}^{E_{ij}} (1 - \hat{\alpha})^{N_{ij} - E_{ij}} + (1 - \rho) \hat{\beta}^{E_{ij}} (1 - \hat{\beta})^{N_{ij} - E_{ij}}}. \tag{23.25}$$

With $\hat{Q}_{ij}$, we have an expression for the probability of an edge given our observations and parameters. Intuitively, Eq. (23.25) makes sense, and notice that if our data includes an unobserved edge, one where $N_{ij} = E_{ij} = 0$, we have $\hat{Q}_{ij} = \hat{\rho}$, our prior estimate for overall network density. The expression is consistent with our priors—exactly what we want. (The case of no data also motivates the need for ρ.)

Fitting algorithm Between our parameter estimates and $\hat{Q}_{ij}$, we have the pieces we need for inference. To fit to data, we use expectation–maximization, which alternates between two steps:

1. (E-step) Compute $\hat{Q}_{ij}$ for all i, j using the observations and our estimated parameters $\hat{\alpha}, \hat{\beta}$, and $\hat{\rho}$ in Eq. (23.25).

2. (M-step) Update parameter estimates $\hat{\alpha}, \hat{\beta}$, and $\hat{\rho}$ using the observations and $\hat{Q}_{ij}$ in Eqs. (23.20) and (23.21).

(Initially, our parameters are randomly drawn from their priors.) Iterate E- and M-steps until convergence (within a tolerance). These steps will converge, but not necessarily to a global optimum [127].

23.3.1 Application: temporal contact network

Let's use the independent edge observer model to estimate the network structure of the Malawi Sociometer Network, treating the sociometers as our "edge observers." For the most part, until now, we have used the weighted version of this focal network. But if we suspect there may be either noise in the data or just an overabundance of weak

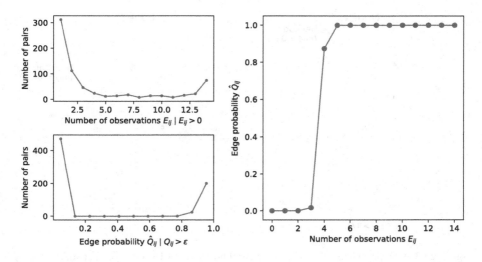

Figure 23.2 Fitting the edge observer model to the Malawi Sociometer Network. To prevent zeroes from visually swamping out the distribution, in the lower-left panel we condition Q_{ij} using $\epsilon = 10^{-8}$.

ties, can we use the edge observer to extract the most meaningful, underlying network (Sec. 10.5)?

In Ch. 15 we studied the time dynamics of this network by aggregating the event representation by days. We use that version here, and we count for each node pair i, j the number of days an edge was observed, E_{ij}, out of the $N_{ij} = 14$ days of observations.[7] (Notice that this uses no additional temporal information nor does it consider multiple observations within a single day.) These counts serve as our input data.

First, in Fig. 23.2 we plot the distribution of E_{ij} over all node pairs where $E_{ij} > 0$. Most edges that were observed tended to be observed only a few times ($E_{ij} < 5$) while a handful of the (presumably) strongest edges were observed on all or nearly all days ($E_{ij} = 13$ or 14). That said, a small portion of observed edges are distributed roughly uniformly between these extremes ($5 < E_{ij} < 13$); although not too bad, this makes it difficult to impose a global cutoff E^* (i.e., retain all edges $E_{ij} > E^*$; Ch. 10), if we wished, to extract the "true" edges from the noise.

Now, we fit the model by iterating on Eqs. (23.20), (23.21), and (23.25) until convergence,[8] which was fast, usually within 13–15 steps. After fitting, we compute $\hat{Q}_{ij}$ for each node pair. This probability admits a natural cutoff for when to infer an edge, $Q_{ij} \geq 1/2$, and this is confirmed in the remaining panels of Fig. 23.2. In particular, we see an immediate jump in Q_{ij}, sharply separating the edges between $E_{ij} = 3$ and $E_{ij} = 4$.

Examining our fitted parameters, starting with density, we have $\hat{\rho} = 0.0303$, meaning

---

[7] An important next step would be to explore aggregation other than daily; Ch. 7.

[8] Specifically, we iterate EM steps s until $\max\{|\hat{\alpha}^{(s)} - \hat{\alpha}^{(s-1)}|, |\hat{\beta}^{(s)} - \hat{\beta}^{(s-1)}|, |\hat{\rho}^{(s)} - \hat{\rho}^{(s-1)}|\} < 10^{-6}$.

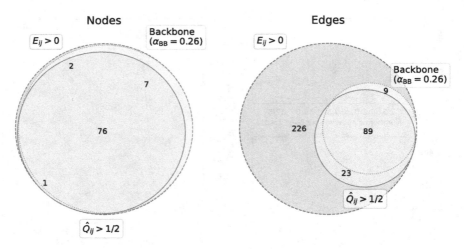

Figure 23.3 Comparing the disparity filter (backbone) and edge observer network recovery methods to the Malawi Sociometer Network.

the network is quite sparse. In comparison, the density of the raw data, using any i, j with $E_{ij} > 0$, is 0.0949, meaning the model has removed about two-thirds of the observed edges. For the other parameters, we have $\hat{\alpha} = 0.737$ and $\hat{\beta} = 0.00690$. These translate to a false negative rate (edges actually present that we fail to detect) $1 - \hat{\alpha}$ of about 25% and a (very low) false positive rate: only 0.7% of the time should we expect to note an edge exists when in fact it does not. We can also quantify performance by measuring how often the model's positive predictions are wrong, the *false discovery rate*, FDR = FP/(TP + FP), where FP is the number of false positives and TP is the number of true positives. For the independent edge observer,

$$\text{FDR} = \frac{(1 - \rho)\beta}{\rho\alpha + (1 - \rho)\beta}, \tag{23.26}$$

which, with our estimated parameters, gives FDR = 0.2301. All said, these are plausible values given the nature of the experiment, including the context of the social interactions and the precision of the sociometer badges. When the model rules out an edge, it is probably correct, but if it confirms an edge is present, it will be wrong roughly 1 in 4 times.

Lastly, it's instructive to compare the results of the edge observer model with our earlier analysis using the disparity filter on the weighted network, which we considered in Sec. 10.5.5 (Fig. 10.2). To compare the two, we simply examine the sets of nodes and edges remaining in the networks extracted with the disparity filter and the edge observer. Figure 23.3 uses Venn diagrams to illustrate the sets of all nodes and edges compared to the sets of nodes and edges remaining in the two networks. Both methods preserve nearly all the nodes, while both remove the majority of edges, which is to be expected. Interestingly, the sets of edges retained by the two methods are nearly identical. At least for this network, our filtering methods are quite consistent.

One disadvantage of the disparity filter over the edge observer is that the disparity filter requires tuning its backbone cutoff parameter α_{BB} (Eq. (10.1)), whereas the edge observer admits a natural cutoff of $Q \geq 1/2$. Previously we found $\alpha_{BB} = 0.26$ to work well on this network (Fig. 10.2) and we used that value here. It is not too difficult to determine the disparity filter's threshold, but it is still not automatic like we have with the edge observer.

Overall, the edge observer model is a simple and often effective statistical model for network data where repeat edge measurements are taken. We focused on the simplest formulation, where edges are iid, but we can relax this assumption if needed. The natural next step [337] is to allow for nodes i to have individual true- and false-positive rates, $\alpha \rightarrow \alpha_i$ and $\beta \rightarrow \beta_i$. Expressing this model is straightforward, noting that it requires $E_{ij} \neq E_{ji}$. Not only can this capture asymmetry in edge formation, the model can also capture measures of data quality on a per-node basis, which the independent edge observer cannot. With all that said, more complex dependencies, such as transitivity, become more difficult to express, leading researchers to consider approaches such as those we describe in the next section.

23.4 Other modeling approaches

Here we discuss three additional statistical models for networks. The first two are suitable when a single instance of the network is observed and were developed specifically for social network analysis. The third can be used when the network is unknown but data for each node, such as time series, are available, and we wish to find the network from relationships between the data.

23.4.1 Exponential random graphs

Suppose you wish to build a probability model for whether edge i, j is present in the network. This probability may depend on a variety of other features (covariates) and you would like to capture that in your model, along with parameters describing which covariates matter that you can infer by fitting the model to data. Ideally, the covariates and their parameters can even be interpreted, giving us inferential insights.

Since we are building a model for a probability given covariates, what may immediately come to mind is logistic regression (Sec. 16.2),

$$\Pr(A_{ij} = 1) = \frac{1}{Z} \exp \left[\sum_k \beta_k g_k(i, j) \right], \tag{23.27}$$

where $g_k(i, j)$ are covariates and β_k are parameters.

The difficulty, however, comes from interactions, dependencies between covariates of different edges, which may or may not be coincident on the same node. Specifying a probability model here requires covariates that capture such dependencies across the configuration of the network, not just on a per-edge basis. Without it, the model could never capture triangles, for instance.

These dependencies need to be introduced as constraints in the probability model. For example, if we need the model to capture the total number of edges, the number of two-paths, the number of triangles, and such, we need constraints for each.

To see these constraints in our model, let's step back and look at the probability $\Pr(G)$ for the entire network G. This should satisfy $\sum_G \Pr(G) = 1$ with the sum running over all possible networks[9] with N nodes. If we know some statistic $s(G)$ from the network, such as the number of edges, and we want the model to reproduce the expected value $\langle s \rangle$ of that statistic, then

$$\sum_G s(G) \Pr(G) = \langle s \rangle \qquad (23.28)$$

acts to constrain $\Pr(G)$. In general we will have a set of statistics $s_k(G)$, each constraining $\Pr(G)$. This set may be large, but it will not fully specify the model as the number of possible networks on N nodes is enormous: $2^{\binom{n}{2}} = 2^{n(n-1)/2}$. Which model then to choose? Among all possible models that meet these constraints, many will have additional constraints or assumptions that we didn't intend to consider and for which we do not have evidence. What we really want is to specify the most "random," least constrained model that meets the assumptions we do want (see also Sec. 23.5). We can approach this by quantifying how random the model is and picking the most random model that still meets our constraints.

The entropy h of a probability distribution,

$$h\left(\Pr(G)\right) = -\sum_G \Pr(G) \log \Pr(G), \qquad (23.29)$$

serves to measure its randomness. Intuitively, it tells us how "surprised" we are by values that are drawn from the distribution. If the distribution is very predictable due to constraints, h will be low. Conversely, an unconstrained, highly random distribution will display high h. Thus we seek a $\Pr(G)$ that maximizes h while still being constrained by our statistics.

We can find the form of $\Pr(G)$ by maximizing $\mathcal{L}$ (not to be confused with the likelihood), a function called the *Lagrangian* that combines h with the equality constraints using *Lagrange multipliers*:

$$\mathcal{L} = h - \beta_0 \left(1 - \sum_G \Pr(G) \right) - \sum_k \beta_k \left[\langle s_k \rangle - \sum_G s_k(G) \Pr(G) \right]. \qquad (23.30)$$

Here β_0 is the Lagrange multiplier that introduces the normalization constraint while β_k is the multiplier for network statistic s_k. Differentiating Eq. (23.30) with respect to $\Pr(G)$ for a particular G and setting equal to zero gives

$$-\log \Pr(G) - 1 + \beta_0 + \sum_k \beta_k s_k(G) = 0, \qquad (23.31)$$

which we solve to find our choice of $\Pr(G)$,

$$\Pr(G) = \exp\left[-1 + \beta_0 + \sum_k \beta_k s_k(G) \right] = \frac{1}{Z} \exp\left[\sum_k \beta_k s_k(G) \right] = \frac{e^{H(G)}}{Z}, \qquad (23.32)$$

---

[9] Undirected networks without self-loops or multi-edges.

where $Z = e^{1-\beta_0}$ and $H(G) = \sum_k \beta_k s_k(G)$. Notice we see a model that looks like a logistic regression (Eq. (23.27)) but the statistics are network-wide, not per-edge.[10]

Challenges remain, as this model is specified over all possible G and computation becomes intractable. For example, the normalization term $Z = \sum_G e^{H(G)}$ cannot be computed except for very small networks or very simple models. The same holds for the β_k although, if we did know Z, we could write down the expected values of our statistics by differentiating $\log Z$ with respect to their β:

$$\langle s_k \rangle = \sum_G s_k(G) \Pr(G) = \frac{1}{Z} \sum_G s_k e^{H(G)} = \frac{1}{Z} \sum_G s_k e^{\sum_k \beta_k s_k(G)}$$

$$= \frac{1}{Z} \frac{\partial}{\partial \beta_k} \sum_G e^{\sum_k \beta_k s_k(G)} = \frac{1}{Z} \frac{\partial Z}{\partial \beta_k} = \frac{\partial \log Z}{\partial \beta_k}. \tag{23.33}$$

To move forward, researchers have worked on simplifying this general model (Eq. (23.32)). Frank and Strauss [168] derive the form for such a probability model if the network is random up to a Markov property, meaning that the presence or absence of two edges is conditionally independent given the rest of the network, unless those edges are coincident at a node.[11] The idea is that this can still capture relationships between edges, such as triadic closure, while greatly simplifying the probability model. Conditional independence places constraints on the probability model as edges become dependent when they participate in two-paths and triangles. Frank and Strauss show that a random network can satisfy the Markov property if and only if its probability distribution can be written as

$$\Pr(G) = \frac{1}{Z} \exp\left(\sum_{k=1}^{N-1} \beta_k s_k(G) + \tau T(G) \right). \tag{23.34}$$

The statistics s_k and T are

$$s_1(G) = \sum_{i<j} A_{ij} \qquad \text{the number of edges,} \tag{23.35}$$

$$s_k(G) = \sum_i \binom{\sum_j A_{ij}}{k} \qquad \text{the number of "k-stars" } (k \geq 2), \tag{23.36}$$

$$T(G) = \sum_{i<j<u} A_{ij} A_{iu} A_{ju} \qquad \text{the number of triangles.} \tag{23.37}$$

Here a k-star is a set of nodes $i, j_1, j_2, \ldots, j_k$ where $A_{ij_t} = 1$ for each j_t. Essentially, it is the degree distribution of the network. (A single edge is a 1-star.)

If we specialize this model by taking $\beta_2 = \beta_3 = \cdots = \beta_{N-1} = \tau = 0$, we are left with

$$\Pr(G) = \frac{e^{\beta_1 M}}{\sum_G e^{\beta_1 M}}, \tag{23.38}$$

---

[10] A distribution of this form is often encountered in statistical physics. We have found an example of Boltzmann's distribution. The Z is known as the *partition function* and $H(G)$ is the *graph Hamiltonian*.

[11] More explicitly, for an undirected network and four distinct nodes i, j, u, v, edges a_{ij} and a_{uv} are independent, conditional on all other variables a_{st}.

where $M = \sum_{i<j} A_{ij}$. The normalization becomes

$$Z = \sum_G \exp\left(\beta_1 \sum_{i<j} A_{ij}\right) = \sum_G \prod_{i<j} e^{\beta_1 A_{ij}}$$

$$= \prod_{i<j} \sum_{A_{ij}=0,1} e^{\beta_1 A_{ij}} = \prod_{i<j}\left(1 + e^{\beta_1}\right) = \left(1 + e^{\beta_1}\right)^{\binom{N}{2}}. \qquad (23.39)$$

Recall that $\partial \log Z / \partial \beta_1 = \langle s_1 \rangle = \langle M \rangle$. Applying this to Eq. (23.39) gives

$$\langle M \rangle = \binom{N}{2} \frac{\partial}{\partial \beta_1} \log(1 + e^{\beta_1}) = \binom{N}{2} \frac{1}{1 + e^{-\beta_1}} \qquad (23.40)$$

and solving for β_1 we get

$$\beta_1 = \log \frac{\langle M \rangle}{\binom{N}{2} - \langle M \rangle}. \qquad (23.41)$$

Now, what is the probability for a single edge u, v? This is given by the expected value of A_{uv}, $\langle A_{uv} \rangle = \Pr(A_{uv} = 0) \times 0 + \Pr(A_{uv} = 1) \times 1 = \Pr(A_{uv} = 1)$, or

$$\langle A_{uv} \rangle = \frac{\sum_{A_{uv}=0,1} A_{uv} e^{\beta_1 A_{uv}}}{\sum_{A_{uv}=0,1} e^{\beta_1 A_{uv}}} = \frac{e^{\beta_1}}{1 + e^{\beta_1}}. \qquad (23.42)$$

We know the value of β_1 from Eq. (23.41) so this becomes,

$$\Pr(A_{uv} = 1) = \frac{1}{1 + e^{-\beta_1}} = \frac{\langle M \rangle}{\binom{N}{2}}. \qquad (23.43)$$

In other words, this specialized model captures a constant probability for edges based just on the expected density. Erdős–Rényi (or the Bernoulli model), among others, is thus a special case of Eq. (23.34).

Extending beyond the special case allows for a probability model to capture higher-order dependencies, which is why they are quite popular for modeling social networks, where homophily and other social phenomena drive triadic closure and other network features. The model Eq. (23.34) can be generalized to directed networks and to arbitrary statistics $s(G)$ (including node-level attributes; Ch. 9), giving

$$\Pr(G = g) = \exp\left[\boldsymbol{\beta}^\mathsf{T} \mathbf{s}(G) - \psi(\boldsymbol{\beta})\right], \qquad (23.44)$$

where $\psi(\boldsymbol{\beta}) = \log Z$ ensures normalization. Written in this form we see the model falls into the exponential family and it is therefore called the *exponential random graph model* (ERGM).

Fitting ERGMs to data is challenging practically and, more importantly, over the years researchers have slowly discovered catastrophic, possibly fatal problems with them. In terms of fitting, early approaches (using pseudo-likelihoods) were found to have flaws and eventually it was determined that Markov Chain Monte Carlo (MCMC) methods were preferred for sampling graphs from the ERGM and for estimating parameters. However, it was those very MCMC methods that revealed serious, (mostly) overlooked problems with how ERGMs are specified.

Roughly speaking, the flaw is that the statistics, Eqs. (23.35)–(23.37), are not independent of one another: triangles involve two-paths, k-stars involve triangles, and so forth. Suppose you change one edge in the model. This will change the number of edges, but it can also change the number of triangles. Those triangles would increase the T statistic, which could increase the probabilities for other edges to form. A cascade of changes could begin, driving the model towards the complete graph. All from a single edge! This is *model degeneracy*, where small changes to a fit parameter can drastically change the probability assigned to different network configurations—suddenly a model that places all its probability on the complete graph will shift entirely over to the nearly empty graph. This *phase transition*[12] is unlikely to reflect a real network—how could one new friendship cause every possible friendship to exist?—and the model's instability should make us extremely skeptical of the robustness of our inferences.

On the bright side, revealing the degeneracy in the model means research can focus on addressing it. Snijders et al. [436] propose a new set of statistics, replacing Eqs. (23.35)–(23.37), intended to prevent the "change cascade" just discussed. For example, they note that models with positive parameters for k-stars will put high probability onto graph configurations containing high-degree nodes based on their subgraph counts. Thus, a possible solution is to use a statistic that decreases the weight on higher degrees. Snijders et al. suggest using geometrically decreasing weights. A similar argument leads to different ways of capturing transitivity and two-paths.

Overall, these new statistics are very interesting and helpful, but they do not completely eliminate degeneracy from ERGMs. Improvements and alternatives to ERGMs remain an important area of research.

23.4.2 Latent space models

Another approach to modeling network edges statistically that has some advantages over ERGMs is *latent space models* [216]. Here each node i is associated with a coordinate z_i in a latent space (or, an embedding space; Ch. 26) and the probability for an edge i, j will depend on their distance $d_{ij} = \|z_i - z_j\|$ in the space, along with other observed covariates x_{ij} and parameters β. In other words, our probability model for the network's adjacency matrix is

$$\Pr(\mathbf{A} \mid \mathbf{X}, \mathbf{Z}, \beta) = \prod_{i \neq j} \Pr(A_{ij} \mid x_{ij}, z_i, z_j, \beta), \tag{23.45}$$

where $\mathbf{X}$ is known but $\mathbf{Z}$ and β are unknown and must be estimated (and we consider $A_{ij} \neq A_{ji}$, otherwise we take the product over $i < j$).

A convenient way to incorporate d_{ij} is by parameterizing the model as a logistic regression, meaning we take the *log-odds* for an edge to be a linear combination of our features, which include d_{ij}:

$$\eta_{ij} = \log \frac{P(A_{ij} = 1 \mid x_{ij}, z_i, z_j, \alpha, \beta)}{1 - P(A_{ij} = 1 \mid x_{ij}, z_i, z_j, \alpha, \beta)} \tag{23.46}$$

$$= \alpha + \beta^\mathsf{T} x_{ij} - \|z_i - z_j\|. \tag{23.47}$$

---

[12] In the language of statistical physics, the problem can be described as the model undergoing spontaneous symmetry breaking [359].

We previously attempted to build the ERGM model with logistic regression (Eq. (23.27)) but it was not so simple to capture transitivity and reciprocity. Here, the great benefit of the latent space is that it is intrinsically reciprocal and transitive, inheriting these properties from the distance metric on $\mathbf{Z}$.

Unlike an ERGM, the log-likelihood of this model is relatively simple,

$$
\begin{aligned}
\log \Pr(\mathbf{A} \mid \eta) &= \sum_{i \neq j} \left[A_{ij} \log \Pr(A_{ij} \mid \eta_{ij}) + (1 - A_{ij}) \log \left(1 - \Pr(A_{ij} \mid \eta_{ij}) \right) \right] \\
&= \sum_{i \neq j} \left[A_{ij} (\eta_{ij} - \log(1 + e^{\eta_{ij}})) - (1 - A_{ij}) \log(1 + e^{\eta_{ij}}) \right] \\
&= \sum_{i \neq j} \left[A_{ij} \eta_{ij} - \log(1 + e^{\eta_{ij}}) \right],
\end{aligned}
\tag{23.48}
$$

where η is a function of the model parameters, latent coordinates, and possible known covariates. This log-likelihood makes latent space models amenable to inference methods such as maximum likelihood estimation or Bayesian inference.

The latent space model as described so far has been based on distances, but a model can also be constructed based on *projections* of the latent positions. The distinction is that distances will be symmetric whereas projections need not be, which is useful for capturing asymmetric edge probabilities, $\Pr(A_{ij}) \neq \Pr(A_{ji})$. Suppose node pairs are more likely to have an edge when the angle between their positions is small and less likely when the angle is large, meaning that edge formation is related to alignment in the latent space. We can represent this with $\mathbf{z}_i^{\mathsf{T}} \mathbf{z}_j / |\mathbf{z}_j|$, which is the signed magnitude of the projection of $\mathbf{z}_i$ in the direction of $\mathbf{z}_j$. We can think of this as measuring the amount of shared characteristics between i and j. This projection can be included in the logistic parameterization (Eq. (23.47)) in place of $-d_{ij}$,

$$
\eta_{ij} = \alpha + \boldsymbol{\beta}^{\mathsf{T}} \mathbf{x}_{ij} + \frac{\mathbf{z}_i^{\mathsf{T}} \mathbf{z}_j}{|\mathbf{z}_j|}.
\tag{23.49}
$$

Here positive alignment increases the odds of an edge, anti-alignment decreases the odds, and orthogonality indicates no change in the odds. Notice also that the projection of $\mathbf{z}_i$ in the direction of j and the projection of $\mathbf{z}_j$ in the direction of i are not equal, unless $|\mathbf{z}_j| = |\mathbf{z}_i|$. Therefore, the projection-based model can capture asymmetries in edge formation that the distance-based model does not by varying the magnitudes of the latent vectors; for example, larger $|\mathbf{z}|$ correspond to nodes with greater overall edge formation rates.

There are some difficulties when performing inference that Hoff et al. [216] overcome. The first is that the log-likelihood is not concave in the set of positions, because the log-odds are not affine. Hoff et al. suggest finding a preliminary set of distances, not necessarily Euclidean, that maximizes the likelihood, which is a convex problem. These distances can be transformed to positions using multidimensional scaling [257] which can then initialize a nonlinear optimization method.

The second difficulty that Hoff et al. overcome is that points in a Euclidean latent space are invariant under rotation, reflection, and translation. This means that, for any given set of positions, there will be an infinite number of other positions with equal likelihood. They propose an algorithm to address this based on a Procrustes transformation [187] of the latent positions, which can make equal any two sets of latent positions that differ only by rotation, reflection and/or translation.

With these difficulties addressed, Hoff et al. show that the model can be very effective on real data.

Latent space models are an example of an *embedding method*, where the nodes are embedded in a vector space such that similarities in the space approximate similarities in the network. This can be helpful as the vector space may be more amenable to analysis than the network itself, which is exactly what we saw when considering the logistic model here compared to that of the ERGM. Embedding methods for networks using machine learning are now an active area and we discuss them further in Ch. 26.

23.4.3 Sparse inference of Gaussian graphs

The *precision matrix* (Ch. 25) $\Sigma^{-1} := \Theta$ (inverse covariance matrix) is a useful representation of a graph G_X of n nodes that underlies a set of n variables structured in $X \in \mathbb{R}^{m \times n}$. For example, time series measurements of n nodes can be arranged into X. Zeroes in Θ show conditional independence between variables (Sec. 25.1.5), assuming the data follow a multivariate normal distribution. Therefore, we can capture the conditional dependencies between our n variables by defining the Gaussian graph G_X that contains an edge i, j if $\Theta_{ij} \neq 0$; otherwise, i, j is not an edge.

Since edges are present or absent based on the zeros of the precision matrix, we are motivated to look for sparse estimates of Θ given X. Sparse inference is now well developed, with methods spanning statistics, machine learning, and signal processing. One of the most celebrated methods is LASSO regression.

Like OLS regression, LASSO seeks to solve a linear system of equations, but now we seek regression coefficients β that minimize both the OLS sum-of-squared-errors and are "norm-constrained." We discussed LASSO in Ch. 16 (Sec. 16.4). In the context of Gaussian graph inference, a method known as Graphical LASSO has been very successful.

Graphical LASSO is motivated by earlier approaches that applied LASSO to this problem. The first, by Meinshausen and Bühlmann [304], is quite simple. Perform a separate LASSO regression on each variable x_i using the remaining $n - 1$ variables x_j ($j \neq i$) as predictors. Entries of $(\Sigma^{-1})_{ij}$ are taken as nonzero if either the LASSO coefficient of variable i on j or j on i is nonzero. Meinshausen and Bühlmann show that this method will (asymptotically) consistently estimate the nonzero $(\Sigma^{-1})_{ij}$.

Graphical LASSO follows along these lines but better exploits the relationships between the repeated LASSO regressions and the Gaussian likelihood first described by Banerjee et al. [34]. To describe Graphical LASSO, first consider the log-likelihood of m variables drawn from an n-dimensional normal $\mathcal{N}(\mu, \Sigma)$,

$$\ell(\mu, \Sigma) = c - \frac{m}{2} \log |\Sigma| - \frac{1}{2} \sum_{i=1}^{m} \left(x^{(i)} - \mu\right)^{\top} \Theta \left(x^{(i)} - \mu\right), \qquad (23.50)$$

where c is a constant. We can put this more concisely by rewriting the sum using properties of the trace[13] and identifying the sample covariance S, to get

$$\ell(\mu, \Sigma) = c - \frac{1}{2} \left\{ m \log |\Sigma| + \sum_{i=1}^{m} \left(x^{(i)} - \mu\right)^{\mathsf{T}} \Theta \left(x^{(i)} - \mu\right) \right\}$$

$$= c + \frac{1}{2} \left\{ m \log |\Theta| - \sum_{i=1}^{m} \text{tr} \left[\left(x^{(i)} - \mu\right) \left(x^{(i)} - \mu\right)^{\mathsf{T}} \Theta \right] \right\}$$

$$= c + \frac{m}{2} \left(\log |\Theta| - \text{tr} \left[S\Theta\right] \right)$$

$$\propto \log |\Theta| - \text{tr} \left(S\Theta\right), \tag{23.51}$$

where $S = \frac{1}{m} \sum_{i=1}^{m} \left(x^{(i)} - \mu\right) \left(x^{(i)} - \mu\right)^{\mathsf{T}}$ is the sample covariance matrix. For Graphical LASSO, the goal is to maximize the *penalized* log-likelihood,

$$\max_{\Theta > 0} \ \log |\Theta| - \text{tr} \left(S\Theta\right) + \lambda \|\Theta\|_1. \tag{23.52}$$

This is an example of semidefinite programming (SDP), a convex optimization where we maximize over a set of positive semidefinite matrices ($\Theta > 0$). But, except for the penalty term, Eq. (23.52), at first glance, this doesn't really look like a LASSO problem.

Let's see why LASSO is relevant. Suppose we write $W = \Theta^{-1}$ as a partition by taking one row and one column out of W to make W_{11}: $W = [W_{11}, w_{12}; w_{12}^{\mathsf{T}}, w_{22}]$. This satisfies

$$W\Theta = \begin{pmatrix} W_{11} & w_{12} \\ w_{12}^{\mathsf{T}} & w_{22} \end{pmatrix} \begin{pmatrix} \Theta_{11} & \theta_{12} \\ \theta_{12}^{\mathsf{T}} & \theta_{22} \end{pmatrix} = \begin{pmatrix} I & 0 \\ 0^{\mathsf{T}} & 1 \end{pmatrix}. \tag{23.53}$$

Writing out the upper-right block gives $W_{11}\theta_{12} + w_{12}\theta_{22} = 0$ or $w_{12} = -W_{11}\theta_{12}/\theta_{22} = W_{11}\beta$, where $\beta = -\theta_{12}/\theta_{22}$.

The maximum of Eq. (23.52) occurs when its gradient equals 0:

$$\Theta^{-1} - S - \lambda \, \text{sign}(\Theta) = \begin{pmatrix} W_{11} & w_{12} \\ w_{12}^{\mathsf{T}} & w_{22} \end{pmatrix} - \begin{pmatrix} S_{11} & s_{12} \\ s_{12}^{\mathsf{T}} & s_{22} \end{pmatrix} - \lambda \, \text{sign} \begin{pmatrix} \Theta_{11} & \theta_{12} \\ \theta_{12}^{\mathsf{T}} & \theta_{22} \end{pmatrix} = 0. \tag{23.54}$$

The upper-right block of this gives

$$w_{12} - s_{12} - \lambda \, \text{sign}(\theta_{12}) = W_{11}\beta - s_{12} + \lambda \, \text{sign}(\beta) = 0. \tag{23.55}$$

This is an estimation equation for a LASSO problem[14] with coefficients β that we arrive at, from the partition, by regressing one variable on the rest. This, along with efficient optimization strategies, motivates the idea of breaking down the maximization into LASSO sub-problems which are then recursively solved, which is the basis of the Graphical LASSO algorithm:

---

[13] The trace is linear and the trace of a product is invariant to cyclic permutations.

[14] To see this, take the LASSO objective function Q, differentiate with respect to β, then set equal to 0 for the estimation equation of $\hat{\beta}$:

$$\min Q = \frac{1}{2}(y - \beta)^2 + \lambda |\beta| \Rightarrow Q' = \beta - y + \lambda \, \text{sign}(\beta) = 0.$$

1. Initialize $\mathbf{W} = \mathbf{S} + \lambda\mathbf{I}$. The diagonal of $\mathbf{W}$ will not be updated.
2. For each $j = 1, 2, \ldots, n$:
   (a) Permute and partition $\mathbf{W}$ so target variable j is the last row and column.
   (b) Solve the LASSO problem to find $\hat{\boldsymbol{\beta}}$ using current $\mathbf{W}_{11}$ and $\mathbf{s}_{12}$.
   (c) Update the corresponding row and column of $\mathbf{W}$ with $\mathbf{w}_{12} = \mathbf{W}_{11}\hat{\boldsymbol{\beta}}$.
3. Repeat from 2 until convergence.

As mentioned, solving the LASSO problem at each step requires permuting rows and columns to make the current variable the last. When introducing Graphical LASSO, Friedman et al. [170] discuss a coordinate descent strategy to exploit this for efficiency.

Upon convergence, the Graphical LASSO algorithm estimates $\mathbf{W} = \hat{\boldsymbol{\Sigma}}$, not the precision matrix $\hat{\boldsymbol{\Sigma}}^{-1}$. Friedman et al. also note the following strategy to invert the result efficiently by exploiting the partitioning (Eq. (23.53)) and computations made along the way. From Eq. (23.53) we have

$$
\begin{aligned}
\mathbf{W}_{11}\boldsymbol{\theta}_{12} + \mathbf{w}_{12}\theta_{22} &= \mathbf{0}, \\
\mathbf{w}_{12}^{\mathsf{T}}\boldsymbol{\theta}_{12} + w_{22}\theta_{22} &= 1,
\end{aligned}
\tag{23.56}
$$

which solves for

$$
\begin{aligned}
\boldsymbol{\theta}_{12} &= -\mathbf{W}_{11}^{-1}\mathbf{w}_{12}\theta_{22}, \\
\theta_{22} &= 1 \Big/ \left(w_{22} - \mathbf{w}_{12}^{\mathsf{T}}\mathbf{W}_{11}^{-1}\mathbf{w}_{12} \right).
\end{aligned}
\tag{23.57}
$$

In these we still have an inverse to compute, but notice that we already have $\hat{\boldsymbol{\beta}} = \mathbf{W}_{11}^{-1}\mathbf{w}_{12}$ which we can substitute into Eq. (23.57). Therefore, we can save the LASSO coefficients $\hat{\boldsymbol{\beta}}$ for each of the n problems, and efficiently compute $\boldsymbol{\Sigma}^{-1}$ after convergence, which was our ultimate goal.

Efficiently estimating sparse precision matrices allows Graphical LASSO and related methods to scale up to networks of thousands of nodes. While the assumption of Gaussianity is endemic to using precision matrices, in many problems it remains either justified in the data or at least still serves as a reasonable modeling choice. Inference of non-Gaussian graphical models along these lines remains an active area of research.

23.5 Ensembles

In general, whenever stochasticity is invoked, a network model does not represent a network but an entire family of networks—an *ensemble*. Take the Erdős–Rényi model with parameters N (number of nodes) and p (probability for a pair of nodes to be connected). This model defines an ensemble of networks, the set of all networks that satisfy these conserved properties or *constraints*. A central tenet of statistical mechanics, and information theory, is that the most likely model ensemble is the one that subject to the constraints is otherwise the most random—the principle of maximum entropy.

Understanding ensembles allows us to better capture properties of a network dataset. Is it plausible that these data came from that ensemble? If we gathered more data, how different can we expect the network to be? A positive answer to the first question gives us information about the second question: if we understand the ensemble the network comes from, we can reason about how different a network drawn from that same ensemble will be from our original sample, and this gives us some confidence in addressing the second question.

Ensembles connect to both mechanistic (Ch. 22) and statistical network models. The stochastic block model, ERGM, and other models discussed here all define ensembles of networks, as do the growing and random graph models from Ch. 22. Indeed, the difficulties inherent in understanding network models often boil down to challenges working with their ensembles, either describing them mathematically or drawing samples from them computationally.

23.6   Summary

Statistical models for network data are both promising and challenging. Many approaches exist, from the stochastic block model and its generalizations to the edge observer, the exponential random graph model, and the Graphical LASSO. All these models help us understand our data but, as we saw, using them can be challenging, either computationally or mathematically. Often the model must be specified with great care, lest it seize on a drastically unexpected network property (Fig. 23.1) or fall victim to degeneracy (Sec. 23.4.1). Or the model must make implausibly strong assumptions, such as conditionally independent edges, leading us to question its applicability to our problem. Or even the data we have may simply be too large for the inference method to handle efficiently. The search continues for better, more tractable statistical models and more efficient, more accurate inference algorithms for network data.

Bibliographic remarks

The stochastic block model has a long and storied history, having been first introduced by Holland et al. [219] by extending non-stochastic block models [73, 489]. *A posteriori* blocking, where the block matrix is inferred from data, was first pursued by Snijders and Nowicki [437] and Nowicki and Snijders [350]. A wealth of research has followed in the intervening years; see Lee and Wilkinson [266] for a recent review.

The "edge observer" model (our name) and inference procedure was introduced by Newman [337]. However, it has a clear antecedent outside the context of networks in the seminal work of Dawid and Skene [122], an early application of the EM algorithm to noisy inferences, soon after the EM algorithm was introduced. The Dawid–Skene model, as it is now commonly called, is central to crowdsourcing [225], where large groups of people provide data to, for example, train machine learning models [234, 25].

Exponential random graph models (ERGMs), also known as p^* models, have a long history: see Robins et al. [398] for a review. The latent space models we discussed were introduced by Hoff et al. [216] as an alternative to avoid some of the problems that arise when fitting ERGMs. Latent space models are an example of an embedding method, and we will encounter such methods again in Ch. 25 and, in particular, Ch. 26. Finally, the Graphical LASSO method was introduced by Friedman et al. [170] to leverage sparsified or penalized regression techniques [206] for one kind of network inference. It has seen success, in particular, in bioinformatics problems (e.g., Cao et al. [93]).

Many statistical network models are fundamentally Bayesian and readers interested in learning more on Bayesian inference may wish to consult Wasserman [484] for an introduction or Gelman et al. [179] for an in-depth treatment.

Exercises

23.1 Produce some sketches of the stochastic block model membership matrix $\mathbf{M}$ for different network structures: a bipartite network, a network with four equally sized communities, a network with two communities each containing three equally sized sub-communities, and a network with core–periphery structure.

23.2 What would $\mathbf{M}$ look like for a (bipartite) network exhibiting nestedness (Ch. 12)?

23.3 Given the parameters $k, \mathbf{z}, \mathbf{M}$, where $\mathbf{M}$ specifies the probability of connection, what is the expected number of edges in a realization of the stochastic block model?

23.4 In the degree-corrected stochastic block model, what should be the relationship between γ_i and degree k_i for node i?

23.5 (**Focal network**) Implement the edge observer and reproduce Fig. 23.2. Is the inferred network connected? If not, how many connected components does it find?

23.6 (**Focal network**) Implement the disparity filter ([424], Ch. 10) and apply it to the (weighted) Malawi Sociometer Network. Going beyond the rudimentary results of Fig. 23.3, use techniques from Ch. 14 to compare the "backbone" found with the disparity filter to the network found with the edge observer model.

Chapter 24

Uncertainty quantification and error analysis

As discussed in Ch. 10, network data are necessarily imperfect. Missing and even spurious nodes and edges can create uncertainty in what the observed data tell us about the original network. In this chapter, we dive deeper into tools that allow us to quantify such effects and probe more deeply into the nature of an unseen network from our observations of it.

24.1 Computational and mathematical approaches

We can understand the effects of errors and missing data by computational methods or mathematical models. The computational approach is quite similar in spirit to the null models discussed in Sec. 11.6. There we applied some form of randomization algorithm to G to generate G_{null} whose properties we could compare to G. Now, instead of randomizing a network, we can apply a sampling or error algorithm to the network, then compare statistics of the sampled network to those same statistics on the original network. For example, Martin and Niemeyer [294] perform experiments looking at the random removal or addition of nodes or edges to see how robust different centrality measures are to such errors, finding for instance that degree centrality is quite robust to such errors while eigenvector centrality is more affected by errors such as missing nodes, especially when the nodes that are missing had high centrality. Borgatti et al. [67] and Frantz et al. [169] perform computational studies along similar lines.

Mathematical approaches, on the other hand, model the sampling or error mechanism probabilistically, which can give further insight into the problems we face. Mathematical *uncertainty quantification* enables us to understand the uncertainty in model parameters and summary statistics computed from data. These models can even provide guidance on important questions, such as whether further data collection is necessary. We discuss a variety of mathematical approaches in this chapter.

The advantage of the computational approach is that it can give us intuition about how different statistics are affected by errors and sampling, especially complicated error modes that are difficult to study mathematically. The disadvantage is that they require starting from a known "true" network. Still, these approaches can give us intuition about the statistical estimators themselves which we can use when calculating those statistics on our data.

The advantage of the mathematical approach is that it gives tractable insights that are not as easily seen computationally. Further, those insights can allow us to build statistical models that can extrapolate in various ways from the observed data to the unseen network. The disadvantage is that the probabilistic models are usually limited to simpler forms of errors, for example that edges are independently observed or missing from the data. While these assumptions and approximations are limiting in many ways, the insights and extrapolations they provide are still quite useful.

In general, we encourage practitioners to consider both computational and mathematical approaches to network uncertainty quantification.

24.2   Missing data and its effects

Suppose that nodes present in the true network are absent in the data we have available. What effect does this have on our understanding of the network? What does the degree distribution look like and how does it compare to the original? What about the overall network structure?

One tool we can use for these questions is *percolation*. Percolation usually models simple random sampling, where nodes or links are randomly retained in the network, or equivalently, randomly removed, although it can be extended in various ways to capture bias (for instance, that high-degree nodes are more likely, or less likely, to be observed in the data than other nodes). In fact, percolation has been used in the context of network resilience, where nodes or edges are not missing but damaged or non-functioning, and can even allow us to understand what happens to a network under attack. But for our data analysis purposes, missingness and damage are effectively equivalent.

One interesting conclusion that percolation shows us is that networks undergo *phase transitions* (see also Ch. 22) based on the level of sampling. We discuss one now.

Is the observed network globally connected?

A network is globally connected when it has a giant component, a connected component containing the majority of nodes. We can determine the level of sampling necessary for a giant component to exist by following a simple heuristic: the network is globally connected when a random node i, whose neighbor j belongs to the giant component, is also connected to at least one other node. If this is not the case, the network is globally fragmented. To find the minimum point where this occurs, we write this condition as

$$\langle k_i \mid i \leftrightarrow j \rangle = \sum_{k_i} k_i P(k_i \mid i \leftrightarrow j) = 2, \qquad (24.1)$$

where $P(k_i \mid i \leftrightarrow j)$ is the probability that i has degree k_i given it is connected to j. Let's simplify this by using the degree distribution $P(k)$ instead of this conditional probability. From Bayes' theorem and the joint probability $P(k_i, i \leftrightarrow j)$ we have

$$P(k_i \mid i \leftrightarrow j) = \frac{P(k_i, i \leftrightarrow j)}{P(i \leftrightarrow j)} = \frac{P(i \leftrightarrow j \mid k_i)P(k_i)}{P(i \leftrightarrow j)}. \tag{24.2}$$

If we assume the network is uncorrelated and sparse (meaning, we neglect loops), then $P(i \leftrightarrow j \mid k_i) = k_i/(N-1)$ and $P(i \leftrightarrow j) = \langle k \rangle /(N-1)$. Substituting into Eq. (24.2),

$$P(k_i \mid i \leftrightarrow j) = \frac{k_i P(k_i)}{\langle k \rangle}. \tag{24.3}$$

Finally, applying this to Eq. (24.1) we arrive at a succinct expression for global connectivity:

$$\frac{1}{\langle k \rangle} \sum_{k_i} k_i^2 P(k_i) = \frac{\langle k^2 \rangle}{\langle k \rangle} := \kappa = 2. \tag{24.4}$$

Now, what happens when nodes are missing? Assume a fraction p of nodes are removed independently from the network (i.e., each node is independently sampled with probability $1 - p$). A node with degree k_0 in the original network will have a new degree k due to sampling, on average,

$$\binom{k_0}{k}(1-p)^k p^{k_0-k}. \tag{24.5}$$

Applied to all nodes, this modifies the original degree distribution into the new distribution

$$P'(k) = \sum_{k_0=k}^{\infty} P(k_0)\binom{k_0}{k}(1-p)^k p^{k_0-k}. \tag{24.6}$$

(Primes denote quantities after sampling.) Now, let's compute the first and second moments, $\langle k \rangle'$ and $\langle k^2 \rangle'$, for this new distribution in terms of the original moments. For the first moment,

$$\langle k \rangle' = \sum_k k P'(k)$$

$$= \sum_k k \sum_{k_0=k}^{\infty} P(k_0)\binom{k_0}{k}(1-p)^k p^{k_0-k}$$

$$= \sum_{k_0=0}^{\infty} P(k_0) \underbrace{\sum_k k \binom{k_0}{k}(1-p)^k p^{k_0-k}}_{\text{mean of a binomial distribution}}$$

$$= \sum_{k_0=0}^{\infty} P(k_0)k_0(1-p)$$

$$= \langle k_0 \rangle (1-p). \tag{24.7}$$

A similar calculation for $\langle k^2 \rangle'$ using the second moment of a binomial distribution gives $\langle k^2 \rangle' = \langle k_0^2 \rangle (1-p)^2 + \langle k_0 \rangle (1-p)p$. The sampled network is globally connected when $p \le p_c$ such that

$$\frac{\langle k^2 \rangle'}{\langle k \rangle'} = \frac{\langle k_0^2 \rangle}{\langle k_0 \rangle}(1 - p_c) + p_c = 2 \tag{24.8}$$

holds, or

$$1 - p_c = \frac{1}{\langle k_0^2 \rangle / \langle k_0 \rangle - 1} = \frac{1}{\kappa_0 - 1}, \tag{24.9}$$

where κ_0 is computed using the original network's degree distribution.

The "critical" sampling value $1 - p_c$ from Eq. (24.9) allows us to understand how much random sampling disconnects a network, before sampling occurs. This p_c is known as the *percolation threshold*, where (in the limit of large sizes) networks undergo a phase transition from globally fragmented to globally connected. And, while this result is somewhat limited by strong assumptions, it allows us to understand how sampling changes the average degree $\langle k \rangle$ (Eq. (24.7)) and even the degree distribution $P(k)$ (Eq. (24.6)).

We now consider a percolation argument in a more complex setting.

24.3   Community structure

In Ch. 12 we discussed communities, how some networks are organized into densely connected groups of nodes called communities, clusters, or modules. One facet of this structure that has attracted interest is overlapping communities (Sec. 12.7), where nodes may belong to multiple groups. (The community structure is now a cover of the nodes, not a partition.)

How does network uncertainty affect our discovery of overlapping communities? Here we explore this question using a mathematical model for overlapping communities along with a (relatively simplistic) modeling assumption of how an overlapping community detection method may perform when nodes are missing.

24.3.1   Modeling overlapping communities

Overlapping communities can be well modeled with a bipartite graph, also known as an affiliation network [485]. This graph contains two types of nodes representing the nodes and the communities in the network. Undirected links represent which node belongs to which community. The graph is characterized by two degree distributions, r_m and s_n, governing the fraction of nodes that belong to m communities and the fraction of communities that contain n nodes, respectively [341, 339]. For simplicity, links are distributed randomly between "node nodes" and "community nodes" following these degree distributions. The average number of communities per node is $\sum_m m r_m := \mu$ and the average number of nodes per community is $\sum_n n s_n := \nu$. We then derive two networks from the bipartite graph by projecting onto either the nodes or the communities. One is the original network between nodes, while the other is a network where each node represents a community and two communities are linked if they overlap, i.e., they share at least one node.

We model missing nodes by (prior to projection) retaining nodes independently with probability p; otherwise, they are removed with probability $1 - p$. Meanwhile, we also model the effects of this sampling on a (hypothetical) community detection algorithm. We assume the algorithm fails to find a community if fewer than a critical fraction f_c of its original nodes remain, the idea being that detection will fail when too little of the original community remains in the sample. These undetected communities are removed from the community network but any member nodes that were sampled are not removed from the node network (Fig. 24.1). A percolation analysis can show us the effects that sampling has on both the network structure and its communities. The giant component in the original network disappears when, due to missing nodes, the sampled network lacks global connectivity; in the community network, it vanishes when the communities become uncoupled (non-overlapping). Can we always detect the overlap, all the way down to the percolation threshold? Or does the overlapping structure disappear (well) before that sampling point?

Before proceeding with analysis, it's worth noting that this model makes two assumptions about the communities: that all interactions within each community exist and are equal, and that there are no differences between individual nodes that share a community—there are no "captains" or "team leaders." While simplistic, these nevertheless can form the basis for more complex analyses as needed.

To understand whether sampling hides overlapping communities, disconnects the network, or both, we determine $S(p)$, the fraction of observed nodes within the giant component, as a function of p, for both the node and community networks. We use four generating functions (Sec. 22.4.1):

$$f_0(z) = \sum_{m=0}^{\infty} r_m z^m, \quad f_1(z) = \frac{1}{\mu} \sum_{m=0}^{\infty} m r_m z^{m-1},$$

$$g_0(z) = \sum_{n=0}^{\infty} s_n z^n, \quad g_1(z) = \frac{1}{\nu} \sum_{n=0}^{\infty} n s_n z^{n-1}. \tag{24.10}$$

These functions generate probabilities for (f_0) a randomly chosen node to belong to m communities, (f_1) a random node within a randomly chosen community to belong to m other communities, (g_0) a random community to contain n nodes, and (g_1) a random community of a randomly chosen node to contain n other nodes.

To analyze this model we now separately study the two projections (the node and community networks) of the original bipartite graph.

Network

Consider a randomly chosen node A that belongs to a group of size n. Let $P(k \mid n)$ be the probability that A still belongs to a connected cluster of k nodes (including itself) in this group after sampling:

$$P(k \mid n) = \binom{n - 1}{k - 1} p^{k-1} (1 - p)^{n-k}. \tag{24.11}$$

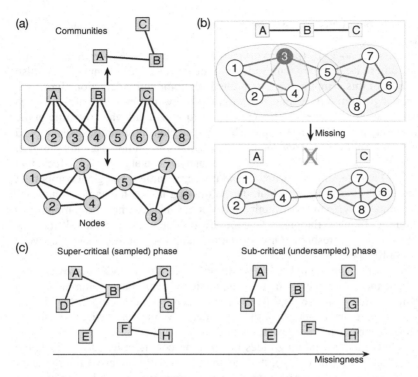

Figure 24.1   Sampling in a network model with overlapping communities. (**a**) Using an affiliation network, we analyze two networks, one representing the linkages between network nodes and a second detailing the overlapping connectivity between the communities themselves. (**b**) Missing nodes (node 3) may prevent communities (community B) from being detected. (**c**) With sufficient missingness, we transition from a well-sampled to an undersampled phase. This can cause (Fig. 24.2) the community network to become disconnected, preventing us from detecting the overlapping community structure, even though the network itself remains connected.

The generating function for the number of other nodes connected to A within this group is

$$h_n(z) = \sum_{k=1}^{n} P(k \mid n)z^{k-1} = (zp + 1 - p)^{n-1}.$$ (24.12)

Averaging over community size:

$$h(z) = \frac{1}{\nu} \sum_{n=0}^{\infty} n s_n h_n(z) = g_1(zp + 1 - p).$$ (24.13)

The total number of nodes that A is connected to, from all communities it belongs to, is then generated by

$$G_0(z) = f_0(h(z)).$$ (24.14)

Likewise, the total number of nodes that a randomly chosen neighbor of A is connected to is generated by

$$G_1(z) = f_1(h(z)).\qquad(24.15)$$

Before determining S, we first identify the critical sampling point p_c where the giant component emerges. This happens when the expected number of nodes two steps away from a random node exceeds the number one step away, or

$$\partial_z G_0(G_1(z))\big|_{z=1} - \partial_z G_0(z)\big|_{z=1} > 0.\qquad(24.16)$$

Substituting Eqs. (24.14) and (24.15) into (24.16) gives $f_0'(1)h'(1)[f_1'(1)h'(1)-1] > 0$ or $f_1'(1)h'(1) > 1$, making the condition for a giant component to exist, since $h'(1) = pg_1'(1)$, be

$$pf_1'(1)g_1'(1) > 1.\qquad(24.17)$$

For constant network degrees, $r_m = \delta(m,\mu)$ and $s_n = \delta(n,\nu)$, where $\delta(a,b) = 1$ if $a = b$ and 0 otherwise, this gives $p(\mu - 1)(\nu - 1) > 1$. If $\mu = 3$ and $\nu = 3$, for example, then the transition occurs at $p_c = 1/4$.

To find S, consider the probability u for node A not to belong to the giant component. A is not a member of the giant component only if all of A's neighbors are also not members, so u satisfies the self-consistency condition $u = G_1(u)$. The size of the giant component is then $S = 1 - G_0(u)$.

Communities

For the community network, we proceed with a similar calculation.

Consider a random community C and then a random member node A. Let $Q(\ell \mid m)$ be the probability that C is connected to ℓ communities, including itself, through node A, who was originally connected to m communities including C:

$$Q(\ell \mid m) = \binom{m-1}{\ell-1}q_1^{\ell-1}(1-q_1)^{m-\ell},\qquad(24.18)$$

where

$$q_1 = \frac{1}{\nu}\sum_{n=0}^{\infty} ns_n \sum_{i=x}^{n}\binom{n-1}{i-1}p^{i-1}(1-p)^{n-i}.\qquad(24.19)$$

(Notice that $q_1 = 1$ when $x(n) := \lceil nf_c \rceil = 1$ for all n.) The generating function j_m for the number of communities that C is connected to, including itself, through A is

$$j_m(z) = \sum_{\ell=1}^{m} Q(\ell \mid m)z^{\ell-1} = (zq_1 + 1 - q_1)^{m-1}.\qquad(24.20)$$

Once again, averaging j_m over memberships gives

$$j(z) = \frac{1}{\mu}\sum_{m=0}^{\infty} mr_m j_m(z) = f_1(zq_1 + 1 - q_1).\qquad(24.21)$$

The total number of communities that C is connected to is *not* generated by $g_0(j(z))$ but by $\tilde{g}_0(j(z))$, where the $\tilde{g}_i$ are generating functions for community sizes in the sampled network:

$$\tilde{g}_0(z) = \sum_{n=0}^{\infty} \tilde{s}_n z^n, \qquad\qquad \tilde{g}_1(z) = \frac{\sum_{n=0}^{\infty} n\tilde{s}_n z^{n-1}}{\sum_{n=0}^{\infty} n\tilde{s}_n}. \qquad (24.22)$$

The probability $\tilde{s}_k$ to have k nodes remaining within a community after sampling is

$$\tilde{s}_k = \frac{\sum_n \binom{n}{k} p^k (1-p)^{n-k} s_n}{\sum_n \sum_{k'=x}^{n} \binom{n}{k'} p^{k'} (1-p)^{n-k'} s_n}. \qquad (24.23)$$

The denominator in $\tilde{s}_k$ is necessary for normalization since the community detection algorithm cannot observe communities with fewer than $\lceil nf_c \rceil$ members. Notice that $\tilde{s}_n = s_n$ when $s_n = \delta(n, \nu)$ and $\lceil nf_c \rceil = n = \nu$. Finally, the total number of communities connected to C through any member node is generated by $F_0(z) = \tilde{g}_0(j(z))$ and the total number of communities connected to a random neighbor of C is generated by $F_1(z) = \tilde{g}_1(j(z))$. As before, the community network has a giant component when $\partial_z F_0(F_1(z))|_{z=1} - \partial_z F_0(z)|_{z=1} > 0$ and $S = 1 - F_0(u) = 1 - \tilde{g}_0(j(u))$, where u satisfies $u = F_1(u) = \tilde{g}_1(j(u))$.

24.3.2   Missing data reveals an *inference gap*

For the uniform case with $\mu = 3$, $\nu = 3$, and $f_c > 2/3$, the critical point for the community network is $p_c = 1/2$, a considerably higher threshold than for the node network ($p_c = 1/4$) discussed above. Figure 24.2 shows S for $\mu = 3$ and $\nu = 6$. The *inference gap*, the difference between the critical points for the node and community networks, grows as the community method's detection cutoff increases, covering a significant range of p for the larger values of f_c. Intuitively this makes sense: a high detection cutoff means an algorithm is sensitive and small changes to the community will lead to detection failure. But even if a method can succeed when half of a community is missing, which is impressive, we still see a non-trivial inference gap in Fig. 24.2.

Of course, realistic networks do not have constant degrees. What do these results look like for scale-free networks? Here we take $r_m = \delta(m, \mu)$ as before, but now $s_n \sim n^{-\lambda}$, with $\lambda \geq 2$. (Scale-free group sizes also model scale-free networks, as the degree distribution after projection remains scale-free, with the same exponent, although the maximum degree may increase.) It is known that the global connectivity of scale-free networks is robust to sampling errors when $2 < \lambda < 3$ (meaning that $p_c \to 0$). However, this result also requires that the maximum value K of the degree distribution be large ($K \gg 1$) [111]. Indeed, as we lower λ, we discover that, while our network is more robust under sampling, we are actually *less robust* when detecting the communities (Fig. 24.3)—overlapping structure vanishes earlier for smaller λ! Interestingly, increasing the maximum size of a community $N = \max \{n \mid s_n > 0\}$ does not make the overlapping structure more robust to node sampling.

So that's where partitions come from?   When one ponders mechanisms for how community structure can appear in a network, it becomes clear that we should expect

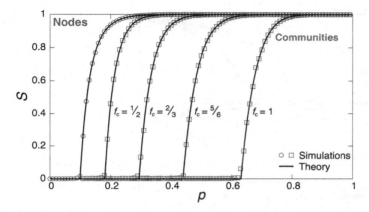

Figure 24.2 The size of the giant component S for the node and community networks as a function of node sampling rate p. Theory and simulations confirm that the network undergoes a transition from coupled to non-overlapping communities well before it loses global connectivity. Symbols represent node ($\odot$) and community ($\square$) networks. Here we used $r_m = \delta(m, \mu)$, $s_n = \delta(n, \nu)$, with $\mu = 3$ and $\nu = 6$. Simulations used 10^6 nodes.

many types of networks to exhibit overlapping communities. Yet in network data we often find high-quality non-overlapping communities. The inference gap revealed here can in part explain this: the overlapping structure is present in the original network but not so easily seen in the sampled network.

24.4 Uncertain networks as probabilistic graphs

Thus far, we have focused most of our attention on the problem of missing nodes (although many site (node) percolation arguments translate to the related bond (edge) percolation problem). Here we go beyond missing nodes or edges by allowing for edges to be uncertain using *probabilistic graphs*. Such probabilistic graphs, while making simplifying assumptions, can capture both missing and spurious edges.

In a probabilistic graph, each edge $e = (i, j)$ is associated with a probability[1] $P(e)$ which we can use to reason about our uncertainties in edges. Assuming edges are independent, we arrive at an expression for the probability of the entire graph which we've encountered several times before,

$$P(G) = \prod_{e \in E} P(e) \prod_{e \notin E} [1 - P(e)]. \tag{24.24}$$

Simple edge sampling can generate a graph, call it G_s, by choosing to include each edge e with probability $P(e)$. Under such a process, what kind of graph statistics can we expect? What is the prevalence of triangles, for instance? What is the *expected* average shortest path length (ASPL)?

---

[1] A good source for $P(e)$ would be the posterior probabilities for edges estimated from the edge observer model; Sec. 23.3.

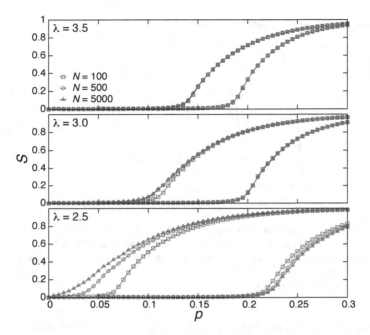

Figure 24.3 Sampling and the community structure of scale-free networks. Here $r_m = \delta(m, 3)$, $s_n \sim n^{-\lambda}$, $f_c = 1/2$, and $N := \max\{n \mid s_n > 0\}$. Increasing N and decreasing λ, measures known to improve the robustness of scale-free networks [111], actually *magnify* the inference gap. Simulations used 10^5 nodes. Figure from [28].

Shortest paths are emphasized in network measures for two reasons. First, they are easy to calculate in a given network using an algorithm such as breadth-first search or (for weighted networks) Dijkstra's algorithm. More importantly, they draw on the belief that heavily used, important paths—say, important for information flow—are short. But for an uncertain graph, we need to account for both the length of the path and *whether it actually exists*. Let ρ_{ij} denote a path between nodes i and j and $E(\rho)$ denote the set of edges comprising a path ρ. Then the probability $P(\rho_{ij})$ that ρ_{ij} exists in our probabilistic graph is

$$P(\rho_{ij}) = \prod_{e \in E(\rho_{ij})} P(e). \tag{24.25}$$

This says nothing about multiple paths; indeed, we expect more than one path can exist between a given pair of nodes. The *most probable path* between nodes i and j is $\rho_{ij}^{(\mathrm{MP})} = \arg\max_{\rho_{ij}} P(\rho_{ij})$. Treating the $P(e)$ as edge weights, we can in principle find ρ_{ij} using Dijkstra's algorithm on the probabilistic graph.

Finding the most probable path can tell us about the existence of paths (although it is a point estimate only), but what about whether the path is actually used? The prior belief is that shorter paths are more likely to be used than longer paths. We can capture this notion by assuming there is a constant "transmission rate" per edge, β, and transmissions along a path of length $|\rho| = \ell$ will then occur at rate β^ℓ, which we call the "permeability." Thus, instead of focusing only on most probable paths, we combine with permeability[2] to find $\rho_{ij}^{(\text{MPP})} = \arg\max_{\rho_{ij}} \left(P(\rho_{ij}) \beta^{|\rho_{ij}|} \right)$. Most Probable and Permeable (MPP) paths are also found efficiently with Dijkstra's algorithm using these different weights.

With MPP paths we now have an analog for shortest paths in an uncertain network. We can use the average MPP path length in place of the ASPL, and we can use MPP paths for betweenness centrality, allowing us to rank the centralities of nodes in probabilistic graphs.

Lastly, let's consider triangles in a probabilistic graph. Usually these are quantified with transitivity or the clustering coefficient (Ch. 12). The (deterministic) clustering coefficient for node a is given by $c_a = \Delta_a / \binom{k_a}{2}$. For a probabilistic graph, a triangle between nodes a, b, and c will occur with probability $P(e_{ab})P(e_{bc})P(e_{ca})$. Similarly, the probability for a two-path on those nodes through a is $P(e_{ab})P(e_{ac})$. We can use these to define an *expected* clustering coefficient for the probabilistic graph:

$$c_a = \mathbb{E}\left[\frac{T_a}{\tau_a}\right] \approx \frac{\mathbb{E}[T_a]}{\mathbb{E}[\tau_a]}, \tag{24.26}$$

where

$$\mathbb{E}[T_a] = \sum_{\substack{b,c \in N_a, \\ b \neq c}} [P(e_{ab})P(e_{bc})P(e_{ca})],$$

$$\mathbb{E}[\tau_a] = \sum_{\substack{b,c \in N_a, \\ b \neq c}} [P(e_{ab})P(e_{ac})], \tag{24.27}$$

(and a better approximation than that given in Eq. (24.26) would incorporate the variances of T and τ, and their covariance).

24.5 Size estimation

One challenge with real network data is trying to use a limited sample of the network to learn about the unseen remainder of the network. Here we describe some strategies for inferring the total number of links,[3] what we call *size estimation*, when only a subgraph has been observed. This is useful both for understanding scientifically how *much* network we're dealing with, and logistically for marshalling our resources—if we are going to pay for experiments, it's good to know if we can expect to find, say, 1% of the data, or half the data.

Suppose we have a sample network $G_{\text{samp}} = (V_{\text{samp}}, E_{\text{samp}})$ and we wish to understand the complete network $G_{\text{full}} = (V_{\text{full}}, E_{\text{full}})$ (with $V_{\text{samp}} \subseteq V_{\text{full}}$ and $E_{\text{samp}} \subseteq E_{\text{full}}$) from the sample. Our sample G_{samp} is the subgraph of G_{full} induced by V_{samp}. We assume G_{full} is generated by some statistical model $P_\theta(G_{\text{full}})$ parameterized by θ and

---

[2] Pfeiffer and Neville [371] refer to most probable and permeable paths as maximum likelihood handicapped (MLH) paths.

[3] We focus on edges, but estimating the number of nodes is interesting too!

that G_{samp} is then sampled from it with probability

$$P_\theta(G_{\text{samp}}) = \sum_{G_{\text{full}} \supseteq G_{\text{samp}}} P_p(G_{\text{samp}} \mid G_{\text{full}})P_\theta(G_{\text{full}}), \qquad (24.28)$$

where p refers to a parameter of the sampling mechanism, which we assume is independent of the network's generating model.

Let us assume for now we know N_{full} and wish to estimate M_{full}. Such a situation is common when we know *a priori* the network nodes but it is prohibitive to confirm, observationally or experimentally, all possible interactions, i.e., all elements of the $N_{\text{full}} \times N_{\text{full}}$ **A**. (If we did not assume N_{full} in Eq. (24.28), we would need to sum over more than all networks of N_{full} nodes.)

Suppose sampling depends only on the nodes and not on how they are connected. Then the sampling mechanism that appears in Eq. (24.28) factors into $P_p(G_{\text{samp}} \mid G_{\text{full}}) = Q_p(N_{\text{samp}})n(G_{\text{samp}}, G_{\text{full}})$, where $Q_p(N_{\text{samp}})$ is the probability of sampling the observed nodes and $n(G_{\text{samp}}, G_{\text{full}})$ is the number of ways that the N_{samp} nodes can be sampled from N_{full}. Enumerating the ways the nodes can be sampled is tricky. If nodes are unlabeled and are all degree zero, it becomes simple, $q(G_{\text{samp}}, G_{\text{full}}) = \binom{N_{\text{full}}}{N_{\text{samp}}}$, but obviously this is a huge oversimplification. However, suppose every node is *identifiable*, uniquely labeled and distinguishable (and the labels are the same in G_{samp} and every possible G_{full}). Then there is only one way to choose the observed nodes, and $n(G_{\text{samp}}, G_{\text{full}}) = 1$. This assumption is not universally true, but quite reasonable: proteins, for instance, are all well identified by their open reading frames (ORFs), and we can expect the assumption to hold in many other contexts such as (some) social networks.

Under the assumptions of connection independence and node identifiability, the sampling mechanism is entirely described by $Q_p(N_{\text{samp}})$. For independent node sampling, meaning we now interpret the sampling parameter p as every node is independently sampled with probability p, we have $Q_p(N_{\text{samp}}) = p^{N_{\text{samp}}}(1 - p)^{N_{\text{full}} - N_{\text{samp}}}$. Solving $\partial Q/\partial p|_{p=\hat{p}} = 0$ for $\hat{p}$ gives us the MLE $\hat{p} = N_{\text{samp}}/N_{\text{full}}$, which is quite intuitive.

Next, because nodes are identifiable (and not because they are independently sampled) the conditional probability

$$P_\theta(G_{\text{full}} \mid G_{\text{samp}}) = \frac{n(G_{\text{samp}}, G_{\text{full}})P_\theta(G_{\text{full}})}{\sum_{G'_{\text{full}} \supseteq G_{\text{samp}}} n(G_{\text{samp}}, G'_{\text{full}})P_\theta(G'_{\text{full}})}, \qquad (24.29)$$

which comes from using Bayes' theorem on the factored $P_p(G_{\text{samp}} \mid G_{\text{full}})$, does not depend on p. It only depends on the model for G_{full}. Therefore, we need to make some assumptions about how G_{full} is generated.

Suppose π, the probability for an edge to appear in G_{full}, is well approximated by the density of the sample, i.e.,

$$\pi \approx \hat{\pi} = \frac{2M_{\text{samp}}}{N_{\text{samp}}(N_{\text{samp}} - 1)}. \qquad (24.30)$$

From this, we have $M_{\text{samp}}/\binom{N_{\text{samp}}}{2} \approx M_{\text{full}}/\binom{N_{\text{full}}}{2}$, and we can solve for an estimate of the unseen network's size,

$$\hat{M}_{\text{full}} = M_{\text{samp}} \frac{N_{\text{full}}(N_{\text{full}} - 1)}{N_{\text{samp}}(N_{\text{samp}} - 1)}. \tag{24.31}$$

While the assumptions so far are quite strong, and we should be skeptical, nevertheless Eq. (24.31) gives us a straightforward way to perform size estimation. Let's see it in an application.

Example   Let's take these results and apply them to HuRI. This network, after removing self-loops, has $N_{\text{samp}} = 8{,}272$ nodes and $M_{\text{samp}} = 52{,}068$ edges. Luck et al. [283] build their experimental protocol around a PPI screening space of approximately 17,500 protein-coding genes: "To increase interactome coverage and generate a reference map of human binary PPIs, we expanded the ORFeome collection to encompass $\sim 90\%$ of the protein-coding genome." Indeed, the GENCODE Release 42 Human dataset lists 19,379 protein-coding genes. We use this value for N_{full}, yielding $\hat{M}_{\text{full}} = 285{,}787$ from Eq. (24.31). According to this, HuRI captures $M_{\text{samp}}/M_{\text{full}} = 18.22\%$ of the human interactome!

It's worth exploring some of the limitations of this estimation procedure.

Uncertainty in the number of nodes   Some problems may give you information on the nodes of the full network independent of the sample, but many will not. How might our estimates of M_{full} change if we don't know N_{full}? Suppose our uncertainty in the now unknown N_{true} compared to N_{full} is $\epsilon \ll 1$ such that

$$N_{\text{full}} = (1 \pm \epsilon)N_{\text{true}}. \tag{24.32}$$

Then, $\hat{p} = N_{\text{samp}}/N_{\text{full}}$ becomes

$$\tilde{p} = \frac{N_{\text{samp}}}{N_{\text{true}}} = \frac{N_{\text{samp}}}{(1 \pm \epsilon)N_{\text{full}}} \approx (1 \mp \epsilon)\hat{p}. \tag{24.33}$$

(The approximation comes from a Taylor series for small ϵ: $(1 \pm \epsilon)^{-1} = 1 \mp \epsilon + O(\epsilon^2)$.) Using $\tilde{p}$ instead of $\hat{p}$ in

$$\hat{M}_{\text{full}} = M_{\text{samp}} \frac{N_{\text{full}}(N_{\text{full}} - 1)}{N_{\text{samp}}(N_{\text{samp}} - 1)} \approx M_{\text{samp}} \left(\frac{N_{\text{full}}}{N_{\text{samp}}} \right)^2 = \frac{M_{\text{samp}}}{\hat{p}^2} \tag{24.34}$$

shows us the effect of the error ϵ is

$$\tilde{M} = \frac{M_{\text{samp}}}{\tilde{p}^2} = \frac{M_{\text{samp}}}{(1 \mp \epsilon)^2 \hat{p}^2} \approx (1 \pm 2\epsilon) \frac{M_{\text{samp}}}{\hat{p}^2}. \tag{24.35}$$

In other words, an uncertainty of ϵ in the true network's number of nodes leads to an uncertainty of roughly 2ϵ in the estimated number of edges.

Uncertainty in measurements of edges Probably your observations will contain errors such as false positives (reported edges not actually present) and false negatives (non-reported edges actually present). The *true* number of edges $M = M_{TP} + M_{FN}$, where M_{TP} is the number of true positive edges and M_{FN} is the number of false negative edges. Likewise, the *observed* number of edges $\tilde{M} = M_{TP} + M_{FP}$, where M_{FP} is the number of false positive edges. Suppose your experimental or observational process has been validated so it has known *precision* and *recall*:

$$\text{Precision} = \frac{M_{TP}}{M_{TP} + M_{FP}}, \qquad \text{Recall} = \frac{M_{TP}}{M_{TP} + M_{FN}}. \tag{24.36}$$

The precision (or positive predicted value) tells us many edges detected by our observations were true while recall (or true positive rate) tells us how many true edges were detected by our observations. These relate nicely to M and $\tilde{M}$, allowing us to estimate $\hat{M} \approx M$ given $\tilde{M}$ and the observation process's precision and recall:

$$\frac{\text{Precision}}{\text{Recall}} = \frac{M_{TP} + M_{FN}}{M_{TP} + M_{FP}} = \frac{M}{\tilde{M}} \quad \Rightarrow \quad M \approx \hat{M} = \frac{\text{Precision}}{\text{Recall}} \tilde{M}. \tag{24.37}$$

Other sampling mechanisms What if nodes are not independently sampled at a constant rate? We can model a sampling mechanism where nodes are sampled independently but non-uniformly. Let p_i be the probability that node i is sampled and assume the values of p_i for different nodes i are not equal but are drawn from the same distribution,

$$p_i \sim D(\beta) \ \forall i, \tag{24.38}$$

where β is some parameter(s) for the sampling rate distribution D. Since nodes are still sampled independently, the probability for an edge to be sampled is now $\pi_{ij} = p_i p_j$. Under these assumptions, the expected value of $\hat{p}$ converges to the expected value of p_i and the variance of $\hat{p} \to 0$ for $M_{\text{full}} \to \infty$ making $\hat{p}$ an unbiased and consistent (converges to the true value) estimator. A similar argument holds for π_{ij}, allowing us to proceed with inference. We can even relax this further by assuming that the rate at which node i is sampled depends in some way on i. We capture this by assuming $p_i \sim D_i(\gamma_i; \beta)$, where γ_i parameterizes how the information related to i changes the distribution of p_i. Such information could be related to i's network properties, such as the degree or clustering, or it could be related to non-network attributes. If we assume the network is uncorrelated given these parameters such that $P(\gamma_i, \gamma_j) = P(\gamma_i)P(\gamma_j)$ for edge i, j and nodes are drawn independently given the probabilities p_i, then we can once again show [448] estimator $\hat{p}$ is unbiased and consistent.

Much work continues on estimating the size of networks. In PPI networks such as HuRI, for instance, it's common to estimate the unseen network's size by equating the densities (Eq. (24.30)) but only across edges of a very thoroughly studied subgraph. This is commonly done using *literature curated* data, the set of edges extracted by analyzing preexisting studies; the argument being that those interactions are more heavily investigated and replicated. For HuRI, Luck et al. [283] include this PPI network, which they call "Lit-BM." Luck et al. [283] themselves provide an estimate of 2–11% coverage for HuRI, less coverage than our estimate of 18% but we are not terribly far off. It is on the whole much more probable to overestimate coverage than underestimate it, and we should in general be prepared for such.

24.5.1 Size estimation from capture–recapture

(This approach to size estimation works when you have access to multiple networks, such as from different experiments or temporal snapshots.)

Capture–recapture (also called mark and recapture) is an idea used for population estimation in ecological studies [14, 61]. Imagine you are trying to count the number of animals that live in a given area. You can go out and capture them with traps, but it will not be possible to capture all of them, especially at the same time. How then can you count the population?

At first glance, this sounds like an impossible problem, but there is a lovely way to address it. Suppose we have a population of unknown size M and we *capture* a sample of M_1 individuals from that population. We tag each individual somehow with a marker that we assume will stay affixed, then release the captured sample. Later, we repeat the capture process exactly as before and capture a second sample of size M_2. Let the number of individuals tagged in sample 1 who were *recaptured* in sample 2 be M_{12}. If we make some assumptions, like that tags don't fall of individuals but also that the two captures are independent from one another, then we have a way to infer M.

The *Lincoln–Petersen* (L–P) *estimator*, which was the impetus for capture–recapture, recognizes that if individuals are equally likely to appear in either sample, then the proportion of tagged individuals found in sample 2 should be equal to the proportion of the total population that was tagged in sample 1, or

$$\frac{M_{12}}{M_2} = \frac{M_1}{M} \quad \Rightarrow \quad \hat{M} = \frac{M_1 M_2}{M_{12}}, \tag{24.39}$$

which we solved for M to estimate the unknown population size $\hat{M} \approx M$. If we also want to compute confidence intervals, it's helpful to have the variance of the estimator [14],

$$\text{Var}\left(\hat{M}\right) = \frac{(M_1 + 1)(M_2 + 1)(M_1 - M_{12})(M_2 - M_{12})}{(M_{12} + 1)^2 (M_{12} + 2)}. \tag{24.40}$$

An estimate with a 95% confidence interval, say, can now be given by $\hat{M} \pm 1.96\sqrt{\text{Var}\left(\hat{M}\right)}$.

Equation (24.39) is a simple, brilliant idea. It can be extended and generalized in many ways, for example going from 2 to K measurements, and considerable work has focused on understanding and improving upon it, especially for small sample sizes. It doesn't work perfectly, though; its assumptions can be restrictive and difficult to validate. For the estimate to be accurate, we need four ingredients: captures are independent, all individuals are equally likely to be captured, population size is constant during captures, and tags remain affixed. These seem innocuous, but imagine yourself a soaking-wet biologist, stomping through the woods in a downpour, trying to find an elk who just ran off with a loose tag—you may be quite skeptical of iid capture probability![4]

---

[4] Indeed, wildlife biologists and ecologists have long debated the accuracy of such estimates [14].

What can the L–P estimator tell us about networks? Our notation in Eq. (24.39) is suggestive. Imagine each edge is a member of the population we are estimating. We observe network edges in one experiment, then repeat the experiment and reobserve them. Using the intersection of the edges, we can estimate the total number of edges M from Eq. (24.39). Thus we have another protocol for size estimation.

⚡ **Example**   Let's illustrate the L–P estimator using the Malawi Sociometer Network. This is a dynamic network (Ch. 15) and we can use the edge events (Sec. 15.2) to "simulate" two samples. First, divide the full set of edges into two sets:

$$E_1 = \{(u, v) \mid (u, v, t_i) \in \text{events}, t_i \le t_*\},$$
$$E_2 = \{(u, v) \mid (u, v, t_i) \in \text{events}, t_i > t_*\}, \tag{24.41}$$

where we take $t_* = \max_i t_i/2$. Then, using $M_1 = |E_1|$, $M_2 = |E_2|$, $M_{12} = |E_1 \cap E_2|$ in Eqs. (24.39) and (24.40) gives a size estimate of $\hat{M} = 396.97 \pm 22.12$ edges. In terms of the full number of observed edges, this means we estimate[5] the Malawi Sociometer Network data to capture $87.41 \pm 4.87\%$ of edges. Good coverage.

Probably the most pressing concern for using the L–P estimator on networks is the assumption that all edges are equally likely to be observed. (That edges are observed independently is also important.) This has been shown to not hold in real-world problems, such as estimating edges in the Internet topology [403]. To overcome this, Roughan et al. [403] introduce a "stratified" model by assuming that edges fall into one of C *classes* and the capture probability is different between classes but the same for all edges within a class. We describe this approach now.

First, suppose we know the class of a given edge. If we take K iid measurements, then the model's probability that we observe an edge in class j a total of k times is

$$\Pr(k \mid K, p_j) = \binom{K}{k} p_j^k (1 - p_j)^{K-k}, \tag{24.42}$$

where p_j is the observation probability for a j-class edge. The observation probability could be estimated with the MLE

$$\hat{p}_j = \frac{\sum_{i \in C_j} k_i}{|C_j| K}, \tag{24.43}$$

where k_i is the number of times edge i was observed and C_j is the set of links in class j.

But Eq. (24.43) breaks down because in practice we only have the j-class edges that were observed at least once. If we knew all the edges in class j, we would have already solved the size estimation problem!

In other words, we actually have the conditional distribution

$$\Pr(k \mid k > 0, K, p_j) = \binom{K}{k} \frac{p_j^k (1 - p_j)^{(K-k)}}{1 - (1 - p_j)^K}, \tag{24.44}$$

---

[5] Here we use error propagation [455] on a ratio $r = M/(\hat{M} \pm \delta M)$ to get $r \pm \delta r$.

known as the truncated binomial distribution. If we estimate its parameter, we can then estimate the number of hidden ($\Pr(k = 0)$) edges. The MLE $\hat{p}_j$ for Eq. (24.44) satisfies [396]

$$M^{(\text{obs})} K p_j = \left(1 - (1 - p_j)^K\right) \sum_{i=1}^{M^{(\text{obs})}} k_i, \tag{24.45}$$

which we can solve numerically, where $M^{(\text{obs})}$ is the number of observed edges. With $\hat{p}_j$ solved from Eq. (24.45), we can estimate the total number of edges with the MLE

$$\hat{M}_j = \frac{M^{(\text{obs})}}{1 - (1 - \hat{p}_j)^K}. \tag{24.46}$$

Lastly, we can iterate through each class j to derive our total estimate.

So far, we assume the class of a given edge is known, which is not realistic. We can address this using expectation–maximization (EM) to simultaneously estimate edge class and class parameters.

The EM method, which we also saw in the edge observer model (Sec. 23.3), iterates between (E-step) averaging a latent (hidden) variable (in our case, edge classes) and (M-step) finding model parameters by maximizing a likelihood. To use EM, we first extend the model to capture how edges fall into classes. We do this with two new statistical parameters: w_j, the proportion of edges in class j, and $c_j^{(i)}$, the probability that edge i belongs to class j. We describe the model in terms of EM in Alg. 24.1. After fitting, the model can also categorize the edges: we estimate the class of edge i to be $\arg\max_j c_j^{(i)}$, the most probable class. And the number of edges in class j is given by $\hat{M}_j = M_j^{(\text{obs})} / \left(1 - (1 - \hat{p}_j)^K\right)$, and we have our size estimates.

One wrinkle. The number of classes C is now a free parameter. The more classes we have, the more complex the model is, up to the extreme of a single class for every edge. Thus we are forced into a tradeoff between model fit and model simplicity. This is a classic problem in model selection, and one way to tackle it is with the Akaike

---

Algorithm 24.1 The EM algorithm for estimating the parameters of the multi-class model, where C is the number of classes. Here we use a uniform initialization, but the choice is not too important. For a convergence condition, continue iterating until the total change in $\hat{p}_j$ from one iteration to the next is less than, say, 10^{-6}.

1: $\hat{p}_j \leftarrow j/(C+1)$, $w_j \leftarrow 1/C$ ▷ *Initialization (uniform)*
2: **while** (not converged) **do**
3:      $c_j^{(i)} \leftarrow \hat{w}_j \Pr(k_i \mid K, \hat{p}_j)$ (Eq. (24.42)) ▷ *E-step*
4:      **for** $j = 1$ to C **do** ▷ *Start M-step*
5:          **while** (not converged) **do**
6:             $\hat{p}_j \leftarrow \left(1 - (1 - \hat{p}_j)^K\right) \sum_i k_i c_j^{(i)} \Big/ \left(K \sum_i c_j^{(i)}\right)$
7:          $\hat{w}_j \leftarrow \sum_i c_j^{(i)} / \left(M^{(\text{obs})}(1 - (1 - \hat{p}_j)^K)\right)$
         ▷ *End M-step*

Information Criterion (AIC), given by

$$\text{AIC} = 2\eta - 2\ln(\hat{L}), \tag{24.47}$$

where η is the number of parameters and $\hat{L}$ is the maximum likelihood of the model. We can then compare models by plotting AIC as a function of η or, in our case, as a function of C (as C determines the total number of parameters).

In general, size estimation in networks is quite interesting, and attempting to probe beyond the certainty of a limited sample into the unknown is exciting. Unfortunately, progress is usually made by taking on some heavy assumptions, and these should make us skeptical. For example, we usually assume edges are sampled independently. But it may be the sampling mechanism is biased towards motifs or other structures; it's certainly the case that edges won't exist independently in the network. That said, experiments such as assays testing for protein interactions (à la HuRI), may meet this assumption. But, in general, don't be too surprised if your estimates (and CIs) are off by noticeable factors.

24.6   Other approaches

The edge observer model described in Sec. 23.3 also fits nicely into our suite of network error analysis methods. (Indeed, such models are known as *observer error models*.) It attempts to estimate the most likely set of edges from repeated, noisy observations, and its estimates of the probabilities of edges can serve as the base for the probabilistic graphs described in Sec. 24.4. Lastly, like the multi-class capture–recapture method in Sec. 24.5.1, it uses the EM algorithm for parameter fitting.

But the EM algorithm gives us point estimates for our parameters. We may, when interrogating uncertainty, be better served with a full Bayesian treatment—in essence, beginning from the same model but then generating samples from the posterior, $P(\mathbf{A}, \theta \mid \text{data})$ (Eq. (23.8)) using, for example, MCMC. The main difference is that instead of one estimate for the posterior probability for an edge to exist, (Q_{ij}, Eq. (23.25)), we would have a distribution, and we can assess our per-edge uncertainty based on how wide or narrow that distribution turned out to be. Young et al. [503] pursue this in depth.

24.7   Summary

The fundamental challenge of measurement error in network data is capturing the error-producing mechanism accurately and then inferring the unseen network from the (imperfectly) observed data. Computational approaches can give us clues and insights, as can mathematical models. Mathematical models can also build up methods of statistical inference, whether in estimating parameters describing a model of the network or estimating the network's structure itself. But such methods quickly become intractable without taking on some possibly serious assumptions, such as edge independence. Even without addressing the full problem of network inference, we can still explore features of the unseen network, such as its size, using the available data.

Bibliographic remarks

For an introduction to uncertainty quantification in general, we can recommend Smith [435] and Sullivan [451]. A long thread of computational studies of network measures under uncertainty includes Borgatti et al. [67], Frantz et al. [169], and Martin and Niemeyer [294] but sociologists have been concerned for far longer; see, for example, Granovetter [189]. Percolation has been one of the main mathematical tools for understanding missing nodes and edges. The percolation condition for a random network (Eq. (24.9)) was first derived by Cohen et al. [111]. Likewise, the calculation showing how sampling may make an overlapping community structure appear to be non-overlapping was studied in Bagrow et al. [28] based on the model and calculation of Newman and Park [339]. Pfeiffer and Neville [371] introduced the probabilistic graph analysis we describe in Sec. 24.4.

Size estimation has been of interest to researchers studying PPI networks for some time [448, 413]. The approach we describe here was introduced by Stumpf et al. [448]. Readers interested in learning more about capture–recapture sampling, which we applied to a dynamic network to estimate our coverage of edges, may start with Amstrup et al. [14] or Böhning et al. [61]. Basic capture–recapture can be extended, and we describe the model of Roughan et al. [403] who allow for different edges to be captured with different probabilities. Their model was proposed in the context of estimating the size of the Internet (at the Autonomous Systems level), a particularly interesting application of size estimation.

Exercises

24.1 What is the probability that a node with degree $k = 1$ appears to have degree $k = 0$ in a network with iid node sampling and a p sampling rate? What can this tell us about missing data?

24.2 Suppose every edge e in a network is sampled with constant probability $P(e) = p$. What is the expected transitivity (Eq. (12.9))?

24.3 Size estimation with capture–recapture requires data from independent experiments or from different time periods. If we take a single network, randomly divide it into two parts in an attempt to simulate two capture periods, and apply the Lincoln–Petersen estimator, we will fail to find useful size estimates. Why?

24.4 (**Focal network**) Use a modularity optimization method to find a high-quality partition of the Malawi Sociometer Network. (For simplicity, take the weighted version and ignore edge weights.)

    (a) Report modularity Q, the number of communities, and the mean and median community size.

    (b) Now, suppose the network is under-observed, meaning edges are missing. Simulate this missingness with iid edge sampling and a $p = 1/2$ sampling rate. Make a computer function subsample(G,p) that takes the network and the edge sampling rate as input and returns a subsampled copy of the network.

(c) Apply subsample independently to the original network 100 times. For each sampled network, reapply the modularity optimization and record its Q. Report the distribution of Q over the subsampled realizations. How does the modularity of the sampled networks compare to that of the original?

24.5 (**Focal network**) Repeat the analysis of Ex. 24.4 but vary the edge sampling rate p and report the mean Q as a function of p. Interpret the dependence between Q and p.

Chapter 25

Ghost in the matrix: spectral methods for networks

Every network (graph) has a corresponding matrix representation. This is powerful. We can leverage tools from linear algebra within network science, and doing so brings great insights. The branch of graph theory concerned with such connections is called *spectral graph theory* and we'll introduce some of its central principles as we explore tools and techniques that use matrices and spectral analysis to work with network data.

25.1 Networks as matrices

First, we tour the various matrices that represent or relate to networks. Some matrices, such as the adjacency matrix (which we've encountered before) or the graph Laplacian, exactly capture the nodes and edges of the network—they are equivalent definitions of the network. Other matrices that we'll describe, such as the precision matrix, describe non-network data in such a way that a network can be extracted. Lastly, matrices such as the modularity matrix do not strictly define the network structure but instead capture statistical patterns related to that structure, patterns we can use to better understand the structure.

25.1.1 Adjacency matrix

The most common matrix representation of a dyadic network is its *adjacency matrix*. For a network with N nodes, the adjacency matrix $\mathbf{A}$ is an $N \times N$ matrix where each row i corresponds to a node i and each column j corresponds to a node j. The elements of $\mathbf{A} = [A_{ij}]$ denote the presence or absence of an edge: $A_{ij} = 1$ if $i, j \in E$ and zero otherwise. $\mathbf{A}$ is symmetric for an undirected network, but not necessarily so in a directed network. We illustrate $\mathbf{A}$ for some example networks in Fig. 25.1.

In a directed network, A_{ij} usually refers to the directed edge from j to i, although defining the other way around is also possible. The reason to use A_{ij} to represent an edge from j to i, but not i to j, is because this convention makes writing various graph

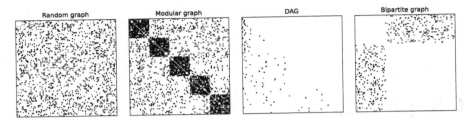

Figure 25.1 Adjacency matrices of different types of graphs (when arranged to reveal their structure).

operations easier. As an example, let's look at a simple graph with just two nodes and one directed edge $1 \rightarrow 2$. Imagine a process where some information flows[1] through the directed edges. By using this convention, we can simply write this process as a matrix multiplication of the adjacency matrix $\mathbf{A}$ and a vector $\mathbf{u}$ that captures the information. Now the product[2] $\mathbf{Au}$ describes the process where a unit of something is passed from node 1 to node 2 through the directed edge from node 1 to node 2:

$$\mathbf{Au} = \begin{bmatrix} 0 & 0 \\ 1 & 0 \end{bmatrix} \begin{bmatrix} \text{"info"} \\ 0 \end{bmatrix} = \begin{bmatrix} 0 \\ \text{"info"} \end{bmatrix}. \tag{25.1}$$

The flow of information from node 1 to 2 is captured by a matrix–vector multiplication!

For weighted networks, the A_{ij} represents the edge weights: $A_{ij} = w_{ij}$ where w_{ij} is the weight on edge i, j. In this case, the adjacency matrix is no longer binary. Although usually $A_{ii} = 0$ because of the absence of *self-edges (self-loops)*, A_{ii} may capture the number or weight of the self-edges if we wish.

How about a bipartite or multipartite network? A bipartite network can be represented in mainly two different ways. First, we can consider a potentially rectangular adjacency matrix where the rows represent one set of nodes V_1 and the columns represent the other set of nodes V_2. In this case, we will not be able to represent any edges within either of the node sets (which is what we want for this type of network). Second, we can sort the nodes based on which node set they belong to and treat it as if it is a unipartite network. (This case is shown in Fig. 25.1.) While the two node sets are not explicitly distinguished (into rows and columns), it will allow having edges within each node set.

Degree and walks from the adjacency matrix Sums over the adjacency matrix are meaningful: $\sum_{i=1}^{N} A_{ij} = k_j$ is the degree of node j. For a directed network, the row and column sums correspond to the in- and out-degree, respectively, when we follow the convention defined above:

$$k_i^{\text{in}} = \sum_{j=1}^{N} A_{ij} \quad \text{and} \quad k_j^{\text{out}} = \sum_{i=1}^{N} A_{ij}. \tag{25.2}$$

---

[1] In fact, as we'll soon see, the adjacency matrix is *not* the most natural way to represent such dynamics.
[2] It's also fine to use the other convention, we'd just write $\mathbf{u}^{\mathsf{T}}\mathbf{A}$, which is slightly less convenient.

We can also sum over all elements to obtain the number of edges (edge stubs). In an unweighted graph, $\sum_i \sum_j A_{ij} = \sum_i k_i$ captures the number of edge stubs or two times the number of edges. In other words,

$$M = |E| = \frac{1}{2} \sum_i k_i = \frac{1}{2} \sum_i \sum_j A_{ij}. \tag{25.3}$$

In a directed graph, because each direction of an edge is distinct, we do not need the factor of $\frac{1}{2}$:

$$M = |E| = \sum_i \sum_j A_{ij}. \tag{25.4}$$

The adjacency matrix can also compute the number of *walks* in a network. Could you think about how to do this? Let's say we are looking for the number of possible walks of length 4 from node j to i. How can we express this number using the adjacency matrix?

The first thing we can notice is that the existence of any walk can be expressed by multiplying elements of the adjacency matrix. For instance, the existence of a walk from node j to i of length 1 can be expressed as A_{ij}. This is 1 if there is an edge from j to i and 0 otherwise. Likewise, the existence of a walk of length 2 *via another node* k can be expressed as $A_{ik}A_{kj}$. How about the *total* number of walks with length 2 from j to i? We simply consider all possible intermediary nodes k and sum over them: $\sum_k A_{ik}A_{kj}$. Look familiar? Yes! This is exactly the definition of an element (i, j) in the matrix obtained by multiplying $\mathbf{A}$ with itself: $\left[\mathbf{A}^2\right]_{ij}$.

This notion can be generalized to walks of arbitrary length. For length 3, the number of walks from j to i is $\sum_k \sum_l A_{ik}A_{kl}A_{lj}$, which is exactly the element $\left[\mathbf{A}^3\right]_{ij}$. Likewise, the number of walks of length ℓ is $\left[\mathbf{A}^\ell\right]_{ij}$. Also note that this gives us the number of *walks* not *paths* because the multiplication cannot exclude any repeated edges (the walk $i \rightarrow j \rightarrow i \rightarrow j$ is length 3 even though it only covers one step on the graph).

Adjacency matrix and classes of graphs A good way to build our intuition about the adjacency matrix is to imagine the patterns that it will exhibit for different types of graphs, such as those shown in Fig. 25.1. For instance, what would an Erdős–Rényi graph's adjacency matrix look like? Because all edges are random, the adjacency matrix will not have any pattern—just random edges are filled. What about a network with a strong *community structure* or even multiple, separate *components*? If we identify them and sort the rows and columns of the adjacency matrix in an informative way, we will see the block structure, where each block captures dense connections and the lack of edges between blocks captures the separability of the clusters or components.[3]

How about a DAG (directed acyclic graph)? If we arrange the adjacency matrix well, what patterns would it show? Remember that a DAG has a clear ordering between nodes. We can always find the node(s) that do not have any incoming edges and that are at the top of the hierarchy. If we put this node at the top of the adjacency matrix, the first row of the matrix will not have any nonzero elements (following the convention

---

[3] Although the pattern can become complicated when we have overlapping communities.

that $\mathbf{A}_{ij}$ represents an edge from j to i), whereas the first column can have nonzero elements. If we extend this idea, we can find that the adjacency matrix of a DAG can be represented as a *lower triangular matrix*.

Finally, let's also examine the case of multipartite networks, assuming that we are sorting the node sets and using the normal adjacency matrix notation. Then we can quickly see that a bipartite network can be represented as a inverted version of a network with two components. Instead of having two blocks that are arranged along the diagonal, we will have two off-diagonal blocks, which capture the edges that run between two node sets.

A preview of what's to come—spectral revelation   Here is a curious thing. The adjacency matrix is a matrix, right? And taking powers of it tells us something about walks? Taking powers of a matrix leads us to start thinking about the eigenvectors of the matrix, since the power method[4] (power iteration) is an efficient algorithm to turn a random vector into the leading eigenvector $\mathbf{v}_1$: just keep multiplying it by the matrix ($\mathbf{r}_{t+1} = c\mathbf{Ar}_t$ with suitable constant c will converge to $\mathbf{v}_1$). Well, what can eigenvectors say about a network?

As an example, let's take the 5-group modular network shown in the second panel of Fig. 25.1 and find its eigenvectors. (Because $\mathbf{A}$ is real and symmetric, these are guaranteed to exist and all N will be linearly independent.) In Fig. 25.2 we draw the network and to the right we do something curious. We plot the eigenvectors against one another, or more precisely, we plot $\mathbf{v}_4$ against $\mathbf{v}_5$, where $\mathbf{v}_j$ is the eigenvector of the jth largest eigenvalue. Look at what happened: we see very clear clustering of points, mirroring the modules of the network itself.

Why did this happen? And how on earth did we decide to plot the fourth and fifth eigenvector? Stay tuned to find out!

25.1.2   Graph Laplacian

While the adjacency matrix is the most basic matrix representation of a network, another matrix is in fact almost always more important and more useful, as we will see. That

---

[4] The power method is a simple algorithm for approximating the eigenvector of a (diagonalizable) matrix $\mathbf{A}$. Start from a random vector $\mathbf{b}$ and left-multiply (and rescale) this vector by $\mathbf{A}$: $\mathbf{b}_2 = \mathbf{Ab}/\|\mathbf{Ab}\|$, where $\|\cdot\|$ is the 2-norm of the vector and serves to rescale $\mathbf{b}_2$. We repeat this left-multiply-and-rescale step on $\mathbf{b}_2$ to get $\mathbf{b}_3$, and continue:

$$\mathbf{b}_{k+1} = \frac{\mathbf{Ab}_k}{\|\mathbf{Ab}_k\|} = \frac{\mathbf{A}^k\mathbf{b}_1}{\|\mathbf{A}^k\mathbf{b}_1\|}. \tag{25.5}$$

If we repeatedly apply this, $\mathbf{b}_k$ will converge to the "leading" eigenvector, the eigenvector corresponding to the dominant (largest magnitude) eigenvalue of $\mathbf{A}$. From the second equality in Eq. (25.5), we can see where the algorithm gets its name: we are repeatedly taking powers of $\mathbf{A}$, hence the power method or power iteration. The power method shows us an algorithm, based on matrix multiplication, for computing the leading eigenvector of a matrix.

Convergence to the leading eigenvector, while not necessarily fast, is guaranteed if: only one eigenvalue has maximum magnitude (the largest magnitude eigenvalue $\mathbf{A}$ has multiplicity 1) and the random initial vector is not orthogonal to the leading eigenvector.

The power method can work well for efficiently finding the leading eigenvector of a large matrix. To find more of the leading eigenvectors, which we will often need, we can use the Lanczos algorithm [184], which extends the power method beyond the top eigenvector.

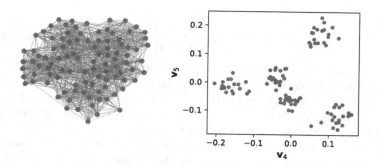

Figure 25.2 Spectral revelation. Plotting some of the eigenvectors of **A** reveals clusters in the network. Each point (v_{4i}, v_{5i}) in the plot corresponds to a node i in the network. Why does this happen?

matrix is the *graph Laplacian* (we often just call it the "Laplacian"), defined as:

$$\mathbf{L} = \mathbf{D} - \mathbf{A}, \tag{25.6}$$

where **D** is a diagonal matrix with diagonal elements equal to the degrees of nodes ($D_{ii} = k_i$ for $i = 1, \ldots, N$).

For example, for this small network of four nodes,

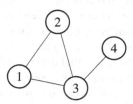

the graph Laplacian can be written as:

$$\mathbf{L} = \begin{bmatrix} 2 & -1 & -1 & 0 \\ -1 & 2 & -1 & 0 \\ -1 & -1 & 3 & -1 \\ 0 & 0 & -1 & 1 \end{bmatrix}. \tag{25.7}$$

Note that any row-wise or column-wise sum of the Laplacian matrix is zero.

Motivation The Laplacian is fundamental to understanding network structure and dynamics on networks, but where does this come from? Why is it defined this way? The information contained in the graph Laplacian is equivalent to the adjacency matrix, so why should we bother to use it?

One way to understand the Laplacian matrix is to consider it as an analogue to the (smooth) Laplacian operator Δ (or ∇^2) defined for a scalar function f on Euclidean space:

$$\Delta f = \nabla^2 f = \nabla \cdot \nabla f = \sum_{i=1}^{n} \frac{\partial^2 f}{\partial x_i^2}. \tag{25.8}$$

If we consider only a one-dimensional space where a function $f(x)$ is defined, then the Laplacian operator is simply the second derivative of the function: $\nabla^2 f(x) = f''(x)$.

Let us examine the meaning of the Laplacian operator by thinking about a heat diffusion process. Assume that the function f is time-varying ($f(x, t)$) and it describes the temperature of a one-dimensional metal rod. If we heat up (or cool down) a small part of the rod, how will the temperature profile across the rod change? Intuitively, if $f(x, t)$—temperature at x—is hotter than its neighbors, then some heat will be transferred to the neighboring region of the rod and $f(x, t)$ will decrease. On the other hand, if $f(x, t)$ is colder than its neighbors, then some heat will be transferred from the neighboring regions of the rod and $f(x, t)$ will increase.

This is exactly what the Laplacian operator captures: is the value of a function in the neighboring region larger or smaller than the value here? Its value will become negative when the first derivative (slope of the function) is decreasing—where f is concave or "peaking"; $f''(x)$ will be positive when the first derivative is *increasing*—where the original function is convex or "plunging." This intuition can be extended to higher dimensions as well. The Laplacian operator computes how much larger the average value of the function in the adjacent region is than the current location.

Now, to think about the graph Laplacian, let's discretize our space. Instead of a smooth, continuous space and a function defined on this space, we have a discrete space. When space is discretized, the derivative of a function at a certain point can be approximated as the difference of the value between two adjacent points. If we consider two neighboring points to x, namely x_l ($x_l < x$) and x_r ($x < x_r$), the first derivative around x can be approximated by $f(x) - f(x_l)$ and $f(x_r) - f(x)$. The second derivative is then the difference between these two first derivatives: $f(x_r) - 2f(x) + f(x_l)$.

If we consider this as a graph—where the point x has two neighbors—and $f(x)$ as a function defined on the graph, now we can see that the second derivative is the same as the term at x in the product $-\mathbf{Lf}$:

$$-\mathbf{Lf} = - \begin{bmatrix} & & \vdots & & \\ \cdots & -1 & 2 & -1 & \cdots \\ & & \vdots & & \end{bmatrix} \begin{bmatrix} \vdots \\ f(x_l) \\ f(x) \\ f(x_r) \\ \vdots \end{bmatrix}. \tag{25.9}$$

Can you see now why the graph Laplacian looks like this?

In other words, the graph Laplacian $\mathbf{L}$ is a *(negative) discretization of the Laplacian operator* Δ. When applied to a scalar function defined over all nodes, the graph Laplacian returns the difference between the function's value at each node and the average value of each node's neighbors, scaled by the node's degree:

$$[\mathbf{Lf}]_i = k_i \left(f_i - \sum_j f_j A_{ij} \right). \tag{25.10}$$

Imagine a diffusion process on the network. A large value in $\mathbf{Lf}$ means the node is a "peak" and the value will diffuse to its neighbors; a negative value would mean that the node is a "valley" and there will be a flow of something from the neighbors to the node. Notice that, if we make an analogy to the Euclidean Laplacian operator, the sign of the graph Laplacian is the opposite to the Euclidean Laplacian operator.

As we will see, the graph Laplacian appears in a variety of data-driven network analyses, including understanding the structural properties of the network's topology (Ch. 12). Often different ways to *normalize* $\mathbf{L}$ appear, and we will encounter those matrices as well. We will talk more about the spectrum of the graph in later sections, especially when we discuss community detection and spectral clustering.

By far the most important matrix for spectral analysis is the graph Laplacian, moreso than the adjacency matrix, but many other matrices appear as well. We briefly discuss a few.

25.1.3 Incidence matrix

Another interesting matrix is the *incidence matrix* $\mathbf{B}$. Unlike $\mathbf{A}$ and $\mathbf{L}$, the incidence matrix is not square. Instead of being $N \times N$, the incidence matrix is $N \times M$, with a row for each node and a column for each edge. Its entries are defined by

$$B_{ij} = \begin{cases} 1 & \text{if node } i \text{ is connected to edge } e_j, \\ 0 & \text{otherwise,} \end{cases} \qquad (25.11)$$

where we use the notation e_j to index edge $e = (u, v)$ into column j of $\mathbf{B}$. Technically, Eq. (25.11) describes the unoriented incidence matrix; for a directed graph, the *oriented incidence matrix* has entries

$$B_{ij} = \begin{cases} 1 & \text{if node } i \text{ is the target of edge } e_j, \\ -1 & \text{if node } i \text{ is the source of edge } e_j, \\ 0 & \text{otherwise.} \end{cases} \qquad (25.12)$$

The incidence matrix appears in problems of dynamics, such a circuit flow. For example, Kirchoff's first law, that the sum of currents flowing into a node equals the sum of currents flowing out of that node, can be expressed as $\mathbf{By} = 0$, where $\mathbf{y}$ is an $M \times 1$ vector of unknowns.

The incidence matrix also acts as a factorization of $\mathbf{L}$, as $\mathbf{L} = \mathbf{BB}^\mathsf{T}$. This allows us to quickly demonstrate that $\mathbf{L}$ is positive semi-definite, i.e.,

$$\mathbf{x}^\mathsf{T} \mathbf{Lx} = \mathbf{x}^\mathsf{T} \mathbf{BB}^\mathsf{T} \mathbf{x} = \|\mathbf{B}^\mathsf{T} \mathbf{x}\|^2 \geq 0, \qquad (25.13)$$

for all $N \times 1$ vectors $\mathbf{x}$, a useful property for applications.

25.1.4 Non-backtracking matrix

A rather curious matrix is the *non-backtracking matrix* $\mathbf{H}$, also known as the *Hashimoto matrix* [204].

This matrix is different from the others as both rows and columns represent edges, not nodes. For a network with M edges, $\mathbf{H}$ has size $2M \times 2M$. Its elements are defined based on two-paths, tracking collisions between pairs of edges. Specifically,

$$H_{(i \to j),(u \to v)} = \begin{cases} 1 & \text{if } j = u \text{ and } i \neq v, \\ 0 & \text{otherwise.} \end{cases} \tag{25.14}$$

Notice that the indices are oriented even if the network is undirected. The same two-path in reverse, $(v \to u) \to (j \to i)$ is not the same element even though the edges reversed are the same. The idea behind the non-backtracking matrix is that it lets us track (random) walks but not walks that go back on themselves, allowing the matrix to distinguish more than the adjacency matrix can (Sec. 25.1.1) when considering flows and dynamics on top of a network.

25.1.5   Precision matrix

(The precision matrix appears in the Graphical LASSO inference method discussed in Sec. 23.4.3.)

In statistical problems with structured data $\mathbf{X} \in \mathbb{R}^{m \times n}$, relationships between variables $\mathbf{x}_i$ and $\mathbf{x}_j$, $i, j = 1, \ldots, n$, are captured in part by the covariance matrix Σ with elements

$$\Sigma_{ij} = \text{Cov}\left(\mathbf{x}_i, \mathbf{x}_j\right) = \mathbb{E}\left[(\mathbf{x}_i - \mathbb{E}[\mathbf{x}_i])\left(\mathbf{x}_j - \mathbb{E}[\mathbf{x}_j]\right)\right], \tag{25.15}$$

assuming each $\mathbf{x}_i, i = 1, \ldots, n$, has finite mean and variance.

The *precision matrix* is the inverse covariance matrix, Σ^{-1} (assuming it exists). The elements of the precision matrix are often taken to represent a graph $G_{\mathbf{X}}$ that governs the variables of $\mathbf{X}$, where each node i in this graph corresponds to the variable $\mathbf{x}_i$. An edge i, j exists in $G_{\mathbf{X}}$ when $[\Sigma^{-1}]_{ij} := \Sigma_{ij}^{-1} \neq 0$; otherwise, no edge exists. Thus, the sparsity of Σ^{-1} directly maps the sparsity of $G_{\mathbf{X}}$.

But why is this? What does the inverse of the covariance matrix tell us? The answer comes from investigating the conditional independence relations of the variables comprising $\mathbf{X}$.

Recall, two random variables x_1 and x_2 are *conditionally independent*, given a third variable[5] x_3 if

$$P(x_1, x_2 \mid x_3) = P(x_1 \mid x_3)P(x_2 \mid x_3). \tag{25.16}$$

Rewrite this in terms of joint probabilities,

$$\frac{P(x_1, x_2, x_3)}{P(x_3)} = \frac{P(x_1, x_3)}{P(x_3)} \frac{P(x_2, x_3)}{P(x_3)}$$

$$P(x_1, x_2, x_3) = \frac{P(x_1, x_3)P(x_2, x_3)}{P(x_3)}$$

$$= f(x_1, x_3)g(x_2, x_3), \tag{25.17}$$

---

[5] For x_1 and x_2 conditionally independent given $x_3, x_4, \ldots, x_n$, the definition remains, with the conditioning on $x_3, x_4, \ldots, x_n$ instead of just x_3.

and we see that the conditional independence requires the joint probability P to factorize into a function f that does not depend on x_2 and a function g that does not depend on x_1.

Now, suppose $\mathbf{x} = x_1, \ldots, x_n$ are distributed according to a multivariate normal (or Gaussian) distribution

$$P(\mathbf{x}) = P(x_1, \ldots, x_n) = \frac{1}{\sqrt{(2\pi)^n |\Sigma|}} \exp\left(-\frac{1}{2} (\mathbf{x} - \boldsymbol{\mu})^\top \Sigma^{-1} (\mathbf{x} - \boldsymbol{\mu})\right), \qquad (25.18)$$

with $n \times 1$ mean vector $\boldsymbol{\mu}$ and $n \times n$ covariance matrix Σ. (Notice how the precision matrix appears.) Taking the log of $P(\mathbf{x})$ we are left with

$$\log P(\mathbf{x}) = c - \frac{1}{2} (\mathbf{x} - \boldsymbol{\mu})^\top \Sigma^{-1} (\mathbf{x} - \boldsymbol{\mu}), \qquad (25.19)$$

where c is a constant with respect to $\mathbf{x}$. Let's write out explicitly the products in Eq. (25.19):

$$(\mathbf{x} - \boldsymbol{\mu})^\top \Sigma^{-1} (\mathbf{x} - \boldsymbol{\mu}) = \sum_{i=1}^{n} \sum_{j=1}^{n} (x_i - \mu_i)(x_j - \mu_j) \Sigma_{ij}^{-1}. \qquad (25.20)$$

Suppose x_i and x_j are conditionally independent given the remaining x_k ($k \neq i, j$) This means that the joint probability $P(\mathbf{x})$ cannot contain terms that depend on both x_i and x_j, except in the factored form we saw in Eq. (25.17) which, after taking the log, becomes a sum. But Eq. (25.20) shows that every pair x_i, x_j will occur in a term of the sum—unless $\Sigma_{ij}^{-1} = 0$. Thus zeros in Σ^{-1} indicate conditional independence relationships.

This result, at least under the assumption of Gaussianity, motivates the search for sparse precision matrices, and has led to a long thread of research on inferring $G_{\mathbf{X}}$ from $\mathbf{X}$ using sparsifying method. We explore some approaches in Ch. 23 (Sec. 23.4.3).

25.1.6 Modularity matrix

Recall from Ch. 12 that modularity, a measure of the quality of a community partition, is defined by

$$Q = \frac{1}{2M} \sum_{i,j} \left(A_{ij} - \frac{k_i k_j}{2M}\right) \delta(g_i, g_j), \qquad (25.21)$$

where g_i is the group (community) containing node i and $\delta(g_i, g_j) = 1$ if nodes i and j are in the same group and zero otherwise. The *modularity matrix* $\mathbf{B} = [B_{ij}]$ contains the elements within the sum:

$$B_{ij} = A_{ij} - \frac{k_i k_j}{2M}. \qquad (25.22)$$

Let $\mathbf{k} = [k_i]$ be the vector of node degrees and we can express $\mathbf{B}$ compactly as

$$\mathbf{B} = \mathbf{A} - \frac{1}{2M} \mathbf{k} \mathbf{k}^\top. \qquad (25.23)$$

This real symmetric matrix is helpful because it lets us rewrite Eq. (25.21) in matrix form. We discuss using $\mathbf{B}$ to find high-modularity communities in Sec. 25.6.

25.2   Spectral properties reflect network properties

Spectral graph theory arose from the realization that graph properties such as connectedness appear in the spectral properties of the graph's matrix. Let's discuss the graph Laplacian.

Having established the analogy between the graph Laplacian L and the Euclidean Laplacian (Sec. 25.1.2), we can also think about the eigenvectors and eigenvalues of L—the so-called *spectrum of the graph*. This eigendecomposition captures a wealth of information about the network. One obvious eigenvalue is zero[6] ($\lambda_1 = 0$), associated with a constant eigenvector ($v_1 = (1, 1, \ldots, 1)^\top$). In general, we number the eigenvalues of L in non-decreasing order[7] $0 = \lambda_1 \leq \lambda_2 \leq \cdots \leq \lambda_N$ and the eigenvectors are ordered likewise; v_i is the eigenvector associated with eigenvalue λ_i. The "ones" eigenvector v_1 corresponds to the case of *equilibrium*. For instance, when the whole space has the same temperature—no gradient—there will be no flow of heat and thus the Laplacian of the temperature function will be zero everywhere. This is true when we consider any quantity that is constant across the whole network.

What if we have multiple disconnected components in the graph? Then we will have multiple eigenvectors v with zero eigenvalue because Lv can be zero as long as we have a constant value for *each* component. This is similar to thinking about a space that is completely disconnected into multiple pieces, where each piece has its own constant temperature. Due to this property, the dimension of the *nullspace* (or the *algebraic multiplicity* of the zero eigenvalue) of the Laplacian is equal to the number of connected components in the graph. In fact, the spectra of the graph Laplacian contain a lot of information about the graph's topology [108]. Another simple example is the number of edges: $\sum_i \lambda_i = 2M$ because the diagonal of L contains the degrees of nodes, so $\sum_i \lambda_i = \operatorname{tr} L = \sum_i k_i = 2M$.

A corollary to the multiplicity of $\lambda = 0$ is that the second smallest eigenvalue $\lambda_2 > 0$ if and only if G is connected. This eigenvalue and its associated eigenvector is quite interesting, as we'll discuss below.

Many more connections between graph topology and spectra are known, beyond what we can treat here. Some results are general, others apply to special cases, such as k-regular graphs (where $k_i = k$ for all i). Interested readers are encouraged to consult Chung [108] and van Mieghem [472].

25.3   Some spectral applications

A variety of network-related problems can be tackled using spectral methods. These include centrality measures, graph partitioning, community detection, and the general problem of spectral clustering. Here we discuss each in turn.

---

[6] We can see this algebraically by looking at row/column sums of L. Every row and column sum is zero by construction. And multiplying a matrix by the ones vector 1 computes that sum, so 1 must be an eigenvector with eigenvalue 0.

[7] Because L is symmetric, its eigenvalues are all real and non-negative.

25.4 Centrality

We briefly discussed spectral centralities in Ch. 12. Here we go into more detail.

Spectral measures of centrality are motivated by a kind of "inheritance" argument. Suppose we have a centrality value c for each node (higher c is more central) and two nodes i and j both have the same value, $c_i = c_j$. This makes those nodes equally central according to c, but it may still be that one node is more central than the other based on who they are connected to: i may have many neighbors with high values of c while j may have few central neighbors. If this were to happen, we may want to update c accordingly, by allowing nodes to "inherit" centrality from their neighbors. One possibility is to assign a new centrality c' that is the sum of the centralities of neighbors:

$$c'_i = \sum_{j=1}^{N} A_{ij} c_j. \tag{25.24}$$

Here we've used a sum that runs over all nodes j and introduced the adjacency matrix $\mathbf{A}$ to remove non-neighbors. This summation can be written for all nodes i using a matrix–vector multiplication:

$$\mathbf{c}' = \mathbf{A}\mathbf{c}, \tag{25.25}$$

where $\mathbf{c}$ and $\mathbf{c}'$ are column vectors whose ith component is c_i and c'_i, respectively.

However, given we have updated our centrality everywhere, all the neighbor centralities will have changed, and we may want to update the centrality *again*. Suppose our original centrality vector is $\mathbf{c}^{(0)}$ and our first update $\mathbf{c}' = \mathbf{c}^{(1)}$. A further update can be written by applying the previous sum again:

$$\mathbf{c}^{(2)} = \mathbf{A}\mathbf{c}^{(1)} = \mathbf{A}\mathbf{A}\mathbf{c}^{(0)} = \mathbf{A}^2 \mathbf{c}^{(0)}, \tag{25.26}$$

where the second expression follows from recursively substituting in the same expression for $\mathbf{c}^{(1)}$. But this update is another change, and we should update again to $\mathbf{c}^{(3)}$. Now we are trapped, updating $\mathbf{c}^{(t)}$ to $\mathbf{c}^{(t+1)}$ forever.

Can we escape? If an update step does not change the centrality vector then we know no further changes will occur and we are done. This will happen if $\mathbf{c}$ satisfies $\mathbf{c} = \mathbf{A}\mathbf{c}$. This condition is an eigenvalue equation of the form $\mathbf{A}\mathbf{v} = \lambda\mathbf{v}$ where the eigenvalue $\lambda = 1$.[8] This means that *the eigenvectors of the adjacency matrix* **are a centrality measure**, and we actually don't need to perform this iterative updating process to find $\mathbf{c}$; we simply compute the (leading) eigenvector $\mathbf{v}_1$ of $\mathbf{A}$ directly and use the ith component for c_i:

$$c_i = v_{i1}. \tag{25.27}$$

This is known as *eigenvector centrality*.

Eigenvector centrality is very useful but it has some problems, particularly for directed networks. A node that has no incoming edges will receive no centrality and

---

[8] In fact, we are being a little loose with our derivation. A better choice for the update step is $\mathbf{c}^{(t+1)} = \lambda^{-1}\mathbf{A}\mathbf{c}^{(t)}$ meaning some constant is used to modify the sum from Eq. (25.24). With this in place, the convergence condition becomes a general eigenvalue equation $\lambda\mathbf{c} = \mathbf{A}\mathbf{c}$. Multiple eigenvectors will in principle satisfy this but the most important one in terms of centrality will be the one with the largest λ.

likewise can contribute no centrality to other nodes. This will happen if the network is not strongly connected, which is quite common for real-world directed networks. An easy fix that introduces some elegant effects is to change our update rule (Eq. (25.24)) to always give the node a small amount of centrality β even when it has no connections:

$$c_i = \alpha \sum_j A_{ij} c_j + \beta, \qquad (25.28)$$

or, written for all nodes using matrices:

$$\mathbf{c} = \alpha \mathbf{A} \mathbf{c} + \beta \mathbf{1}, \qquad (25.29)$$

where $\mathbf{1}$ is a column vector of all 1's. We have also, for the sake of generality, changed the update to include a positive constant α which we can use to balance the neighbor contribution against the constant contribution given by β. Solving this for $\mathbf{c}$ with some matrix operations gives $\mathbf{c} = \beta (\mathbf{I} - \alpha \mathbf{A})^{-1} \mathbf{1}$. Written in this way, we see that β only acts as an overall scaling of c_i, so we may as well ignore it by setting $\beta = 1$. The other constant α acts as a free parameter, although it is related to the eigenvalues of $\mathbf{A}$. This centrality measure is known as *Katz centrality* [237] and can be used for undirected as well as directed networks.

A final set of spectral centrality measures are motivated by incorporating node degree into the update equation:

$$c_i = \alpha \sum_j A_{ij} \frac{c_j}{k_j^{\text{out}}} + \beta. \qquad (25.30)$$

The idea is that nodes with many neighbors are "spread out" compared to nodes with few neighbors. If i has a high-degree neighbor j, the centrality j contributes to i should count less when we update i's centrality than another neighbor with fewer other ties, because j is spreading their centrality to many more nodes than the low-degree neighbor. (We use out-degree in Eq. (25.30) because, if the network is directed, centralities update outward from a neighbor j.) The c_i that satisfies Eq. (25.30) with $\alpha = 1 - \beta$ is called *PageRank*, and it was a major component in the early success of Google's search engine.[9]

25.5 Partitioning

Another area where spectral methods are used is graph partitioning, trying to cut the network apart by deleting edges. We can always reduce a network to multiple connected components by deleting enough edges—the problem of partitioning becomes interesting when (1) we seek the minimum number of cuts, sometimes referred to as *sparsest cuts*

---

[9] PageRank has a "random web surfer" interpretation and that gives us another motivation for dividing by out-degree. Imagine the directed network represents hyperlinks between web pages. The update equation will tell us the probability for a web surfer who randomly clicks on links to arrive at the web page corresponding to node i, assuming that every so often (i.e., with probability β) they get bored and pick a web page uniformly at random instead of, with probability $1 - \beta$, clicking a link. Dividing by out-degree captures the fact that a surfer landing at our high-degree neighbor is less likely to move on a link from that neighbor to us than if the surfer had landed on our low-degree neighbor.

(2) we try to avoid trivial results like breaking the network into two by disconnecting a single node. We will focus on *bisection*, a special case of partitioning into two, roughly equally sized (balanced) connected components. More than two components can be handled by recursive bisection: bisect the network, then bisect each of the components we made, then bisect *those* components, and so forth.

Why partition?   Partitioning helps measure the robustness of a network by discovering bottlenecks. If only a few edges are keeping a road network or power grid, for example, globally connected, the networks are vulnerable when outages occur. Finding those edges gives us a target for what to add to the network to improve its robustness. Partitioning also helps with optimizations where we wish to group nodes together with many edges inside groups and few between. Parallel processing, image segmentation, the design of circuit boards (VLSI), and community detection are all classic examples of such optimization problems [82].

What does cutting edges from a network have to do with the graph spectrum? Good question! A clue: recall that the number of connected components in a network is given by the multiplicity of zero as an eigenvalue of $L = D - A$, the graph Laplacian. Spectral partitioning begins from this idea.

Some definitions. Suppose we divide the N nodes of our (connected) network into two groups, $S \subset V$ and $S^c = V \setminus S$, the set of nodes in V but not in S. If we want to make each group its own component, how many edges[10] do we need to cut? The *cut number* of a non-empty subset of nodes $S \subset V$ is the number of edges that span the boundary of S:

$$\text{cut}(S) = \sum_{i \in S, j \in S^c} A_{ij}. \tag{25.31}$$

Notice that $\text{cut}(S) = \text{cut}(S^c)$. To separate S and S^c, we will have to delete $\text{cut}(S)$ edges.

Suppose we define an $N \times 1$ vector $\mathbf{x}$ with elements,

$$x_i = \begin{cases} c & \text{if } i \in S, \\ -c & \text{otherwise,} \end{cases} \tag{25.32}$$

_____

[10] We'll focus on undirected, unweighted networks for clarity. For weighted networks, it's straightforward to follow these steps using the total weight as the cut, so long as weights are non-negative. For directed networks, we can proceed along similar lines, taking care about double-counting to ensure we capture both directions of any edge if both are present. However, directed networks are more challenging for spectral methods as the matrices involved are no longer symmetric.

for some constant $c \neq 0$. What happens if we compute $\mathbf{x}^\mathsf{T} \mathbf{L} \mathbf{x}$? We have,

$$
\begin{aligned}
\mathbf{x}^\mathsf{T} \mathbf{L} \mathbf{x} &= \sum_{i,j} L_{ij} x_i x_j = \sum_{i=j} k_i x_i^2 - \sum_{i \neq j} A_{ij} x_i x_j \\
&= \sum_{i=j} k_i x_i^2 - \sum_{i,j \in E} 2 x_i x_j \\
&= \sum_{i,j \in E} \left(x_i^2 + x_j^2 - 2 x_i x_j \right) = \sum_{i,j \in E} \left(x_i - x_j \right)^2 \\
&= \sum_{i \in S, j \in S^c} A_{ij} \, (2c)^2 \\
&= 4 c^2 \mathrm{cut}(S).
\end{aligned}
\tag{25.33}
$$

In other words, the quadratic form $\mathbf{x}^\mathsf{T} \mathbf{L} \mathbf{x}$ gives, up to a constant, our partition's cut number!

The connection between the cut number and a quadratic form involving $\mathbf{L}$ is yet another reason why $\mathbf{L}$ is such an important matrix. It's also what lets us bring in spectral optimization methods to find partitions.

If we want to find the partition that minimizes $\mathrm{cut}(S)$, what does $\mathbf{L}$ tell us? A lot. First, as we mentioned before, the ones vector $\mathbf{1}$ is an eigenvector[11] of $\mathbf{L}$ with eigenvalue $\lambda_0 = 0$. If we choose $\mathbf{x} = c\mathbf{1}$, we will have $\mathbf{x}^\mathsf{T} \mathbf{L} \mathbf{x} = (c\mathbf{1}^\mathsf{T}) \mathbf{L} (c\mathbf{1}) = c^2 \mathbf{1}^\mathsf{T} \mathbf{L} \mathbf{1} = c^2 \mathbf{1}^\mathsf{T} \lambda_0 \mathbf{1} = 0$. This corresponds to the trivial partition where all nodes are placed in S, so it's not surprising that we get zero. What about a more useful partition?

Let's consider an $\mathbf{x}$ that is orthogonal to $\mathbf{1}$, $\mathbf{x}^\mathsf{T} \mathbf{1} = 0$. Because it is still subject to Eq. (25.32), to be orthogonal to $\mathbf{1}$ means we should have equal numbers[12] of c and $-c$ elements—we are enforcing a *balanced* partition. This will push us away from the trivial partition $c\mathbf{1}$. In other words, we want to solve

$$
\begin{aligned}
\text{minimize} \quad & \mathbf{x}^\mathsf{T} \mathbf{L} \mathbf{x} \\
\text{subject to} \quad & x_i \in \{c, -c\} \text{ for all } i, \\
& \mathbf{x}^\mathsf{T} \mathbf{1} = 0.
\end{aligned}
\tag{25.34}
$$

This problem is hard, very hard: it is NP-complete [174], leaving us in a sticky situation.

A way to proceed is to relax somewhat the constraint on $\mathbf{x}$. This will no longer (perfectly) balance the partition, but it makes finding the partition much easier. As $\mathbf{1}$ is an eigenvector of $\mathbf{L}$, and since $\mathbf{L}$ is a real, symmetric matrix, it can be diagonalized, $\mathbf{L} = \mathbf{V} \mathbf{\Lambda} \mathbf{V}^\mathsf{T}$, and its eigenvectors $\mathbf{v}_i$, which are the columns of the real orthogonal matrix $\mathbf{V}$, are orthonormal. Thus, we can suspect that another eigenvector, which will be orthonormal to $\mathbf{v}_0 = \mathbf{1}$, is somehow related to $\mathbf{x}$.

Let's use the diagonalized $\mathbf{L}$ to express our cut number. First, write $\mathbf{x}$ in terms of

---

[11] Usually we'll consider unit eigenvectors, so technically $1/\sqrt{N}$ is the eigenvector.
[12] We're also assuming an even number of nodes in G. Let's ignore this for now, as we'll relax the constraint on $\mathbf{x}$ shortly anyway.

the eigenvectors,[13]

$$\mathbf{x} = \sum_{i=1}^{N} \alpha_i \mathbf{v}_i, \text{ where } \alpha_i = \mathbf{x}^\mathsf{T} \mathbf{v}_i. \tag{25.35}$$

Substituting in,

$$\mathbf{x}^\mathsf{T} \mathbf{L} \mathbf{x} = \left(\sum_i \alpha_i \mathbf{v}_i^\mathsf{T} \right) \mathbf{L} \left(\sum_i \alpha_i \mathbf{v}_i \right)$$

$$= \sum_i \alpha_i \mathbf{v}_i^\mathsf{T} \sum_i \alpha_i \mathbf{L} \mathbf{v}_i$$

$$= \sum_i \alpha_i \mathbf{v}_i^\mathsf{T} \sum_i \alpha_i \lambda_i \mathbf{v}_i$$

$$= \sum_i \lambda_i \alpha_i^2 \quad (\text{since } \mathbf{v}_j^\mathsf{T} \mathbf{v}_i = \delta_{ij}), \tag{25.36}$$

or, written more compactly using $\mathbf{x} = \mathbf{V}\alpha$, where $[\alpha]_i = \alpha_i$,

$$\mathbf{x}^\mathsf{T} \mathbf{L} \mathbf{x} = \alpha^\mathsf{T} \mathbf{V}^\mathsf{T} \mathbf{L} \mathbf{V} \alpha$$

$$= \alpha^\mathsf{T} \mathbf{V}^\mathsf{T} \mathbf{V} \Lambda \mathbf{V}^\mathsf{T} \mathbf{V} \alpha$$

$$= \alpha^\mathsf{T} \Lambda \alpha \quad (\text{since } \mathbf{V}^\mathsf{T} \mathbf{V} = \mathbf{I})$$

$$= \sum_i \lambda_i \alpha_i^2. \tag{25.37}$$

Since $\lambda_1 = 0$, we can drop that term from the sum. And since we've ordered the eigenvalues, $0 = \lambda_1 \le \lambda_2 \le \cdots \le \lambda_N$, we can bound $\mathbf{x}^\mathsf{T} \mathbf{L} \mathbf{x}$ from either side,

$$\lambda_2 \sum_{i=2}^{N} \alpha_i^2 \le \mathbf{x}^\mathsf{T} \mathbf{L} \mathbf{x} \le \lambda_N \sum_{i=2}^{N} \alpha_i^2$$

$$\lambda_2 \mathbf{x}^\mathsf{T} \mathbf{x} \le \mathbf{x}^\mathsf{T} \mathbf{L} \mathbf{x} \le \lambda_N \mathbf{x}^\mathsf{T} \mathbf{x}$$

$$\lambda_2 \le \frac{\mathbf{x}^\mathsf{T} \mathbf{L} \mathbf{x}}{\mathbf{x}^\mathsf{T} \mathbf{x}} \le \lambda_N \tag{25.38}$$

since $\sum_i \alpha_i^2 = \alpha^\mathsf{T} \alpha = \mathbf{x}^\mathsf{T} \mathbf{V}^\mathsf{T} \mathbf{V} \mathbf{x} = \mathbf{x}^\mathsf{T} \mathbf{x}$.

This result, Eq. (25.38), is known as the *Courant–Fischer theorem*, and terms of the form $\mathbf{x}^\mathsf{T} \mathbf{M} \mathbf{x} / \mathbf{x}^\mathsf{T} \mathbf{x}$, where $\mathbf{M}$ is a Hermitian matrix, are called *Rayleigh quotients*. It's not surprising to encounter a Rayleigh quotient here, they occur in many optimization problems.

In our problem, using Eq. (25.33), this gives us bounds on the cut number,

$$\frac{\lambda_2 N}{4} \le \mathrm{cut}(S) \le \frac{\lambda_N N}{4}, \tag{25.39}$$

---

[13] To see where this comes from, suppose it is true for some $\{\alpha_i\}$ and compute

$$\mathbf{v}_j^\mathsf{T} \mathbf{x} = \mathbf{v}_j^\mathsf{T} \sum_i \alpha_i \mathbf{v}_i = \sum_i \alpha_i \mathbf{v}_j^\mathsf{T} \mathbf{v}_i = \alpha_j \mathbf{v}_j^\mathsf{T} \mathbf{v}_i = \alpha_j \Rightarrow \alpha_j = \mathbf{x}^\mathsf{T} \mathbf{v}_j,$$

where we used the fact that the eigenvectors are orthonormal, $\mathbf{v}_j^\mathsf{T} \mathbf{v}_i = \delta_{ij}$.

where we've used $x^T x = Nc^2$. Nicely, the value of c is now irrelevant.[14]

If we seek to minimize our cut number, how can we choose x to match this lower bound? Well, the minimum of the Rayleigh quotient will be when $x \propto v_2$, as all terms in $\sum_i \lambda_i \alpha_i^2$ for $i > 2$ will be zero. Of course, this won't satisfy our constraint on the values of x_i, but we can use this to find an *approximate* solution if we relax a constraint on our minimization problem (Eq. (25.34)).

Specifically, instead of Eq. (25.34) let us simply constrain the norm of x, giving this relaxed optimization problem:

$$\begin{aligned} \text{minimize} \quad & x^T L x \\ \text{subject to} \quad & x^T x = N, \\ & x^T 1 = 0. \end{aligned} \qquad (25.40)$$

This problem is always solved by $x = \sqrt{N} v_2$. To translate this to a **graph bisection algorithm**, we use the elements of v_2 to decide which group to place each node into. The simplest way is based on the sign: if $v_{2i} < 0$ put i in S, otherwise put i in S^c. Other thresholds may be used. Thresholding in this way can often give good partitions, but it doesn't guarantee they are balanced. (One idea, proposed by Pothen et al. [377], is to threshold at the median of v_2.) Nor does it tell us what to do with i where $v_{2i} = 0$, which has led to much debate. (When many elements of v_2 are zero, it is a warning sign that the graph is not easy to bisect.)

Fiedler vector and algebraic connectivity   This derivation has highlighted the importance of the second eigenvalue and second eigenvector of L. These are famously known as the algebraic connectivity and Fiedler vector. Fiedler proved [158] for unweighted graphs (that are not the complete graph) that the vertex connectivity, the minimum number of nodes that must be removed to disconnect a connected graph, is at least λ_2. The edge connectivity, the number of edges that must be cut to disconnect the graph, is never less than the vertex connectivity, so it too is bounded by λ_2. (This bound differs from $\lambda_2 N/4$ because it doesn't address a balanced cut.) These are powerful, inspiring results.

25.5.1   Alternative balancing

Many other specific ways to balance the partition have been pursued. One is called the *normalized cut* of S or $\text{NCut}(S)$ [428]. The normalized cut seeks to balance the cut number of S with the size or "volume" $\text{vol}(S)$ of S,

$$\text{vol}(S) = \sum_{i \in S} k_i, \qquad (25.41)$$

which is the total degree of nodes in S. If we wish to partition into groups that, like communities, are both internally dense and externally sparse, then we seek an objective

---

[14] Assuming the same c is used for both S and S^c. There can be situations where you want imbalanced groups, and one way to promote that is to choose c_1 for S and $c_2 \neq c_1$ for S^c.

function that minimizes cut(S) and maximizes vol(S). Normalized cut addresses this by minimizing a combination of cut(S) and $1/\text{vol}(S)$:

$$\text{NCut}(S) = \frac{\text{cut}(S)}{\text{vol}(S)} + \frac{\text{cut}(S^c)}{\text{vol}(S^c)} = \text{cut}(S)\left(\frac{1}{\text{vol}(S)} + \frac{1}{\text{vol}(S^c)}\right). \qquad (25.42)$$

Our problem is again to find the S that minimizes NCut(S): $\arg\min_{S \subset V} \text{NCut}(S)$. Unfortunately, this is difficult; in fact, it's known to be NP-complete [428]. However, we can proceed in a very similar manner to the previous method, defining a constrained indicator vector $\mathbf{x}$ and relaxing the constraint to find a spectral algorithm.

To find the approximate solution, make a clever choice of $\mathbf{x}$:

$$x_i = \begin{cases} 1/\text{vol}(S) & \text{if } i \in S, \\ -1/\text{vol}(S^c) & \text{otherwise.} \end{cases} \qquad (25.43)$$

Proceeding as before, if we apply $\mathbf{x}$ to $\mathbf{L}$, we get

$$\mathbf{x}^{\mathsf{T}}\mathbf{L}\mathbf{x} = \left(\frac{1}{\text{vol}(S)} + \frac{1}{\text{vol}(S^c)}\right)^2 \text{cut}(S). \qquad (25.44)$$

Notice how close this is to Eq. (25.42)? We can do more. Dividing by

$$\mathbf{x}^{\mathsf{T}}\mathbf{D}\mathbf{x} = \sum_i k_i x_i^2 = \sum_{i \in S} \frac{k_i}{\text{vol}(S)^2} + \sum_{i \in S^c} \frac{k_i}{\text{vol}(S^c)^2}$$

$$= \frac{1}{\text{vol}(S)} + \frac{1}{\text{vol}(S^c)} \qquad (25.45)$$

gives

$$\text{NCut}(S) = \frac{\mathbf{x}^{\mathsf{T}}\mathbf{L}\mathbf{x}}{\mathbf{x}^{\mathsf{T}}\mathbf{D}\mathbf{x}}, \qquad (25.46)$$

which we can rewrite using $\mathbf{y} := \mathbf{D}^{1/2}\mathbf{x}$ to get,

$$\text{NCut}(S) = \frac{\mathbf{y}^{\mathsf{T}}\mathbf{D}^{-1/2}\mathbf{L}\mathbf{D}^{-1/2}\mathbf{y}}{\mathbf{y}^{\mathsf{T}}\mathbf{D}^{-1/2}\mathbf{D}\mathbf{D}^{-1/2}\mathbf{y}} = \frac{\mathbf{y}^{\mathsf{T}}\mathbf{L}_{\text{sym}}\mathbf{y}}{\mathbf{y}^{\mathsf{T}}\mathbf{y}} \qquad (25.47)$$

In other words, to minimize NCut(S) we can seek the Fiedler vector of the *normalized* graph Laplacian,

$$\mathbf{L}_{\text{sym}} = \mathbf{D}^{-1/2}\mathbf{L}\mathbf{D}^{-1/2},$$

$$[\mathbf{L}_{\text{sym}}]_{ij} = \begin{cases} 1 & \text{if } i = j, \\ -\frac{1}{\sqrt{k_i k_j}} & \text{if } A_{ij} = 1 \text{ and } i \neq j, \\ 0 & \text{otherwise.} \end{cases} \qquad (25.48)$$

The normalized Laplacian $\mathbf{L}_{\text{sym}}$ also appears in many spectral methods [108]. (We encounter another normalized Laplacian in Sec. 25.7.)

25.5.2   Worst case bounds on sparsest cuts

We can go further in our understanding of sparsest cuts using spectral bisection. And again, our focus is on edge cuts but many similar results hold for vertex cuts (removing nodes to disconnect the network) [108]. When asking whether a sparse cut exists, can we determine bounds on the cut number, both lower and upper bounds? A lower bound we already have, from the algebraic connectivity, λ_2. A fascinating result is one that provides an upper bound, which gives a clue about worst case scenarios ("it won't take more than X cuts to disconnect this network...").

From Sec. 25.5.1, we have a lower bound on the cut number,

$$
\begin{aligned}
\lambda_2 &\leq \mathrm{cut}(S)\left(\frac{1}{\mathrm{vol}(S)} + \frac{1}{\mathrm{vol}(S^c)}\right) \\
&\leq \frac{2\,\mathrm{cut}(S)}{\min\{\mathrm{vol}(S), \mathrm{vol}(S^c)\}} \\
&= 2h_G(S),
\end{aligned}
\tag{25.49}
$$

where $h_G(S)$ is the *Cheeger constant* of S. This inequality holds for any S, including the optimal S, $h_G := \min_S h_G(S)$, which is the Cheeger constant of the graph G. Therefore,

$$
2h_G \geq \lambda_2.
\tag{25.50}
$$

In perhaps the most profound result in spectral graph theory, the *Cheeger inequality* also provides a bound on the cut number in the opposite direction. We derive this now. Let $\mathbf{x} = \mathbf{v}_2$ be the eigenvector that corresponds to λ_2, meaning that $\sum_i x_i k_i = 0$ and

$$
R(\mathbf{x}) := \frac{\mathbf{x}^\mathsf{T} \mathbf{L} \mathbf{x}}{\mathbf{x}^\mathsf{T} \mathbf{x}} = \frac{\sum_{i,j \in E}(x_i - x_j)^2}{\sum_i x_i^2 k_i} = \lambda_2.
\tag{25.51}
$$

Suppose we order the nodes by $\mathbf{x}$ in descending order,

$$
x_1 \geq x_2 \geq \cdots \geq x_N.
\tag{25.52}
$$

Define the ordered partitions $S_i = \{1, \ldots, i\}$ and let $\alpha_G = \min_i h_{S_i}$. As h_G is the global minimum, we have $\alpha_G \geq h_G$. Let r be the largest integer such that $\mathrm{vol}(S_r) \leq \mathrm{vol}(G)/2$. Notice that

$$
\sum_i x_i^2 k_i = \min_c \sum_i (x_i - c)^2 k_i \leq \sum_i (x_i - x_r)^2 k_i.
\tag{25.53}
$$

Consider the nodes that sort before and after r by defining the positive and negative parts of $\mathbf{x}$,

$$
x_{(+)i} = \begin{cases} x_i - x_r & \text{if } x_i \geq x_r, \\ 0 & \text{otherwise,} \end{cases}
$$
$$
\tag{25.54}
$$
$$
x_{(-)i} = \begin{cases} x_r - x_i & \text{if } x_i \leq x_r, \\ 0 & \text{otherwise,} \end{cases}
$$

respectively. We now apply these to $\lambda_2 = R(\mathbf{x})$. First, we have

$$\lambda_2 = \frac{\Sigma_{i,j\in E}(x_i - x_j)^2}{\Sigma_i x_i^2 k_i} \geq \frac{\Sigma_{i,j\in E}(x_i - x_j)^2}{\Sigma_i (x_i - x_r)^2 k_i}. \tag{25.55}$$

Notice, for the denominator, that $(x_i - x_r)^2 = \left(x_{(+)i} - x_{(-)i}\right)^2 = x_{(+)i}^2 + x_{(-)i}^2 - 2x_{(+)i}x_{(-)i} \leq x_{(+)i}^2 + x_{(-)i}^2$ and, for the numerator,

$$\begin{aligned}
x_i - x_j &= x_i - x_r - (x_j - x_r) \\
&= x_{(+)i} - x_{(-)i} - \left(x_{(+)j} - x_{(-)j}\right) \\
&= x_{(+)i} - x_{(+)j} - \left(x_{(-)i} - x_{(-)j}\right),
\end{aligned} \tag{25.56}$$

$$\begin{aligned}
(x_i - x_j)^2 &= \left(x_{(+)i} - x_{(+)j}\right)^2 + \left(x_{(-)i} - x_{(-)j}\right)^2 \\
&\quad - 2\left(x_{(+)i} - x_{(+)j}\right)\left(x_{(-)i} - x_{(-)j}\right).
\end{aligned} \tag{25.57}$$

The cross term is non-negative regardless of the ordering of x_i, x_j and x_r, so we know

$$(x_i - x_j)^2 \geq \left(x_{(+)i} - x_{(+)j}\right)^2 + \left(x_{(-)i} - x_{(-)j}\right)^2 \tag{25.58}$$

and therefore

$$\lambda_2 \geq \frac{\Sigma_{i,j\in E}\left(\left(x_{(+)i} - x_{(+)j}\right)^2 + \left(x_{(-)i} - x_{(-)j}\right)^2\right)}{\Sigma_i \left(x_{(+)i}^2 + x_{(-)i}^2\right)k_i}. \tag{25.59}$$

Without loss of generality, we assume $R\left(\mathbf{x}_{(+)}\right) \leq R\left(\mathbf{x}_{(-)}\right)$ and therefore, since $(a + b)/(c + d) \geq \min(a/c, b/d)$,

$$\begin{aligned}
\lambda_2 &\geq R(\mathbf{x}_{(+)}) \\
&= \frac{\Sigma_{i,j\in E}\left(x_{(+)i} - x_{(+)j}\right)^2}{\Sigma_i x_{(+)i}^2 k_i} \\
&= \frac{\Sigma_{i,j\in E}\left(x_{(+)i} - x_{(+)j}\right)^2}{\Sigma_i x_{(+)i}^2 k_i} \frac{\Sigma_{i,j\in E}\left(x_{(+)i} + x_{(+)j}\right)^2}{\Sigma_{i,j\in E}\left(x_{(+)i} + x_{(+)j}\right)^2}.
\end{aligned} \tag{25.60}$$

We again bound the numerator and denominator. First, the numerator, by the Cauchy–Schwarz inequality,

$$\sum_{i,j\in E}\left(x_{(+)i} - x_{(+)j}\right)^2 \sum_{i,j\in E}\left(x_{(+)i} + x_{(+)j}\right)^2 \geq \left(\sum_{i,j\in E}\left(x_{(+)i}^2 - x_{(+)j}^2\right)\right)^2. \tag{25.61}$$

We also have, once more using Cauchy–Schwarz,

$$\begin{aligned}
\sum_{i,j\in E}\left(x_{(+)i} + x_{(+)j}\right)^2 &\leq 2\sum_{i,j\in E}\left(x_{(+)i}^2 + x_{(+)j}^2\right) \\
&\leq 4\sum_{i,j\in E} x_{(+)i}^2 \\
&= 2\sum_i x_{(+)i}^2 k_i.
\end{aligned} \tag{25.62}$$

Putting these together, we now have

$$\lambda_2 \geq \frac{\left(\sum_{i,j \in E} \left(x^2_{(+)i} - x^2_{(+)j}\right)\right)^2}{2 \left(\sum_i x^2_{(+)i} k_i\right)^2}. \tag{25.63}$$

Next, we rewrite the numerator in terms of consecutive differences, revealing the presence of $\mathrm{cut}(S_i)$:

$$\sum_{i,j \in E} \left(x^2_{(+)i} - x^2_{(+)j}\right) = \sum_{i<j} A_{ij} \left(x^2_{(+)i} - x^2_{(+)j}\right)$$

$$= \sum_{i<j} A_{ij} \sum_{i \leq u < j} \left(x^2_{(+)u} - x^2_{(+)u+1}\right)$$

$$= \sum_{u=0}^{r} \left(x^2_{(+)u} - x^2_{(+)u+1}\right) \sum_{\substack{i \leq u, \\ j > u}} A_{ij}$$

$$= \sum_{u=0}^{r} \left(x^2_{(+)u} - x^2_{(+)u+1}\right) \mathrm{cut}(S_u). \tag{25.64}$$

Accordingly, we have

$$\lambda_2 \geq \frac{\left(\sum_i \left(x^2_{(+)i} - x^2_{(+)i+1}\right) \mathrm{cut}(S_i)\right)^2}{2 \left(\sum_i x^2_{(+)i} k_i\right)^2}. \tag{25.65}$$

The cut number is related to α_G (and therefore h_G), by $\mathrm{cut}(S_i) \geq \min_i \mathrm{cut}(S_i) = \alpha_G \mathrm{mvol}(S_i)$, where $\mathrm{mvol}(S) := \min(\mathrm{vol}(S), \mathrm{vol}(S^c))$. Applying this,

$$\lambda_2 \geq \frac{\alpha_G^2}{2} \frac{\left(\sum_i \left(x^2_{(+)i} - x^2_{(+)i+1}\right) \mathrm{mvol}(S_i)\right)^2}{\left(\sum_i x^2_{(+)i} k_i\right)^2}$$

$$= \frac{\alpha_G^2}{2} \frac{\left(\sum_i x^2_{(+)i} \left[\mathrm{mvol}(S_i) - \mathrm{mvol}(S_{i-1})\right]\right)^2}{\left(\sum_i x^2_{(+)i} k_i\right)^2}$$

$$= \frac{\alpha_G^2}{2} \frac{\left(\sum_i x^2_{(+)i} k_i\right)^2}{\left(\sum_i x^2_{(+)i} k_i\right)^2}. \tag{25.66}$$

Finally, given Eq. (25.50) and since $\alpha_G \geq h_G$, we arrive at Cheeger's inequality,

$$2h_G \geq \lambda_2 \geq \frac{h_G^2}{2} \tag{25.67}$$

or

$$\frac{\lambda_2}{2} \leq h_G \leq \sqrt{2\lambda_2}. \tag{25.68}$$

The power of Cheeger's inequality is both that it bounds the size of the sparse cut (G has a natural bisection if and only if λ_2 is small) and it tells us (approximately) how to find it by partitioning based on the sign of the second eigenvector of **L**. Even for very large, sparse networks, this eigenvector can be computed reasonably quickly.

The quantity $cut(S)/vol(S)$ is also known as the *conductance* of S, and, like NCut(S), serves to balance the cut.

25.5.3 Partitioning beyond two groups

Our discussion has focused on bisection, where we seek only S and S^c. If we're interested in finding more than two groups, which we often are, we have some options. The first as we mentioned is recursive bisection: find S and S^c in the original graph, then bisect the subgraph induced by S and the subgraph induced by S^c. Now we have four groups, and we can continue further if we choose. Another alternative is using information beyond the Fiedler vector. The higher eigenvectors and eigenvalues are also helpful for these problems, and even analytical quantities such as Cheeger's inequality, which connects only with the algebraic connectivity, can be extended to higher-order inequalities that consider eigenvalues λ_k, $k > 2$ [279, 267]. We'll see where this higher-order information comes into play in subsequent sections.

25.6 Community detection

A problem closely related to graph partitioning is community detection. The goal is (usually) to find dense groups of nodes within the network that are sparsely connected to the rest of the network. Like graph partitioning, these groups are easily cut out of the network. Spectral methods, including bisection methods, are commonly used for community detection as well, although the emphasis is usually on multi-way partitions (Sec. 25.5.3).

Here we focus on a popular spectral method due to Newman [334] that uses the modularity matrix **B**, Eq. (25.22) (Sec. 25.1.6), with elements

$$B_{ij} = A_{ij} - \frac{k_i k_j}{2M}. \tag{25.69}$$

Notice that **1** is always an eigenvector of **B**, with eigenvalue zero.

For two groups, define the vector **s** with elements $s_i = +1$ if i belongs to the first group and $s_i = -1$ if i belongs to the second group. Two nodes i and j are in the same group when $\frac{1}{2}(s_i s_j + 1) = 1$ and 0 otherwise. Substituting into Eq. (25.21),

$$Q = \frac{1}{4M} \sum_{i,j} \left(A_{ij} - \frac{k_i k_j}{2M} \right) (s_i s_j + 1) = \frac{1}{4M} \sum_{i,j} \left(A_{ij} - \frac{k_i k_j}{2M} \right) s_i s_j. \tag{25.70}$$

The second expression contains $s_i s_j$ instead of $s_i s_j + 1$ because $\sum_{i,j} A_{ij} = \sum_i k_i = 2M$ and $\sum_{i,j} k_i k_j / (2M) = \sum_i k_i \sum_j k_j / (2M) = 2M$. We can now write Q as a quadratic form,

$$Q = \frac{1}{4M} \mathbf{s}^\mathsf{T} \mathbf{B} \mathbf{s}. \tag{25.71}$$

Similar to spectral bisection where we minimized the cut number, expressing Q in this form shows us that Q can be maximized by choosing $\mathbf{s}$ based on the eigenvectors of $\mathbf{B}$.[15]

Because $\mathbf{B}$ is a real, symmetric matrix, by the spectral theorem it is diagonalizable and its eigenvectors form an orthonormal basis for $\mathbb{R}^N$. Let $\mathbf{v}_i$ be the ith (normalized) eigenvector of $\mathbf{B}$. We express $\mathbf{s}$ in the eigenvector basis of $\mathbf{B}$,

$$\mathbf{s} = \sum_i \alpha_i \mathbf{v}_i, \quad \alpha_i = \mathbf{v}_i^\mathsf{T} \mathbf{s}, \tag{25.72}$$

and rewrite Q to give,

$$Q = \frac{1}{4M} \sum_i \alpha_i \mathbf{v}_i^\mathsf{T} \mathbf{B} \sum_j \alpha_j \mathbf{v}_j = \frac{1}{4M} \sum_i \lambda_i (\mathbf{v}_i^\mathsf{T} \mathbf{s})^2, \tag{25.73}$$

where λ_i is the ith eigenvalue of $\mathbf{B}$ and we assume the eigenvalues are labeled such that $\lambda_1 \geq \lambda_2 \geq \cdots \geq \lambda_N$.

If we chose $\mathbf{s}$ parallel to $\mathbf{v}_1$ we would maximize Q, as we would concentrate all our weight on the largest term, that with λ_1, in the sum for Q. All other terms would be zero because the eigenvectors are all orthogonal. Unfortunately, the problem is not so simple, as $\mathbf{s}$ cannot be proportional to $\mathbf{v}_1$ because its elements are constrained to take values ± 1. However, if we proceed and try to make $\mathbf{s}$ as close to parallel to $\mathbf{v}_1$ as possible, we see that we can achieve this by setting $s_i = 1$ when $v_{1i} > 0$ or $s_i = -1$ when $v_{1i} < 0$.[16] The algorithm to find the maximum modularity bisection is, similar to the spectral bisection before, to find the leading eigenvector of $\mathbf{B}$, and partition nodes based on the sign of the eigenvector's elements.

As remarked by Newman [334], partitioning in this way has the advantage that we do not need to specifically balance the size of the two groups, like we did with NCut or the Cheeger constant (conductance). The density of positive elements of $\mathbf{v}_1$ naturally expresses how big the two groups are, and if all (or none) of the $v_{1i} > 0$, then so be it—the algorithm has told us (correctly, hopefully) that the network should not be partitioned.

As a brief example, we apply the partitioning techniques we have learned to Zachary's Karate Club. This club split after a disagreement between the club members [505], so it provides a nice case study to benchmark such methods. In Fig. 25.3 we compare the observed split to ones found by graph bisection and spectral modularity maximization. We also compute the modularity Q for each partition. All three partitions are quite similar, but the computational methods actually discover slightly higher values of modularity Q than the observed split.

---

[15] And like the null partition of spectral bisection, the $\mathbf{1}$ eigenvector reflects that $Q = 0$ when all nodes are placed into one group.

[16] A good question becomes what to do if $v_{1i} = 0$.

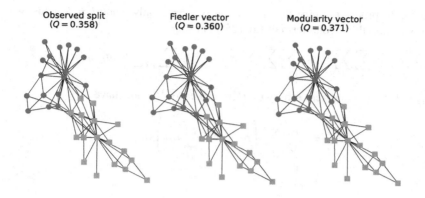

Figure 25.3   Partitioning Zachary's Karate Club into two groups.

Of course, it's rare to seek only two communities. Ideally we'd want the algorithm to inform us how many communities will be in the network. (Note that this is not strictly possible with modularity due to the resolution limit, Sec. 12.7.) For spectral bisection, we can proceed by recursion, repeated partitions of the original graph into halves, quarters, and so forth. At first, this also seems like a straightforward strategy for maximizing Q: find two groups, cut the edges between them, then repeat to maximize Q inside of each group separately. But it is not appropriate, as the edges cut between the groups will contribute to Q and so removing them will alter the objective function we are trying to maximize.

Instead of removing edges and recursively maximizing modularity, a better approach is to express how Q changes after one of the groups is further partitioned in two (bisected) while remaining in the network. Suppose the group g with n_g nodes is bisected and the $n_g \times 1$ vector $\tilde{\mathbf{s}}$ indicates the memberships of those nodes. How does Q change due to this? The groups outside of g are unchanged, so any $i, j \notin g$ terms of Q are the same and we need only look at nodes $i, j \in g$. Before g was bisected, all i, j were in the same group, so we lose all their B_{ij} terms. On the other hand, we gain terms where $i, j \in g$ have the same membership in the new groups, $\tilde{s}_i = \tilde{s}_j$. Putting these together, the change in Q due to bisecting the group g is

$$
\Delta Q = \frac{1}{2M} \sum_{i,j \in g} B_{ij} \frac{(\tilde{s}_i \tilde{s}_j + 1)}{2} - \frac{1}{2M} \sum_{i,j \in g} B_{ij}
$$

$$
= \frac{1}{2M} \left(\frac{1}{2} \sum_{i,j \in g} B_{ij} (\tilde{s}_i \tilde{s}_j + 1) - \sum_{i,j \in g} B_{ij} \right)
$$

$$
= \frac{1}{4M} \left(\sum_{i,j \in g} B_{ij} \tilde{s}_i \tilde{s}_j - \sum_{i,j \in g} B_{ij} \right). \tag{25.74}
$$

Now it looks like we're stuck. To continue leveraging spectral methods, we want to factor Eq. (25.74) so that $\tilde{s}_i \tilde{s}_j$ is on all terms, letting us express the double sum as $\tilde{\mathbf{s}}^\top \mathbf{M} \tilde{\mathbf{s}}$

for some $\mathbf{M}$. To proceed, notice that $1 = \tilde{s}_i^2$ but $\tilde{s}_i \tilde{s}_j = 1$ only when $i = j$. Thus, we introduce $\tilde{s}_i \tilde{s}_j$ into the second term of Eq. (25.74) using

$$\sum_i \sum_j B_{ij} = \sum_i \sum_k B_{ik} = \sum_i \sum_j \sum_k B_{ij} \delta_{ij} = \sum_i \sum_j \sum_k B_{ij} \delta_{ij} \tilde{s}_i \tilde{s}_j, \quad (25.75)$$

where $\delta_{ij} = 1$ if $i = j$ and zero otherwise (Kronecker δ). We now have

$$\Delta Q = \frac{1}{4M} \sum_{i,j \in g} \left(B_{ij} - \delta_{ij} \sum_{k \in g} B_{ik} \right) \tilde{s}_i \tilde{s}_j$$

$$= \frac{1}{4M} \tilde{s}^\mathsf{T} \mathbf{B}^{(g)} \tilde{s}, \quad (25.76)$$

written in terms of the new $n_g \times n_g$ matrix $\mathbf{B}^{(g)}$ with elements

$$B_{ij}^{(g)} = B_{ij} - \delta_{ij} \sum_{k \in g} B_{ik}. \quad (25.77)$$

Just as with the calculation for Q, here we correctly have $\Delta_Q = 0$ if g is not bisected.

To build a full algorithm based on repeatedly bisecting groups, we will need to know when to stop. The expression for ΔQ provides an answer: continue bisecting while $\Delta Q > 0$. When bisecting a group no longer increases ΔQ, leave it undivided and move onto other groups, continuing to divide until all groups are left undivided and the final communities have been revealed.

We briefly discuss alternatives to repeated (in-place) bisection. Another way to establish a spectral method that goes from two groups to C is to generalize the group indicator vector $\mathbf{s}$ to a matrix $\mathbf{S}$ of size $N \times C$, with elements

$$S_{ig} = \begin{cases} 1 & \text{if node } i \text{ belongs to group } g, \\ 0 & \text{otherwise.} \end{cases} \quad (25.78)$$

Because we seek a partition of the nodes, columns of $\mathbf{S}$ are orthogonal, rows sum to 1, and $\operatorname{tr} \mathbf{S}^\mathsf{T} \mathbf{S} = N$. To indicate whether nodes i and j belong to the same group, we have

$$\delta(g_i, g_j) = \sum_{g=1}^{C} S_{ig} S_{jg}. \quad (25.79)$$

Therefore, modularity can be written as,

$$Q = \frac{1}{2M} \sum_{i,j} \sum_{g} B_{ij} S_{ig} S_{jg} = \operatorname{tr} \left(\mathbf{S}^\mathsf{T} \mathbf{B} \mathbf{S} \right). \quad (25.80)$$

(In fact, we only need $C-1$ variables to identify a node's membership in one of C groups, so this formulation can be condensed.) Written in this way, diagonalizing $\mathbf{B} = \mathbf{V} \mathbf{\Lambda} \mathbf{V}^\mathsf{T}$ points to methods that use more eigenvectors than the leading eigenvector, since

$$Q = \sum_j \sum_g \lambda_j (\mathbf{V}_{*j}^\mathsf{T} \mathbf{S}_{*k})^2, \quad (25.81)$$

where $\lambda_j = \Lambda_{jj}$ is the jth eigenvalue of $\mathbf{B}$, $\mathbf{V}_{*j}$ is the jth eigenvector, and $\mathbf{S}_{*k}$ is the kth column of $\mathbf{S}$. We can maximize this expression by loading our vectors (columns) of $\mathbf{S}$ on the positive eigenvalues of $\mathbf{B}$. The number of positive eigenvalues will tell us how many communities we have.[17] However, remember that the elements of $\mathbf{S}$ must be constrained to be in $\{0, 1\}$, which complicates matters. Approaches to finding those elements include vector partitioning [333] and spectral clustering (Sec. 25.7).

Beyond the modularity matrix

The non-backtracking matrix (Sec. 25.1.4) has also found applications in spectral detection of communities. The modularity matrix suffers when trying to infer groups in sparse, homogeneous networks. Optimal methods can find communities all the way down to the detectability limit (Sec. 23.2.5), but spectral methods fail before that point. This failure occurs when the largest eigenvalue of $\mathbf{B}$ becomes indistinguishable from the "bulk" or remaining, random eigenvalues. Krzakala et al. [259] argue that the spectrum of the non-backtracking matrix is better behaved in this regard, with its C top eigenvalues tending to remain outside the bulk, allowing detection to occur. While this comes at a price, the non-backtracking matrix is larger and thus computing its eigenvectors can be more expensive, optimizations are possible.

25.7 Spectral clustering

Spectral clustering, which we briefly discussed in Ch. 14 (Sec. 14.3.2), is a general technique that uses spectral information to build a point cloud that we can then cluster with an algorithm such as k-means (which we discuss below). Often in general data problems, we do not have a ready representation of our data in $\mathbb{R}^d$, but we can compare observations quantitatively, so we proceed by building an $N \times N$ similarity matrix, where S_{ij} tells us how similar the ith and jth data observations are. For spectral clustering, this matrix is then converted into a similarity network whose graph Laplacian is then analyzed; for all the reasons discussed above, the graph Laplacian is packed with useful spectral information about the network. We use this information to "embed" the data into $\mathbb{R}^k$ and then find k clusters of points with k-means (or another method if we wish). One of the advantages of spectral clustering is that it makes few assumptions on how clusters of points are shaped. Spectral clustering is known to work well when the data cluster into groups but those groups are non-convex, nested, or otherwise distributed such that the center coordinate of a group is not a meaningful way to distinguish different groups.

While spectral clustering is a general method, it is especially handy for networks. In our case, we usually already have a network, so we will skip directly to the network's Laplacian, but remember that this network extraction step (Ch. 7) turns spectral clustering into a general purpose method.

---

[17] The number of communities is one greater than the number of positive eigenvalues because we must ensure the normalization condition on $\mathbf{S}$, meaning we can only have $C - 1$ linearly independent columns of $\mathbf{S}$.

Before we introduce the major spectral clustering methods, we briefly describe k-means, the "classic" data clustering method.

25.7.1   Clustering with k-means

The k-means algorithm is the prototypical clustering method. Clustering, unlike graph partitioning or community detection, is not about grouping nodes in a network but about grouping points in a d-dimensional space such that points within a cluster are "close" by some measure of distance while points in different clusters are "far." For networks, clustering becomes important to us when we perform embedding (Ch. 26) where the nodes in the network *become* points in a d-dimensional space. We can then cluster those points and thus group those nodes.

The basic idea with k-means is to represent k clusters as additional points called *centroids*. We first place the centroids at random among the N points in our space and then move the centroids to, well, the center of each cluster. This is done by computing distances between each of the N points and the nearest of the k centroids, then setting the centroid to the average position of points that are closer to that centroid than any other. This needs to be repeated since moving the centroids will change which points are closest, so we iterate until the centroids no longer move. The full algorithm is described in Alg. 25.1.

Algorithm 25.1 k-means clustering.

1: Input: Data $\mathbf{X} \in \mathbb{R}^{N \times d}$, number k of clusters
2: Initialize centroid positions $\boldsymbol{\mu}_1, \boldsymbol{\mu}_2, \ldots, \boldsymbol{\mu}_k \in \mathbb{R}^d$ randomly[18]
3: **repeat**
4:        **for** $i = 1, \ldots, N$ **do**   ▷ *find centroid closest to ith point*
5:              $c_i \leftarrow \arg\min_j \|\mathbf{x}_i - \boldsymbol{\mu}_j\|^2$ ▷ $\mathbf{x}_i$ *is the ith point (row of* $\mathbf{X}$*),* $\|\cdot\|^2$ *is the squared Euclidean distance*
6:        **for** $j = 1, \ldots, k$ **do**   ▷ *move jth centroid to average position of closest points*
7:              $\boldsymbol{\mu}_j \leftarrow \dfrac{\sum_{i=1}^{N} \mathbf{x}_i \mathbb{1}(c_i = j)}{\sum_{i=1}^{n} \mathbb{1}(c_i = j)}$
8: **until** convergence
9: Return: Cluster assignments $\{c_i\}_{i=1}^{N}$

We illustrate k-means with a toy example in Fig. 25.4, showing the true clusters and the first few iterations.

The expensive step for k-means is determining for each point the closest centroid. In principle, for one update pass, this takes $O(Nk)$ distance calculations (efficient data structures can improve this; see remarks). Futhermore, as described, the basic k-means algorithm uses Euclidean distances and the vector mean to define the "center" of a group of points. Both can be generalized. For example, k-medoids uses the median instead of the mean.

---

[18] One common strategy is to pick k data points uniformly at random and place the centroids at those same coordinates.

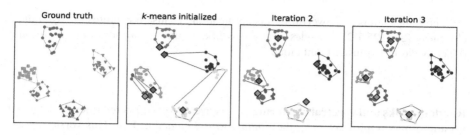

Figure 25.4 Applying k-means to a five-cluster toy dataset in $\mathbb{R}^2$. We illustrate the ground truth, the random initialization, and the first two iterations of k-means. Diamonds ($\diamond$) represent the cluster centroids and convex hulls illustrate the cluster groups.

An important question is how to choose the value of k. This is essentially a model selection question and we can overfit if k is too high. Many methods exist and are worth exploring in practical applications [478]. For our purposes we will discuss a spectral heuristic shortly; see Sec. 25.7.3 for other ways to choose the number of clusters.

We will focus in the remainder on the basic k-means algorithm, but clustering is an entire cottage industry—many methods are available and, while k-means is the best choice to start with, other methods can perform better in specific circumstances. See remarks for more.

25.7.2   Spectral clustering

With k-means established, we now describe spectral clustering. The method, parameterized by k, consists of finding the k leading eigenvectors of a Laplacian matrix, treating the elements of those vectors as coordinates to turn the nodes in the network into a "point cloud" in $\mathbb{R}^k$, and finding groups of nodes by clustering those coordinates using k-means. As described by Alg. 25.2, there are three main methods, differing only by what version of the Laplacian is used.

Algorithm 25.2 Spectral clustering.

1: Input: (Weighted) network, number k of clusters
2: Compute the unnormalized Laplacian $\mathbf{L} = \mathbf{D} - \mathbf{A}$
3: Compute the first k eigenvectors $\mathbf{v}_1, \ldots, \mathbf{v}_k$ for one of:
   - $\mathbf{L}$ ▷ *unnormalized spectral clustering*
   - $\mathbf{L}_{\text{rw}} = \mathbf{D}^{-1}\mathbf{L}$ ▷ *normalized spectral clustering*
   - $\mathbf{L}_{\text{sym}} = \mathbf{D}^{-1/2}\mathbf{L}\mathbf{D}^{-1/2}$ (Eq. (25.48))   ▷ *symmetric normalized spectral clustering*

   Let $\mathbf{V} \in \mathbb{R}^{N \times k}$ be the matrix containing $\mathbf{v}_1, \ldots, \mathbf{v}_k$ as columns
   - If using $\mathbf{L}_{\text{sym}}$, normalize rows of $\mathbf{V}$: $v_{ij} \leftarrow v_{ij} / \left(\sum_k v_{ik}^2 \right)^{1/2}$

   Let $\mathbf{x}_i \in \mathbb{R}^k, i = 1, \ldots, N$, be the ith row of $\mathbf{V}$ ▷ *point cloud*
4: Group the points $(\mathbf{x}_i)_{i=1}^N$ into clusters $c_1, \ldots, c_k$ using k-means
5: Return: Clusters $g_1, \ldots, g_k$ where $g_j = \{i \mid \mathbf{x}_i \in c_j\}$

The Laplacian is so powerful in this application because it captures the dynamical closeness (Sec. 25.1.2) of nodes in the network, leading to a more meaningful and typically easier to cluster point cloud.

Random walks and spectral clustering   We have also encountered a new normalized version of the graph Laplacian, the *random walk Laplacian*, $\mathbf{L}_{rw}$. This version of the Laplacian provides some useful insights.

Imagine a *random walker* placed on a node i and then randomly hopping to a neighboring node j, with each neighbor being an equally likely destination. Suppose the walker continues to move randomly over the network for many time steps. What is the probability it will be on node i at time $t + 1$, given it was on j at time t? For a random walker, this transition probability is a constant, $\Pr(i \mid j) = A_{ij}/k_j := P_{ij}$ (note the order of the indices i, j). As a matrix, this probability is $\mathbf{P} = \mathbf{D}^{-1}\mathbf{A}$, which is closely related to $\mathbf{L}_{rw} = \mathbf{I} - \mathbf{P}$.

The transition probabilities tells us about the random walk movement, but not the walker's location. The *occupation probability* $p_i(t)$ is the probability to be at node i at time t. These probabilities evolve according to the transition probability, given by summing over any of the possible neighbors j that could lead the walker to i, $p_i(t) = \sum_j \Pr(i \mid j)p_j(t-1) = \sum_j P_{ij}p_j$ or, for all nodes, $\mathbf{p}(t) = \mathbf{P}\mathbf{p}(t-1)$. Suppose the random walker has moved so much that it reaches equilibrium, meaning the occupation probability $\mathbf{p}(t) \to \pi$ no longer changes. If this occurs, the stationary distribution π satisfies $\pi = \mathbf{P}\pi$, an eigenvalue equation. This distribution, if it exists, is unique and has elements $\pi_i = k_i/\sum_j k_j = k_i/2M$

Many properties of a network's structure are reflected in this equilibrium distribution. One is a relationship with graph bisection, allowing us to interpret what $\mathbf{L}_{rw}$ means in terms of cutting the network into disjoint components S and S^c. Instead of asking how many edges must be cut to disconnect a component, we now ask *what is the probability a random walker can get into or out of a component*, a closely related question.

Denote a random walker's position at time t with X_t. Suppose the random walker is located on node $i \in S$ following the stationary distribution. What is the probability $\Pr(X_1 \in S^c \mid X_0 \in S)$ it will move to a node $j \in S^c$? First, the joint probability,

$$
\begin{aligned}
\Pr(X_0 \in S, X_1 \in S^c) &= \sum_{i \in S, j \in S^c} \Pr(X_0 = i, X_1 = j) \\
&= \sum_{i \in S, j \in S^c} P_{ji}\pi_i \\
&= \sum_{i \in S, j \in S^c} \frac{A_{ji}}{k_i} \frac{k_i}{2M} \\
&= \frac{\mathrm{cut}(S)}{2M}.
\end{aligned}
\tag{25.82}
$$

For our original question, we obtain

$$\Pr(X_1 \in S^c \mid X_0 \in S) = \frac{\Pr(X_0 \in S, X_1 \in S^c)}{\Pr(X_0 \in S)}$$

$$= \frac{\mathrm{cut}(S)}{2M} \left(\frac{\sum_{i \in S} k_i}{2M} \right)^{-1}$$

$$= \frac{\mathrm{cut}(S)}{\mathrm{vol}(S)}. \tag{25.83}$$

This is very interesting as it tells us how to interpret NCut(S) (Eq. (25.42)) and therefore normalized spectral clustering:

$$\mathrm{NCut}(S) = \mathrm{cut}(S) \left(\frac{1}{\mathrm{vol}(S)} + \frac{1}{\mathrm{vol}(S^c)} \right)$$

$$= \Pr(X_1 \in S^c \mid X_0 \in S) + \Pr(X_1 \in S \mid X_0 \in S^c). \tag{25.84}$$

In other words, by minimizing NCut we are looking for a cut through the graph that a random walker is unlikely to move across.

Why three methods? Which to use? In a highly influential paper, von Luxburg [478] makes a strong argument in favor of $\mathbf{L}_{\mathrm{rw}}$. Using different analyses, von Luxburg shows that all three versions work well when the clusters are clearly distinguishable. When they are not distinguishable, the normalized versions $\mathbf{L}_{\mathrm{sym}}$ and $\mathbf{L}_{\mathrm{rw}}$ are better behaved. In particular, normalized versions seek to maximize both the difference between nodes in different clusters and the similarity between nodes in the same cluster, while unnormalized versions seek only the former. Between $\mathbf{L}_{\mathrm{sym}}$ and $\mathbf{L}_{\mathrm{rw}}$, von Luxburg recommends $\mathbf{L}_{\mathrm{rw}}$ as it is more robust to the negative effects of low-degree nodes. Recall the eigenvectors of $\mathbf{L}_{\mathrm{sym}}$ are $\mathbf{D}^{1/2}\mathbf{v}$, which we saw when we rewrote NCut(S) in Eq. (25.47). When a node has low degree, that factor of $\mathbf{D}^{1/2}$ will make the corresponding entry in the eigenvector small, which makes mis-clustering more likely. This is why the rows of $\mathbf{V}$ are rescaled when using $\mathbf{L}_{\mathrm{sym}}$, although this doesn't eliminate the problem [478].

25.7.3 Choosing the number of clusters and resolving the spectral revelation

As with community detection and other clustering problems, we are left to determine the number of clusters ourselves. Many methods can be employed (see below). For spectral clustering we have a particularly nice heuristic, the *eigengap*. Just as the algebraic connectivity tells us how easy it is to disconnect the network, additional eigenvalues tell us how "clusterable" the network will be. If the network has k clusters, when we plot the eigenvalues sorted by magnitude, we should see an obvious gap between the first k smallest eigenvalues and the remaining bulk, exactly what we show in Fig. 25.5. When the clusters are clear, this works very well. When the clusters are not-so-clear, the gap will shrink and k may be hard to determine. But in a sense this is a good thing, as we will learn how clear the clusters are from this simple plot, something we cannot

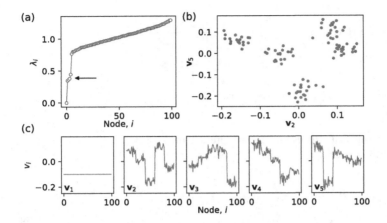

Figure 25.5 Resolving the spectral revelation of Fig. 25.2. Examining the eigendecomposition of $\mathbf{L}_{rw}$ shows how the leading eigenvectors capture the modular structure. (a) The eigenvalues ordered by magnitude. The *eigengap*, a useful diagnostic, clearly shows we should use $k = 5$. (b) The second and fifth eigenvectors plotted against one another, highlighting the clusters. (c) Elementwise plots of each of the leading eigenvectors. The plateaus seen in v_2 through v_5 demonstrate that cluster information is embedded in multiple vectors, motivating spectral clustering.

learn so easily trying to visualize the clusters themselves (except perhaps when $k = 2$ or 3). We recommend always looking at the sorted eigenvalues when applying spectral clustering.

There are some practical considerations when clustering data. The first is efficiency. When data are large or high-dimensional, many distance comparisons are needed. Efficient data structures, particularly k-d trees [53], can help with this. See Kanungo et al. [233] for more. The second and often more pressing concern is choosing the number of clusters. Some techniques to help choose the number of clusters include the elbow method, a heuristic which dates back to Thorndike [458]; silhouettes, introduced by Rousseeuw [404]; gap statistics [459]; and information criteria such as AIC or BIC [367].

25.8   Summary

The topology of a network can be represented one-to-one with its adjacency matrix. This is powerful, as it brings forth connections to powerful ideas from linear algebra, one of the most important areas of mathematics. Over time, the study of these connections has given rise to a dedicated area, *spectral graph theory*.

Many matrices appear in different cases when studying networks, including the modularity matrix, non-backtracking matrix, and the precision matrix. But one matrix stands out—the graph Laplacian. Not only does it capture dynamical processes unfolding over a network's structure, its spectral properties have deep connections to that structure. We found many relationships between the Laplacian's eigendecomposition and graph bisection and optimal partitioning tasks, including Cheeger's inequality, perhaps the most important result in spectral graph theory. Combining the dynamical information and the connections with partitioning also motivates spectral clustering, a powerful and successful way to find groups of data in general. Spectral clustering works by first converting the data into a meaningful network, then using k eigenvectors of that network's Laplacian to "embed" the nodes of the network into $\mathbb{R}^k$.

This idea of network embedding, converting network elements to points in a meaningful vector space, is now a common and important data analysis technique. We turn our attention in the next chapter to embedding and its interplay with machine learning.

Bibliographic remarks

Spectral graph theory is a rich area of mathematics. Chung [108] and van Mieghem [472] are wonderful general resources while Spielman [441] and Nica [345] are great introductions. Linear algebra underlies all of this, of course, and recommended references are Strang [447] and, for exciting connections with machine learning, Strang [446]. Meanwhile, see Golub and Van Loan [184] for more on general and efficient algorithms for numerical linear algebra, critical for applications to large datasets.

Spectral approaches to network centrality have a long history, going back to Katz [237]. Google's PageRank was memorably called the "$25,000,000,000 eigenvector" in an evocative review by Bryan and Leise [80].

Graph bisection problems arise in the study of network flows [121]. The work of Kernighan and Lin [239] provides an early heuristic view on graph bisection. Spectral properties of graphs were explored by Donath and Hoffman [132] and the especially influential work of Fiedler [158], where algebraic connectivity was introduced. The second eigenvector of the graph Laplacian is named the "Fiedler vector" in his honor. Barnes [39] provides another early look at graph partitioning. Many more results on spectral bisection and extensions have followed in Fiedler's work, including connections with Markov chains [433], efficient numerical methods [377] and Cheeger's inequality, which originally comes from Cheeger's work on Riemannian manifolds [100]. Chung [106] provides an interesting overview of different proofs for Cheeger's inequality, and highlights connections with Markov chains and PageRank. Recent work on higher-order Cheeger inequalities that go beyond the second-smallest eigenvalue include Louis et al. [279] and Lee et al. [267]. See Buluç et al. [82] for a recent overview on graph partitioning.

Community detection has emerged as a close sibling of graph partitioning, and many community methods are also spectral. The modularity maximization technique we presented, due to Newman [334, 333], is perhaps the most popular. See Newman [336] for an analysis showing some interesting parallels between modularity maximization and graph partitioning.

As discussed, the graph Laplacian is fundamental to dynamics evolving over network topologies. For general references on network dynamics—such as random walks, diffusion, and other spreading processes—and connections with the Laplacian, see, e.g.,

Porter and Gleeson [375], Masuda et al. [298], and, for a general introduction from a dynamical systems and physics perspective, Nolte [349].

Data clustering, of which spectral clustering is one instance, is a vast area of research, with many methods and many useful applications; it is one of the most important problems in the domain of unsupervised learning. See, for example, Kaufman and Rousseeuw [238], Bouveyron et al. [71], Hennig et al. [211], and Hastie et al. [205] for introductions and reference. We focused on k-means clustering, the prototypical form of clustering, but many other methods exist and can perform better in different situations. In particular, k-means makes assumptions about the size and shape of the clusters. Methods that avoid this, or make less strong assumptions, include DBSCAN [150] and the spectral clustering algorithms we described here. Spectral clustering has a long history dating back to Donath and Hoffman [132] and Fiedler [158]. The normalized variants of spectral clustering were popularized by Shi and Malik [428] and Ng et al. [344]. For more on spectral clustering, see the influential tutorial by von Luxburg [478] and references therein.

Exercises

25.1 Show how to compute the number of edges and the number of triangles from the adjacency matrix.

25.2 A bipartite network can be projected onto one of its two node sets. Given the original adjacency matrix, describe this projection with matrix operations.

25.3 For a network of k disconnected components, show that the k leading eigenvectors of **L** are *indicator variables* for the clusters.

25.4 (**Focal network**) Implement recursive bisection, apply it to the Karate Club and report the number of groups and average group size you find.

25.5 (**Focal network**) Repeat Ex. 25.4 but with repeated spectral modularity maximization. Report the number of groups found using the $\Delta Q > 0$ heuristic and compare to bisection.

25.6 (**Focal network**) *Heuristics beyond the spectra.* Not all graph partitioning and community detection methods are spectral. In fact, most are not. These different kinds of methods can coexist, however. For instance, a spectral method can be used as an efficient initialization for a more expensive method.

Devise a heuristic method, perhaps based on swapping nodes between groups, for optimizing the modularity of an existing partition, and apply this to one of the partitions of the Karate Club shown in Fig. 25.3. Do we improve Q and, if so, by how much?

Chapter 26

Embedding and machine learning

Machine learning, especially neural network methods, is increasingly important in network analysis. As discussed in Ch. 16, much of the success of modern machine learning is thanks to a neural network's power to obtain *useful representations—embeddings—* of data. In this chapter, we will discuss the theoretical aspects of network embedding methods and graph neural networks.

26.1   Embeddings as representations

Network embedding refers to a variety of methods that represent each node (or edge, or even a whole network) as a point in a space. Formally, this means we seek to learn a function $V \to \mathbb{R}^d$ that maps each node[1] to a d-dimensional vector that meaningfully captures the network's structure. In other words, network embedding methods *embed* a network into a (vector) *space*.

You may be wondering: "But doesn't the adjacency matrix itself represent each node as a vector?" Very clever! Indeed, if we look at each row of the adjacency matrix, the row is a vector with a fixed dimension (N) that captures information about the node's neighbors. Then, why do we apply embedding methods? Don't we already have an "embedding"? Usually, embedding methods aim to find representations that are (1) compact,[2] (2) continuous, and (3) dense.

To understand why we need these three properties, let's think about the adjacency matrix as an embedding. First, assuming that the network is sparse like most real-world networks, each node representation vector would have few nonzero entries; the representation vectors are *sparse*. If we pick two random nodes from a large network, they are likely to share no common neighbors. In other words, they are completely orthogonal from the perspective of adjacency matrix embedding and we don't have *any*

---

[1] Or each edge $E \to \mathbb{R}^d$ or even network $G \to \mathbb{R}^d$.

[2] We can also think of embedding methods as dimensionality reduction methods, because the compactness leads to a reduced number of dimensions, often significantly reduced.

information about their relationship. This would be the case for most pairs of nodes for many real-world networks. Even if they share some common neighbors, it is very likely that the amount of information that can be gleaned from the shared nodes is negligible for most pairs.

Second, the adjacency vectors—assuming a non-weighted network—treats all neighbors exactly the same, which is quite limiting. For instance, suppose i has two neighbors u and v, where i and u share numerous common neighbors and i and v do not share any. It is reasonable to assume that the relationship between i and u is much stronger than that between i and v. Yet, the adjacency embedding—which only considers immediate neighbors—does not tell us anything about such relationships. The two neighbors are identical. A better representation would capture the difference.

Lastly, the adjacency vectors are big, of dimension N. We are not compressing the information in the original adjacency matrix into a compact representation. It not only produces undesirable sparsity and discontinuity, it is also inefficient, and high-dimensional spaces are notoriously poor settings for building predictive models. We need compactness.

What's a simple alternative to using the adjacency matrix? A low-rank approximation, found with matrix factorization:

Laplacian Eigenmaps Although the term "embedding" is popular recently, thanks to the advancement of deep learning and neural network methods, the idea of low-dimensional representation has a long history. One notable method is *Laplacian Eigenmaps* [48], which leverages the spectral properties of networks.

In fact, we have encountered this embedding method before, as it is spectral clustering (Sec. 25.7.2). The graph Laplacian is diagonalized and its k leading eigenvectors are used to represent the nodes in a k-dimensional space (specifically, for a node i, its jth embedding coordinate is the ith element of eigenvector $\mathbf{v}_j$). The only difference from earlier was that spectral clustering limited the dimensionality to the number of clusters being inferred. For a Laplacian eigenmap, we can take the dimensionality as a "hyperparameter" of the method.

Of course, many other approaches exist, such as non-negative matrix factorization (Sec. 16.7.1).

What can we do with embedded representations?

Embeddings are helpful for exploratory tasks such as visualization, but one major reason we seek embedded versions of nodes in a network is that these representations can help with subsequent *machine learning tasks*. We may want to train a node classifier, for instance, and a classifier that uses node vectors instead of the network structure may be easier to set up, less costly to train, and may work better at predicting node classes. Node classes can be considered a form of attribute (Ch. 9) and we can even train a model to impute missing attributes by learning a combined representation of nodes and attributes. Another example is link prediction, which is essentially a node-pair classification task. And we can even in principle learn to generate synthetic networks by embedding the entire network and then learning a "decoder" mechanism that can translate an embedding back to a network structure. In general, dense vectors may be more "learnable" than sparse network structure. Logistic regression (Sec. 16.2.1) with the embedding vectors as features is often used.

Transductive and inductive learning

One important distinction we should mention involves the kinds of predictions we make. Will our model make predictions within a network it has already seen, or will it be expected to accommodate an unseen network? For the former, as an example, suppose we have our network G and a $N \times p$ node attribute matrix $\mathbf{X}$. One of our attributes, say the last one, is incomplete for some nodes: X_{ip} is missing for some nodes $i \in V_{\text{missing}}$. Can we train a model to predict X_{ip} for those nodes? Our model is able to see all of G, all of $\mathbf{X}$ for nodes $i \in V \setminus V_{\text{missing}}$, and the non-missing columns of $\mathbf{X}$ for nodes $i \in V_{\text{missing}}$. We can use all this information to supervise the training and validation of our model, then apply it to the same network to impute X_{ip}.

This imputation problem is an example of *transductive learning* (we also can call it a *semi-supervised* problem). Because the model will be trained on the same network it will make its predictions on, we can't expect it to generalize to an entirely different network. Doing that is called *inductive learning*, which is a more difficult problem. Many other problems can be either transductive or inductive depending on the setting, such as link prediction in a seen or unseen network.

26.2 Language models and word2vec—embeddings come of age

Often but not always when we now say "embedding," researchers now refer to neural network methods. To explain common embedding methods, it is useful to first visit the idea of *language models* and *word embedding* methods, upon which many network embedding methods build.

The *word2vec* method [307] is one of the early neural network methods that heralded the recent boom of deep learning. Word2vec and related models build on the classical idea of "language models." To understand where the idea came from, let's imagine a basic approach to obtain word representations. Let us simply assume that the properties of any given word can be represented by a bunch of numbers—a vector. Each element of this vector answers a particular question about the property of the word. Say, the first element is about whether the word is a verb or not. The second element quantifies how concrete (vs. abstract) the word is. We can imagine preparing a bunch of questions and building a vector for each word. Once we build these vectors—representations—for every word, we can estimate the similarity between words based on the similarity (e.g., cosine similarity) of the vectors and do all sorts of other tasks. But the list of questions can be arbitrary and it will not be trivial to build these vectors. The question is: can we build a *meaningful* vector of a word so that it captures the *meaning* of every word coherently? But then, how do we know the *meaning* of a word and how can we represent it quantitatively?

Language models assume that the meaning of a word comes from the *contexts* in which that word appears. This idea can be traced to Ferdinand de Saussure, who lays an important foundation for modern linguistics. Saussure argued [123] that the meaning of a "sign" converges onto the same concept through the averaging of the speakers who use the sign to communicate:

> Among all the individuals that are linked together by speech, some sort of average will be set up: all will reproduce—not exactly of course, but approximately—the same signs united with the same concepts.

This idea was later more formalized into the *distributional hypothesis* by Zellig S. Harris, who argued [202], as summarized by Pantel [357]: "*words that occur in the same contexts tend to have similar meaning.*" John R. Firth succinctly captures this idea by saying [159],

> You shall know a word by the company it keeps.

Therefore, we can study large corpora of natural language to understand the meaning of those words.

Language models   In other words, the idea is that we can figure out the meaning of a word by examining the other words around it. Let's look at these three sentences:

> "The quick brown _____ jumps over the lazy dog."
> "He is cunning as a _____."
> "The _____ was already in the hen house."

Just by examining the words around the blank, we can see that the word that can fill the blank should represent something brown, quick, cunning, something that can jump and likes to go into hen houses. And you may be already thinking about a *fox*.

Let's approach it more formally. A language model is a statistical model of natural language, in a sense answering the question, "can we distinguish a *natural* document from gibberish?" Mathematically, this question can be written as: can we accurately estimate the probability of an observed sequence of words $(\Pr(\{w_1, w_2, \ldots, w_T\}))$? If we can assign high probability to *actual* sentences and low probability to *random* sequences of words, then we have a *good* language model. Note that this joint probability can be broken down into a product of conditional probabilities using the chain rule:

$$\Pr(w_1, w_2, \ldots, w_T) = \Pr(w_T \mid w_1, w_2, \ldots, w_{T-1}) \cdots \Pr(w_3 \mid w_1, w_2)$$
$$\times \Pr(w_2 \mid w_1) \Pr(w_1). \quad (26.1)$$

In other words, if we can accurately estimate $\Pr(\text{target word} \mid \text{context words})$, we can construct a good language model. And this conditional probability captures the conceptual idea of language models—given the context (previous words), can we predict the word that comes next?[3]

---

[3] Also note that we can rearrange context and target words: "what would be the missing word in the middle, given the words around it?"

N-gram model   However, estimating the conditional probabilities directly from data is not feasible due to the exploding number of word combinations. A common simplification to address this issue is the *n-gram* model, where we only consider the previous n words, not all previous words. Namely, we assume that

$$\Pr(w_t \mid w_1, \ldots, w_{t-1}) \simeq \Pr(w_t \mid w_{t-n}, \ldots, w_{t-1}). \qquad (26.2)$$

If we set n small enough, we can find most n-gram combinations and count their actual occurrences. Then we can estimate the conditional probabilities by using our counts. This is the n-gram language model.

Word2vec takes a different approach. Instead of estimating the conditional probabilities directly, we assume that there exists a meaningful vector representation for every word that allows us to estimate the conditional probabilities without counting the actual occurrences.

Imagine every word has two vector representations that capture their meaning very well. For convenience, we call them *query* and *key* vectors[4] and denote as $\mathbf{q}_i$ and $\mathbf{k}_i$ (for word i). Now we assume that

$$\Pr(w_t \mid w_{t-n}, \ldots, w_{t-1}) \simeq f(\mathbf{k}_{w_t}, \mathbf{q}_{w_{t-n}}, \ldots, \mathbf{q}_{w_{t-1}}). \qquad (26.3)$$

In other words, instead of directly estimating $\Pr(w_t \mid w_{t-n}, \ldots, w_{t-1})$ from the data, we imagine a function f that can estimate the conditional probability based on the *representations of the words (vectors)*. A database is a nice analogy. The *query* to the database is the sequence of n-gram context words. When the query matches the *key* of a target word, we return this target word. Unlike a real database, where we identify a single perfect and unique match, everything here is probabilistic. Our training goal becomes learning the vectors from data (the corpus) to make this work.

Skip-gram model   The word2vec method suggested a further simplification to the n-gram model. It asks, why don't we just decompose the n-gram conditional probability into a product of 1-gram conditional probabilities (and also consider both preceding and following contexts)? This gives

$$\Pr(w_t \mid w_{t-l}, \ldots, w_{t-1}, w_{t+1}, \ldots, w_{t+l}) \simeq \prod_{t-l \leq i \leq t+l} \Pr(w_t \mid w_i). \qquad (26.4)$$

This is called the "skip-gram" model, where we *skip* all the other words but one from the n-gram formulation. Then, we can focus on a much simpler function with just two vectors: $f(\mathbf{k}_{w_i}, \mathbf{q}_{w_j})$. Specifically, the word2vec model proposes to use the *softmax* function of the dot product $\mathbf{k} \cdot \mathbf{q} = \mathbf{k}^{\mathsf{T}} \mathbf{q}$ between the two vectors:

$$f(\mathbf{k}_{w_t}, \mathbf{q}_{w_c}) = \frac{\exp(\mathbf{k}_{w_t} \cdot \mathbf{q}_{w_c})}{\sum_i \exp(\mathbf{k}_{w_i} \cdot \mathbf{q}_{w_c})}. \qquad (26.5)$$

---

[4] They are analogous to the query and key vectors in the self-attention mechanism used in Transformer models like BERT and GPT [277].

Then,

$$\Pr(w_1, \ldots, w_T) \simeq \prod_t \prod_{c \in C_t} \Pr(w_t \mid w_c)$$

$$= \prod_t \prod_{c \in C_t} \frac{\exp(\mathbf{k}_{w_t} \cdot \mathbf{q}_{w_c})}{\sum_i \exp(\mathbf{k}_{w_i} \cdot \mathbf{q}_{w_c})}, \tag{26.6}$$

where C is the set of words that are considered as context words (usually $w_{t-1}, \ldots, w_{t+l}$ not including w_t, with l being the size of the context window). The skip-gram model tries to maximize the log probability $\log \Pr(w_t \mid w_i)$.

Hierarchical softmax and negative sampling   Once we can compute $\Pr(w_t \mid w_c)$, we can initialize every word vector randomly and then go through actual natural sentences to generate context–target word pairs. For every context–target word pair from our data, we can estimate $\Pr(w_t \mid w_c)$ and use the backpropagation algorithm to learn the vectors. All standard stuff at this point. Yet, this is computationally challenging because we have to compute $\sum_i \exp(\mathbf{k}_{w_i} \cdot \mathbf{q}_{w_c})$ every time for all possible words. The word2vec model proposes another simplifying innovation here. Instead of directly computing this summation, the authors suggest two methods to drastically reduce the computational complexity: *hierarchical softmax* and *negative sampling* (NS). The hierarchical softmax method is a tree-based data structure that allows us to compute the summation in $\log(n)$ time, where n is the number of words. Although it is a very clever algorithm, it is not as popular as NS. NS is a simple idea that belongs to the class of methods called *"contrastive learning,"* which suggests that, instead of computing the probability directly, we can solve *another problem*, that of distinguishing the actual word w_t from a small set of randomly sampled words. If our representation and conditional probability function can distinguish the actual answer very well from random noise, then we can argue that the representation is good (recall the basic idea of language models). Formally, we can write the NS objective as

$$\log \sigma(\mathbf{k}_{w_t} \cdot \mathbf{q}_{w_c}) + \sum_{i=1}^b \mathbb{E}_{w_i \sim \Pr_n(w)} \left[\log \sigma(-\mathbf{k}_{w_i} \cdot \mathbf{q}_{w_c}) \right], \tag{26.7}$$

where $\sigma(x) = 1/(1+e^{-x})$ is the sigmoid function and b is the number of negative pairs we sample from the "noise" or "negative" distribution $\Pr_n$. (Mikolov et al. [307] used $\Pr_n(w) \propto n_w^{3/4}$, where n_w is the (unigram) count of occurrences of word w.) This is called the *"negative sampling"* model or *skip-gram negative sampling*. We seek vectors that align the word and context (first term) and oppose the negative word and context (second term; note the minus sign on the dot product). It allows us to learn by sampling a few random words rather than computing the softmax over all possible words.

26.3   From writing to walking: embedding networks

What does all this have to do with networks? Once word2vec achieved great success at natural language processing tasks, particularly at capturing *analogical relationships* between words, many researchers recognized the generalizability of the word2vec model and the idea of language modeling itself. Note that in the word2vec model, all we need is a fixed vocabulary[5] and lots of example sentences, where a sentence is simply a sequence of "words" in the vocabulary. There isn't, however, any restriction about the nature of the "words." Any set of entities where we can find their natural *sequences* can be considered as "words" and "sentences."

Not surprisingly, networks were one such generalization. Networks consist of nodes and they can be considered as "words." How can we get the natural sequences of nodes? One place to start is a *random walk*. If we perform a random walk on a network, we will have a sequence of nodes that organically captures the structural information of the network. If two nodes are close in a network, they will likely appear nearby in more random walk trajectories than another pair of nodes that are far apart. If two nodes share many neighbors together, they are more likely to co-appear in random walk trajectories than another pair of nodes that do not share any neighbors.

The first model that applies this line of thinking is "DeepWalk" [369]. The idea is exactly as laid out above. We generate many random walks from a network and then feed them as "sentences" to the word2vec model. Soon, another model called "node2vec" [193] was published with a similar idea, but an additional twist of using a biased random walk. Node2vec argued that we can modulate the nature of the random walks to obtain different representations that focus on different aspects of the network structure. For instance, if we want to capture the *community structure* of a network, we can bias the random walk so that it is more likely to stay in the same community (like BFS); if we want to capture the *hierarchical structure* of a network, we can bias the random walk so that it explores the network more (like depth-first search). These two models became foundational for many following models that adopt the paradigm of language models.

There was an interesting, subtle difference in the implementation of DeepWalk and node2vec. As we discussed, the softmax function is difficult to calculate and word2vec proposed using either hierarchical softmax or negative sampling to speed it up. DeepWalk's implementation adopted hierarchical softmax while node2vec used negative sampling. It turned out that this choice was actually an important one that changed how the two behave and perform.

The hidden bias of negative sampling Analysis has shown that negative sampling has an implicit bias [325, 249]. To see this, we need to first look at a similar, yet unbiased model called *"noise contrastive estimation"* (NCE) [197]. NCE is a general contrastive estimator that allows us to estimate a probability model $\Pr_m(x)$ of the following form:

$$\Pr_m(x) = \frac{f(x;\theta)}{\sum_{x' \in X} f(x';\theta)}, \tag{26.8}$$

where f is a non-negative function of x and θ is a parameter vector. The word2vec model can be considered a special case of NCE where $f(x) = \exp(x)$ and $x = \mathbf{k}_i^\top \mathbf{q}_j$. NCE tries to solve the same logistic regression problem as the negative sampling model, but

---

[5] How about new words? It is still possible to *inductively* learn the vector representations of new words based on those of existing words.

using Bayesian inference. Given one positive example and b randomly sampled negative examples, we take as prior probabilities for the positive and negative samples:

$$\Pr(Y_j = 1) = \frac{1}{b+1}, \quad \Pr(Y_j = 0) = \frac{b}{b+1}. \tag{26.9}$$

Here, the positive example is sampled from $\Pr_m(j)$ and negative examples are sampled from a noise distribution $p_0(j)$,

$$\Pr(j \mid Y_j = 1) = \Pr_m(\mathbf{q}_i \cdot \mathbf{k}_j), \quad \Pr(j \mid Y_j = 0) = p_0(j). \tag{26.10}$$

Based on Bayes' rule, the posterior probability of the positive example is

$$\Pr_{\text{NCE}}(Y_j = 1 \mid j) = \frac{\Pr(j \mid Y_j = 1) \Pr(Y_j = 1)}{\sum_{y \in \{0,1\}} \Pr(j \mid Y_j = y) \Pr(Y_j = y)}$$

$$= \frac{\Pr_m(\mathbf{q}_i \cdot \mathbf{k}_j)}{\Pr_m(\mathbf{q}_i \cdot \mathbf{k}_j) + b p_0(j)}, \tag{26.11}$$

which can be written in the form of a sigmoid function:

$$\Pr_{\text{NCE}}(Y_j = 1 \mid j) = \frac{1}{1 + b p_0(j)/\Pr_m(\mathbf{q}_i \cdot \mathbf{k}_j)}$$

$$= \frac{1}{1 + \exp\left[-\ln f(\mathbf{q}_i \cdot \mathbf{k}_j) + \ln p_0(j) + c\right]}. \tag{26.12}$$

The negative sampling estimator $\Pr_{\text{NS}}$ is similar to the NCE estimator, and can be written as

$$\Pr_{\text{NS}}(Y_j = 1 \mid j) = \frac{1}{1 + \exp(-\mathbf{q}_i \cdot \mathbf{k}_j)}$$

$$= \frac{1}{1 + \exp\left[-\left(\mathbf{q}_i \cdot \mathbf{k}_j + \ln p_0(j) + c\right) + \ln p_0(j) + c\right]}$$

$$= \frac{1}{1 + \exp\left[-\ln f'(\mathbf{q}_i \cdot \mathbf{k}_j) + \ln p_0(j) + c\right]}. \tag{26.13}$$

By comparing the two estimators, we can see that the negative sampling estimator is a special case of the NCE estimator with $f'(\mathbf{q}_i \cdot \mathbf{k}_j) = \exp(\mathbf{q}_i \cdot \mathbf{k}_j + \ln p_0(j) + c)$. This means that the negative sampling word2vec model is an unbiased estimator for the following probability model:

$$\Pr_{\text{w2v-NS}}(j \mid i) = \frac{p_0(j) \exp(\mathbf{q}_i \cdot \mathbf{k}_j)}{\sum_{j'} p_0(j') \exp(\mathbf{q}_i \cdot \mathbf{k}_{j'})}. \tag{26.14}$$

In other words, the negative sampling model of the word2vec model is a biased estimator of the original word2vec model, but an unbiased estimator of a modified model where the word similarity represents the *deviation* from $p_0(j)$, or the information about the words that is *not* captured by the noise distribution, which is the frequency of each word.

The implicit bias of the negative sampling model has a profound impact when applied to random walks on networks. The random walk is naturally biased towards the nodes with higher degrees, because whenever we sample a *neighbor* we preferentially sample a node with higher degree (the friendship paradox; Ch. 21). This degree bias is, however, captured by the noise distribution $p_0(j)$. Because of this, the resulting embedding of the negative sampling model captures the deviation from what is expected based on the degree alone. In other words, the bias of the negative sampling model exactly *negates* the bias of the random walk, producing embedding vectors that are free of the degree bias!

Although DeepWalk and node2vec methods look like more or less the same method when we don't apply any biased random walk for node2vec, the subtle difference in the choice of estimation method in their implementations has a profound impact on the resulting embedding vectors, usually producing much better results for node2vec.

Other embedding methods

There exist a plethora of other methods for graph embedding and it will be impossible to cover all of them in this chapter. Let us mention a few of the most popular methods with unique ideas.

LINE (Large-scale Information Network Embedding) [453], a simpler special case of DeepWalk, aims to encode the "proximity" between nodes into a dense embedding. LINE adopts the ideas of word2vec and seeks to model from random walks the probability for a directed edge $\Pr(j \mid i) = \exp(\mathbf{u}_j \cdot \mathbf{v}_i)/\sum_{s=1}^{N} \exp(\mathbf{u}_s \cdot \mathbf{v}_i)$, where $\mathbf{v}_i$ and $\mathbf{u}_j$ are node vectors and context vectors, respectively. Crucially, the same node gets different vectors depending on whether it is treated as a context for a walk or the target of the walk itself. LINE's overall objective is to learn the vectors which maximize $\sum_{i,j \in E} A_{ij} \log p(j \mid i)$. It also adopts NS, leading to an objective function for each edge i, j that mirrors Eq. (26.7) with $\mathbf{k} \rightarrow \mathbf{u}$ and $\mathbf{q} \rightarrow \mathbf{v}$. The negative distribution $\Pr_n(v) = k_v^a/\sum_{v'} k_{v'}^a$,[6] accounts for the overall degree distribution of the network.

The ComplEx method [463] shows success at link prediction by using a matrix factorization technique similar to many others (such as Laplacian Eigenmaps), but with the twist of allowing for complex-valued vectors as the representations. Among other motivations, directed networks lead to non-symmetric matrices, a problem we mostly avoided in Ch. 25, and we may encounter complex eigenvalues and eigenvectors. Embracing complex values and using this for link prediction was very successful.

ComplEx was tailored for *knowledge graphs* (Ch. 27) as are many other methods. In a knowledge graph, links are semantic triples ("Rome *IsA* City") that represent factual statements. Nodes are identified by words or phrases, and this brings in many ways to use word embeddings, either as the node embeddings or, more often, as *part* of learning the node embeddings. The TransE [65] method, for example, was designed around translation of words to learn embeddings; the factual relationships should hold across languages. Other knowledge graph methods include DistMult [501], which shares some similarities with TransE, and RESCAL [346], which uses tensor factorization instead of matrix factorization.

---

[6] The authors followed word2vec and used $a = 3/4$; $a = 1$ is also common.

It's common to treat the space we embed in as Euclidean but there is no need to do so. Indeed, *hyperbolic* spaces have properties of interest specifically for modeling networks. A metric defined on a hyperbolic geometry can incorporate heterogeneous degree distributions, transitive closure, and hierarchy more naturally than in Euclidean space [254, 60]. This has led to researchers pursuing embedding methods specific to hyperbolic spaces [347, 98].

Finally, a large class of methods called graph neural networks incorporate embedding as part of their function. We discussed these briefly in Sec. 16.7.3 and we'll return to them in Sec. 26.5.

26.4 Embedding as matrix factorization

Soon after word2vec emerged, researchers seeking to understand what it was calculating showed that it was implicitly performing a matrix factorization [269]. This is powerful to know. Matrix factorization is very useful for data analysis, as we see whenever we use singular value decomposition (SVD), and many other embedding approaches use it explicitly. Indeed, in the context of natural language processing, SVD powers latent semantic analysis (LSA) a classical NLP technique [261]. In a way, it's both surprising (because it looks so different) and not surprising that the more advanced word2vec method does something similar.

Let's derive the factorization occurring in word2vec. Then we'll discuss some network-specific embedding methods in this context.

Negative sampling as implicit matrix factorization

First, we summarize an influential discovery by Levy and Goldberg [269]: the sampling strategy used by word2vec (and adopted for networks by DeepWalk and its descendents) is implicitly factorizing a matrix **M**.

Recall that a language model seeks to understand the co-occurrence between words $w \in V_W$ and contexts $c \in V_C$, the surrounding words; for a word w_i, the surrounding L-sized context is $w_{i-L}, \ldots, w_{i-1}, w_{i+1}, \ldots, w_{i+L}$. Let D be the multiset of observed word–context pairs and use $\#(w, c)$ to denote the number of occurrences of pair $(w, c) \in D$. Marginalizing gives counts for w and c, $\#(w) = \sum_{c' \in V_C} \#(w, c')$ and $\#(c) = \sum_{w' \in V_W} \#(w', c)$, respectively. Our embedding goal is to find vector representations for words and contexts. Let $\mathbf{w} \in \mathbb{R}^d$ be the vector representing word $w \in V_W$ and likewise $\mathbf{c} \in \mathbb{R}^d$ for context $c \in V_C$, where d is the embedding dimension. (Previously we used **k** and **q**.) Generally only the word vectors are used for subsequent NLP tasks, but both are necessary for optimization.

From Eq. (26.7), the negative sampling objective is

$$
\begin{aligned}
\ell &= \sum_{w \in V_W} \sum_{c \in V_C} \#(w, c) \left(\log \sigma(\mathbf{w} \cdot \mathbf{c}) + b \, \mathbb{E}_{c_N \sim \mathrm{Pr}_D} \left[\log \sigma \left(-\mathbf{w} \cdot \mathbf{c}_N \right) \right] \right) \\
&= \sum_{w \in V_W} \sum_{c \in V_C} \#(w, c) \log \sigma(\mathbf{w} \cdot \mathbf{c}) + \sum_{w \in V_W} \#(w) \left(b \, \mathbb{E}_{c_N \sim \mathrm{Pr}_D} \left[\log \sigma \left(-\mathbf{w} \cdot \mathbf{c}_N \right) \right] \right),
\end{aligned}
$$

$$(26.15)$$

where b is the number of negative samples and c_N is the sampled context drawn from the empirical distribution $\Pr_D(c) := \#(c)/|D|$.[7] Next, pull the true context out of the NS expectation:

$$\mathbb{E}_{c_N \sim \Pr_D} [\log \sigma(-\mathbf{w} \cdot \mathbf{c}_N)] = \sum_{c_N \in V_C} \frac{\#(c_N)}{|D|} \log \sigma(-\mathbf{w} \cdot \mathbf{c}_N)$$

$$= \frac{\#(c)}{|D|} \log \sigma(-\mathbf{w} \cdot \mathbf{c}) + \sum_{c_N \in V_C \setminus \{c\}} \frac{\#(c_N)}{|D|} \sigma(-\mathbf{w} \cdot \mathbf{c}_N). \tag{26.16}$$

For sufficiently large d, we can assume each product $\mathbf{w} \cdot \mathbf{c}$ takes on a value independently of the others, letting us treat ℓ as a function of independent $\mathbf{w} \cdot \mathbf{c}$ terms. Using this and Eq. (26.16), the term specific to the pair (w, c) is

$$\ell(w, c) = \#(w, c) \log \sigma(\mathbf{w} \cdot \mathbf{c}) + b \, \#(w) \frac{\#(c)}{|D|} \log \sigma(-\mathbf{w} \cdot \mathbf{c}). \tag{26.17}$$

Because we seek to optimize this objective, we find the partial derivative with respect to $x := \mathbf{w} \cdot \mathbf{c}$:

$$\frac{\partial \ell}{\partial x} = \#(w, c) \, \sigma(-x) - b \, \#(w) \frac{\#(c)}{|D|} \sigma(x). \tag{26.18}$$

Simplifying and setting equal to zero gives

$$e^{2x} - \left(\frac{\#(w, c)}{b \, \#(w) \frac{\#(c)}{|D|}} - 1 \right) e^x - \frac{\#(w, c)}{b \, \#(w) \frac{\#(c)}{|D|}} = 0. \tag{26.19}$$

Letting $y := e^x$, this equation becomes a quadratic of y, which has two solutions. The first, $y = -1$, is invalid (since $e^x > 0$) while the second (using $\mathbf{w} \cdot \mathbf{c} = \log(y)$) is

$$\mathbf{w} \cdot \mathbf{c} = \log \left(\frac{\#(w, c)}{b \, \#(w) \frac{\#(c)}{|D|}} \right) = \log \left(\frac{\#(w, c)|D|}{\#(w)\#(c)} \right) - \log b. \tag{26.20}$$

We can now see better what is happening. The expression

$$\log \left(\frac{\frac{\#(w, c)}{|D|}}{\frac{\#(w)}{|D|} \frac{\#(c)}{|D|}} \right) = \log \left(\frac{\#(w, c)|D|}{\#(w)\#(c)} \right) \tag{26.21}$$

is the *pointwise mutual information* (PMI) for (w, c), estimated from the corpus D. It tells us how strongly associated w and c are by comparing their joint distribution to the joint distribution if they were independent. And lastly, the matrix being factorized by NS, since this is equal to a dot product for each term, has elements defined by the PMI: $M_{ij} = PMI(w_i, c_j) - \log b$. For $b > 1$, we can think of this as a *shifted* PMI matrix.

---

[7] The original word2vec and DeepWalk models drew negative contexts from $\propto \#(c)^{3/4}$ instead of $\propto \#(c)$. This difference does not substantially change our results [269] and avoiding the 3/4 exponent also lets us make clear graph-theoretic connections we'll use shortly.

However, we assumed every $\mathbf{w}_i \cdot \mathbf{c}_j$ was independent. If this is not true, the loss for a given pair (Eq. (26.17)) depends on the observed counts of the pair together ($\#(w, c)$) versus the expected number of negative samples ($b\,\#(w)\#(c)/|D|$) and we can instead think of NS as performing a *weighted* factorization, where the solution is biased in favor of more frequent pairs, the same bias discussed above.

Having a better understanding of what NS is doing "behind-the-scenes" motivates spectral methods, the original approach to embedding, which can be more computationally efficient than using stochastic gradient descent and more theoretically tractable. Indeed, Levy and Goldberg [269] compare NS to SVD on the shifted PPMI[8] and show that it achieves better optimization of the loss function than NS. That said, NS performed better at subsequent linguistic tasks, such as finding word analogies, probably due to the weighted factorizing, as PMI solutions are known to be over-affected by rare observations. It is good to understand what NS is doing here because it motivates in NLP the search for better weighted factorization methods [269].

Network embedding as factorization

Network embedding methods such as DeepWalk and node2vec follow the spirit of word2vec closely, so it stands to reason that they too are implicitly performing matrix factorization. Indeed, Qiu et al. [382] show that this is exactly the case.

For brevity, we focus our discussion on LINE [453], which is a special case of DeepWalk. As mentioned, LINE follows the word2vec model and uses NS to learn vector representations that help predict node associations during random walks. The objective function they derive is nearly identical to that of word2vec:

$$\ell = \sum_{i \in V} \sum_{j \in V} A_{ij} \left(\log \sigma(\mathbf{v}_i \cdot \mathbf{u}_j) + b\mathbb{E}_{s \sim \mathrm{Pr}_n} \left[\log \sigma(-\mathbf{v}_i \cdot \mathbf{u}_s) \right] \right), \qquad (26.22)$$

where, as discussed in Sec. 26.4, for simplicity, we now take $\mathrm{Pr}_n(v) = k_v/2M$. (For a weighted network, k_v and M are the total edge weights for a node v and the entire graph, respectively.) Our previous analysis for word2vec transfers over almost exactly given the similar objective function, meaning that LINE also performs an implicit matrix factorization [382]:

$$\mathbf{v}_i \cdot \mathbf{u}_j = \log \left(\frac{2M A_{ij}}{b k_i k_j} \right) \qquad (26.23)$$

or, in matrix form,

$$\mathbf{V}^\top \mathbf{U} = \log(2M\mathbf{D}^{-1}\mathbf{A}\mathbf{D}^{-1}) - \log b, \qquad (26.24)$$

where $\mathbf{V}$ and $\mathbf{U}$ contain the node and context vectors as columns, $\mathbf{D}$ is the diagonal degree matrix, and the log works element-wise. Notice the resemblance between the matrix $\mathbf{D}^{-1}\mathbf{A}\mathbf{D}^{-1}$ and some of the matrices we encountered in Ch. 25.

LINE is a special case of DeepWalk, and Qiu et al. [382] show that DeepWalk factorizes $\log \left(2M \left(L^{-1} \sum_{r=1}^{L} \left(\mathbf{D}^{-1}\mathbf{A} \right)^r \right) \mathbf{D}^{-1} \right) - \log b$. (We recover LINE when the

---

[8] The *positive* PMI matrix is defined as $PPMI = \max(PMI, 0)$. This is often used in NLP because the PMI matrix estimated from a corpus will have entries $\log 0 = -\infty$ for any pairs not observed in the corpus. The PPMI ensures the matrix is well-defined and sparse. (Other fixes, such as Dirichlet smoothing, will not ensure sparsity.)

context window $L = 1$.) Qiu et al. [382] also derive the implicit factorization for node2vec and use these results to demonstrate a superior method called NetMF that explicitly factorizes these matrices using SVD instead of implicitly using NS.

26.5 Graph neural networks

The 2010s conclusively demonstrated that neural networks had finally fulfilled their promise, addressing and in some cases even solving longstanding problems of computer vision, speech recognition, and natural language processing. And now neural networks are also making inroads with network data (Ch. 16).

> ℹ️ We're going to start using "graphs" more consistently when referring to our data to avoid any confusion with the neural networks.

Just as neural network methods have wildly succeeded in NLP, so have neural methods proliferated for graphs. These networks also learn representations—embeddings— but often also seek to incorporate node and link attributes (Ch. 9) into their representations. Using these as features can enable more and better predictive models, including *inductive* models that work on entirely unseen graphs.

The challenge with graphs, unlike other forms of data such as written text or spoken audio, is that we must be *permutation-invariant*: shuffling the neighborhood of a node should change nothing, but shuffling the context of a word should change (or destroy) the meaning of the writing. This need for permutation invariance extends to graph isomorphism (Sec. 12.6), and two isomorphic graphs (including attributes) should ideally lead to the same representations; we return to this point later.

Recall the basic feedforward neural network propagates data through a collection of layers where linear combinations of values at one layer are passed through a nonlinear activation function when arriving at the next layer, that is,

$$\mathbf{a}^{(l)} = \mathbf{W}^{(l-1)}\mathbf{h}^{(l-1)}, \quad \mathbf{h}^{(l)} = \sigma(\mathbf{a}^{(l)}), \quad (26.25)$$

where $\mathbf{W}^{(l-1)}$ represents the matrix of weights connecting layers $l - 1$ and l, $\mathbf{h}^{(l)}$ is the vector of activations at layer l, and $\sigma(\cdot)$ is a nonlinear function, often a sigmoid but not always. Bias units for each layer, which act like the intercept in a linear model, are absorbed into the weight matrices. Our input data $\mathbf{x}$ serve as the original activations, $\mathbf{h}^{(0)} = \mathbf{x}$. The parameters $\{\mathbf{W}\}$ of the network are learned with optimization, often by "backpropagating" errors on training data using stochastic gradient descent and possibly various regularization techniques. Again, for practitioners, all standard stuff.

Modern graph neural networks (GNNs) have coalesced around a framework of learning functions that iteratively update a node's representation by aggregating it with the representations of its neighbors in the graph, including possibly higher-order neighbors such as next-nearest neighbors. Let $\mathbf{x}_i$ be the attribute vector for node i and $\mathbf{h}_i^{(l)}$ be the representation (or activation) of i at the lth layer of the network, with $\mathbf{h}_i^{(0)} = \mathbf{x}_i$ (we feed the original features into the neural network). (Note in principle the dimensionality of $\mathbf{h}$ may be different for different layers.) Broadly, a GNN's lth iteration for node i comes from two learned functions:

$$\mathbf{a}_i^{(l)} = \text{AGGREGATE}^{(l)}\left(\left\{\mathbf{h}_j^{(l-1)} \,\middle|\, j \in N_i\right\}\right), \tag{26.26}$$

$$\mathbf{h}_i^{(l)} = \text{COMBINE}^{(l)}\left(\mathbf{h}_i^{(l-1)}, \mathbf{a}_i^{(l)}\right), \tag{26.27}$$

where N_i are the (possibly higher-order) neighbors of i. Choices for AGGREGATE include mean, max, and sum, which can encapsulate many different GNN architectures, while vector concatenation is often used for COMBINE. Lastly, when seeking a representation $\mathbf{h}_G$ for the entire graph, a permutation-invariant READOUT function is applied to the final iteration L:

$$\mathbf{h}_G = \text{READOUT}\left(\left\{\mathbf{h}_i^{(L)} \,\middle|\, i \in V\right\}\right). \tag{26.28}$$

Some authors have also pursued more complex pooling READOUT functions [21] but a simple sum is often used.

Many architectures are described by these functions, including graph convolutional networks (GCNs) that explicitly feed the graph's adjacency matrix into the neural network. For example, Kipf and Welling [242] use the following layer-wise propagation step:

$$\mathbf{H}^{(l)} = \sigma\left(\tilde{\mathbf{D}}^{-1/2}(\mathbf{I}_N + \mathbf{A})\tilde{\mathbf{D}}^{-1/2}\mathbf{H}^{(l-1)}\mathbf{W}^{(l-1)}\right), \tag{26.29}$$

where $\mathbf{H}^{(l)} \in \mathbb{R}^{N \times d}$ are the layer representations, $\mathbf{W}^{(l)}$ are the corresponding network weights, and $\sigma(\cdot)$ is a nonlinear (not necessarily sigmoid) activation function. Here the adjacency matrix has been augmented with self-loops ($\mathbf{I}_N$) which are included in the diagonal rescaling matrices $[\tilde{\mathbf{D}}]_{ii} = k_i + 1$. (Notice the similarities with the Laplacian matrices we saw in Ch. 25.) By "hitting" the output of each layer of the neural network with the adjacency matrix, the network is forced into accounting for the graph structure, a form of *masking*. However, by explicitly including the adjacency matrix, the GCN is unable to accommodate an unseen network structure, unlike follow-up methods like GraphSAGE [200] whose aggregate methods only sample features from neighborhoods (usually first and second neighbors) and thus can handle novel graphs and perform *inductive* learning.

One of the problems faced with NLP methods and the network methods they inspire is learning long-range relationships. The solutions[9] devised for language, the attention mechanism and transformers, have been wildly successful, and have been adopted for graph structured data. *Graph attention networks* [475] (GATs; here we described what is called GATv2 [77]) learn a scoring function $e : \mathbb{R}^d \times \mathbb{R}^d \to \mathbb{R}$ for every edge i, j which captures how important the features of neighbor j are to node i (we omit the layer index l):

$$e(\mathbf{h}_i, \mathbf{h}_j) = \mathbf{a}^\top \text{LeakyReLU}\left(\mathbf{W}\left[\mathbf{h}_i \| \mathbf{h}_j\right]\right), \tag{26.30}$$

where $\mathbf{a}$ and $\mathbf{W}$ are learned parameters whose dimensionalities depends on the network architecture, $[\cdot\|\cdot]$ denotes concatenation, $\text{LeakyReLU}(y) = y$ if $y > 0$, βy otherwise,[10]

---

[9] Recurrent neural networks (RNNs) have long struggled with this, with LSTMs being a solution. Attention works even better, allowing networks to learn long-range associations without the long-range computational paths of RNNs, speeding up training and avoiding vanishing and exploding gradients.

[10] A common choice in GAT is $\beta = 0.2$.

and bias units are again omitted for brevity. Attention scores come from a softmax of the edge scores,

$$\alpha_{ij} = \frac{\exp(e(\mathbf{h}_i, \mathbf{h}_j))}{\sum_{j' \in N_i} \exp(e(\mathbf{h}_i, \mathbf{h}_{j'}))}, \tag{26.31}$$

and the next layer's representations are computed from an attentive weighted average,

$$\mathbf{h}'_i = \sigma \left(\sum_{j \in N_i} \alpha_{ij} \mathbf{W} \mathbf{h}_j \right). \tag{26.32}$$

For simplicity we described a single attention mechanism, but common practice is to learn K parallel attention "heads" which are then concatenated or averaged. Many other important details such as hyperparameter values and choice of regularizer will depend on the researcher and the task at hand, but GAT networks have proven successful at transductive and inductive prediction tasks.

Expressiveness and isomorphism

As we mentioned, neural networks intended for graph-structured data have to deal with permutation invariance in ways that networks working with images or other data do not—the set of neighboring nodes in a graph is order-invariant, the set of neighboring pixels in an image is not. The *graph isomorphism* (GI) problem, understanding whether two graphs are the same regardless of how we order or label the nodes, is thus highly relevant to learning GNNs: GNNs trained on two apparently different but actually isomorphic graphs should converge on the same learned functions. We often need to ask whether the GNN architecture is expressive enough to accomplish this.

GI is an interesting computational problem. For many years, it had its own location, called "GI," in the hierarchy of complexity classes, not in P and not in NP. Recently, it has been shown to be in quasi-polynomial time, with complexity $2^{O((\log N)^c)}$, worse than polynomial time but not as bad as exponential [18, 19]. An influential algorithm that tests for isomorphism is called the *Weisfeiler–Lehman* (WL) algorithm [487]. Roughly speaking, WL works by "coloring" nodes of the graph, then updating the coloring of a node using the colors of its neighbors. These colors are found with hash functions that ensure different colorings are represented differently.

The color-passing idea of WL is very closely connected to what GNNs do, passing the node representations as messages during learning. Indeed, many researchers have used this connection as the foundation to understand what GNNs can and cannot do [293, 320, 497, 32]. For example, many of the GNNs we have encountered above cannot distinguish graphs that WL can distinguish, meaning they are less expressive or else powerful. Work continues [32] to improve upon this.

Open problems and the future

GNNs struggle in several key areas. One is computational complexity. Often the methods require significant training time and memory. Many methods are currently limited to small networks or small subgraphs of larger networks. As an extreme example, the

Graphormer, a recently introduced, successful application of the Transformer natural language network to GNNs, is intended for networks of at most only dozens of nodes. GNNs currently struggle with larger and more complex graphs and attributes, especially heterogeneous attributes.

Because GNNs work by passing updates between nodes, they suffer when it comes to long-range signals. Skip connections, which allow for information to bypass some layer updates, can be a remedy, but the problem remains substantial. Bottlenecks (cf. Sec. 25.5) often prevent appropriate learning [12]. This leads to "over-smoothing" where the network fails to retain differences in representations between obviously different nodes—as we move outward in a graph, the exponentially growing number of nodes lead to information that must be squashed by the network into a representation of fixed dimensionality. This remains a fundamental challenge. Natural language networks struggled with similar problems for decades, before recent progress led away from recurrent networks to attention and Transformer models. While the problems with graphs are different, the hope is that, like with natural language, the problems can be overcome.

Lastly, robustness and interpretability are concerns for users of GNNs just as with other neural network models. Adversarial data and even noise (Ch. 10) can be a particular problem for GNNs [119, 514], leading us to worry about our robustness. Likewise, in practice, GNNs can be sensitive to choices of architecture and hyperparameter settings. Even design choices, like how to incorporate graph features such as centrality measures or shortest path lengths into the GNN, are often approached in an ad hoc manner. And these models are far over to one side of the interpretability–flexibility curve (Chs. 3 and 16): like many complex machine learning methods, they are black boxes that we can struggle to understand.

26.6   Summary

Embedding network nodes and edges is big business. Embeddings should be compact, continuous and dense, and these properties allow for novel ways to work with network data. We've already encountered ways to embed networks, such as the spectral methods of Ch. 25 but more and better embedding techniques continue to be introduced.

Machine learning and embedding are closely aligned. Translating network elements to embedding vectors and sending those vectors as features to a subsequent predictive model often leads to a simpler, more performant model than designing a model that works directly with the network. Embeddings help with network learning tasks, from node classification to link prediction. We can even embed entire networks and then use models to summarize and compare networks.

But the relationship is also a two-way street. Not only does machine learning benefit from embeddings, but embeddings benefit from machine learning. Inspired by the incredible recent progress with natural language data, embeddings *created* by predictive models are becoming increasingly important and useful. Often these embeddings are produced by neural networks of various flavors.

Neural networks are a major area of machine learning and graph neural networks are currently a very fast moving area of research. Many types exist and many more are sure to come. Currently, they work well at many tasks but significant challenges remain. Mathematically, our understanding of these neural networks is still nascent. Computationally, they are very costly and often struggle to scale up to very large networks. Yet neural networks have shown great success in many previously intractable settings, and it's likely they'll make their mark on graphs as well.

Bibliographic remarks

Although dense vector representations have long been pursued in NLP [261, 51, 357] and network analysis (Ch. 25), word2vec [307] has had a large influence on natural language processing. DeepWalk [369] brought this idea directly to networks (graphs), leading to an explosion of interest in learned embeddings; see Goyal and Ferrara [188] and Xu [498] for recent reviews.

The connections between random walk embeddings, pioneered by Perozzi et al. [369], and matrix factorization were noted by Qiu et al. [382] following the word2vec analysis of Levy and Goldberg [269]. Kojaku et al. [249] noted the mechanism that negative sampling introduced (and effectively removed) which led to a general way (residual2vec) to remove the bias from graph embeddings.

Graph neural networks, proposed by Scarselli et al. [415], are now a very large and fast-evolving area. See Zhou et al. [508] and Chami et al. [99] for recent overviews. Grohe and Schweitzer [192] give a recent review of the graph isomorphism, one of the tools we use to understand the representational power of different graph machine learning methods.

Exercises

26.1 (**Focal network**) Use a Laplacian Eigenmap to embed the Zachary Karate Club. In two dimensions, visualize the embedding in a meaningful manner and interpret what you see in terms of the club's known community structure.

26.2 How might graph neural networks help with some of the problems discussed in Sec. 9.5?

26.3 What are AGGREGATE($\cdot$) and COMBINE($\cdot$) for the basic GCN (Eq. (26.29))?

26.4 (**Focal network**) Implement the basic GCN (Eq. (26.29)), making some choices for hyperparameters and other details as necessary. Apply to the Malawi Sociometer Network and use its embeddings as input to a logistic regression link prediction method. Validate your predictive model as per Ex. 16.5.

26.5 (**Focal network**) Same as Ex. 26.4 but try changing the propagation mechanism to use the graph Laplacian or other variation. Can you improve your predictions?

26.6 *The devil is in the details.* You may have noticed that our treatments of different graph neural networks omitted many important, practical details such as hyperparameter values, fitting methods, choice of regularization technique (if any), validation procedures, and more.

(a) Find the original paper introducing one of the specific GNNs we described, such as GAT, and identify all the specific implementation details employed in the paper. Use the paper's references if necessary. Describe all the details.

(b) Having determined all this information, are you able to reproduce exactly the original study? If not, what is missing?

26.7 **(Focal network)** Gather GO terms (Sec. 9.1) as node attributes for HuRI. What opportunities are there to predict new or missing GO terms? If the attributes support it, build a GNN-based node classifier to predict GO terms. Validate using held-out attributes or other means and interpret your performance.

Chapter 27

Big data and scalability

Networks get big. Really big. In this chapter, we discuss some of the challenges you face and solutions you can try when scaling up to massive networks. These range from implementation details to new algorithms and strategies to reduce the burden of such big data.

27.1 Do you *really* have big data?

Before talking about how to scale the network analysis, the very first thing to remember is that chances are *your network is probably not that big* to necessitate the following practices and that you may really want to *avoid using them* unless it is absolutely necessary or the big data infrastructure is already set up for you. Most methods and practices that we will discuss here come with substantial overhead and extra cost, and you want to avoid prematurely jumping into a fancy, complex technology.

The first questions you need to ask are: what kinds of data do we need to address our question? Is our data really that *big*? Can we reduce the data enough so that we don't need "big data" technology? For instance, even if your network has many millions of nodes, you may be able to answer your particular question by carefully *sampling* a small fraction of nodes or by reducing the network with the methods explained in Ch. 10. You may even be able to address your question simply by using a more efficient library or more resource-aware programming language.

Remember: premature optimization is the root of all evil. Try to tackle your problems in the simplest ways possible. Only if that fails should you turn to some of the technologies and methods described here.

27.2 When networks become large

Computers are so powerful nowadays, with so much storage. We shouldn't have to worry about our data becoming too big to work with, right? Yes and no. Although many networks are large, they are often still manageable with powerful computers and large

447

disks. They are not so massive that we need to turn to specialized storage systems and algorithms. On the other hand, there are areas where networks are absolutely massive, forcing us to worry about more efficient storage systems, ways to interact with the network efficiently, and using specialized algorithms for analysis.

In some sense, network data are simpler than other forms of data. Representing a network (Ch. 8) will not be as data-intensive as, say, storing all the videos on YouTube, or all the pictures shared on Instagram. However, if a network becomes dense, the set of links will become quadratic, meaning $M = O(N^2)$, as any node can in principle be linked to any other node. Fortunately, we almost never encounter real, large networks close to that level of density—sparsity saves us. But working with large networks requires data structures that may encounter such density, even if only locally (a fully connected subgraph for example). The storage of one YouTube video will never depend on the contents of other videos.[1]

Examples of large networks.   Three problem domains in particular stand out as focusing on massive networks:

The web graph The network of web pages connected by directed hyperlinks. This evolving network is the focus of web search engines from major companies like Google that spend considerable resources to index and re-index, or *crawl* the web. PageRank (Secs. 12.9 and 25.4) was introduced to tame the chaos of the web by giving a centrality measure to rank important web pages. A 2014 analysis of a Common Crawl[2] dataset found the web page graph contains 3.5 billion nodes and 128 billion links [306]. The public web graph is highly dynamic, however, with many pages coming and going, and estimating its size will depend on when and for how long the crawl occurred. A recent data release of the Common Crawl project, for instance, which crawled only from 26 January to 9 February 2023, contained an estimated 3.15 billion web pages and 1.3 billion new URLs, not seen in previous crawls.

Online social networks Major online platforms deal with huge amounts of social connections. For example, researchers at Facebook released a 2015 paper describing a graph storage system capable of handling up to 1 trillion edges [103]. Of course, such a graph will contain far more than a social network, since businesses and other organizations exist on such platforms. Further, nodes in the graph represent individual pieces of content such as posts, comments, and likes. Twitter, a less popular social platform, probably still comprises a graph structure of 10^8 nodes and 10^9 or more edges.

Knowledge graphs Some of the largest sources of network data are knowledge graphs. These networks, or *ontologies*, contain huge numbers of semantic *triples* meant to represent factual statements as links. For example, "Leonardo da Vinci is a painter" is the triple "Leonardo da Vinci" (subject), "IsA" (predicate), "painter"

---

[1] That said, there are lots of networks related to features on platforms like YouTube, such as the user subscriber lists and comment threads.

[2] https://commoncrawl.org

(object). The nodes are subjects and objects while the links can be categorized or arranged into layers based on the predicates. Such representations are meant to enable computational reasoning by enabling network algorithms to work with facts and statements. These can perhaps one day imbue AI systems with commonsense reasoning. As you can imagine, the set of all possible triples is large and very large knowledge graphs have been created. For example, DBpedia,[3] an open knowledge graph extracted from Wikimedia projects such as Wikipedia, contains over 21 billion triples [215]. Meanwhile, proprietary knowledge graphs, such as those created by Facebook or Google, are expected to be at least as large; a 2020 post from Google [450] described their knowledge graph as having "amassed over 500 billion facts about five billion entities" gathered from hundreds of online sources. Working with knowledge graphs needs significant computing infrastructure.

Problems and strategies

Big data can bring us back to the drawing board. We need scalable methods (see below). Often the solutions we have used so far take details for granted that are no longer applicable.

For example, the network can become so large it no longer fits into memory. If we have relied on, say, a Python dictionary to store a network as an adjacency set (Ch. 8), we have to rethink our data store. We can move to a database that works on disk, but this will make working with the network far slower. And we may need to be more efficient with how we represent the network, even to the level of choosing node identifiers more appropriately. We need methods to compress the graph.

Even if we can be space-efficient storing the graph, in principle it can get so big it won't fit on a single computer. We then need a *distributed system* where parts of the graph can be stored on different computers and we can efficiently access those different parts. Given how the network structure pulls together disparate regions of the network, this can get tricky. If many nodes stored on one machine are connected to many nodes on another machine, we are likely to require more data transmission between the machines as we work with the network, leading to wasted bandwidth and even reliability issues when synchronizing changes to the data.

Sometimes a network is so big relative to our compute resources that we have no chance of storing and reanalyzing the data. Or perhaps we could store the network but we don't have time to reanalyze the data—we need to finish the calculation in real-time for a customer-facing feature. (A practical example of a (nearly) real-time calculation is verifying a credit card transaction as non-fraudulent using a large knowledge graph.) Imagine a *stream of data* flying past, where we can only read it once. And maybe we can't even read it completely, only a subset of it. How can we implement algorithms on a graph stream? Even generating a random sample of nodes can be challenging.

Lastly, many network algorithms have high complexity. If we want to compute the diameter of a network, say, we need to run breadth-first search between every pair of nodes. That's $O(N^2)$, which is no longer feasible when we have a billion or more

---

[3] https://www.dbpedia.org

nodes. This challenge demands approximations. Can we estimate the diameter from a sample, lowering the cost but introducing a (hopefully acceptable) degree of error into our estimate? Even simpler calculations than the diameter are challenging: how can we efficiently estimate how many *distinct* nodes are present in a stream of dynamic graph events so large we cannot store the stream in memory? Approximations and local algorithms rule the day when it comes to massive networks.

27.3   What do we mean by "scalability"?

What does it mean for a network analysis algorithm or technique to be *scalable*? Scalability usually means it can handle large networks or "scale up" to that large scale. But whether a network is large-scale or not depends on many factors; there is not a clear dividing line between small-scale and large-scale. One of the major factors is the computer hardware being used. A network of a million nodes is probably considered as medium size today, but 20 or 30 years ago at the dawn of modern network science it was large. Earlier in time, it would be considered absolutely massive.

To skip past the details of hardware means, just like with any algorithmic analysis, we consider the computational complexity of the algorithm either in time (number of calculations) or space (amount of memory or storage). For our purposes, a scalable method must be cheaper than quadratic. An $O(N^2)$ or $O(M^2)$ algorithm will be inaccessible to us. Of course, this too depends on various other factors. An algorithm may have good average complexity but poor worse-case complexity. Likewise, one algorithm may have better scaling than another but worse constant factors, which may make a difference in practice.

In other words, roughly speaking, a network analysis tool is scalable when its complexity is sub-quadratic in the size of the network.[4]

Note that network size includes both nodes and links. An algorithm with complexity $O(N)$ or $O(N \log N)$ is usually scalable; algorithms with complexity $O(N^2)$ or $O(M^2)$ or $O(NM)$ are not.

Sometimes, an algorithm is scalable for some networks and not others. The most common cause is density. An algorithm can be very efficient when the network is sparse but slows down as it becomes dense. We can see this in the complexity because a dense network is defined as one where most edges exist; there are $N(N-1)/2 \sim N^2$ possible edges, so if most exist then $M = O(N^2)$. An algorithm that is scalable, for instance with complexity $O(N + M)$, will now be too costly: $O(N + M) \to O(N + N^2) = O(N^2)$. Such a method will not scale to a large, dense network.

27.4   Compressing, distributing, and streaming graphs

Here we discuss problems and strategies for operating with very large amounts of graph data. Keep in mind our earlier admonishment that hardware continues to advance and some of these problems can be solved not by complex data engineering or involved algorithms but just by brute-force hardware. A modern multi-core, shared-memory

---

[4] Of course, anything worse than polynomial is completely out the window when it comes to scalability.

machine can store graphs with billions of edges in memory. Compress, distribute, stream when you need to, not before: premature optimization is the root of all evil.

27.4.1 Compressing graphs

We have already seen in Ch. 8 how data structures such as adjacency sets can save space. We can store the network as an edgelist $(i, j_1), (i, j_2), \ldots$ but we are repeating the node i for each of i's edges. Instead, store i's neighbors altogether: $(i, j_1, j_2, \ldots), \ldots$

We can take this further. If each neighbor j_s is represented with an identifier, we want to use as few bits as possible to store those identifiers. Compression algorithms on text data, for instance, work by assigning smaller identifiers (fewer bits) to more common words or letters. We can do something similar for networks, giving high-degree nodes shorter identifiers. But networks give us more opportunities: we can exploit the connectivity patterns of the network to save more space.

Consider this subset of a directed adjacency set, written as a table (node 17 has degree 0) [62]:

⋮	⋮
15:	13, 15, 16, 17, 18, 19, 23, 24, 203, 315, 1034
16:	15, 16, 17, 22, 23, 24, 315, 316, 317, 3041
17:	
18:	13, 15, 16, 17, 50
⋮	⋮

Notice how often we have a consecutive run $n, n+1, \ldots$ of IDs? We can encode this information more efficiently by using *gaps* (also called *deltas*):

⋮	⋮
15:	3, 1, 0, 0, 0, 0, 3, 0, 178, 111, 718
16:	1, 0, 0, 4, 0, 0, 290, 0, 0, 2723
17:	
18:	9, 1, 0, 0, 32
⋮	⋮

Here an entry of zero in a neighbor list indicates that the node is one more than the preceding node. If $S(x) = (s_1, \ldots s_k)$ are the neighbors (successors) of node x, the gap representation stores[5] them as $(s_1 - x, s_2 - s_1 - 1, s_3 - s_2 - 1, \ldots, s_k - s_{k-1} - 1)$. In practice, of course, the benefit of gaps will depend on how nodes are identified, and often a reordering of the nodes needs to be computed.

Gaps exploits consecutivity. We can also exploit similarity. Two nodes i and j may have nearly identical adjacency lists, if their neighborhoods strongly overlap. We can

---

[5] All these gaps will be positive except possibly the first one ($s_1 - x$). If we are dealing with enough data that we care about this, we probably do not want to allocate space for storing the sign of an ID so we want to avoid negative entries. The solution is to encode the first element $y = s_1 - x$ using the map $v(y) = 2y$ if $y \geq 0$, $2|y| - 1$ otherwise.

exploit this by storing $S(i)$ and then, instead of $S(j)$, store a reference from j to i along with a list of differences between $S(j)$ and $S(i)$. It will require computation to find the differences, but the space savings can be substantial.

Many algorithmic questions still need to be addressed to combine these strategies together into a workable, large scale system. See our remarks for references with more details. For brevity, we discuss one more common compression strategy: virtual nodes.

Some network motifs (Sec. 12.6) are highly dense and regular. If they occur frequently, it is not worth storing every instance. Instead, we can remove all the links within the motif and replace it with a node that represents the entire motif, then link the nodes within the motif to this new virtual node. The savings can be substantial for completely dense subgraphs:

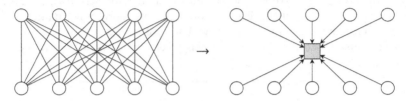

For a clique of n nodes, we go from n nodes and $n(n-1)/2$ edges to $n+1$ nodes and n edges. An efficient algorithm called virtual node mining was introduced by Buehrer and Chellapilla [81] to find those nodes.

Virtual nodes are one example of a grouping compression (and summarization) technique [278]. In general, these compression strategies, like all data compression algorithms, have a tradeoff: more space can be saved but at the expense of more computation. For very large networks, it can become costly to find all the redundancy possible, so we often settle for "good enough" compression.

Random data generally compress poorly and the same holds for graphs. Some graphs are inherently more compressible than others. The web graph in particular compresses surprisingly well. In part, this is because of redundancy in how links are set up. Many links on a web page repeat, for example, a navigation bar on every page within a site will lead to a highly redundant set of links. Social networks also show compressibility but generally less so [64].

27.4.2   Distributing and streaming graphs

For exceptionally large networks, it becomes untenable to store them in memory (even on disk) on a single machine. The network needs to be distributed to multiple machines. Computations on the network also need to be distributed. While we have techniques and standards for distributed computation, such as the MapReduce framework [124], they may not be suitable for graph-structured data.

For example, suppose we wish to compute shortest paths on a weighted graph. We can run Dijkstra's method but, because not all of the graph is on a single machine or single disk, the algorithm must be distributed across all the data machines. If we have distributed the graph poorly, such that many edges exist between nodes on different machines, we may find poor performance: the distributed computation will need to pass many messages between those different machines as the shortest paths are computed, and inter-machine communication is far slower than intra-machine communication.

Minimizing cross-talk between data machines is another motivation for graph partitioning (Sec. 25.5). We can use (scalable) methods to distribute subgraphs between our data machines such that few edges exist between machines. Many graph databases have algorithms for this *sharding* or *horizontal distribution* problem, but it remains a difficult feat of data engineering. Often, careful choices of schema for a graph database are needed to ensure queries of the data can run efficiently. One technique for improving efficiency is by incorporating overlap: allow the "boundary" of groups to exist on *multiple* machines. If chosen with care, this can lower significantly the need for inter-machine messaging, at the expense of more storage. Consistency is also an issue here (as it is always a concern with distributed data): if the data shards are changing after they have been distributed, we may run into scenarios where the boundaries duplicated across machines are no longer isomorphic. Errors may creep in.

Distributed computation engines A problem closely associated with distributing the graph is distributing computations on the graph. Frameworks such as MapReduce [124] and its successor implementations (Apache Spark, Beam, etc.) have served this purpose well for traditional relational data stores. But graphs bring distinct challenges as their connectivity patterns demand more from parallel calculations. (It is difficult to apply a divide-and-conquer parallelism strategy to a small-world network, for instance.) Graph-specific distributed computational models have been proposed, including Pregel [289, 103], GraphLab [280, 185, 281], and GraphX [496].

Beyond distributing graph data to multiple machines, we can consider the problem of *streaming graphs*. A data stream is a read-once sequence of data and we generally have no control over the order we see items from the sequence. The idea being that the data come by in such a high volume or velocity that computationally we are not able to store the data, or at best we only have memory capacity that is sublinear in the length of the data stream. Often we do not even know how long the stream will be.

We can imagine a graph data stream where each item in the stream is an edge in the graph, or perhaps a subgraph of some kind. What calculations can we perform when we are forced to interact with the graph in such a piecemeal manner [300]?

27.4.3 Sampling large streams

Reservoir sampling [476] is an interesting algorithm for sampling uniformly from a large stream of data, so large that you cannot store it in memory and you don't even know how big the stream is. Without knowing the number of observations n it becomes difficult to define the uniform sample probability $1/n$.

We wish to sample k items uniformly at random from a stream of items we observe one at a time. The length of the stream is unknown. We want to guarantee at any given sampling step, if the stream ends, the probability for each item to be sampled is the same. The reservoir sampling algorithm is quite simple:

1. Store the first k elements of the stream in a buffer (the reservoir).

2. Every time we look at an element $i > k$, we decide to keep it with probability k/i. If kept, select an item from within the buffer uniformly at random and replace it with the ith item.

Does this really work? We need to show that this sampling strategy leads to a uniform sample of the stream. Specifically, because we sampled k items, we need the probability that element i is sampled to be k/n for all i. We study two cases, $i \leq k$ and $i > k$:

- For $i \leq k$: i has already been selected. What is the probability it is not replaced? The probability that element $j > k$ does not replace i is 1 minus the probability j is selected and i is chosen as the replacement location in the reservoir, or

$$\Pr(j > k \text{ does not replace } i) = 1 - \frac{k}{j}\frac{1}{k} = \frac{j-1}{j}. \tag{27.1}$$

From this, the probability that *every* $k < j \leq n$ does not replace i is given by

$$\frac{k}{k+1}\frac{k+1}{k+2} \cdots \frac{n-1}{n} = \frac{k}{n}. \tag{27.2}$$

- For $i > k$: now we ask, what is the probability i is selected and then not subsequently replaced? Element i is selected with probability k/i. The probability it is not replaced later is, like before, $\frac{i}{i+1}\frac{i+1}{i+2} \cdots \frac{n-1}{n} = i/n$. Put together, the probability that i is selected and not replaced is k/n.

Both cases show the correct probability, and therefore we know reservoir sampling generates a uniform sample of the data stream. All while not knowing the value of n. Amazing!

One immediate application for our purposes is computing statistics like the assortativity of a graph stream. Use reservoir sampling to generate a sample of k edges, then compute the correlation between the attributes or degrees (assuming we can efficiently compute or estimate the degrees) of their constituent nodes. Note, of course, that sampling edges uniformly at random is not the same as sampling nodes uniformly at random (Chs. 12 and 21). And with further effort, reservoir sampling-like streaming algorithms can be introduced for sampling triangles and other motifs [231, 301]. Indeed, many algorithms have been proposed for graph streams beyond subgraph enumeration, including finding minimum spanning trees, checking for bipartiteness, thinning and sparse cuts, and more [300].

27.5 Approximations and local methods

Graph streaming brings us more broadly to problems and solutions that work either locally or by making approximations. Approximations trade off computational complexity for accuracy, allowing breathtakingly efficient solutions if we are willing for answers to have some (known) degree of error. We first discuss probabilistic counting, a general technique that we will then use for network-specific data structures.

27.5.1 Probabilistic counting

Suppose we are faced with a huge data stream, far too large for our memory, and we want to count the unique items in the stream. For example, how many *unique* edges are present in a large graph stream of edges. This is called *cardinality estimation* and it is often encountered in big data applications. Of course, we can store all the edges we've seen before in a data structure, but this will consume considerable space and even time. Can we estimate cardinality without using a large in-memory hashing data structure such as a Python dict or resorting to slow out-of-memory storage like a disk? Yes, if we are willing to have some errors in the count.

We discuss HyperLogLog (HLL) [161], one of the best solutions to large-scale cardinality estimation. HyperLogLog (and its ancestors [160, 137]) begin from the following intuition. Suppose each item x in our stream can be mapped to a string of bits using a *hash function* $h(x)$ such that (approximately) all bit strings are equally likely; each bit in the string is equal to zero with probability 1/2. If this is the case and you observe in your stream a bit string that begins with $\rho - 1$ zeros before the first 1, then it stands to reason the cardinality n of the stream is at least 2^ρ. As we encounter items in our stream, we can simply track the largest value of ρ we observe, using very little memory.

Unfortunately, as you can imagine, this solution alone will be extremely error-prone (it also turns out to be biased). We may get unlucky in a small stream and by chance encounter a large ρ, wildly affecting our estimate. To solve this, HLL takes two courses of action. One, it splits the stream into substreams that are averaged over, helping with noise. Two, it uses a harmonic mean which is less susceptible to outliers. The authors of HLL also identify biases and corrective factors in their analysis of the algorithm.

Algorithm 27.1 describes the HyperLogLog algorithm. Each hashed item is placed into a substream based on its first few bits, while cardinality is estimated using the remaining bits. The storage requirements for HLL are extremely modest: when estimating cardinalities $\leq N$, only a collection of "registers," each of which using $O\left(\log_2 \log_2 N\right)$ bits, are needed.

How does this method work? If n is the unknown cardinality, each substream should have cardinality approximately n/m, where m is the number of registers. Then the maximum ρ stored in the register for that substream should be approximately $\log_2(n/m)$ and the harmonic mean (times m) of 2^M should be of the order n/m. Therefore $m^2 Z$ should be of the order n and we have our cardinality estimate. The term α_m corrects a multiplicative bias in $m^2 Z$ that was shown from the analysis by Flajolet et al. [161]. The authors also prove that the relative standard error between the estimated cardinality E and n is $\leq 1.04/\sqrt{m}$.

To illustrate the power of HLL, the authors describe that "using $m = 2048$, hashing on 32 bits, and short bytes of 5 bit length each: cardinalities till values over $N = 10^9$ can be estimated with a typical accuracy of 2% using 1.5 kB (kilobyte) of storage" [161].

With a tool like HLL,[6] we can solve otherwise intractable enumeration questions. Because the counters are so small, we can have very many of them, and we can now, for example, enumerate motifs or shortest paths in a very large network.

---

[6] Other probabilistic data structures can be used for similar problems, including Bloom filters [58] and MinHashes [76].

Algorithm 27.1 HyperLogLog cardinality estimation [161].

1: Input: Data stream (multiset) S, hash function h, $m = 2^b$ with $b \in \mathbb{Z}^+$
2: Initialize m registers $M[1], \ldots, M[m] \leftarrow -\infty$
3: **for** $v \in S$ **do**
4:         $x \leftarrow h(v)$
5:         $j \leftarrow 1 + \langle x_1 x_2 \cdots x_b \rangle_2$ ▷ *the binary address determined by the first b bits of x*
6:         $r \leftarrow \rho(x_{b+1} x_{b+2} \cdots)$ ▷ *$\rho(s)$ is the position of the leftmost 1 in s*
7:         $M[j] \leftarrow \max(M[j], r)$

8: $Z \leftarrow \left(\displaystyle\sum_{j=1}^{m} 2^{-M[j]} \right)^{-1}$ ▷ *Harmonic mean*

9: Return: $E \leftarrow \alpha_m m^2 Z$, where

$$\alpha_m := \left(m \int_0^\infty \left(\log_2 \left(\frac{2+u}{1+u} \right) \right)^m du \right)^{-1} \qquad (27.3)$$

27.5.2 Approximate Neighborhood Functions

The *neighborhood function* (NF) counts how many pairs of nodes i, j in a graph are reachable within t hops (i.e., the shortest path distance $\ell_{ij} \leq t$) as a function of t. This function contains a wealth of information about the graph topology (see also: network portraits, Secs. 13.3 and 14.2.2; graph distances and connectedness, Sec. 12.10). From the NF, we can compute quantities like the typical distance between nodes, and the diameter. But computing the NF requires global breadth-first search (usually) calculations, which are expensive and memory-intensive.

A highly space-efficient algorithm [355, 63] to approximate the NF can be constructed by recognizing that the NF can be determined by taking set unions and that probabilistic counting registers (Sec. 27.5.1) can estimate the cardinality of a set union. Specifically, let $B(i, r)$ be the set of all nodes within distance r from node i. This "ball" satisfies $B(i, 0) = \{i\}$ and

$$B(i, r) = \bigcup_{j \in N_i} B(j, r - 1). \qquad (27.4)$$

We can compute the NF (and generalizations) from these sets. Given two sets A and B, the cardinality of $A \cup B$ can be estimated from HyperLogLog counters for A and B by maximizing register-by register (assuming the counters have the same bit length), i.e., $M_{A \cup B}[i] = \max(M_A[i], M_B[i])$.[7]

These two ingredients combine to make the *approximate* NF (ANF) method. The sets $B(i, r)$ are never stored as they will become massive for large enough networks. Probabilistic counters are used instead to track the cardinalities of their unions, saving

---

[7] In principle, we could also use this to estimate the set intersection from $|A \cap B| = |A| + |B| - |A \cup B|$ as we have counters for all three. In practice, the errors scale poorly for this strategy and more sophisticated methods are employed [460].

considerable space. The ANFs are built by scanning over the adjacency set of the network, applying HLL counters to the unions (Eq. (27.4)). When finished, meaning when the HLL counters have converged (and they must be run to convergence for the estimates to be accurate), we have the information necessary to approximate the NF. The basic HyperANF algorithm is given in Alg. 27.2.

Algorithm 27.2 HyperANF without optimizations [63].

1: Input: Network G as adjacency sets $\{i : N_i\}$

2: **procedure** UNION(M, N)
3:     **for each** $i < m$ **do**
4:         $M[i] \leftarrow \max(M[i], N[i])$

5: Initialize $c[-]$, an array of N HLL counters
6: **for** $v \in G$ **do**
7:     Add v to $c[v]$
8: $t \leftarrow 0$
9: **while** $c[-]$ has not converged **do**
10:     $NF(t) \leftarrow \sum_v \text{size}(c[v])$ ▸ *The neighborhood function*
11:     **for** $v \in G$ **do**
12:         $m_v \leftarrow c[v]$
13:         **for** $w \in N_v$ **do**
14:             $m_v \leftarrow$ UNION($c[w], m_v$)   ▸ *Estimator from Eq. (27.4)*
15:         Store (v, m_v)
16:     Recall $\{(v, m_v) \mid v \in G\}$ and update array $c[-]$
17:     $t \leftarrow t + 1$

Several further optimizations make the HyperANF method highly efficient (and these are what really distinguish it from the earlier ANF method [355]). For one, the outer loop over G can be parallelized. Second, the HLL union, which is performed many times, can be parallelized using clever bitwise operations [63]. Third, we can track which c's have not changed and avoid maximizing them. This last improvement can be taken even further by allowing nodes to "signal" backwards to predecessors when their counter changes, and we can later skip over nodes whose successors have unmodified counters.

With ANFs we can derive a variety of useful network statistics [355, 63], the most prominent being the "effective" diameter. Computing the cumulative distance distribution $H_G(t) := NF(t)/\max_t NF(t)$, the effective diameter is the distance t such that $H_G(t) < \alpha$ (usually, $\alpha = 0.9$). By modifying Alg. 27.2, we can also devise other statistics related to cut numbers (Sec. 25.5) and similar quantities by tracking subsets of the NF, such as $NF_i(t)$ for individual nodes i or even sets of nodes [355]. Thus we have a way to measure how big the network is, and other properties, even when interacting with it is challenging due to its scale.

27.5.3   Application: The small world of Facebook

As an application of these technologies, we briefly describe a 2012 study by Backstrom et al. [20] to estimate social distance (small world; Sec. 22.5) in the Facebook social network. They used both graph compression (Sec. 27.4.1) and approximate neighbor functions (Sec. 27.5.2) with HyperLogLog counters (Sec. 27.5.1) and other optimization to make the calculation feasible. With this infrastructure in place, they analyze social networks for individual countries and the entire platform, at the time consisting of over 700 million active users and nearly 70 billion friendship edges. They found the average distance (number of hops) between individuals was 4.74 or 3.74 "degrees of separation," lower than the famous six degrees of separation and lower than the 4.4–5.7 degrees famously found by Milgram [309]. Regional networks showed consistently smaller distances still, which are more directly comparable to Milgram's work which was itself restricted to a single country. Perhaps with richer data to analyze, or perhaps with more venues for communication, our small world is smaller still.

27.5.4   Community detection at scale

Fast community detection with label propagation

Community detection (Secs. 12.7, 23.2, and 25.6) on large networks is challenging. Here we discuss an efficient algorithm using label propagation (LP), proposed by Raghavan et al. [384]. Nearly all the previous community methods we've studied, from modularity maximization to inference of the stochastic block model, are too costly for very large networks. The LP method works by transforming the global community detection problem into a local updating rule, akin to an eigenvector centrality update but without the need for matrix operations.

The LP algorithm is beautifully simple. Suppose each node is given a label denoting which community it belongs to. The labels propagate by a majority rule: each node i looks at every neighbor $j \in N_i$ and adopts the most common label among the neighbors. Initially, every node has a unique label, corresponding to N communities all of size 1. As labels propagate, dense groups will reach consensus on a label, which crowds out uncommon labels and spreads until colliding with the labels of other dense groups, until propagation stops. Updating stops naturally at consensus, when every node has the same label as the majority of its neighbors.

This spreading process can be implemented with synchronous updating steps, where all nodes update their labels at the same time, or with asynchronous updating, where one node updates its label, then another, etc. and later updates can be affected by earlier updates. Raghavan et al. [384] show that asynchronous updating has an advantage because, with synchronous updates, certain network structures (bipartite subgraphs) can cause "blinking" where labels oscillate back and forth instead of converging. Asynchronous updating avoids this, and cycling through nodes in a random order for each update helps speed up mixing.

The LP algorithm is as follows. Let $C_i(t)$ be the label of node i at step t.

1. Initialize the labels, giving each node a unique label: $C_i(0) = i$.
2. Set $t = 1$.

3. Arrange the nodes in a random order L.
4. For each $i \in L$, chosen in the order of L, let

$$C_i(t) = f\left(C_{i_1}(t), \ldots, C_{i_u}(t), C_{i_{u+1}}(t-1), \ldots, C_{i_k}(t-1)\right), \tag{27.5}$$

where $i_1, \ldots, i_u \in N_i$ are neighbors of i who have already updated (preceded i's position in L) and $i_{u+1}, \ldots, i_k \in N_i$ are neighbors who have not, and $f(\cdot)$ returns its most frequent argument, breaking ties at random.
5. If every node has the label that is most common among its neighbors, terminate the algorithm and return the $\{C_i\}$. Otherwise, set $t = t + 1$ and continue from 3.

There is one wrinkle after the algorithm converges. While usually rare, it is possible for disjoint groups to independently converge on the same label. This occurs when two or more neighbors of a node adopt its label and then pass that label in different directions, which eventually leads to separated communities adopting that same label. The fix to distinguish different groups with the same label is to find the connected components of the subgraph induced by each label, which can be found by one breadth-first search per component in $O(N + M)$ time.

During each iteration, finding $C_i(t)$ for each node requires looking at the k neighbors of i, at a cost of $O(k_i)$. For all the nodes, this gives a total cost of $\sum_i k_i = O(M)$—each iteration is linear in the number of edges. The algorithm tends to converge very quickly, even when there is no community structure (Raghavan et al. [384] report that typically 95% of nodes reach consensus after five iterations), although the number of iterations needed is not known theoretically. This fast convergence in practice famously gives the algorithm a near-linear complexity.

Local community methods

Another problem of interest in big data is finding a local community only, and not worrying about partitioning the full network. Local methods can do this, finding a single community belonging to a starting node by spreading outward from that node. We discussed local community methods in Sec. 12.7.

27.6 Updating schemes for network statistics

Dynamic updating refers to tracking the evolution of a network summary statistic s as the data are changing [452]. Suppose, for instance, we wish to know the degree assortativity of our very large network. Stopping to compute this will be intractable, it is a global calculation. Instead, we can design our data system to track the necessary components of the statistic, updating them efficiently on any changes to the network (node insertion, edge insertion, node deletion, edge deletion). Then, when needed, we can perform the final calculation to derive s. While there can be a space tradeoff depending on s, often it is minor compared to the time complexity of the global calculation.

We now develop dynamic updating schemes for several important network statistics. Some are trivially straightforward, such as updating the degree. Others require more bookkeeping. We focus on node degree, clustering coefficient, assortativity, and modularity.

Recall the clustering coefficient (Eq. (12.8)):

$$C_i = \begin{cases} \frac{2\Delta_i}{k_i(k_i-1)} & \text{if } k_i \geq 2, \\ 0 & \text{otherwise,} \end{cases} \quad (27.6)$$

where Δ_i is the number of triangles that contain i. Then the average clustering coefficient[8] of the whole network is simply the average of all C_i's:

$$C = \frac{1}{N} \sum_i C_i. \quad (27.7)$$

Meanwhile, we now write the assortativity coefficient (Eq. (12.10)) as,

$$r := \frac{8M \sum_{(i,j)\in E} k_i k_j - \left[\sum_{(i,j)\in E} (k_i + k_j)\right]^2}{4M \sum_{(i,j)\in E} \left(k_i^2 + k_j^2\right) - \left[\sum_{(i,j)\in E} (k_i + k_j)\right]^2}$$

$$= \frac{8Mu - v^2}{4Mw - v^2}, \quad (27.8)$$

where

$$u := \sum_{(i,j)\in E} k_i k_j,$$

$$v := \sum_{(i,j)\in E} (k_i + k_j), \quad (27.9)$$

$$w := \sum_{(i,j)\in E} \left(k_i^2 + k_j^2\right),$$

and modularity (Eqs. (12.15) and (12.16)) as

$$Q = \frac{1}{2M} \sum_{i,j} \left(A_{ij} - \frac{k_i k_j}{2M}\right) \delta(g_i, g_j) = \frac{1}{2M} \left[S_A - \frac{1}{2M} S_P\right], \quad (27.10)$$

where $\delta(g_i, g_j) = 1$ if nodes i and j are in the same group and zero otherwise, and

$$S_A := \sum_{i,j} A_{ij} \delta(g_i, g_j), \quad S_P := \sum_{i,j} k_i k_j \delta(g_i, g_j). \quad (27.11)$$

27.6.1   Updating schemes

Connecting a new node

First, consider connecting a new node. We can decompose this into two successive operations, adding a new isolated node to the network, then adding an edge from that new node to an existing node. (Multiple connections can then be handled using the updating scheme for connecting edges between existing nodes; see below.) We use $\widetilde{\ }$ to denote updated statistics.

---

[8] Remember that transitivity is a better global measure; Ch. 12 and Ex. 27.5.

With a new node and no new edges, we have the following update relations:

$$\widetilde{N} = N + 1, \quad \widetilde{M} = M, \quad \widetilde{E} = E, \tag{27.12}$$

and[9]

$$\widetilde{A}_{ij} = \begin{cases} A_{ij} & \text{if } i \neq N + 1 \text{ and } j \neq N + 1, \\ 0 & \text{otherwise.} \end{cases} \tag{27.13}$$

The other statistics follow likewise:

$$\widetilde{k}_i = k_i, \ i \neq N + 1; \quad \widetilde{k}_{N+1} = 0; \tag{27.14}$$

and

$$\widetilde{C}_i = C_i, \ i \neq N + 1; \quad \widetilde{C}_{N+1} = 0. \tag{27.15}$$

Similarly, $\widetilde{r} = r$ since $\widetilde{u} = u$, $\widetilde{v} = v$, and $\widetilde{w} = w$; and $\widetilde{Q} = Q$ since $\widetilde{S}_A = S_A$, and $\widetilde{S}_P = S_P$. Deleting a degree-zero node follows likewise.

Adding a new edge between existing nodes

Suppose nodes p and q are not connected ($A_{pq} = 0$). Here we derive how adding edge p, q changes our statistics. First, since we inserted an edge we have

$$\widetilde{E} = E \cup \{(p, q), (q, p)\}, \tag{27.16}$$
$$\widetilde{M} = M + \Delta^+ M = M + 1, \tag{27.17}$$

and

$$\widetilde{A}_{ij} = A_{ij} + \Delta^+ A_{ij} = A_{ij} + \delta_{ip}\delta_{jq} + \delta_{iq}\delta_{jp}, \tag{27.18}$$

where we use the "update delta" Δ^+ to denote the change in statistic after adding an edge to the existing network. Using these expressions, we can now derive efficient updating schemes for network statistics.

Degree The degree update is simple, as only p and q are affected:

$$\widetilde{k}_i = k_i + \Delta^+ k_i = k_i + \delta_{ip} + \delta_{iq}. \tag{27.19}$$

If we wish to study the network's degree distribution, we can use this update to track how many nodes currently have a given degree by, after an update $k \rightarrow k + 1$, decrementing the number of nodes with degree k and incrementing the number with $k + 1$.

---

[9] We use **A** for presenting calculations, but it will serve as a placeholder for how the network data system is actually implemented, which is unlikely to be with an adjacency matrix, even if the network is very sparse.

Clustering coefficient To compute the new clustering coefficient of each node, and thus the whole network, we need the updated number of triangles at node i:

$$\widetilde{\Delta}_i = \begin{cases} \Delta_i & \text{if } i \notin \{p,q\} \cup N_{pq}, \\ \Delta_i + 1 & \text{if } i \in N_{pq}, \\ \Delta_i + |N_{pq}| & \text{if } i \in \{p,q\}, \end{cases} \tag{27.20}$$

where $N_{ij} := N_i \cap N_j$ is the shared neighborhood of nodes i and j. Combining this with Eq. (27.19) and $\Delta_i = \frac{1}{2} C_i k_i(k_i - 1)$, from Eq. (27.6), we have

$$\widetilde{C}_i = \begin{cases} C_i & \text{if } i \notin \{p,q\} \cup N_{pq}, \\ C_i + \frac{2}{k_i(k_i-1)} & \text{if } i \in N_{pq}, \\ \frac{k_i-1}{k_i+1} C_i + \frac{2|N_{pq}|}{k_i(k_i+1)} & \text{if } i \in \{p,q\}. \end{cases} \tag{27.21}$$

(Whenever the denominator of a fraction is zero, we define the fraction to be zero, in Eq. (27.21) and throughout. This ensures $C_i = 0$ if $k_i < 2$.) Finally, the average clustering coefficient C becomes

$$\widetilde{C} = C + \Delta^+ C = C + \frac{2}{N} \left[\sum_{i \in N_{pq}} \frac{1}{k_i(k_i-1)} + \sum_{i \in \{p,q\}} \left(\frac{|N_{pq}|}{k_i(k_i+1)} - \frac{C_i}{k_i+1} \right) \right], \tag{27.22}$$

where

$$\Delta^+ C = \frac{2}{N} \left[\sum_{i \in N_{pq}} \frac{1}{k_i(k_i-1)} + \sum_{i \in \{p,q\}} \left(\frac{|N_{pq}|}{k_i(k_i+1)} - \frac{C_i}{k_i+1} \right) \right]. \tag{27.23}$$

Notice that updating the average clustering coefficient requires keeping C_i for each i, at $O(N)$ space complexity.

Degree assortativity To compute $\widetilde{r}$, we need $\widetilde{u}$, $\widetilde{v}$, and $\widetilde{w}$. The update for u is

$$\widetilde{u} = \sum_{(i,j)\in\widetilde{E}} \widetilde{k}_i \widetilde{k}_j = \sum_{(i,j)\in E} \widetilde{k}_i \widetilde{k}_j + 2(k_p+1)(k_q+1)$$

$$= \sum_{(i,j)\in\widetilde{E}} k_i k_j + 2\sum_{i\in N_p} k_i(k_p+1) + 2\sum_{i\in N_q} k_i(k_q+1)$$

$$+ 2(k_p+1)(k_q+1)$$

$$= u + 2\left(\sum_{i\in N_p} k_i + \sum_{i\in N_q} k_i \right) + 2(k_p+1)(k_q+1)$$

$$= u + \Delta^+ u. \tag{27.24}$$

Here $\widehat{E} = E \setminus \{(p,q),(q,p)\}$ is the edge set that contains all edges in E but (p,q) and (q,p) and

$$\Delta^+ u = 2\left(\sum_{i\in N_p} k_i + \sum_{i\in N_q} k_i \right) + 2(k_p+1)(k_q+1). \tag{27.25}$$

Similarly, the update formula for v and w are

$$\widetilde{v} = \sum_{(i,j)\in\widetilde{E}} (\widetilde{k}_i + \widetilde{k}_j) = v + \Delta^+ v = v + 4(k_p + k_q + 1), \tag{27.26}$$

and

$$\widetilde{w} = \sum_{(i,j)\in\widetilde{E}} (\widetilde{k}_i^2 + \widetilde{k}_j^2) = w + \Delta^+ w, \tag{27.27}$$

where

$$\Delta^+ w = 6 \left[k_p(k_p + 1) + k_q(k_q + 1) \right] + 4. \tag{27.28}$$

Finally, the new assortativity coefficient can be updated using

$$\widetilde{r} = r + \Delta^+ r = \frac{8\widetilde{M}\widetilde{u} - \widetilde{v}^2}{4\widetilde{M}\widetilde{w} - \widetilde{v}^2} = \frac{8(M+1)(u+\Delta^+ u) - (v+\Delta^+ v)^2}{4(M+1)(w+\Delta^+ w) - (v+\Delta^+ v)^2}. \tag{27.29}$$

Modularity For modularity, we assume that after connecting the nodes p and q, the partitions g_i do not change for any node i.[10] Then the new modularity measure will be

$$\widetilde{Q} = \frac{1}{2\widetilde{M}} \left[\widetilde{S}_A - \frac{1}{2\widetilde{M}} \widetilde{S}_P \right]. \tag{27.30}$$

We already have $\widetilde{M} = M + 1$, we now derive updating formulas for S_A and S_P. By Eq. (27.11), we have

$$\widetilde{S}_A = S_A + \Delta^+ S_A = \sum_{i,j} \widetilde{a}_{ij}\delta(g_i, g_j)$$

$$= \sum_{i,j} \left(a_{ij} + \delta_{ip}\delta_{jq} + \delta_{iq}\delta_{jp}\right) \delta(g_i, g_j)$$

$$= S_A + 2\delta(g_p, g_q) \tag{27.31}$$

and

$$\widetilde{S}_P = S_P + \Delta^+ S_P = \sum_{i,j} \widetilde{k}_i\widetilde{k}_j\delta(g_i, g_j)$$

$$= \sum_{i,j} \left(k_i + \delta_{ip} + \delta_{iq}\right) \left(k_j + \delta_{jp} + \delta_{jq}\right) \delta(g_i, g_j)$$

$$= S_P + 2\sum_i k_i\left[\delta(g_i, g_p) + \delta(g_i, g_q)\right] + 2\left[\delta(g_p, g_q) + 1\right]. \tag{27.32}$$

However, computing the sum in Eq. (27.32) after every update is expensive. To avoid this, we introduce the following auxiliary statistics:

$$K_g := \sum_i k_i\delta(g_i, g) \tag{27.33}$$

---

[10] We actually already encountered the updating scheme for a partition change in Sec. 25.6.

with updating scheme

$$\widetilde{K}_g = K_g + \Delta^+ K_g = K_g + \delta(g_p, g) + \delta(g_q, g), \tag{27.34}$$

giving

$$\widetilde{S}_P = S_P + \Delta^+ S_P = S_P + 2\left(K_{g_p} + K_{g_q}\right) + 2\left[\delta(g_p, g_q) + 1\right], \tag{27.35}$$

where $\Delta^+ S_P = 2\left(K_{g_p} + K_{g_q}\right) + 2\left[\delta(g_p, g_q) + 1\right]$. Finally, combining Eq. (27.31) and Eq. (27.35) with Eq. (27.30) gives the updating scheme for Q:

$$\widetilde{Q} = Q + \Delta^+ Q = \frac{1}{2(M+1)}\left[S_A + 2\delta(g_p, g_q)\right.$$
$$\left. - \frac{1}{2(M+1)}\left(S_P + 2[K_{g_p} + K_{g_q}] + 2[\delta(g_p, g_q) + 1]\right)\right]. \tag{27.36}$$

From Eq. (27.36) we can predict whether Q increases or decreases given the existing partition and the edge to be added. For example, if there is a preexisting partition of the network into two groups, then if a new edge is added in between the groups, $\Delta^+ Q < 0$. On the other hand, if a new edge connects nodes in the same group, then the modularity is sure to increase *only if* the edge is added to the group with smaller total degree. Perhaps surprisingly, adding an edge within the group does not necessarily increase Q if the edge is added into the group with larger total degree.

Removing an existing edge

Now we focus on updates when an existing edge is deleted. These updates can also be used for the removal of a node, since removing a node requires deleting its edges then deleting the now-disconnected node.

Suppose $p \neq q$ and p, q are connected, and we delete this edge, $(p, q) \cup (q, p)$, from our edge set E. We'll use $\widehat{\cdot}$ to represent the updated statistics. First, we immediately have

$$\widehat{E} = E \setminus \{(p, q), (q, p)\}, \tag{27.37}$$
$$\widehat{M} = M - 1, \tag{27.38}$$
$$\widehat{A}_{ij} = A_{ij} + \Delta^- A_{ij} = A_{ij} - \delta_{ip}\delta_{jq} - \delta_{iq}\delta_{jp}, \tag{27.39}$$

where Δ^- will be used to denote change for statistics upon deleting an existing edge.

Degree　　The change in degree for node i is simply

$$\widehat{k}_i = k_i + \Delta^- k_i = k_i - \delta_{ip} - \delta_{iq}. \tag{27.40}$$

Clustering coefficient   For the new clustering coefficient, as before, we start with the updating scheme for the number of triangles containing node i:

$$\widehat{\Delta}_i = \begin{cases} \Delta_i & \text{if } i \notin \{p,q\} \cup N_{pq}, \\ \Delta_i - 1 & \text{if } i \in N_{pq}, \\ \Delta_i - |N_{pq}| & \text{if } i \in \{p,q\}. \end{cases} \tag{27.41}$$

Then we obtain the scheme for updating C_i:

$$\widehat{C}_i = \begin{cases} C_i & \text{if } i \notin \{p,q\} \cup N_{pq}, \\ C_i - \frac{2}{k_i(k_i-1)} & \text{if } i \in N_{pq}, \\ \frac{k_i}{k_i-2}C_i - \frac{2|N_{pq}|}{(k_i-1)(k_i-2)} & \text{if } i \in \{p,q\}. \end{cases} \tag{27.42}$$

The average clustering coefficient C is updated by

$$\widehat{C} = C + \Delta^- C = \frac{1}{N}\sum_i \widehat{C}_i$$

$$= C - \frac{2}{N}\left[\sum_{i \in N_{pq}} \frac{1}{k_i(k_i-1)} + \sum_{i \in \{p,q\}} \left(\frac{|N_{pq}|}{(k_i-1)(k_i-2)} - \frac{C_i}{k_i-2}\right)\right]. \tag{27.43}$$

Degree assortativity   The updating formulas for u, v, w are

$$\widehat{u} = u + \Delta^- u = u - 2\left(\sum_{i \in N(p)} k_i + \sum_{i \in N(q)} k_i\right) - 2(k_p - 1)(k_q - 1),$$

$$\widehat{v} = v + \Delta^- v = v - 4(k_p + k_q - 1), \tag{27.44}$$

$$\widehat{w} = w + \Delta^- w = w - 6\left[k_p(k_p - 1) + k_q(k_q - 1)\right] - 4,$$

and the new assortativity coefficient $\widehat{r}$ is

$$\widehat{r} = \frac{8\widehat{M}\widehat{u} - \widehat{v}^2}{4\widehat{M}\widehat{w} - \widehat{v}^2} = \frac{8(M-1)(u + \Delta^- u) - (v + \Delta^- v)^2}{4(M-1)(w + \Delta^- w) - (v + \Delta^- v)^2}. \tag{27.45}$$

Modularity   Once more, we assume that the community partitions g_i are unchanged after disconnecting the edge between p and q. It follows that

$$\widehat{S}_A = S_A + \Delta^- S_A = S_A - 2\delta(g_p, g_q), \tag{27.46}$$

$$\widehat{S}_P = S_P + \Delta^- S_P = S_P - 2\left(K_{g_p} + K_{g_q}\right) + 2\left(\delta(g_p, g_q), +1\right), \tag{27.47}$$

where K_g is now updated using

$$\widehat{K}_g = K_g + \Delta^- K_g = K_g - \delta(g_p, g) - \delta(g_q, g). \tag{27.48}$$

These now define the updating scheme for $\widehat{Q} = \left(\widehat{S}_A - \widehat{S}_P/2\widehat{M}\right)/2\widehat{M}$.

Lastly, in Table 27.1 we compare the computational complexity of these updating schemes to solutions using an adjacency matrix and an edgelist [452]. In all cases, the updating schemes have better (lower) complexity.

Table 27.1 Complexity of updating schemes [452].

Statistic	Adjacency matrix	Edge list	Updating scheme
Degree (one node)	$O(N)$	$O(\langle k \rangle)$	$O(1)$
Degree (network)	$O(N^2)$	$O(\langle k \rangle N)$	$O(1)$
Clustering coefficient (one node)	$O(\langle k \rangle N)$	$O\left(\langle k \rangle^3\right)$	$O(1)/O(\langle k \rangle)$
Clustering coefficient (network)	$O(\langle k \rangle N^2)$	$O\left(\langle k \rangle^3 N\right)$	$O(\langle k \rangle)$
Assortativity	$O(N^2)$	$O(\langle k \rangle N)$	$O(\langle k \rangle)$
Modularity	$O(N^2)$	$O(\langle k \rangle N)$	$O(1)$

27.7 Making graphs

Most graph models are implemented in ways that don't account for scalability. For example, suppose we wish to generate a very large Erdős–Rényi graph $G(N, p)$ on $N \gg 1$ nodes where each edge exists with probability p. The naive, or unscalable way is to take each pair of nodes (i, j), of which there are $\binom{N}{2} = O(N^2)$, generate a random number r uniformly distributed on $[0, 1]$, and insert edge i, j if $r < p$. Because we are looping through the set of nodes twice, we have a quadratic algorithm; fine for small networks, prohibitive for large ones.

This naive double-loop is also wasteful. We individually test every pair of nodes, and (depending on p) many of those tests will fail and no edge will be created. A more efficient approach (see also remarks) is to change how edges are inserted to skip over the node pairs where no edge will be inserted.[11] We describe this now.

Suppose nodes are numbered $0, 1, \dots, N - 1$. We begin at node $u = 0$. Then for each node $v = 1, \dots, N - 1$, the naive algorithm would generate r and insert edge u, v if $r < p$. The process would continue with $u \geq 1, 2, \dots$ and $v = u + 1$. Let's be more efficient. Let $v_1 = u + 1 + \delta$ be the first neighbor with which u forms an edge. Here δ is the number of intervening pairs u, v_i that did not form an edge. Similarly, $v_2 = v_1 + 1 + \delta$ where now δ is the new number of intervening pairs between v_1 and v_2. The steps δ are distributed geometrically: $\Pr(\delta) = (1 - p)^\delta p$. Therefore, instead of considering every v after u, we generate δ and skip ahead that many nodes. Specifically, generate r uniformly on $(0, 1)$ and set $\delta = \lfloor \log(r)/\log(1 - p) \rfloor$, where $\lfloor \cdot \rfloor$ is the floor function.[12] (We take $\delta = 0$ if $p = 1$.) Then we can insert edge $u, v + 1 + \delta$, and continue. We give the full algorithm in Alg. 27.3.

While the naive algorithm has an $O(N^2)$ complexity, Alg. 27.3 has a complexity of only $O(N + M)$. To see this, consider the average number of times the inner loop

---

[11] We can also just generate edges as integer pairs $i, j \sim U[0, N - 1]$, but we focus on the "skipping" method because it is more easily generalizes to the other graph models we will discuss.

[12] This expression comes from using *inverse transform sampling* [130]. We compute the CDF of $\Pr(\delta)$, set it equal to $r \sim U[0, 1]$, then solve for δ. When the CDF $F(x)$ can be inverted, this is a clever way to transform $U[0, 1]$-distributed random numbers to $x \sim f(x)$. In our case, we have $r = \sum_{d=\delta}^{\infty} p(1 - p)^d = (1 - p)^\delta \Rightarrow \delta = \log r / \log(1 - p)$.

Algorithm 27.3 Efficient algorithm for generating Erdős–Rényi graphs. Runs in $O(N + M)$.

1: Input: Number of nodes N, edge probability $0 < p < 1$
2: $E \leftarrow \emptyset$
3: **for** $u = 0$ to $N - 2$ **do**
4:      $v \leftarrow u + 1$
5:      **while** $v < N$ **do**
6:          draw $r \sim U[0, 1]$
7:          $v \leftarrow v + \left\lfloor \frac{\log(r)}{\log(1-p)} \right\rfloor$
8:          **if** $v < N$ **then**
9:              $E \leftarrow E \cup \{(u, v)\}$
10:             $v \leftarrow v + 1$
11: Return: instance of $G(N, p)$ with nodes $0, \ldots, N - 1$

---

executes: $1 + N/\langle \delta \rangle = 1 + O(Np)$, the 1 being the number of unsuccessful times. The outer loop executes N times, giving a total complexity of $N(1 + O(Np)) = O(N + M)$ since $M = \binom{N}{2}p$. This complexity is just about optimal in that we'd need to examine every edge during generation regardless.

Now, we generalize this method from Erdős–Rényi networks to *Chung–Lu* networks [107], networks with a given expected degree sequence. These networks introduce heterogeneity into the degree distribution by using node weights w_u and setting the probability for an edge u, v to be $p_{uv} = w_u w_v / N \bar{w}$, where $\bar{w} = \sum_u w_u / N$.

The obvious, naive $O(N^2)$ algorithm consists of again performing a double loop over all nodes and checking for each edge by generating $r \sim U[0, 1]$ and inserting edge u, v if $r < w_u w_v / N \bar{w}$. We again wish to be efficient by skipping unnecessary node pairs, but this will be more difficult because the probabilities for possible edges will not be fixed. To derive a more efficient algorithm, we first present an alternative to the naive algorithm that can be modified to be efficient.

First, assume the list W of weights for the network is available and sorted in descending order. As we consider every $v = u + 1, \ldots, N - 1$, p_{uv} decreases monotonically, so we avoid recalculating p for each v by setting $p = p_{u,u+1} = w_u w_{u+1} / N \bar{w}$ and skipping each v with probability $1 - p$. Arriving at the first node v_1 we do not skip, we then need to check if v_1 is actually neighbors with u, since $p_{uv_1} \le p$. We calculate $q = p_{uv_1}$ and assign the edge with probability q/p. We then set $p = q$ and continue to discard nodes with probability $1 - p$ until all v have been considered. Then we increment u and repeat the process. The probability that an edge is inserted, $p p_{uv_1} / p = p_{uv_1}$ is as we expect and, like the naive algorithm, this algorithm is $O(N^2)$.

Why organize the naive algorithm in this way? Doing so ensures that p is fixed at each step until a *potential neighbor* is identified, allowing us to identify how to skip over those intervening nodes. Starting with $u = 0$, and setting $p = p_{u,u+1}$, draw $r \sim U[0, 1]$ and find the first potential neighbor $v_1 = u + 1 + \delta$ by generating a step $\delta = \lfloor \log r / \log(1-p) \rfloor$ (taking $\delta = 0$ if $p = 1$). Once v_1 is selected, insert edge u, v_1 with probability p_{uv_1}/p. Then set $p = p_{u,v_1}$ and continue to the second potential neighbor v_2 by generating another value of δ. Since p decreases monotonically for a given u, the expected value of δ increases monotonically. We give the full algorithm for Chung–Lu graphs in Alg. 27.4.

Algorithm 27.4 Efficient algorithm for generating Chung–Lu [107] graphs. Runs in $O(N + M)$.

1: Input: sorted list of N weights $w_0 \geq w_1 \geq \ldots \geq w_{N-1}$
2: $E \leftarrow \emptyset$
3: $S \leftarrow \sum_u w_u$
4: **for** $u = 0$ to $N - 2$ **do**
5: $\quad v \leftarrow u + 1$
6: $\quad p \leftarrow \min(w_u w_v / S, 1)$
7: $\quad$ **while** $v < N$ and $p > 0$ **do**
8: $\quad\quad$ **if** $p \neq 1$ **then**
9: $\quad\quad\quad$ draw $r \sim U[0, 1]$
10: $\quad\quad\quad v \leftarrow v + \left\lfloor \frac{\log(r)}{\log(1-p)} \right\rfloor$
11: $\quad\quad$ **if** $v < N$ **then**
12: $\quad\quad\quad q \leftarrow \min(w_u w_v / S, 1)$
13: $\quad\quad\quad$ draw $r \sim U[0, 1]$
14: $\quad\quad\quad$ **if** $r < q/p$ **then**
15: $\quad\quad\quad\quad E \leftarrow E \cup \{(u, v)\}$
16: $\quad\quad\quad p \leftarrow q$
17: $\quad\quad\quad v \leftarrow v + 1$
18: Return: instance of Chung–Lu graph with nodes $0, \ldots, N - 1$

Algorithm 27.4 has complexity $O(N + M)$. We omit the proof, which is more involved compared to that of Alg. 27.3 due to the dynamic rejection sampling. We refer readers to Miller and Hagberg [310] for details.

27.8 Summary

What should we do when networks get big? Really big? A variety of tools, such as graph databases, probabilistic data structures, and local algorithms, are at our disposal, especially if we are able to accept sampling effects and uncertainty. But remember our admonishment: there is an opportunity cost to these solutions and it is very often the case that they are not needed. Try simpler methods first, then adapt.

Areas where such tools are needed include web crawls, online social networks, and knowledge graphs, and major graph database systems have been proposed for such domains. Systems biology is yet another area where networks will continue to grow: as sequencing methods continue to advance, more networks and larger, denser networks will need to be analyzed. Perhaps not to the scale of trillions of edges, but perhaps so. Some day, big data solutions may be necessary. Either way, infrastructure is in place if and when that scale needs to be confronted.

Bibliographic remarks

For more on "big data," its general challenges and opportunities, see Madden [286], Marx [296], Fan et al. [155], Chen et al. [102], and references therein. For a broad introduction to graph databases see Robinson et al. [399]. An early and very influential graph storage project was the Connectivity Server [54]. Many of its ideas became standard practice for large graph storage and compression, including in the WebGraph Framework [62]. For a recent survey on the related problem of graph summarization, see Liu et al. [278]

Many problems related to graph streams have been studied, see McGregor [300] for a survey. Detecting anomalies in a graph stream is an important, practical problem; see Aggarwal et al. [4], Manzoor et al. [290], and Eswaran et al. [151] for methods and Ranshous et al. [389] for a review. Enumerating subgraphs is another streaming problem, and McGregor et al. [301] discuss algorithms for it. More generally, Al Hasan and Dave [8] give a survey of triangle counting methods that covers both graph streams and systems where random access to the graph is possible.

Space-efficient approximate or probabilistic counting dates back to Morris [321]. The HyperLogLog method for cardinality estimation we describe is the culmination of a long line of research [160, 137, 161]. Heule et al. [212] devised the HyperLogLog++ algorithm to help improve practical application of the method.

Approximate Neighborhood Functions were introduced by Palmer et al. [355]. ANF used Flajolet–Martin counters [160]. Boldi et al. [63] introduced HyperANF by extending ANF with more advanced HyperLogLog counters (which were introduced after ANF) and other programming advancements.

Fast community detection with label propagation (LP) was introduced in a highly influential paper by Raghavan et al. [384]. An interesting variant that aligns LP with modularity by using constraints was introduced by Barber and Clark [38]. A multi-resolution variant of LP was also employed by Boldi et al. [64] for graph compression.

The dynamic updating schemes we presented were introduced by Sun et al. [452].

The algorithms we presented for efficient Erdős–Rényi and Chung–Lu graphs were introduced by Miller and Hagberg [310]. Batagelj and Brandes [45] and Hagberg and Lemons [198] also address this problem for other graph models. Ramani et al. [385] provide a readable summary of efficient graph-making strategies across a variety of models.

Exercises

27.1 HyperANF (Alg. 27.2) assumes you can sequentially access the neighbors of a node (the inner loop). What if you have a graph stream where edges are given one at a time? What problems may this pose with the basic algorithm and with optimizing it?

27.2 Suppose you use HyperANF to compute the neighborhood function for a network. Then that network changes, perhaps some edges have been rewired. How much does this affect your previous computation? Will you need to rerun HyperANF from scratch?

27.3 Modify HyperANF to compute the graph conductance cut(S)/vol(S) (Sec. 25.5) of a set of nodes S (S is an input to the new algorithm and assume it can be stored in memory). What additional data must be tracked during the outer loop?

27.4 Can label propagation be used for local community detection, i.e., to find the community containing a starting node i without needing to examine the (entire) rest of the network? Why or why not?

27.5 Derive the updating schemes for transitivity (Eq. (12.9)) instead of the average clustering coefficient.

27.6 Modify the Miller–Hagberg algorithm (Alg. 27.3) for Erdős–Rényi graphs to generate *bipartite* Erdős–Rényi graphs.

Conclusion

What are the nodes?

What are the links?

These questions are not the start of your work—the upstream task makes sure of that—but they are an inflection point. Keep them front of mind. Your methods, the paths you take to analyze and interrogate your data, all unfold from the answers (plural!) to these questions.

Here we reflect on where we have gone, where we can go for more, and, perhaps, what the future has in store for networks and network data.

What we've learned

With this book, we hope that you have learned the practices and principles to approach, analyze and understand network data. A brief summary of what we have covered:

Part I discussed (i) how ubiquitous network data is (Ch. 1), (ii) how network thinking allows us to understand the world around us (Ch. 2), and (iii) how important it is to think carefully about ethical issues surrounding network data (Ch. 3). We also reviewed basic computational, mathematical, and statistical tools and concepts that are necessary to understand and analyze network data (Ch. 4).

In **Part II**, we focused on the *practice* of working with network data. We began with an overview of the life cycle of a network study (Ch. 5), which involves (i) asking network questions, (ii) gathering data (Ch. 6), (iii) the "upstream task"—extracting networks from data (Ch. 7), (iv) storing and representing networks and their attributes in our computers (Chs. 8 and 9), (v) handling measurement errors (Ch. 10), (vi) exploratory and explanatory analysis (Ch. 11) by using visualization (Ch. 13) or various network analysis methods for understanding structure (Ch. 12), (vii) summarizing and comparing networks (Ch. 14). We also discussed dynamics *on* networks and *of* networks (Ch. 15) and capped the part with the basics of machine learning for networks (Ch. 16).

Part III was all about the *fundamentals* of network data. We opened up with a discussion on the *friendship paradox* (Ch. 21), which highlights how counterintuitive networks can be. We then studied models: more mechanistic models (Ch. 22) and more statistical models (Ch. 23); both have a lot to teach us about networks and network data. Many methods unfold from here, from analyzing noisy and uncertain data (Ch. 24) to spectral methods (Ch. 25) that leverage the vast array of tools from linear algebra, to

471

machine learning and neural networks (Ch. 26). Lastly, big data, and the challenges it brings us, were the focus on Ch. 27

Part II and **Part III** were connected via the **Interlude** about good computing practices. We strongly believe that adopting these practices will help you be a more efficient and effective data scientist.

Where to go for more

Like so many, the data scientist is defined by lifelong learning. Network data demands skills from a wide set of areas and we hope to have kick-started your imagination with all the different subjects you may want to learn. If you would like to know more about the topics covered in this book, please first consult each chapter's bibliographic remarks, where we recommend useful books and articles.

Here are some general reading recommendations:

Networks If you would like to understand more about the theoretical aspects of network science, we recommend the following books. For general introductions, we recommend *A First Course in Network Science* by Menczer et al. [305], *Network Science* by Barabási [35], and *Networks* by Newman [342] as essentials. It's likely some of these are already on your shelf.

Networks, Crowds, and Markets by Easley and Kleinberg [140] is another wonderful read, giving a very nice blend of network science with social science and economics. For social networks, the influential classic *Social Network Analysis* by Wasserman and Faust [485] remains worthwhile. For dynamic networks, we encourage starting with *Temporal Networks* by Holme and Saramäki [221] and *A Guide to Temporal Networks* by Masuda and Lambiotte [297].

Statistics We cannot more highly recommend *All of Statistics* by Wasserman [484]. The balance between breadth and depth is perfect. Another recommended read is *Computer Age Statistical Inference* by Efron and Hastie [142], which takes us through the connections between classical and computational statistics.

For Bayesian inference, we recommend *Doing Bayesian Data Analysis* by Kruschke [256] and, for more technical depth, *Bayesian Data Analysis* by Gelman et al. [179].

Machine learning Machine learning is a huge topic by itself—where to start? *An Introduction to Statistical Learning* by James et al. [230] may be one place, alongside the more technical *The Elements of Statistical Learning* by Hastie et al. [205] for those interested.

Machine learning is fast-moving, especially lately. But the classics are still worth consulting, including *Machine Learning* by Mitchell [312] and *Pattern Recognition* by Bishop [56].

More recently, a number of excellent books have been produced or newly revised, and we recommend *Deep Learning* by Goodfellow et al. [186], *Machine Learning* by Zhou [510], *Artificial Intelligence: A Modern Approach* by Russell and Norvig

[410], and *Probabilistic Machine Learning* by Murphy [324]. A heavy, but highly useful, reading list.

Visualization and communication Visualization is a major tool for working with data. We strongly recommend that scientists become well grounded in using visualization effectively and safely (it's easy to mislead). The perennial classic remains *The Visual Display of Quantitative Information* by Tufte [464]. We also recommend *Better Data Visualizations* by Schwabish [419], *Data Visualization: A Practical Introduction* by Healy [208], and *The Wall Street Journal Guide to Information Graphics* by Wong [495]. For even more on the science and general principles of visualization, *The Data Visualization Handbook* by Koponen and Hildén [250] is a great starting point. Visualization is one form of communication, writing is another, and *Writing Science* by Schimel [417] should be required reading for all scientists.

The future of network data and network science

Science does not stand still, and new network data and methods continue to emerge. Where will things go from here? What challenges and opportunities will there be? Predicting the future is hard, but allow us to speculate a bit.

New data sources

All fields of science are on the verge of exciting new data.

Social media will continue to evolve, driven by algorithmic changes, interaction with new AI [480], and perhaps the rise of new and decentralized platforms. Wearables, smart sensors, and Internet-of-Things devices will give more high-resolution data about ourselves and our environment. High-speed sampling will resolve our view on temporal networks [445, 268]. And human mobility data will push forward transportation research, urban planning, and even epidemiology.

Biology and bioinformatics are poised to make further breakthroughs. Single-cell biology aims to understand cellular heterogeneity and the complex biological processes that occur at the level of individual cells. The sheer volume and complexity of data generated by single-cell biology experiments is a challenge we are now preparing to tackle.

Single-cell biology will naturally lead to the study of networks of cells, and major efforts such as the The Human Tumor Atlas Network [405] are already underway. Examining both physical interactions between cells (e.g., through signal transduction pathways) and functional interactions between cells (e.g., through gene expression networks) can provide insights into the mechanisms of cellular heterogeneity and disease. For example, network analysis can link gene co-expression networks to specific cell states. This may one day identify the key drivers of diseases or enable predictions of disease progression.

While single-cell biology delves deeper into the inner workings of individual cells, connectomics works to understand the complex network of neurons in the brain and its impact on human behavior and cognition. The quality and extent of connectome data is improving rapidly. Connectomics data is huge and complex, presenting an exciting opportunity for neuroscientists, data scientists, and network scientists to collaborate and develop new techniques for constructing and analyzing the networks in the brain.

Models and methods

Alongside new and exciting data are new means to analyze those data.

Recently, much attention has been brought to higher-order networks and hypergraphs. These assume edges in the graph can contain more than two nodes, an interesting generalization. Although any hypergraph can be represented with a (non-hyper) bipartite graph (don't count binary relations out!), we anticipate exciting opportunities, especially in temporal networks, cell biology and neuroscience, where incorporating higher-order data can produce better models, predictions, and understanding.

Statistical inference and machine learning continue to break amazing ground and more is sure to come. Inference with network data remains challenging, as we saw with the difficulties that ERGMs (Sec. 23.4.1) and GNNs (Sec. 26.5) face. But advancements continue to be made. Machine learning tools can also help with data collection and processing. For example, processing text data such as social media posts will further evolve thanks to NLP, and non-text social data such as images and videos will continue becoming amenable to analysis thanks to computer vision.

Lastly, network science struggles in particular with questions of causality [217]. Causal inference with networked data is challenging, and large-scale controlled experiments to study, for instance, social influence and behavior change remain limited in scope. These challenges demand a better understanding of natural or found experiments [136, 372] so we can leverage existing, observational data, and research continues on better inferential methods.

New norms

The landscape of data science is changing, as researchers become increasingly aware of real-world concerns and potential harms. Efforts beginning to gain ground include the use of datasheets in machine learning [177]. For network data specifically, we have proposed a new effort, the *network card* [22], that we hope will continue the trend of better describing and documenting datasets.

Open science and open data have recently taken off, in part due to concerns over the replication crisis [328]. Awareness is spreading [265], and we believe that better and more appropriate use of statistics will complement data documentation and help make science more reliable and reproducible. Of course, education on the practice and theory of data science plays a major role in establishing these norms.

Challenges and opportunities

Network science and data science are inherently interdisciplinary, posing challenges and opportunities for research and teaching. The best research programs are cross-cutting, yet it takes real effort to maintain healthy interdisciplinary collaborations and make sure knowledge and know-how are shared across the silos of disciplines. Likewise, it is fiendishly difficult to teach a highly interdisciplinary topic to students from many disciplines, for student and teacher alike. We simply need far more and better institutional investment, support, and recognition to overcome these barriers facing network science, data science, and other interdisciplinary fields.

Finally, concerns over privacy and ethical practices with data have become more pressing, especially with the recent advances in machine learning. Privacy and data ownership clash in the face of machine learning's incessant hunger for massive training data. Better protection of personal information is needed, and there is a growing call for awareness and for regulating data brokers to ensure that data is being collected and used ethically. Organizations and researchers alike are adopting improved practices for deidentification and implementing differential privacy strategies [138]. We hope to see this trend continue.

Bibliography

[1] Abbe, E. 2018. Community Detection and Stochastic Block Models: Recent Developments. *Journal of Machine Learning Research*, **18**(177), 1–86. (Cited on p. 356.)

[2] Acebrón, J. A., Bonilla, L. L., Pérez Vicente, C. J., Ritort, F., and Spigler, R. 2005. The Kuramoto Model: A Simple Paradigm for Synchronization Phenomena. *Reviews of Modern Physics*, **77**(1), 137–185. (Cited on p. 235.)

[3] Adamic, L. A., and Adar, E. 2003. Friends and Neighbors on the Web. *Social Networks*, **25**(3), 211–230. (Cited on p. 131.)

[4] Aggarwal, C. C., Zhao, Y., and Yu, P. S. 2011. Outlier Detection in Graph Streams. Pages 399–409 of: *2011 IEEE 27th International Conference on Data Engineering*. Hannover, Germany: IEEE. (Cited on p. 469.)

[5] Ahn, Y.-Y., Bagrow, J. P., and Lehmann, S. 2010. Link Communities Reveal Multiscale Complexity in Networks. *Nature*, **466**(7307), 761–764. (Cited on pp. 184, 201.)

[6] Ahn, Y.-Y., Ahnert, S. E., Bagrow, J. P., and Barabási, A.-L. 2011. Flavor Network and the Principles of Food Pairing. *Scientific Reports*, **1**(1), 196. (Cited on pp. 9, 10, 22, 24, 26.)

[7] Airoldi, E. M., Blei, D., Fienberg, S., and Xing, E. 2008. Mixed Membership Stochastic Blockmodels. In: *Advances in Neural Information Processing Systems*, vol. 21. Curran Associates, Inc. (Cited on p. 358.)

[8] Al Hasan, M., and Dave, V. S. 2018. Triangle Counting in Large Networks: A Review. *WIREs Data Mining and Knowledge Discovery*, **8**(2), e1226. (Cited on p. 469.)

[9] Albert, R., Jeong, H., and Barabási, A.-L. 2000. Error and Attack Tolerance of Complex Networks. *Nature*, **406**(6794), 378–382. (Cited on p. 167.)

[10] Allen, A. 2007. *Vaccine: The Controversial Story of Medicine's Greatest Lifesaver*. New York, NY: W.W. Norton. (Cited on p. 67.)

[11] Almeida-Neto, M., Guimarães, P., Guimarães Jr, P. R., Loyola, R. D., and Ulrich, W. 2008. A Consistent Metric for Nestedness Analysis in Ecological Systems: Reconciling Concept and Measurement. *Oikos*, **117**(8), 1227–1239. (Cited on p. 202.)

[12] Alon, U., and Yahav, E. 2021. On the Bottleneck of Graph Neural Networks and

Its Practical Implications. In: *International Conference on Learning Representations*. Virtual Event, Austria: OpenReview.net. (Cited on p. 444.)

[13] Alstott, J., Bullmore, E., and Plenz, D. 2014. Powerlaw: A Python Package for Analysis of Heavy-Tailed Distributions. *PLoS ONE*, **9**(1), e85777. (Cited on pp. 156, 163.)

[14] Amstrup, S. C., McDonald, T. L., and Manly, B. F. J. (eds.). 2005. *Handbook of Capture-Recapture Analysis*. Princeton, NJ: Princeton University Press. (Cited on pp. 391, 395.)

[15] Ashburner, M., Ball, C. A., Blake, J. A. et al. 2000. Gene Ontology: Tool for the Unification of Biology. *Nature Genetics*, **25**(1), 25–29. (Cited on p. 116.)

[16] Ashmore, R., Calinescu, R., and Paterson, C. 2021. Assuring the Machine Learning Lifecycle: Desiderata, Methods, and Challenges. *ACM Computing Surveys*, **54**(5), 111:1–111:39. (Cited on p. 31.)

[17] Atmar, W., and Patterson, B. D. 1993. The Measure of Order and Disorder in the Distribution of Species in Fragmented Habitat. *Oecologia*, **96**(3), 373–382. (Cited on p. 202.)

[18] Babai, L. 2016. Graph Isomorphism in Quasipolynomial Time [Extended Abstract]. Pages 684–697 of: *Proceedings of the Forty-Eighth Annual ACM Symposium on Theory of Computing*. STOC '16. New York, NY: Association for Computing Machinery. (Cited on p. 443.)

[19] Babai, L. 2019. Canonical Form for Graphs in Quasipolynomial Time: Preliminary Report. Pages 1237–1246 of: *Proceedings of the 51st Annual ACM SIGACT Symposium on Theory of Computing*. STOC 2019. New York, NY: Association for Computing Machinery. (Cited on p. 443.)

[20] Backstrom, L., Boldi, P., Rosa, M., Ugander, J., and Vigna, S. 2012. Four Degrees of Separation. Pages 33–42 of: *Proceedings of the 4th Annual ACM Web Science Conference*. Evanston, IL: Association for Computing Machinery. (Cited on pp. 341, 458.)

[21] Baek, J., Kang, M., and Hwang, S. J. 2022. Accurate Learning of Graph Representations with Graph Multiset Pooling. In: *International Conference on Learning Representations*. Virtual Event, Austria: OpenReview.net. (Cited on p. 442.)

[22] Bagrow, J., and Ahn, Y.-Y. 2022. Network Cards: Concise, Readable Summaries of Network Data. *Applied Network Science*, **7**(1), 1–17. (Cited on pp. 225, 232, 474.)

[23] Bagrow, J. P. 2008. Evaluating Local Community Methods in Networks. *Journal of Statistical Mechanics: Theory and Experiment*, **2008**(05), P05001. (Cited on p. 184.)

[24] Bagrow, J. P. 2012. Communities and Bottlenecks: Trees and Treelike Networks Have High Modularity. *Physical Review E*, **85**(6), 066118. (Cited on p. 179.)

[25] Bagrow, J. P. 2020. Democratizing AI: Non-Expert Design of Prediction Tasks. *PeerJ Computer Science*, **6**(Sept.), e296. (Cited on p. 374.)

[26] Bagrow, J. P., and Bollt, E. M. 2005. Local Method for Detecting Communities. *Physical Review E*, **72**(4), 046108. (Cited on pp. 184, 185, 201.)

[27] Bagrow, J. P., and Bollt, E. M. 2019. An Information-Theoretic, All-Scales Approach to Comparing Networks. *Applied Network Science*, **4**(1), 45. (Cited on pp. 24, 216, 228.)

[28] Bagrow, J. P., Lehmann, S., and Ahn, Y.-Y. 2015. Robustness and Modular Structure in Networks. *Network Science*, **3**(4), 509–525. (Cited on pp. 386, 395.)

[29] Bagrow, J. P., Danforth, C. M., and Mitchell, L. 2017. Which Friends Are More Popular than You?: Contact Strength and the Friendship Paradox in Social Networks. Pages 103–108 of: *Proceedings of the 2017 IEEE/ACM International Conference on Advances in Social Networks Analysis and Mining 2017*. Sydney, Australia: Association for Computing Machinery. (Cited on p. 323.)

[30] Bagrow, J. P., Liu, X., and Mitchell, L. 2019. Information Flow Reveals Prediction Limits in Online Social Activity. *Nature Human Behaviour*, **3**(2), 122–128. (Cited on p. 37.)

[31] Baker, R. J. 2010. *CMOS: Circuit Design, Layout, and Simulation*. 3rd edn. IEEE Press Series on Microelectronic Systems. Piscataway, NJ; Hoboken, NJ: IEEE. (Cited on p. 26.)

[32] Balcilar, M., Renton, G., Héroux, P., Gaüzère, B., Adam, S., and Honeine, P. 2022. Analyzing the Expressive Power of Graph Neural Networks in a Spectral Perspective. In: *International Conference on Learning Representations*. Virtual Event, Austria: OpenReview.net. (Cited on p. 443.)

[33] Ball, P. 2006. *Critical Mass: How One Thing Leads to Another*. New York, NY: Farrar, Straus and Giroux. (Cited on p. 5.)

[34] Banerjee, O., El Ghaoui, L., and d'Aspremont, A. 2008. Model Selection Through Sparse Maximum Likelihood Estimation for Multivariate Gaussian or Binary Data. *The Journal of Machine Learning Research*, **9**(June), 485–516. (Cited on p. 371.)

[35] Barabási, A.-L. 2016. *Network Science*. Cambridge: Cambridge University Press. (Cited on pp. 16, 472.)

[36] Barabási, A.-L., and Albert, R. 1999. Emergence of Scaling in Random Networks. *Science*, **286**(5439), 509–512. (Cited on pp. 344, 346, 349.)

[37] Barabási, A.-L., and Oltvai, Z. N. 2004. Network Biology: Understanding the Cell's Functional Organization. *Nature Reviews Genetics*, **5**(2), 101–113. (Cited on p. 25.)

[38] Barber, M. J., and Clark, J. W. 2009. Detecting Network Communities by Propagating Labels under Constraints. *Physical Review E*, **80**(2), 026129. (Cited on p. 469.)

[39] Barnes, E. R. 1982. An Algorithm for Partitioning the Nodes of a Graph. *SIAM Journal on Algebraic Discrete Methods*, **3**(4), 541–550. (Cited on p. 427.)

[40] Bartomeus, I., Vilà, M., and Santamaría, L. 2008. Contrasting Effects of Invasive Plants in Plant–Pollinator Networks. *Oecologia*, **155**(4), 761–770. (Cited on p. 23.)

[41] Bascompte, J. 2007. Networks in Ecology. *Basic and Applied Ecology*, **8**(6), 485–490. (Cited on p. 25.)

[42] Basser, P. J., Mattiello, J., and LeBihan, D. 1994. Estimation of the Effective Self-Diffusion Tensor from the NMR Spin Echo. *Journal of Magnetic Resonance, Series B*, **103**(3), 247–254. (Cited on p. 8.)

[43] Basser, P. J., Mattiello, J., and LeBihan, D. 1994. MR Diffusion Tensor Spectroscopy and Imaging. *Biophysical Journal*, **66**(1), 259–267. (Cited on p. 8.)

[44] Bassett, D. S., and Sporns, O. 2017. Network Neuroscience. *Nature Neuroscience*, **20**(3), 353–364. (Cited on pp. 25, 82.)

[45] Batagelj, V., and Brandes, U. 2005. Efficient Generation of Large Random Networks. *Physical Review E*, **71**(3), 036113. (Cited on p. 469.)

[46] Batson, J., Spielman, D. A., Srivastava, N., and Teng, S.-H. 2013. Spectral Sparsification of Graphs: Theory and Algorithms. *Communications of the ACM*, **56**(8), 87–94. (Cited on p. 134.)

[47] Battiston, S., Caldarelli, G., May, R. M., Roukny, T., and Stiglitz, J. E. 2016. The Price of Complexity in Financial Networks. *Proceedings of the National Academy of Sciences*, **113**(36), 10031–10036. (Cited on p. 22.)

[48] Belkin, M., and Niyogi, P. 2003. Laplacian Eigenmaps for Dimensionality Reduction and Data Representation. *Neural Computation*, **15**(6), 1373–1396. (Cited on p. 430.)

[49] Belkin, M., Hsu, D., Ma, S., and Mandal, S. 2019. Reconciling Modern Machine-Learning Practice and the Classical Bias–Variance Trade-Off. *Proceedings of the National Academy of Sciences*, **116**(32), 15849–15854. (Cited on p. 277.)

[50] Bender, E. M., Gebru, T., McMillan-Major, A., and Shmitchell, S. 2021. On the Dangers of Stochastic Parrots: Can Language Models Be Too Big? Pages 610–623 of: *Proceedings of the 2021 ACM Conference on Fairness, Accountability, and Transparency*. Virtual Event, Canada: Association for Computing Machinery. (Cited on p. 29.)

[51] Bengio, Y., Ducharme, R., Vincent, P., and Janvin, C. 2003. A Neural Probabilistic Language Model. *The Journal of Machine Learning Research*, **3**(null), 1137–1155. (Cited on p. 445.)

[52] Bentley, J., Knuth, D., and McIlroy, D. 1986. Programming Pearls: A Literate Program. *Communications of the ACM*, **29**(6), 471–483. (Cited on p. 313.)

[53] Bentley, J. L. 1975. Multidimensional Binary Search Trees Used for Associative Searching. *Communications of the ACM*, **18**(9), 509–517. (Cited on p. 426.)

[54] Bharat, K., Broder, A., Henzinger, M., Kumar, P., and Venkatasubramanian, S. 1998. The Connectivity Server: Fast Access to Linkage Information on the Web. *Computer Networks and ISDN Systems*, **30**(1-7), 469–477. (Cited on p. 469.)

[55] Bhattacharya, A., Friedland, S., and Peled, U. N. 2008. On the First Eigenvalue of Bipartite Graphs. *The Electronic Journal of Combinatorics [electronic only]*, **15**(1). (Cited on p. 197.)

[56] Bishop, C. M. 2006. *Pattern Recognition and Machine Learning*. Berlin, Heidelberg: Springer. (Cited on pp. 277, 472.)

[57] Blondel, V. D., Guillaume, J.-L., Lambiotte, R., and Lefebvre, E. 2008. Fast

Unfolding of Communities in Large Networks. *Journal of Statistical Mechanics: Theory and Experiment*, **2008**(10), P10008. (Cited on pp. 179, 358.)

[58] Bloom, B. H. 1970. Space/Time Trade-Offs in Hash Coding with Allowable Errors. *Communications of the ACM*, **13**(7), 422–426. (Cited on p. 455.)

[59] Boguñá, M., Pastor-Satorras, R., and Vespignani, A. 2004. Cut-Offs and Finite Size Effects in Scale-Free Networks. *The European Physical Journal B*, **38**(2), 205–209. (Cited on pp. 163, 173.)

[60] Boguñá, M., Bonamassa, I., De Domenico, M., Havlin, S., Krioukov, D., and Serrano, M. Á. 2021. Network Geometry. *Nature Reviews Physics*, **3**(2), 114–135. (Cited on p. 438.)

[61] Böhning, D., van der Heijden, P. G. M., and Bunge, J. (eds.). 2018. *Capture-Recapture Methods for the Social and Medical Sciences*. Boca Raton, FL: CRC Press/Taylor & Francis Group. (Cited on pp. 391, 395.)

[62] Boldi, P., and Vigna, S. 2004. The WebGraph Framework I: Compression Techniques. Pages 595–602 of: *Proceedings of the 13th International Conference on World Wide Web*. New York, NY: Association for Computing Machinery. (Cited on pp. 451, 469.)

[63] Boldi, P., Rosa, M., and Vigna, S. 2011. HyperANF: Approximating the Neighbourhood Function of Very Large Graphs on a Budget. Pages 625–634 of: *Proceedings of the 20th International Conference on World Wide Web*. WWW '11. New York, NY: Association for Computing Machinery. (Cited on pp. 456, 457, 469.)

[64] Boldi, P., Rosa, M., Santini, M., and Vigna, S. 2011. Layered Label Propagation: A Multiresolution Coordinate-Free Ordering for Compressing Social Networks. Pages 587–596 of: *Proceedings of the 20th International Conference on World Wide Web*. WWW '11. New York, NY: Association for Computing Machinery. (Cited on pp. 452, 469.)

[65] Bordes, A., Usunier, N., Garcia-Duran, A., Weston, J., and Yakhnenko, O. 2013. Translating Embeddings for Modeling Multi-relational Data. In: *Advances in Neural Information Processing Systems*, vol. 26. Curran Associates, Inc. (Cited on p. 437.)

[66] Borgatti, S. P., and Everett, M. G. 2000. Models of Core/Periphery Structures. *Social Networks*, **21**(4), 375–395. (Cited on pp. 194, 195, 201.)

[67] Borgatti, S. P., Carley, K. M., and Krackhardt, D. 2006. On the Robustness of Centrality Measures under Conditions of Imperfect Data. *Social Networks*, **28**(2), 124–136. (Cited on pp. 377, 395.)

[68] Borwein, J., and Bailey, D. 2013. The Reinhart-Rogoff Error – or How Not to Excel at Economics. *The Conversation*, Apr. (Cited on p. 36.)

[69] Bostrom, N. 2014. *Superintelligence: Paths, Dangers, Strategies*. Oxford: Oxford University Press. (Cited on p. 277.)

[70] Boswell, D., and Foucher, T. 2011. *The Art of Readable Code*. Theory in Practice. Sebastopol, CA: O'Reilly Media. (Cited on pp. 293, 300.)

[71] Bouveyron, C., Celeux, G., Murphy, T. B., and Raftery, A. E. 2019. *Model-Based Clustering and Classification for Data Science: With Applications in R.* Cambridge: Cambridge University Press. (Cited on p. 428.)

[72] boyd, d., Golder, S., and Lotan, G. 2010. Tweet, Tweet, Retweet: Conversational Aspects of Retweeting on Twitter. Pages 1–10 of: *2010 43rd Hawaii International Conference on System Sciences.* Honolulu, HI: IEEE. (Cited on p. 84.)

[73] Breiger, R. L., Boorman, S. A., and Arabie, P. 1975. An Algorithm for Clustering Relational Data with Applications to Social Network Analysis and Comparison with Multidimensional Scaling. *Journal of Mathematical Psychology*, **12**(3), 328–383. (Cited on p. 374.)

[74] Brenner, S. 2009. In the Beginning Was the Worm .... *Genetics*, **182**(2), 413–415. (Cited on p. 82.)

[75] Brin, S., and Page, L. 1998. The Anatomy of a Large-Scale Hypertextual Web Search Engine. *Computer Networks and ISDN Systems*, **30**(1), 107–117. (Cited on p. 189.)

[76] Broder, A. 1997. On the Resemblance and Containment of Documents. Pages 21–29 of: *Proceedings. Compression and Complexity of SEQUENCES 1997 (Cat. No.97TB100171).* Salerno, Italy: IEEE. (Cited on p. 455.)

[77] Brody, S., Alon, U., and Yahav, E. 2022. How Attentive Are Graph Attention Networks? In: *International Conference on Learning Representations.* Virtual Event, Austria: OpenReview.net. (Cited on pp. 273, 442.)

[78] Broido, A. D., and Clauset, A. 2019. Scale-Free Networks Are Rare. *Nature Communications*, **10**(1), 1017. (Cited on p. 346.)

[79] Brown, T., Mann, B., Ryder, N. et al. 2020. Language Models Are Few-Shot Learners. Pages 1877–1901 of: *Advances in Neural Information Processing Systems*, vol. 33. Vancouver, BC: Curran Associates, Inc. (Cited on p. 261.)

[80] Bryan, K., and Leise, T. 2006. The $25,000,000,000 Eigenvector: The Linear Algebra behind Google. *SIAM Review*, **48**(3), 569–581. (Cited on p. 427.)

[81] Buehrer, G., and Chellapilla, K. 2008. A Scalable Pattern Mining Approach to Web Graph Compression with Communities. Pages 95–106 of: *Proceedings of the 2008 International Conference on Web Search and Data Mining.* WSDM '08. New York, NY: Association for Computing Machinery. (Cited on p. 452.)

[82] Buluç, A., Meyerhenke, H., Safro, I., Sanders, P., and Schulz, C. 2016. Recent Advances in Graph Partitioning. Pages 117–158 of: Kliemann, L., and Sanders, P. (eds.), *Algorithm Engineering*, vol. 9220. Cham, Switzerland: Springer. (Cited on pp. 409, 427.)

[83] Buolamwini, J., and Gebru, T. 2018. Gender Shades: Intersectional Accuracy Disparities in Commercial Gender Classification. Pages 77–91 of: *Proceedings of the 1st Conference on Fairness, Accountability and Transparency.* PMLR. (Cited on p. 28.)

[84] Burnham, K. P., and Anderson, D. R. 2004. Multimodel Inference: Understanding AIC and BIC in Model Selection. *Sociological Methods & Research*, **33**(2), 261–304. (Cited on p. 162.)

[85] Burt, R. S. 1992. *Structural Holes: The Social Structure of Competition*. Cambridge, MA: Harvard University Press. (Cited on p. 15.)

[86] Busetto, L., Wick, W., and Gumbinger, C. 2020. How to Use and Assess Qualitative Research Methods. *Neurological Research and Practice*, **2**(1), 14. (Cited on p. 226.)

[87] Butts, C. T. 2003. Network Inference, Error, and Informant (in)Accuracy: A Bayesian Approach. *Social Networks*, **25**(2), 103–140. (Cited on p. 134.)

[88] Butts, C. T. 2009. Revisiting the Foundations of Network Analysis. *Science*, **325**(5939), 414–416. (Cited on p. 86.)

[89] Callaway, D. S., Hopcroft, J. E., Kleinberg, J. M., Newman, M. E. J., and Strogatz, S. H. 2001. Are Randomly Grown Graphs Really Random? *Physical Review E*, **64**(4), 041902. (Cited on p. 201.)

[90] Callaway, E. 2022. 'The Entire Protein Universe': AI Predicts Shape of Nearly Every Known Protein. *Nature*, **608**(7921), 15–16. (Cited on p. 17.)

[91] Cameron, A. C., and Trivedi, P. K. 2013. *Regression Analysis of Count Data*. 2nd edn. Econometric Society Monographs. Cambridge; New York, NY: Cambridge University Press. (Cited on pp. 159, 163.)

[92] Cantwell, G. T., Kirkley, A., and Newman, M. E. J. 2021. The Friendship Paradox in Real and Model Networks. *Journal of Complex Networks*, **9**(2), cnab011. (Cited on pp. 317, 324, 325.)

[93] Cao, J., Packer, J. S., Ramani, V. et al. 2017. Comprehensive Single-Cell Transcriptional Profiling of a Multicellular Organism. *Science*, **357**(6352), 661–667. (Cited on p. 375.)

[94] Cao, M., Shu, N., Cao, Q., Wang, Y., and He, Y. 2014. Imaging Functional and Structural Brain Connectomics in Attention-Deficit/Hyperactivity Disorder. *Molecular Neurobiology*, **50**(3), 1111–1123. (Cited on p. 85.)

[95] Caretta Cartozo, C., and De Los Rios, P. 2009. Extended Navigability of Small World Networks: Exact Results and New Insights. *Physical Review Letters*, **102**(23), 238703. (Cited on p. 344.)

[96] Carmi, S., Carter, S., Sun, J., and ben-Avraham, D. 2009. Asymptotic Behavior of the Kleinberg Model. *Physical Review Letters*, **102**(23), 238702. (Cited on p. 344.)

[97] Casey, B. J., Cannonier, T., Conley, M. I. et al. 2018. The Adolescent Brain Cognitive Development (ABCD) Study: Imaging Acquisition across 21 Sites. *Developmental Cognitive Neuroscience*, **32**(Aug.), 43–54. (Cited on pp. 19, 80.)

[98] Chami, I., Wolf, A., Juan, D.-C., Sala, F., Ravi, S., and Ré, C. 2020. Low-Dimensional Hyperbolic Knowledge Graph Embeddings. Pages 6901–6914 of: *Proceedings of the 58th Annual Meeting of the Association for Computational Linguistics*. Online: Association for Computational Linguistics. (Cited on p. 438.)

[99] Chami, I., Abu-El-Haija, S., Perozzi, B., Ré, C., and Murphy, K. 2022. Machine Learning on Graphs: A Model and Comprehensive Taxonomy. *Journal of Machine Learning Research*, **23**(89), 1–64. (Cited on p. 445.)

[100] Cheeger, J. 1969. A Lower Bound for the Smallest Eigenvalue of the Laplacian. Pages 195–200 of: Gunning, R. C. (ed.), *Problems in Analysis: A Symposium in Honor of Salomon Bochner (PMS-31)*. Princeton, NJ: Princeton University Press. (Cited on p. 427.)

[101] Chen, B. L., Hall, D. H., and Chklovskii, D. B. 2006. Wiring Optimization Can Relate Neuronal Structure and Function. *Proceedings of the National Academy of Sciences*, **103**(12), 4723–4728. (Cited on p. 227.)

[102] Chen, M., Mao, S., and Liu, Y. 2014. Big Data: A Survey. *Mobile Networks and Applications*, **19**(2), 171–209. (Cited on p. 469.)

[103] Ching, A., Edunov, S., Kabiljo, M., Logothetis, D., and Muthukrishnan, S. 2015. One Trillion Edges: Graph Processing at Facebook-Scale. *Proceedings of the VLDB Endowment*, **8**(12), 1804–1815. (Cited on pp. 448, 453.)

[104] Christen, P. 2012. *Data Matching: Concepts and Techniques for Record Linkage, Entity Resolution, and Duplicate Detection*. Data-Centric Systems and Applications. Berlin; New York: Springer. (Cited on p. 116.)

[105] Chu, C.-C., and Iu, H. H.-C. 2017. Complex Networks Theory For Modern Smart Grid Applications: A Survey. *IEEE Journal on Emerging and Selected Topics in Circuits and Systems*, **7**(2), 177–191. (Cited on p. 26.)

[106] Chung, F. 2007. Four Proofs for the Cheeger Inequality and Graph Partition Algorithms. Page 378 of: *Proceedings of ICCM*, vol. 2. Providence, RI; Boston, MA: American Mathematical Society & International Press of Boston, for Citeseer. (Cited on p. 427.)

[107] Chung, F., and Lu, L. 2002. Connected Components in Random Graphs with Given Expected Degree Sequences. *Annals of Combinatorics*, **6**(2), 125–145. (Cited on pp. 467, 468.)

[108] Chung, F. R. K. 1997. *Spectral Graph Theory*. Providence, RI: American Mathematical Society. (Cited on pp. 406, 413, 414, 427.)

[109] Clauset, A., Shalizi, C. R., and Newman, M. E. J. 2009. Power-Law Distributions in Empirical Data. *SIAM Review*, **51**(4), 661–703. (Cited on pp. 153, 163.)

[110] Cleveland, W. S., and McGill, R. 1984. Graphical Perception: Theory, Experimentation, and Application to the Development of Graphical Methods. *Journal of the American Statistical Association*, **79**(387), 531–554. (Cited on p. 214.)

[111] Cohen, R., Erez, K., ben-Avraham, D., and Havlin, S. 2000. Resilience of the Internet to Random Breakdowns. *Physical Review Letters*, **85**(21), 4626–4628. (Cited on pp. 167, 384, 386, 395.)

[112] Cohen, R., Erez, K., ben-Avraham, D., and Havlin, S. 2001. Breakdown of the Internet under Intentional Attack. *Physical Review Letters*, **86**(16), 3682–3685. (Cited on p. 167.)

[113] Colizza, V., Flammini, A., Serrano, M. A., and Vespignani, A. 2006. Detecting Rich-Club Ordering in Complex Networks. *Nature Physics*, **2**(2), 110–115. (Cited on p. 202.)

[114] Cong, J., and Liu, H. 2014. Approaching Human Language with Complex Networks. *Physics of Life Reviews*, **11**(4), 598–618. (Cited on p. 26.)

[115] Conover, M. D., Gonçalves, B., Flammini, A., and Menczer, F. 2012. Partisan Asymmetries in Online Political Activity. *EPJ Data Science*, **1**(1), 6. (Cited on p. 84.)

[116] Conte, D., Foggia, P., Sansone, C., and Vento, M. 2004. Thirty Years of Graph Matching in Pattern Recognition. *International Journal of Pattern Recognition and Artificial Intelligence*, **18**(03), 265–298. (Cited on p. 232.)

[117] Cormen, T. H., Leiserson, C. E., Rivest, R. L., and Stein, C. 2022. *Introduction to Algorithms*. 4th edn. Cambridge, MA: MIT Press. (Cited on p. 61.)

[118] Cutler, A. 1991. Nested Faunas and Extinction in Fragmented Habitats. *Conservation Biology*, **5**(4), 496–504. (Cited on p. 202.)

[119] Dai, H., Li, H., Tian, T. et al. 2018. Adversarial Attack on Graph Structured Data. Pages 1115–1124 of: *Proceedings of the 35th International Conference on Machine Learning*. PMLR. (Cited on p. 444.)

[120] Dalton, J. 1808. *A New System of Chemical Philosophy. Part I*. Cambridge: Cambridge University Press. (Cited on p. 4.)

[121] Dantzig, G. B., and Fulkerson, D. R. 1955. *On the Max Flow Min Cut Theorem of Networks*. Santa Monica, CA: RAND corporation. (Cited on pp. 201, 427.)

[122] Dawid, A. P., and Skene, A. M. 1979. Maximum Likelihood Estimation of Observer Error-Rates Using the EM Algorithm. *Journal of the Royal Statistical Society: Series C (Applied Statistics)*, **28**(1), 20–28. (Cited on p. 374.)

[123] de Saussure, F. 1959. *Course in General Linguistics*. New York, NY: Philosophical Library. (Cited on p. 432.)

[124] Dean, J., and Ghemawat, S. 2008. MapReduce: Simplified Data Processing on Large Clusters. *Communications of the ACM*, **51**(1), 107–113. (Cited on pp. 452, 453.)

[125] Decelle, A., Krzakala, F., Moore, C., and Zdeborová, L. 2011. Asymptotic Analysis of the Stochastic Block Model for Modular Networks and Its Algorithmic Applications. *Physical Review E*, **84**(6), 066106. (Cited on pp. 356, 357.)

[126] Decelle, A., Krzakala, F., Moore, C., and Zdeborová, L. 2011. Inference and Phase Transitions in the Detection of Modules in Sparse Networks. *Physical Review Letters*, **107**(6), 065701. (Cited on pp. 356, 357.)

[127] Dempster, A. P., Laird, N. M., and Rubin, D. B. 1977. Maximum Likelihood from Incomplete Data Via the EM Algorithm. *Journal of the Royal Statistical Society: Series B (Methodological)*, **39**(1), 1–22. (Cited on pp. 134, 360, 362.)

[128] Dessimoz, C., and Škunca, N. (eds.). 2017. *The Gene Ontology Handbook*. Methods in Molecular Biology, vol. 1446. New York, NY: Springer New York. (Cited on p. 116.)

[129] Devlin, J., Chang, M.-W., Lee, K., and Toutanova, K. 2019. BERT: Pre-training of Deep Bidirectional Transformers for Language Understanding. Pages 4171–4186 of: *Proceedings of the 2019 Conference of the North American Chapter of the Association for Computational Linguistics: Human Language Technologies, Volume 1 (Long and Short Papers)*. Minneapolis, MN: Association for Computational Linguistics. (Cited on p. 261.)

[130] Devroye, L. 1986. *Non-Uniform Random Variate Generation*. New York, NY: Springer. (Cited on p. 466.)

[131] Dodds, P. S., Muhamad, R., and Watts, D. J. 2003. An Experimental Study of Search in Global Social Networks. *Science*, **301**(5634), 827–829. (Cited on p. 341.)

[132] Donath, W. E., and Hoffman, A. J. 1973. Lower Bounds for the Partitioning of Graphs. *IBM Journal of Research and Development*, **17**(5), 420–425. (Cited on pp. 427, 428.)

[133] Dong, X., Thanou, D., Rabbat, M., and Frossard, P. 2019. Learning Graphs From Data: A Signal Representation Perspective. *IEEE Signal Processing Magazine*, **36**(3), 44–63. (Cited on p. 238.)

[134] Doolittle, W. F. 2013. Is Junk DNA Bunk? A Critique of ENCODE. *Proceedings of the National Academy of Sciences*, **110**(14), 5294–5300. (Cited on p. 122.)

[135] Dunbar, R. I. M. 1993. Coevolution of Neocortical Size, Group Size and Language in Humans. *Behavioral and Brain Sciences*, **16**(4), 681–694. (Cited on p. 181.)

[136] Dunning, T. 2012. *Natural Experiments in the Social Sciences: A Design-Based Approach*. Cambridge: Cambridge University Press. (Cited on p. 474.)

[137] Durand, M., and Flajolet, P. 2003. Loglog Counting of Large Cardinalities. Pages 605–617 of: Di Battista, G., and Zwick, U. (eds.), *Algorithms—ESA 2003*. Lecture Notes in Computer Science. Berlin, Heidelberg: Springer. (Cited on pp. 455, 469.)

[138] Dwork, C., and Roth, A. 2014. The Algorithmic Foundations of Differential Privacy. *Foundations and Trends® in Theoretical Computer Science*, **9**(3–4), 211–407. (Cited on p. 475.)

[139] Eades, P. 1984. A Heuristic for Graph Drawing. *Congressus Numerantium*, **vol.42**, 149–160. (Cited on p. 221.)

[140] Easley, D., and Kleinberg, J. 2010. *Networks, Crowds, and Markets: Reasoning about a Highly Connected World*. Cambridge: Cambridge University Press. (Cited on p. 472.)

[141] Edwards, P. N., Mayernik, M. S., Batcheller, A. L., Bowker, G. C., and Borgman, C. L. 2011. Science Friction: Data, Metadata, and Collaboration. *Social Studies of Science*, **41**(5), 667–690. (Cited on p. 292.)

[142] Efron, B., and Hastie, T. 2016. *Computer Age Statistical Inference*. Cambridge: Cambridge University Press. (Cited on pp. 62, 277, 472.)

[143] Ehrlich, B. 2022. *The Brain in Search of Itself: Santiago Ramón y Cajal and the Story of the Neuron*. New York, NY: Farrar, Straus and Giroux. (Cited on p. 16.)

[144] Einstein, A. 1905. Über Die von Der Molekularkinetischen Theorie Der Wärme Geforderte Bewegung von in Ruhenden Flüssigkeiten Suspendierten Teilchen. *Annalen der Physik*, **322**(8), 549–560. (Cited on p. 327.)

[145] Ellson, J., Gansner, E., Koutsofios, L., North, S. C., and Woodhull, G. 2002. Graphviz—Open Source Graph Drawing Tools. Pages 483–484 of: Mutzel, P.,

Jünger, M., and Leipert, S. (eds.), *Graph Drawing*. Lecture Notes in Computer Science. Berlin, Heidelberg: Springer. (Cited on p. 206.)

[146] Emmons, S. W. 2015. The Beginning of Connectomics: A Commentary on White et al. (1986) 'The Structure of the Nervous System of the Nematode Caenorhabditis Elegans'. *Philosophical Transactions of the Royal Society B: Biological Sciences*, **370**(1666), 20140309. (Cited on p. 82.)

[147] Eom, Y.-H., and Jo, H.-H. 2015. Generalized Friendship Paradox in Complex Networks: The Case of Scientific Collaboration. *Scientific Reports*, **4**(1), 4603. (Cited on p. 324.)

[148] Erdős, P., and Rényi, A. 1959. On Random Graphs I. *Publicationes Mathematicae*, **6**(1), 290–297. (Cited on pp. 163, 349.)

[149] Erdős, P., and Rényi, A. 1960. On the Evolution of Random Graphs. *Publ. Math. Inst. Hung. Acad. Sci*, **5**(1), 17–60. (Cited on pp. 163, 349.)

[150] Ester, M., Kriegel, H.-P., Sander, J., Xu, X., et al. 1996. A Density-Based Algorithm for Discovering Clusters in Large Spatial Databases with Noise. Pages 226–231 of: *Proceedings of the Second International Conference on Knowledge Discovery and Data Mining*. KDD'96, vol. 96. Portland, OR: AAAI Press. (Cited on p. 428.)

[151] Eswaran, D., Faloutsos, C., Guha, S., and Mishra, N. 2018. SpotLight: Detecting Anomalies in Streaming Graphs. Pages 1378–1386 of: *Proceedings of the 24th ACM SIGKDD International Conference on Knowledge Discovery & Data Mining*. KDD '18. New York, NY: Association for Computing Machinery. (Cited on p. 469.)

[152] Euzenat, J., and Shvaiko, P. 2013. *Ontology Matching*. 2nd edn. Berlin, Heidelberg: Springer. (Cited on p. 116.)

[153] Evans, T. S., and Lambiotte, R. 2009. Line Graphs, Link Partitions, and Overlapping Communities. *Physical Review E*, **80**(1), 016105. (Cited on pp. 184, 201.)

[154] Evtushenko, A., and Kleinberg, J. 2023. Node-Based Generalized Friendship Paradox Fails. *Scientific Reports*, **13**(1), 2074. (Cited on p. 324.)

[155] Fan, J., Han, F., and Liu, H. 2014. Challenges of Big Data Analysis. *National Science Review*, **1**(2), 293–314. (Cited on p. 469.)

[156] Feld, S. L. 1991. Why Your Friends Have More Friends Than You Do. *American Journal of Sociology*, **96**(6), 1464–1477. (Cited on p. 324.)

[157] Ferguson, N. 2019. *The Square and the Tower: Networks and Power, from the Freemasons to Facebook*. New York, NY: Penguin. (Cited on p. 16.)

[158] Fiedler, M. 1973. Algebraic Connectivity of Graphs. *Czechoslovak Mathematical Journal*, **23**(2), 298–305. (Cited on pp. 412, 427, 428.)

[159] Firth, J. R. 1957. *Studies in Linguistic Analysis*. Publications of the Philological Society. Oxford: Blackwell. (Cited on pp. 272, 432.)

[160] Flajolet, P., and Nigel Martin, G. 1985. Probabilistic Counting Algorithms for Data Base Applications. *Journal of Computer and System Sciences*, **31**(2), 182–209. (Cited on pp. 455, 469.)

[161] Flajolet, P., Fusy, É., Gandouet, O., and Meunier, F. 2007. HyperLogLog: The Analysis of a near-Optimal Cardinality Estimation Algorithm. Pages 137–156 of: *Discrete Mathematics and Theoretical Computer Science*, vol. AH. Nancy, France: Discrete Mathematics and Theoretical Computer Science. (Cited on pp. 455, 456, 469.)

[162] Ford, L. R., and Fulkerson, D. R. 1956. Maximal Flow Through a Network. *Canadian Journal of Mathematics*, **8**, 399–404. (Cited on p. 201.)

[163] Ford, L. R., and Fulkerson, D. R. 1962. *Flows in Networks*. Princeton, NJ: Princeton University Press. (Cited on p. 201.)

[164] Fortunato, S. 2010. Community Detection in Graphs. *Physics Reports*, **486**(3), 75–174. (Cited on p. 201.)

[165] Fortunato, S., and Barthélemy, M. 2007. Resolution Limit in Community Detection. *Proceedings of the National Academy of Sciences*, **104**(1), 36–41. (Cited on p. 201.)

[166] Fortunato, S., and Newman, M. E. J. 2022. 20 Years of Network Community Detection. *Nature Physics*, **18**(8), 848–850. (Cited on p. 201.)

[167] Fosdick, B. K., Larremore, D. B., Nishimura, J., and Ugander, J. 2018. Configuring Random Graph Models with Fixed Degree Sequences. *SIAM Review*, **60**(2), 315–355. (Cited on p. 163.)

[168] Frank, O., and Strauss, D. 1986. Markov Graphs. *Journal of the American Statistical Association*, **81**(395), 832–842. (Cited on pp. 163, 367.)

[169] Frantz, T. L., Cataldo, M., and Carley, K. M. 2009. Robustness of Centrality Measures under Uncertainty: Examining the Role of Network Topology. *Computational and Mathematical Organization Theory*, **15**(4), 303–328. (Cited on pp. 377, 395.)

[170] Friedman, J., Hastie, T., and Tibshirani, R. 2008. Sparse Inverse Covariance Estimation with the Graphical Lasso. *Biostatistics*, **9**(3), 432–441. (Cited on pp. 373, 374.)

[171] Fruchterman, T. M. J., and Reingold, E. M. 1991. Graph Drawing by Force-Directed Placement. *Software: Practice and Experience*, **21**(11), 1129–1164. (Cited on p. 221.)

[172] Gao, X., Xiao, B., Tao, D., and Li, X. 2010. A Survey of Graph Edit Distance. *Pattern Analysis and Applications*, **13**(1), 113–129. (Cited on p. 232.)

[173] Garcia, D. 2017. Leaking Privacy and Shadow Profiles in Online Social Networks. *Science Advances*, **3**(8), e1701172. (Cited on p. 37.)

[174] Garey, M., Johnson, D., and Stockmeyer, L. 1976. Some Simplified NP-complete Graph Problems. *Theoretical Computer Science*, **1**(3), 237–267. (Cited on p. 410.)

[175] Gauvin, L., Génois, M., Karsai, M. et al. 2022. Randomized Reference Models for Temporal Networks. *SIAM Review*, **64**(4), 763–830. (Cited on p. 243.)

[176] Gawande, A. 2010. *The Checklist Manifesto: How to Get Things Right*. New York, NY: Metropolitan Books. (Cited on p. 285.)

[177] Gebru, T., Morgenstern, J., Vecchione, B. et al. 2020. Datasheets for Datasets. *arXiv:1803.09010 [cs]*, Mar. (Cited on pp. 31, 474.)

[178] Gelman, A., and Hill, J. 2007. *Data Analysis Using Regression and Multi-level/Hierarchical Models*. Analytical Methods for Social Research. Cambridge; New York: Cambridge University Press. (Cited on pp. 116, 134.)

[179] Gelman, A., Carlin, Stern, H. S., Dunson, D. B., Vehtari, A., and Rubin, D. B. 2014. *Bayesian Data Analysis*. 3rd edn. Chapman & Hall/CRC Texts in Statistical Science. Boca Raton, FL: CRC Press. (Cited on pp. 62, 375, 472.)

[180] Gertner, J. 2013. *The Idea Factory: Bell Labs and the Great Age of American Innovation*. London: Penguin Books. (Cited on p. 313.)

[181] Ghosh, S., Das, N., Gonçalves, T., Quaresma, P., and Kundu, M. 2018. The Journey of Graph Kernels through Two Decades. *Computer Science Review*, **27**(Feb.), 88–111. (Cited on p. 232.)

[182] Gilbert, E. N. 1959. Random Graphs. *The Annals of Mathematical Statistics*, **30**(4), 1141–1144. (Cited on pp. 163, 349.)

[183] Girvan, M., and Newman, M. E. J. 2002. Community Structure in Social and Biological Networks. *Proceedings of the National Academy of Sciences*, **99**(12), 7821–7826. (Cited on pp. 185, 201.)

[184] Golub, G. H., and Van Loan, C. F. 2013. *Matrix Computations*. 4th edn. Johns Hopkins Studies in the Mathematical Sciences. Baltimore, MD: The Johns Hopkins University Press. (Cited on pp. 400, 427.)

[185] Gonzalez, J. E., Low, Y., Gu, H., Bickson, D., and Guestrin, C. 2012. Power-Graph: Distributed Graph-Parallel Computation on Natural Graphs. Pages 17–30 of: *Proceedings of the 10th USENIX Conference on Operating Systems Design and Implementation*. OSDI'12. Hollywood, CA: USENIX Association. (Cited on p. 453.)

[186] Goodfellow, I., Bengio, Y., and Courville, A. 2016. *Deep Learning*. Cambridge, MA: MIT Press. (Cited on pp. 277, 472.)

[187] Gower, J. C. 1975. Generalized Procrustes Analysis. *Psychometrika*, **40**(1), 33–51. (Cited on p. 371.)

[188] Goyal, P., and Ferrara, E. 2018. Graph Embedding Techniques, Applications, and Performance: A Survey. *Knowledge-Based Systems*, **151**(July), 78–94. (Cited on p. 445.)

[189] Granovetter, M. 1976. Network Sampling: Some First Steps. *American Journal of Sociology*, **81**(6), 1287–1303. (Cited on p. 395.)

[190] Granovetter, M. S. 1973. The Strength of Weak Ties. *American Journal of Sociology*, **78**(6), 1360–1380. (Cited on pp. 21, 343.)

[191] Gray, J., Liu, D. T., Nieto-Santisteban, M., Szalay, A., DeWitt, D. J., and Heber, G. 2005. Scientific Data Management in the Coming Decade. *ACM SIGMOD Record*, **34**(4), 34–41. (Cited on p. 292.)

[192] Grohe, M., and Schweitzer, P. 2020. The Graph Isomorphism Problem. *Communications of the ACM*, **63**(11), 128–134. (Cited on p. 445.)

[193] Grover, A., and Leskovec, J. 2016. Node2vec: Scalable Feature Learning for Networks. Pages 855–864 of: *Proceedings of the 22nd ACM SIGKDD International Conference on Knowledge Discovery and Data Mining*. KDD '16. New York, NY: Association for Computing Machinery. (Cited on p. 435.)

[194] Groves, R. M. 2011. Three Eras of Survey Research. *Public Opinion Quarterly*, **75**(5), 861–871. (Cited on pp. 65, 68, 71.)

[195] Guimerà, R., and Sales-Pardo, M. 2009. Missing and Spurious Interactions and the Reconstruction of Complex Networks. *Proceedings of the National Academy of Sciences*, **106**(52), 22073–22078. (Cited on p. 134.)

[196] Guimerà, R., Sales-Pardo, M., and Amaral, L. A. N. 2004. Modularity from Fluctuations in Random Graphs and Complex Networks. *Physical Review E*, **70**(2), 025101. (Cited on p. 179.)

[197] Gutmann, M., and Hyvärinen, A. 2010. Noise-Contrastive Estimation: A New Estimation Principle for Unnormalized Statistical Models. In: *Proceedings of the Thirteenth International Conference on Artificial Intelligence and Statistics*. Proceedings of Machine Learning Research. Sardinia, Italy: PMLR. (Cited on p. 435.)

[198] Hagberg, A., and Lemons, N. 2015. Fast Generation of Sparse Random Kernel Graphs. *PLOS ONE*, **10**(9), e0135177. (Cited on p. 469.)

[199] Hall, K. M. 1970. An R-Dimensional Quadratic Placement Algorithm. *Management Science*, **17**(3), 219–229. (Cited on p. 221.)

[200] Hamilton, W., Ying, Z., and Leskovec, J. 2017. Inductive Representation Learning on Large Graphs. In: Guyon, I., Luxburg, U. V., Bengio, S. et al. (eds.), *Advances in Neural Information Processing Systems*, vol. 30. Long Beach, CA: Curran Associates, Inc. (Cited on pp. 273, 442.)

[201] Hanel, R., Corominas-Murtra, B., Liu, B., and Thurner, S. 2017. Fitting Power-Laws in Empirical Data with Estimators That Work for All Exponents. *PLOS ONE*, **12**(2), e0170920. (Cited on p. 163.)

[202] Harris, Z. S. 1954. Distributional Structure. *WORD*, **10**(2-3), 146–162. (Cited on pp. 272, 432.)

[203] Hartle, H., Klein, B., McCabe, S. et al. 2020. Network Comparison and the Within-Ensemble Graph Distance. *Proceedings of the Royal Society A: Mathematical, Physical and Engineering Sciences*, **476**(2243), 20190744. (Cited on p. 232.)

[204] Hashimoto, K.-i. 1989. Zeta Functions of Finite Graphs and Representations of P-Adic Groups. Pages 211–280 of: *Automorphic Forms and Geometry of Arithmetic Varieties*. Advanced Studies in Pure Mathematics, vol. 15. Cambridge, MA: Academic Press. (Cited on pp. 357, 403.)

[205] Hastie, T., Tibshirani, R., and Friedman, J. H. 2009. *The Elements of Statistical Learning: Data Mining, Inference, and Prediction*. 2nd edn. Springer Series in Statistics. New York, NY: Springer. (Cited on pp. 277, 428, 472.)

[206] Hastie, T., Tibshirani, R., and Wainwright, M. 2015. *Statistical Learning with*

Sparsity: The Lasso and Generalizations. New York, NY: Taylor & Francis. (Cited on pp. 277, 375.)

[207] Head, M. L., Holman, L., Lanfear, R., Kahn, A. T., and Jennions, M. D. 2015. The Extent and Consequences of P-Hacking in Science. *PLOS Biology*, **13**(3), e1002106. (Cited on pp. 35, 71.)

[208] Healy, K. 2018. *Data Visualization: A Practical Introduction.* Princeton, NJ: Princeton University Press. (Cited on p. 473.)

[209] Heinrich, J. 2001. *Drug Safety: Most Drugs Withdrawn in Recent Years Had Greater Health Risks for Women.* Tech. rept. GAO-01-286R. US Government Printing Office, Washington, DC. (Cited on p. 28.)

[210] Hellman, M. 2002. An Overview of Public Key Cryptography. *IEEE Communications Magazine*, **40**(5), 42–49. (Cited on p. 313.)

[211] Hennig, C., Meila, M., Murtagh, F., and Rocci, R. (eds.). 2016. *Handbook of Cluster Analysis.* Chapman & Hall/CRC Handbooks of Modern Statistical Methods. Boca Raton, FL; London; New York: CRC Press, Taylor & Francis Group. (Cited on p. 428.)

[212] Heule, S., Nunkesser, M., and Hall, A. 2013. HyperLogLog in Practice: Algorithmic Engineering of a State of the Art Cardinality Estimation Algorithm. Pages 683–692 of: *Proceedings of the 16th International Conference on Extending Database Technology.* EDBT '13. New York, NY: Association for Computing Machinery. (Cited on p. 469.)

[213] Hočevar, T., and Demšar, J. 2014. A Combinatorial Approach to Graphlet Counting. *Bioinformatics*, **30**(4), 559–565. (Cited on p. 175.)

[214] Hodas, N., Kooti, F., and Lerman, K. 2013. Friendship Paradox Redux: Your Friends Are More Interesting Than You. *Proceedings of the International AAAI Conference on Web and Social Media*, **7**(1), 225–233. (Cited on p. 324.)

[215] Hofer, M., Hellmann, S., Dojchinovski, M., and Frey, J. 2020. The New DBpedia Release Cycle: Increasing Agility and Efficiency in Knowledge Extraction Workflows. Pages 1–18 of: Blomqvist, E., Groth, P., de Boer, V. et al. (eds.), *Semantic Systems. In the Era of Knowledge Graphs.* Lecture Notes in Computer Science. Cham, Switzerland: Springer. (Cited on p. 449.)

[216] Hoff, P. D., Raftery, A. E., and Handcock, M. S. 2002. Latent Space Approaches to Social Network Analysis. *Journal of the American Statistical Association*, **97**(460), 1090–1098. (Cited on pp. 369, 370, 371, 374.)

[217] Hofman, J. M., Watts, D. J., Athey, S. et al. 2021. Integrating Explanation and Prediction in Computational Social Science. *Nature*, **595**(7866), 181–188. (Cited on p. 474.)

[218] Holland, P. W., and Leinhardt, S. 1981. An Exponential Family of Probability Distributions for Directed Graphs. *Journal of the American Statistical Association*, **76**(373), 33–50. (Cited on p. 163.)

[219] Holland, P. W., Laskey, K. B., and Leinhardt, S. 1983. Stochastic Blockmodels: First Steps. *Social Networks*, **5**(2), 109–137. (Cited on pp. 163, 374.)

[220] Holme, P. 2019. Rare and Everywhere: Perspectives on Scale-Free Networks. *Nature Communications*, **10**(1), 1016. (Cited on p. 346.)

[221] Holme, P., and Saramäki, J. 2012. Temporal Networks. *Physics Reports*, **519**(3), 97–125. (Cited on pp. 249, 472.)

[222] Holme, P., and Saramäki, J. (eds.). 2019. *Temporal Network Theory*. Computational Social Science. Cham, Switzerland: Springer Naure. (Cited on p. 249.)

[223] Holten, D., and Van Wijk, J. J. 2009. Force-Directed Edge Bundling for Graph Visualization. *Computer Graphics Forum*, **28**(3), 983–990. (Cited on p. 213.)

[224] Hosmer, D. W., Lemeshow, S., and Sturdivant, R. X. 2013. *Applied Logistic Regression*. Hoboken, NJ: John Wiley & Sons. (Cited on p. 256.)

[225] Howe, J., et al. 2006. The Rise of Crowdsourcing. *Wired magazine*, **14**(6), 1–4. (Cited on p. 374.)

[226] Humphreys, I. R., Pei, J., Baek, M. et al. 2021. Computed Structures of Core Eukaryotic Protein Complexes. *Science*, **374**(6573), eabm4805. (Cited on p. 17.)

[227] Hyndman, R. J., and Athanasopoulos, G. 2021. *Forecasting: Principles and Practice*. 3rd edn. Melbourne, Australia: Otexts, Online Open-Access Textbooks. (Cited on p. 249.)

[228] Israel, M., and Hay, I. 2006. *Research Ethics for Social Scientists*. Thousand Oaks, CA: SAGE. (Cited on p. 37.)

[229] Jackson, M. O. 2019. The Friendship Paradox and Systematic Biases in Perceptions and Social Norms. *Journal of Political Economy*, **127**(2), 777–818. (Cited on p. 325.)

[230] James, G., Witten, D., Hastie, T., and Tibshirani, R. 2021. *An Introduction to Statistical Learning: With Applications in R*. 2nd edn. Springer Texts in Statistics. New York, NY: Springer. (Cited on pp. 277, 472.)

[231] Jha, M., Seshadri, C., and Pinar, A. 2015. A Space-Efficient Streaming Algorithm for Estimating Transitivity and Triangle Counts Using the Birthday Paradox. *ACM Transactions on Knowledge Discovery from Data*, **9**(3), 15:1–15:21. (Cited on p. 454.)

[232] Kanare, H. M. 1985. *Writing the Laboratory Notebook*. Washington, DC: American Chemical Society. (Cited on pp. 283, 287.)

[233] Kanungo, T., Mount, D., Netanyahu, N., Piatko, C., Silverman, R., and Wu, A. 2002. An Efficient K-Means Clustering Algorithm: Analysis and Implementation. *IEEE Transactions on Pattern Analysis and Machine Intelligence*, **24**(7), 881–892. (Cited on p. 426.)

[234] Karger, D., Oh, S., and Shah, D. 2011. Iterative Learning for Reliable Crowdsourcing Systems. In: *Advances in Neural Information Processing Systems*, vol. 24. Granada, Spain: Curran Associates, Inc. (Cited on p. 374.)

[235] Karinthy, F. 1929. Chain-Links. *Everything is different*, 21–26. (Cited on p. 15.)

[236] Karrer, B., and Newman, M. E. J. 2011. Stochastic Blockmodels and Community Structure in Networks. *Physical Review E*, **83**(1), 016107. (Cited on p. 355.)

[237] Katz, L. 1953. A New Status Index Derived from Sociometric Analysis. *Psychometrika*, **18**(1), 39–43. (Cited on pp. 189, 408, 427.)

[238] Kaufman, L., and Rousseeuw, P. J. 2009. *Finding Groups in Data: An Introduction to Cluster Analysis*. New York, NY: John Wiley & Sons. (Cited on pp. 232, 428.)

[239] Kernighan, B. W., and Lin, S. 1970. An Efficient Heuristic Procedure for Partitioning Graphs. *The Bell System Technical Journal*, **49**(2), 291–307. (Cited on pp. 201, 427.)

[240] Kernighan, B. W. 2020. *UNIX: A History and a Memoir*. Seattle, WA: Kindle Direct Publishing. (Cited on p. 313.)

[241] Kessler, M. D., Yerges-Armstrong, L., Taub, M. A. et al. 2016. Challenges and Disparities in the Application of Personalized Genomic Medicine to Populations with African Ancestry. *Nature Communications*, **7**(1), 12521. (Cited on p. 28.)

[242] Kipf, T. N., and Welling, M. 2017. Semi-Supervised Classification with Graph Convolutional Networks. In: *International Conference on Learning Representations*. Toulon, France: OpenReview.net. (Cited on pp. 273, 442.)

[243] Kirchgässner, G., Wolters, J., and Hassler, U. 2013. *Introduction to Modern Time Series Analysis*. 2nd edn. Springer Texts in Business and Economics. Berlin, Heidelberg: Springer. (Cited on p. 249.)

[244] Kleinberg, J. M. 1999. Authoritative Sources in a Hyperlinked Environment. *Journal of the ACM*, **46**(5), 604–632. (Cited on p. 189.)

[245] Kleinberg, J. M. 2000. Navigation in a Small World. *Nature*, **406**(6798), 845–845. (Cited on p. 344.)

[246] Kleinberg, J. M., Kumar, R., Raghavan, P., Rajagopalan, S., and Tomkins, A. S. 1999. The Web as a Graph: Measurements, Models, and Methods. Pages 1–17 of: Goos, G., Hartmanis, J., van Leeuwen, J. et al. (eds.), *Computing and Combinatorics*, vol. 1627. Berlin, Heidelberg: Springer. (Cited on pp. 346, 349.)

[247] Knuth, D. E. 1984. Literate Programming. *The Computer Journal*, **27**(2), 97–111. (Cited on p. 301.)

[248] Kojaku, S., and Masuda, N. 2017. Finding Multiple Core-Periphery Pairs in Networks. *Physical Review E*, **96**(5), 052313. (Cited on p. 195.)

[249] Kojaku, S., Yoon, J., Constantino, I., and Ahn, Y.-Y. 2021. Residual2Vec: Debiasing Graph Embedding with Random Graphs. Pages 24150–24163 of: *Advances in Neural Information Processing Systems*, vol. 34. Online: Curran Associates, Inc. (Cited on pp. 435, 445.)

[250] Koponen, J., and Hildén, J. 2019. *Data Visualization Handbook*. Aalto, Finland: Aalto University. (Cited on p. 473.)

[251] Kosinski, M., Stillwell, D., and Graepel, T. 2013. Private Traits and Attributes Are Predictable from Digital Records of Human Behavior. *Proceedings of the National Academy of Sciences*, **110**(15), 5802–5805. (Cited on p. 32.)

[252] Köster, J., and Rahmann, S. 2012. Snakemake—a Scalable Bioinformatics Workflow Engine. *Bioinformatics*, **28**(19), 2520–2522. (Cited on p. 302.)

[253] Krapivsky, P. L., and Redner, S. 2001. Organization of Growing Random Networks. *Physical Review E*, **63**(6), 066123. (Cited on pp. 345, 346, 348, 349.)

[254] Krioukov, D., Papadopoulos, F., Kitsak, M., Vahdat, A., and Boguñá, M. 2010. Hyperbolic Geometry of Complex Networks. *Physical Review E*, **82**(3), 036106. (Cited on p. 438.)

[255] Kruja, E., Marks, J., Blair, A., and Waters, R. 2002. A Short Note on the History of Graph Drawing. Pages 272–286 of: Mutzel, P., Jünger, M., and Leipert, S. (eds.), *Graph Drawing*. Lecture Notes in Computer Science. Berlin, Heidelberg: Springer. (Cited on p. 221.)

[256] Kruschke, J. K. 2015. *Doing Bayesian Data Analysis: A Tutorial with R, JAGS, and Stan*. 2nd edn. Boston, MA: Academic Press. (Cited on pp. 62, 472.)

[257] Kruskal, J. B., and Wish, M. 1978. *Multidimensional Scaling*. Sage University Papers Quantitative Applications in the Social Sciences, no. 11. Newbury Park, CA: SAGE. (Cited on p. 370.)

[258] Kryshtafovych, A., Schwede, T., Topf, M., Fidelis, K., and Moult, J. 2021. Critical Assessment of Methods of Protein Structure Prediction (CASP)—Round XIV. *Proteins: Structure, Function, and Bioinformatics*, **89**(12), 1607–1617. (Cited on p. 17.)

[259] Krzakala, F., Moore, C., Mossel, E. et al. 2013. Spectral Redemption in Clustering Sparse Networks. *Proceedings of the National Academy of Sciences*, **110**(52), 20935–20940. (Cited on pp. 356, 357, 421.)

[260] Kumar, S., Morstatter, F., and Liu, H. 2014. *Twitter Data Analytics*. Springer-Briefs in Computer Science. New York, NY: Springer. (Cited on p. 71.)

[261] Landauer, T. K., Foltz, P. W., and Laham, D. 1998. An Introduction to Latent Semantic Analysis. *Discourse Processes*, **25**(2-3), 259–284. (Cited on pp. 438, 445.)

[262] Lang, T. A., and Altman, D. G. 2015. Basic Statistical Reporting for Articles Published in Biomedical Journals: The "Statistical Analyses and Methods in the Published Literature" or the SAMPL Guidelines. *International Journal of Nursing Studies*, **52**(1), 5–9. (Cited on p. 163.)

[263] Latapy, M., Viard, T., and Magnien, C. 2018. Stream Graphs and Link Streams for the Modeling of Interactions over Time. *Social Network Analysis and Mining*, **8**(1), 61. (Cited on p. 240.)

[264] Lazer, D., Pentland, A., Adamic, L. et al. 2009. Computational Social Science. *Science*, **323**(5915), 721–723. (Cited on p. 25.)

[265] Lazer, D. M. J., Pentland, A., Watts, D. J. et al. 2020. Computational Social Science: Obstacles and Opportunities. *Science*, **369**(6507), 1060–1062. (Cited on p. 474.)

[266] Lee, C., and Wilkinson, D. J. 2019. A Review of Stochastic Block Models and Extensions for Graph Clustering. *Applied Network Science*, **4**(1), 1–50. (Cited on pp. 353, 374.)

[267] Lee, J. R., Gharan, S. O., and Trevisan, L. 2014. Multiway Spectral Partitioning and Higher-Order Cheeger Inequalities. *Journal of the ACM*, **61**(6), 1–30. (Cited on pp. 417, 427.)

[268] Lehmann, S. 2019. Fundamental Structures in Temporal Communication Networks. Pages 25–48 of: Holme, P., and Saramäki, J. (eds.), *Temporal Network Theory*. Computational Social Sciences. Cham, Switzerland: Springer. (Cited on p. 473.)

[269] Levy, O., and Goldberg, Y. 2014. Neural Word Embedding as Implicit Matrix Factorization. In: *Proceedings of the 27th International Conference on Neural Information Processing Systems*, vol. 2. Montreal, Canada: Curran Associates, Inc. (Cited on pp. 272, 438, 439, 440, 445.)

[270] Levy, O., Goldberg, Y., and Dagan, I. 2015. Improving Distributional Similarity with Lessons Learned from Word Embeddings. *Transactions of the Association for Computational Linguistics*, 3(Dec.), 211–225. (Cited on p. 272.)

[271] Lewandowsky, S., Mann, M. E., Bauld, L., Hastings, G., and Loftus, a. E. F. 2013. The Subterranean War on Science. *APS Observer*, 26. (Cited on p. 284.)

[272] Lewis, S. 2019. The Racial Bias Built into Photography. *The New York Times*, Apr. (Cited on p. 29.)

[273] Liben-Nowell, D., and Kleinberg, J. 2007. The Link-Prediction Problem for Social Networks. *Journal of the American Society for Information Science and Technology*, 58(7), 1019–1031. (Cited on p. 134.)

[274] Liggett, T. M. 1999. *Stochastic Interacting Systems: Contact, Voter, and Exclusion Processes*. Grundlehren Der Mathematischen Wissenschaften, no. 324. Berlin; New York: Springer. (Cited on p. 235.)

[275] Liggett, T. M. 2005. *Interacting Particle Systems*. Classics in Mathematics. Berlin; New York: Springer. (Cited on p. 235.)

[276] Lin, J. W.-B., Aizenman, H., Espinel, E. M. C., Gunnerson, K. N., and Liu, J. 2022. *An Introduction to Python Programming for Scientists and Engineers*. Cambridge; New York, NY: Cambridge University Press. (Cited on p. 61.)

[277] Lin, T., Wang, Y., Liu, X., and Qiu, X. 2022. A Survey of Transformers. *AI Open*, 3(Jan.), 111–132. (Cited on p. 433.)

[278] Liu, Y., Safavi, T., Dighe, A., and Koutra, D. 2018. Graph Summarization Methods and Applications: A Survey. *ACM Computing Surveys*, 51(3), 62:1–62:34. (Cited on pp. 232, 452, 469.)

[279] Louis, A., Raghavendra, P., Tetali, P., and Vempala, S. 2012. Many Sparse Cuts via Higher Eigenvalues. Pages 1131–1140 of: *Proceedings of the Forty-Fourth Annual ACM Symposium on Theory of Computing*. New York, NY: Association for Computing Machinery. (Cited on pp. 417, 427.)

[280] Low, Y., Gonzalez, J., Kyrola, A., Bickson, D., Guestrin, C., and Hellerstein, J. 2010. GraphLab: A New Framework for Parallel Machine Learning. Pages 340–349 of: *Proceedings of the Twenty-Sixth Conference on Uncertainty in Artificial Intelligence*. UAI'10. Arlington, VA: AUAI Press. (Cited on p. 453.)

[281] Low, Y., Bickson, D., Gonzalez, J., Guestrin, C., Kyrola, A., and Hellerstein, J. M. 2012. Distributed GraphLab: A Framework for Machine Learning and Data Mining in the Cloud. *Proceedings of the VLDB Endowment*, 5(8), 716–727. (Cited on p. 453.)

[282] Lü, L., and Zhou, T. 2011. Link Prediction in Complex Networks: A Survey. *Physica A: Statistical Mechanics and its Applications*, **390**(6), 1150–1170. (Cited on p. 134.)

[283] Luck, K., Kim, D.-K., Lambourne, L. et al. 2020. A Reference Map of the Human Binary Protein Interactome. *Nature*, **580**(7803), 402–408. (Cited on pp. 24, 84, 130, 389, 390.)

[284] Lusher, D., Koskinen, J., and Robins, G. (eds.). 2013. *Exponential Random Graph Models for Social Networks: Theory, Methods, and Applications*. Structural Analysis in the Social Sciences, no. 32. Cambridge: Cambridge University Press. (Cited on p. 163.)

[285] Macleod, M., Collings, A. M., Graf, C. et al. 2021. The MDAR (Materials Design Analysis Reporting) Framework for Transparent Reporting in the Life Sciences. *Proceedings of the National Academy of Sciences*, **118**(17), e2103238118. (Cited on p. 163.)

[286] Madden, S. 2012. From Databases to Big Data. *IEEE Internet Computing*, **16**(3), 4–6. (Cited on p. 469.)

[287] Makalowski, W. 2003. Not Junk After All. *Science*, **300**(5623), 1246–1247. (Cited on p. 122.)

[288] Makin, T. R., and Orban de Xivry, J.-J. 2019. Ten Common Statistical Mistakes to Watch out for When Writing or Reviewing a Manuscript. *eLife*, **8**(Oct.), e48175. (Cited on p. 163.)

[289] Malewicz, G., Austern, M. H., Bik, A. J. et al. 2010. Pregel: A System for Large-Scale Graph Processing. Pages 135–146 of: *Proceedings of the 2010 ACM SIGMOD International Conference on Management of Data*. SIGMOD '10. New York, NY: Association for Computing Machinery. (Cited on p. 453.)

[290] Manzoor, E., Milajerdi, S. M., and Akoglu, L. 2016. Fast Memory-efficient Anomaly Detection in Streaming Heterogeneous Graphs. Pages 1035–1044 of: *Proceedings of the 22nd ACM SIGKDD International Conference on Knowledge Discovery and Data Mining*. KDD '16. New York, NY: Association for Computing Machinery. (Cited on p. 469.)

[291] Marai, G. E., Pinaud, B., Bühler, K., Lex, A., and Morris, J. H. 2019. Ten Simple Rules to Create Biological Network Figures for Communication. *PLOS Computational Biology*, **15**(9), e1007244. (Cited on p. 221.)

[292] Mariani, M. S., Ren, Z.-M., Bascompte, J., and Tessone, C. J. 2019. Nestedness in Complex Networks: Observation, Emergence, and Implications. *Physics Reports*, **813**(June), 1–90. (Cited on pp. 197, 198, 202.)

[293] Maron, H., Ben-Hamu, H., Shamir, N., and Lipman, Y. 2019. Invariant and Equivariant Graph Networks. In: *International Conference on Learning Representations*. New Orleans, LA: OpenReview.net. (Cited on p. 443.)

[294] Martin, C., and Niemeyer, P. 2019. Influence of Measurement Errors on Networks: Estimating the Robustness of Centrality Measures. *Network Science*, **7**(2), 180–195. (Cited on pp. 377, 395.)

[295] Martin, T., Ball, B., and Newman, M. E. J. 2016. Structural Inference for Uncertain Networks. *Physical Review E*, **93**(1), 012306. (Cited on p. 134.)

[296] Marx, V. 2013. The Big Challenges of Big Data. *Nature*, **498**(7453), 255–260. (Cited on p. 469.)

[297] Masuda, N., and Lambiotte, R. 2020. *A Guide to Temporal Networks*. 2nd edn. Series on Complexity Science, vol. 06. London: World Scientific. (Cited on pp. 238, 240, 241, 242, 249, 472.)

[298] Masuda, N., Porter, M. A., and Lambiotte, R. 2017. Random Walks and Diffusion on Networks. *Physics Reports*, **716–717**(Nov.), 1–58. (Cited on p. 428.)

[299] Matejka, J., and Fitzmaurice, G. 2017. Same Stats, Different Graphs: Generating Datasets with Varied Appearance and Identical Statistics through Simulated Annealing. Pages 1290–1294 of: *Proceedings of the 2017 CHI Conference on Human Factors in Computing Systems*. Denver, CO: Association for Computing Machinery. (Cited on p. 203.)

[300] McGregor, A. 2014. Graph Stream Algorithms: A Survey. *ACM SIGMOD Record*, **43**(1), 9–20. (Cited on pp. 453, 454, 469.)

[301] McGregor, A., Vorotnikova, S., and Vu, H. T. 2016. Better Algorithms for Counting Triangles in Data Streams. Pages 401–411 of: *Proceedings of the 35th ACM SIGMOD-SIGACT-SIGAI Symposium on Principles of Database Systems*. San Francisco, CA: Association for Computing Machinery. (Cited on pp. 454, 469.)

[302] McInnes, L., Healy, J., and Melville, J. 2018. *UMAP: Uniform Manifold Approximation and Projection for Dimension Reduction*. arXiv:1802.03426. (Cited on p. 259.)

[303] Mead, C., and Conway, L. 1980. *Introduction to VLSI Systems*. Reading, MA: Addison-Wesley. (Cited on p. 26.)

[304] Meinshausen, N., and Bühlmann, P. 2006. High-Dimensional Graphs and Variable Selection with the Lasso. *The Annals of Statistics*, **34**(3). (Cited on p. 371.)

[305] Menczer, F., Fortunato, S., and Davis, C. A. 2020. *A First Course in Network Science*. Cambridge: Cambridge University Press. (Cited on pp. 16, 472.)

[306] Meusel, R., Vigna, S., Lehmberg, O., and Bizer, C. 2014. Graph Structure in the Web — Revisited: A Trick of the Heavy Tail. Pages 427–432 of: *Proceedings of the 23rd International Conference on World Wide Web*. WWW '14 Companion. New York, NY: Association for Computing Machinery. (Cited on p. 448.)

[307] Mikolov, T., Sutskever, I., Chen, K., Corrado, G. S., and Dean, J. 2013. Distributed Representations of Words and Phrases and Their Compositionality. In: *Advances in Neural Information Processing Systems*, vol. 26. Lake Tahoe, NV: Curran Associates, Inc. (Cited on pp. 272, 431, 434, 445.)

[308] Milgram, S. 1963. Behavioral Study of Obedience. *The Journal of Abnormal and Social Psychology*, **67**, 371–378. (Cited on p. 341.)

[309] Milgram, S. 1967. The Small-World Problem. *Psychology Today*, **1**, 61. (Cited on pp. 15, 192, 458.)

[310] Miller, J. C., and Hagberg, A. 2011. Efficient Generation of Networks with Given Expected Degrees. Pages 115–126 of: Frieze, A., Horn, P., and Prałat, P. (eds.), *Algorithms and Models for the Web Graph*, vol. 6732. Berlin, Heidelberg: Springer. (Cited on pp. 468, 469.)

[311] Milo, R., Shen-Orr, S., Itzkovitz, S., Kashtan, N., Chklovskii, D., and Alon, U. 2002. Network Motifs: Simple Building Blocks of Complex Networks. *Science*, **298**(5594), 824–827. (Cited on p. 201.)

[312] Mitchell, T. M. 1997. *Machine Learning*. New York, NY: McGraw-Hill. (Cited on pp. 254, 472.)

[313] Molloy, M., and Reed, B. 1995. A Critical Point for Random Graphs With a Given Degree Sequence. *Random Structures & Algorithms*, **6**(2-3), 161–180. (Cited on pp. 163, 333.)

[314] Molloy, M., and Reed, B. 1998. The Size of the Giant Component of a Random Graph with a Given Degree Sequence. *Combinatorics, Probability and Computing*, **7**(3), 295–305. (Cited on pp. 163, 333.)

[315] Molnar, C. 2020. *Interpretable Machine Learning*. Morrisville, NC: Lulu Press, Inc. (Cited on p. 275.)

[316] Moore, P. B., Hendrickson, W. A., Henderson, R., and Brunger, A. T. 2022. The Protein-Folding Problem: Not Yet Solved. *Science*, **375**(6580), 507–507. (Cited on p. 17.)

[317] Moreno, J. L. 1934. *Who Shall Survive?: A New Approach to the Problem of Human Interrelations*. Washington, DC: Nervous and Mental Disease Publishing Co. (Cited on pp. 7, 16, 25, 78, 221.)

[318] Moreno, J. L. 1953. *Who Shall Survive? Foundations of Sociometry, Group Psychotherapy and Socio-Drama*. 2nd edn. Oxford: Beacon House. (Cited on p. 7.)

[319] Moreno, R. 2021. *Words of the Daughter: A Memoir*. Morrisville, NC: Lulu Press, Inc. (Cited on p. 16.)

[320] Morris, C., Ritzert, M., Fey, M. et al. 2019. Weisfeiler and Leman Go Neural: Higher-Order Graph Neural Networks. *Proceedings of the AAAI Conference on Artificial Intelligence*, **33**(01), 4602–4609. (Cited on p. 443.)

[321] Morris, R. 1978. Counting Large Numbers of Events in Small Registers. *Communications of the ACM*, **21**(10), 840–842. (Cited on p. 469.)

[322] Mucha, P. J., Richardson, T., Macon, K., Porter, M. A., and Onnela, J.-P. 2010. Community Structure in Time-Dependent, Multiscale, and Multiplex Networks. *Science*, **328**(5980), 876–878. (Cited on pp. 181, 201.)

[323] Munafò, M. R., Nosek, B. A., Bishop, D. V. M. et al. 2017. A Manifesto for Reproducible Science. *Nature Human Behaviour*, **1**(1), 0021. (Cited on p. 163.)

[324] Murphy, K. P. 2022. *Probabilistic Machine Learning: An Introduction*. Cambridge, MA: MIT Press. (Cited on p. 473.)

[325] Murray, D., Yoon, J., Kojaku, S. et al. 2021 (June). *Unsupervised Embedding of Trajectories Captures the Latent Structure of Mobility*. arXiv:2012.02785. (Cited on p. 435.)

[326] Nadakuditi, R. R., and Newman, M. E. J. 2012. Graph Spectra and the Detectability of Community Structure in Networks. *Physical Review Letters*, **108**(18), 188701. (Cited on p. 357.)

[327] Narayanan, A., Shi, E., and Rubinstein, B. I. P. 2011. Link Prediction by De-Anonymization: How We Won the Kaggle Social Network Challenge. Pages 1825–1834 of: *The 2011 International Joint Conference on Neural Networks*. San Jose, CA: IEEE. (Cited on p. 33.)

[328] National Academies of Sciences, Engineering, and Medicine. 2019. *Reproducibility and Replicability in Science*. Washington, DC: National Academies Press. (Cited on p. 474.)

[329] Nemeth, R. J., and Smith, D. A. 1985. International Trade and World-System Structure: A Multiple Network Analysis. *Review (Fernand Braudel Center)*, **8**(4), 517–560. (Cited on p. 194.)

[330] Newman, E. A., Araque, A., Dubinsky, J. M., Swanson, L. W., King, L. S., and Himmel, E. (eds.). 2017. *The Beautiful Brain: The Drawings of Santiago Ramón y Cajal*. New York, NY: Abrams. (Cited on p. 16.)

[331] Newman, M. E. J. 2002. Assortative Mixing in Networks. *Physical Review Letters*, **89**(20), 208701. (Cited on pp. 111, 201.)

[332] Newman, M. E. J. 2003. Mixing Patterns in Networks. *Physical Review E*, **67**(2), 026126. (Cited on pp. 111, 116, 163, 173, 201.)

[333] Newman, M. E. J. 2006. Finding Community Structure in Networks Using the Eigenvectors of Matrices. *Physical Review E*, **74**(3), 036104. (Cited on pp. 421, 427.)

[334] Newman, M. E. J. 2006. Modularity and Community Structure in Networks. *Proceedings of the National Academy of Sciences*, **103**(23), 8577–8582. (Cited on pp. 201, 417, 418, 427.)

[335] Newman, M. E. J. 2012. Communities, Modules and Large-Scale Structure in Networks. *Nature Physics*, **8**(1), 25–31. (Cited on p. 201.)

[336] Newman, M. E. J. 2013. Spectral Methods for Community Detection and Graph Partitioning. *Physical Review E*, **88**(4), 042822. (Cited on p. 427.)

[337] Newman, M. E. J. 2018. Network Structure from Rich but Noisy Data. *Nature Physics*, **14**(6), 542–545. (Cited on pp. 134, 365, 374.)

[338] Newman, M. E. J., and Girvan, M. 2004. Finding and Evaluating Community Structure in Networks. *Physical Review E*, **69**(2), 026113. (Cited on p. 201.)

[339] Newman, M. E. J., and Park, J. 2003. Why Social Networks Are Different from Other Types of Networks. *Physical Review E*, **68**(3), 036122. (Cited on pp. 380, 395.)

[340] Newman, M. E. J., Strogatz, S. H., and Watts, D. J. 2001. Random Graphs with Arbitrary Degree Distributions and Their Applications. *Physical Review E*, **64**(2), 026118. (Cited on p. 340.)

[341] Newman, M. E. J., Watts, D. J., and Strogatz, S. H. 2002. Random Graph Models of Social Networks. *Proceedings of the National Academy of Sciences*, **99**(suppl. 1), 2566–2572. (Cited on p. 380.)

[342] Newman, M. 2018. *Networks*. Oxford: Oxford University Press. (Cited on pp. 16, 172, 472.)

[343] Newman, M. 2005. Power Laws, Pareto Distributions and Zipf's Law. *Contemporary Physics*, **46**(5), 323–351. (Cited on p. 163.)

[344] Ng, A., Jordan, M., and Weiss, Y. 2001. On Spectral Clustering: Analysis and an Algorithm. *Advances in Neural Information Processing Systems*, **14**. (Cited on p. 428.)

[345] Nica, B. 2018. *A Brief Introduction to Spectral Graph Theory*. EMS Textbooks in Mathematics. Zürich, Switzerland: European Mathematical Society. (Cited on p. 427.)

[346] Nickel, M., Tresp, V., and Kriegel, H.-P. 2011. A Three-Way Model for Collective Learning on Multi-Relational Data. Pages 809–816 of: *Proceedings of the 28th International Conference on International Conference on Machine Learning*. ICML'11. Madison, WI: Omnipress. (Cited on p. 437.)

[347] Nickel, M., and Kiela, D. 2017. Poincaré Embeddings for Learning Hierarchical Representations. In: *Advances in Neural Information Processing Systems*, vol. 30. Long Beach, CA: Curran Associates, Inc. (Cited on p. 438.)

[348] Nilsson, N. J. 2009. *The Quest for Artificial Intelligence*. Cambridge: Cambridge University Press. (Cited on p. 276.)

[349] Nolte, D. D. 2019. *Introduction to Modern Dynamics: Chaos, Networks, Space and Time*. 2nd edn. Oxford; New York, NY: Oxford University Press. (Cited on p. 428.)

[350] Nowicki, K., and Snijders, T. A. B. 2001. Estimation and Prediction for Stochastic Blockstructures. *Journal of the American Statistical Association*, **96**(455), 1077–1087. (Cited on p. 374.)

[351] Obermeyer, Z., Powers, B., Vogeli, C., and Mullainathan, S. 2019. Dissecting Racial Bias in an Algorithm Used to Manage the Health of Populations. *Science*, **366**(6464), 447–453. (Cited on pp. 30, 31.)

[352] O'Neil, C. 2017. *Weapons of Math Destruction: How Big Data Increases Inequality and Threatens Democracy*. New York, NY: Crown. (Cited on pp. 37, 277.)

[353] Ozella, L., Paolotti, D., Lichand, G. et al. 2021. Using Wearable Proximity Sensors to Characterize Social Contact Patterns in a Village of Rural Malawi. *EPJ Data Science*, **10**(1), 46. (Cited on pp. 25, 37, 246.)

[354] Palla, G., Derényi, I., Farkas, I., and Vicsek, T. 2005. Uncovering the Overlapping Community Structure of Complex Networks in Nature and Society. *Nature*, **435**(7043), 814–818. (Cited on pp. 181, 201.)

[355] Palmer, C. R., Gibbons, P. B., and Faloutsos, C. 2002. ANF: A Fast and Scalable Tool for Data Mining in Massive Graphs. Pages 81–90 of: *Proceedings of the Eighth ACM SIGKDD International Conference on Knowledge Discovery and Data Mining*. KDD '02. New York, NY: Association for Computing Machinery. (Cited on pp. 456, 457, 469.)

[356] Panko, R. R. 1998. What We Know About Spreadsheet Errors. *Journal of Organizational and End User Computing*, **10**(2), 15–21. (Cited on p. 36.)

[357] Pantel, P. 2005. Inducing Ontological Co-Occurrence Vectors. Pages 125–132 of: *Proceedings of the 43rd Annual Meeting on Association for Computational Linguistics*. ACL '05. Ann Arbor, MI: Association for Computational Linguistics. (Cited on pp. 432, 445.)

[358] Park, J., Wood, I. B., Jing, E. et al. 2019. Global Labor Flow Network Reveals the Hierarchical Organization and Dynamics of Geo-Industrial Clusters. *Nature Communications*, **10**(1), 3449. (Cited on p. 21.)

[359] Park, J., and Newman, M. E. J. 2004. Solution of the Two-Star Model of a Network. *Physical Review E*, **70**(6), 066146. (Cited on p. 369.)

[360] Park, J., and Newman, M. E. J. 2005. A Network-Based Ranking System for US College Football. *Journal of Statistical Mechanics: Theory and Experiment*, **2005**(10), P10014–P10014. (Cited on p. 216.)

[361] Pascual, M., and Dunne, J. A. 2005. *Ecological Networks: Linking Structure to Dynamics in Food Webs*. Oxford: Oxford University Press. (Cited on p. 25.)

[362] Pastor-Satorras, R., and Vespignani, A. 2001. Epidemic Spreading in Scale-Free Networks. *Physical Review Letters*, **86**(14), 3200–3203. (Cited on p. 333.)

[363] Patterson, B. D., and Atmar, W. 1986. Nested Subsets and the Structure of Insular Mammalian Faunas and Archipelagos. *Biological Journal of the Linnean Society*, **28**(1–2), 65–82. (Cited on p. 202.)

[364] Peel, L., Larremore, D. B., and Clauset, A. 2017. The Ground Truth about Metadata and Community Detection in Networks. *Science Advances*, **3**(5), e1602548. (Cited on p. 107.)

[365] Peixoto, T. P. 2013. Parsimonious Module Inference in Large Networks. *Physical Review Letters*, **110**(14), 148701. (Cited on p. 354.)

[366] Peixoto, T. P. 2014. Hierarchical Block Structures and High-Resolution Model Selection in Large Networks. *Physical Review X*, **4**(1), 011047. (Cited on p. 358.)

[367] Pelleg, D., and Moore, A. W. 2000. X-Means: Extending K-means with Efficient Estimation of the Number of Clusters. Pages 727–734 of: *Proceedings of the Seventeenth International Conference on Machine Learning*. ICML '00. San Francisco, CA: Morgan Kaufmann. (Cited on p. 426.)

[368] Pereira, D. A., and Williams, J. A. 2007. Origin and Evolution of High Throughput Screening. *British Journal of Pharmacology*, **152**(1), 53–61. (Cited on p. 82.)

[369] Perozzi, B., Al-Rfou, R., and Skiena, S. 2014. DeepWalk: Online Learning of Social Representations. Pages 701–710 of: *Proceedings of the 20th ACM SIGKDD International Conference on Knowledge Discovery and Data Mining*. New York, NY: Association for Computing Machinery. (Cited on pp. 435, 445.)

[370] Perry, B. L., Pescosolido, B. A., and Borgatti, S. P. 2018. *Egocentric Network Analysis: Foundations, Methods, and Models*. Cambridge: Cambridge University Press. (Cited on p. 201.)

[371] Pfeiffer, J., and Neville, J. 2021. Methods to Determine Node Centrality and Clustering in Graphs with Uncertain Structure. *Proceedings of the International AAAI Conference on Web and Social Media*, **5**(1), 590–593. (Cited on pp. 387, 395.)

[372] Phan, T. Q., and Airoldi, E. M. 2015. A Natural Experiment of Social Network Formation and Dynamics. *Proceedings of the National Academy of Sciences*, **112**(21), 6595–6600. (Cited on p. 474.)

[373] Pilgrim, C., and Hills, T. T. 2021. Bias in Zipf's Law Estimators. *Scientific Reports*, **11**(1), 17309. (Cited on p. 163.)

[374] Pimentel, J. F., Murta, L., Braganholo, V., and Freire, J. 2019. A Large-Scale Study About Quality and Reproducibility of Jupyter Notebooks. Pages 507–517 of: *2019 IEEE/ACM 16th International Conference on Mining Software Repositories (MSR)*. Montreal, Canada: IEEE. (Cited on p. 302.)

[375] Porter, M., and Gleeson, J. 2016. *Dynamical Systems on Networks: A Tutorial*. Frontiers in Applied Dynamical Systems: Reviews and Tutorials, no. 4. Cham, Switzerland: Springer. (Cited on pp. 235, 428.)

[376] Porter, M. A., Onnela, J.-P., Mucha, P. J., et al. 2009. Communities in Networks. *Notices of the AMS*, **56**(9), 1082–1097. (Cited on p. 201.)

[377] Pothen, A., Simon, H. D., and Liou, K.-P. 1990. Partitioning Sparse Matrices with Eigenvectors of Graphs. *SIAM Journal on Matrix Analysis and Applications*, **11**(3), 430–452. (Cited on pp. 412, 427.)

[378] Price, D. D. S. 1976. A General Theory of Bibliometric and Other Cumulative Advantage Processes. *Journal of the American Society for Information Science*, **27**(5), 292–306. (Cited on p. 349.)

[379] Price, D. J. d. S. 1965. Networks of Scientific Papers: The Pattern of Bibliographic References Indicates the Nature of the Scientific Research Front. *Science*, **149**(3683), 510–515. (Cited on p. 349.)

[380] Proulx, S. R., Promislow, D. E. L., and Phillips, P. C. 2005. Network Thinking in Ecology and Evolution. *Trends in Ecology & Evolution*, **20**(6), 345–353. (Cited on p. 25.)

[381] Pržulj, N. 2007. Biological Network Comparison Using Graphlet Degree Distribution. *Bioinformatics*, **23**(2), e177–e183. (Cited on p. 228.)

[382] Qiu, J., Dong, Y., Ma, H., Li, J., Wang, K., and Tang, J. 2018. Network Embedding as Matrix Factorization: Unifying DeepWalk, LINE, PTE, and Node2vec. Pages 459–467 of: *Proceedings of the Eleventh ACM International Conference on Web Search and Data Mining*. WSDM '18. New York, NY: Association for Computing Machinery. (Cited on pp. 440, 441, 445.)

[383] Radford, A., Narasimhan, K., Salimans, T., and Sutskever, I. 2018. *Improving Language Understanding by Generative Pre-Training*. (Cited on p. 261.)

[384] Raghavan, U. N., Albert, R., and Kumara, S. 2007. Near Linear Time Algorithm to Detect Community Structures in Large-Scale Networks. *Physical Review E*, **76**(3), 036106. (Cited on pp. 458, 459, 469.)

[385] Ramani, A. S., Eikmeier, N., and Gleich, D. F. 2019. Coin-Flipping, Ball-Dropping, and Grass-Hopping for Generating Random Graphs from Matrices of Edge Probabilities. *SIAM Review*, **61**(3), 549–595. (Cited on p. 469.)

[386] Ramón y Cajal, S. 1899. *Comparative Study of the Sensory Areas of the Human Cortex*. Worcester, MA: Clark University. (Cited on p. 5.)

[387] Ramón y Cajal, S. 2004. *Advice for a Young Investigator*. Cambridge, MA: MIT Press. (Cited on p. 16.)

[388] Ramón y Cajal, S., and Cowan, W. M. 1989. *Recollections of My Life*. Cambridge, MA: MIT Press. (Cited on p. 16.)

[389] Ranshous, S., Shen, S., Koutra, D., Harenberg, S., Faloutsos, C., and Samatova, N. F. 2015. Anomaly Detection in Dynamic Networks: A Survey. *WIREs Computational Statistics*, **7**(3), 223–247. (Cited on p. 469.)

[390] Reed, B. J., and Segal, D. R. 2006. Social Network Analysis and Counterinsurgency Operations: The Capture of Saddam Hussein. *Sociological Focus*, **39**(4), 251–264. (Cited on p. 33.)

[391] Reichardt, J., and Bornholdt, S. 2006. Statistical Mechanics of Community Detection. *Physical Review E*, **74**(1), 016110. (Cited on p. 181.)

[392] Reichardt, J., and Bornholdt, S. 2006. When Are Networks Truly Modular? *Physica D: Nonlinear Phenomena*, **224**(1-2), 20–26. (Cited on p. 179.)

[393] Reinhart, C. M., and Rogoff, K. S. 2010. Growth in a Time of Debt. *American Economic Review*, **100**(2), 573–578. (Cited on p. 36.)

[394] Resnik, D. B. 1998. *The Ethics of Science: An Introduction*. London: Routledge. (Cited on p. 37.)

[395] Ribeiro, P., Paredes, P., Silva, M. E. P., Aparicio, D., and Silva, F. 2022. A Survey on Subgraph Counting: Concepts, Algorithms, and Applications to Network Motifs and Graphlets. *ACM Computing Surveys*, **54**(2), 1–36. (Cited on p. 201.)

[396] Rider, P. R. 1955. Truncated Binomial and Negative Binomial Distributions. *Journal of the American Statistical Association*, **50**(271), 877–883. (Cited on p. 393.)

[397] Robins, G., and Alexander, M. 2004. Small Worlds Among Interlocking Directors: Network Structure and Distance in Bipartite Graphs. *Computational & Mathematical Organization Theory*, **10**(1), 69–94. (Cited on p. 194.)

[398] Robins, G., Pattison, P., Kalish, Y., and Lusher, D. 2007. An Introduction to Exponential Random Graph (P*) Models for Social Networks. *Social Networks*, **29**(2), 173–191. (Cited on pp. 163, 374.)

[399] Robinson, I., Webber, J., and Eifrem, E. 2015. *Graph Databases: New Opportunities for Connected Data*. Sebastopol, CA: O'Reilly Media. (Cited on p. 469.)

[400] Rombach, M. P., Porter, M. A., Fowler, J. H., and Mucha, P. J. 2014. Core-Periphery Structure in Networks. *SIAM Journal on Applied Mathematics*, **74**(1), 167–190. (Cited on pp. 195, 202.)

[401] Rota, G.-C. 1964. The Number of Partitions of a Set. *The American Mathematical Monthly*, **71**(5), 498. (Cited on p. 178.)

[402] Roughan, M., and Tuke, J. 2015. The Hitchhikers Guide to Sharing Graph Data. Pages 435–442 of: *2015 3rd International Conference on Future Internet of Things and Cloud*. Rome, Italy: IEEE. (Cited on p. 98.)

[403] Roughan, M., Tuke, S. J., and Maennel, O. 2008. Bigfoot, Sasquatch, the Yeti and Other Missing Links: What We Don't Know about the as Graph. Pages 325–330 of: *Proceedings of the 8th ACM SIGCOMM Conference on Internet Measurement*. Vouliagmeni, Greece: Association for Computing Machinery. (Cited on pp. 392, 395.)

[404] Rousseeuw, P. J. 1987. Silhouettes: A Graphical Aid to the Interpretation and Validation of Cluster Analysis. *Journal of Computational and Applied Mathematics*, **20**(Nov.), 53–65. (Cited on p. 426.)

[405] Rozenblatt-Rosen, O., Regev, A., Oberdoerffer, P. et al. 2020. The Human Tumor Atlas Network: Charting Tumor Transitions across Space and Time at Single-Cell Resolution. *Cell*, **181**(2), 236–249. (Cited on p. 473.)

[406] Rubin, D. B. 1976. Inference and Missing Data. *Biometrika*, **63**(3), 581–592. (Cited on p. 134.)

[407] Rubin, D. B. 1996. Multiple Imputation after 18+ Years. *Journal of the American Statistical Association*, **91**(434), 473–489. (Cited on pp. 116, 134.)

[408] Rumelhart, D. E., Hinton, G. E., and Williams, R. J. 1986. Learning Representations by Back-Propagating Errors. *Nature*, **323**(6088), 533–536. (Cited on p. 258.)

[409] Russell, M. A. 2013. *Mining the Social Web: Data Mining Facebook, Twitter, LinkedIn, Google+, GitHub, and More*. Sebastopol, CA: O'Reilly Media. (Cited on p. 71.)

[410] Russell, S. J., and Norvig, P. 2021. *Artificial Intelligence: A Modern Approach*. New York, NY: Pearson. (Cited on p. 473.)

[411] Russell, S. J., Russell, S., and Norvig, P. 2020. *Artificial Intelligence: A Modern Approach*. New York, NY: Pearson. (Cited on p. 277.)

[412] Salganik, M. J. 2018. *Bit by Bit: Social Research in the Digital Age*. Princeton, NJ: Princeton University Press. (Cited on pp. 25, 65, 68, 71, 82.)

[413] Sambourg, L., and Thierry-Mieg, N. 2010. New Insights into Protein-Protein Interaction Data Lead to Increased Estimates of the S. Cerevisiae Interactome Size. *BMC Bioinformatics*, **11**(1), 605. (Cited on p. 395.)

[414] Sarigol, E., Garcia, D., and Schweitzer, F. 2014. Online Privacy as a Collective Phenomenon. Pages 95–106 of: *Proceedings of the Second ACM Conference on Online Social Networks*. COSN '14. New York, NY: Association for Computing Machinery. (Cited on p. 37.)

[415] Scarselli, F., Gori, M., Tsoi, A. C., Hagenbuchner, M., and Monfardini, G. 2009. The Graph Neural Network Model. *IEEE Transactions on Neural Networks*, **20**(1), 61–80. (Cited on p. 445.)

[416] Schaub, M. T., Delvenne, J.-C., Rosvall, M., and Lambiotte, R. 2017. The Many Facets of Community Detection in Complex Networks. *Applied Network Science*, **2**(1), 1–13. (Cited on p. 201.)

[417] Schimel, J. 2012. *Writing Science: How to Write Papers That Get Cited and Proposals That Get Funded.* New York, NY: Oxford University Press. (Cited on p. 473.)

[418] Schreier, A. A., Wilson, K., and Resnik, D. 2006. Academic Research Record-Keeping: Best Practices for Individuals, Group Leaders, and Institutions. *Academic Medicine,* **81**(1), 42–47. (Cited on pp. 287, 292.)

[419] Schwabish, J. 2021. *Better Data Visualizations: A Guide for Scholars, Researchers, and Wonks.* New York, NY: Columbia University Press. (Cited on p. 473.)

[420] Scopatz, A., and Huff, K. D. 2015. *Effective Computation in Physics: Field Guide to Research with Python.* Sebastopol, CA: O'Reilly Media. (Cited on pp. 102, 312, 313.)

[421] Seal, H. L. 1952. The Maximum Likelihood Fitting of the Discrete Pareto Law. *Journal of the Institute of Actuaries,* **78**(1), 115–121. (Cited on p. 163.)

[422] Seebauer, E. G., and Barry, R. L. 2000. *Fundamentals of Ethics for Scientists and Engineers.* Oxford: Oxford University Press. (Cited on p. 37.)

[423] Serafino, M., Cimini, G., Maritan, A. et al. 2021. True Scale-Free Networks Hidden by Finite Size Effects. *Proceedings of the National Academy of Sciences,* **118**(2), e2013825118. (Cited on p. 346.)

[424] Serrano, M. Á., Boguñá, M., and Vespignani, A. 2009. Extracting the Multiscale Backbone of Complex Weighted Networks. *Proceedings of the National Academy of Sciences,* **106**(16), 6483–6488. (Cited on pp. 126, 134, 375.)

[425] Sheppard, C. 2017. *Tree-Based Machine Learning Algorithms: Decision Trees, Random Forests, and Boosting.* Seattle, WA: CreateSpace. (Cited on p. 275.)

[426] Sherman, R. M., Forman, J., Antonescu, V. et al. 2019. Assembly of a Pan-Genome from Deep Sequencing of 910 Humans of African Descent. *Nature Genetics,* **51**(1), 30–35. (Cited on p. 29.)

[427] Shervashidze, N., Vishwanathan, S. V. N., Petri, T., Mehlhorn, K., and Borgwardt, K. 2009. Efficient Graphlet Kernels for Large Graph Comparison. Pages 488–495 of: *Proceedings of the Twelth International Conference on Artificial Intelligence and Statistics.* Proceedings of Machine Learning Research, vol. 5. Clearwater Beach, FL: PMLR. (Cited on p. 228.)

[428] Shi, J., and Malik, J. 2000. Normalized Cuts and Image Segmentation. *IEEE Transactions on Pattern Analysis and Machine Intelligence,* **22**(8), 888–905. (Cited on pp. 412, 413, 428.)

[429] Shneiderman, B. 2020. Bridging the Gap Between Ethics and Practice: Guidelines for Reliable, Safe, and Trustworthy Human-centered AI Systems. *ACM Transactions on Interactive Intelligent Systems,* **10**(4), 26:1–26:31. (Cited on p. 31.)

[430] Shumway, R. H. 2017. *Time Series Analysis and Its Applications: With R Examples.* New York, NY: Springer Science+Business Media. (Cited on p. 249.)

[431] Simkin, M., and Roychowdhury, V. 2011. Re-Inventing Willis. *Physics Reports,* May, S0370157310003339. (Cited on p. 349.)

[432] Simon, H. A. 1955. On a Class of Skew Distribution Functions. *Biometrika*, **42**(3–4), 425–440. (Cited on pp. 344, 349.)

[433] Sinclair, A., and Jerrum, M. 1989. Approximate Counting, Uniform Generation and Rapidly Mixing Markov Chains. *Information and Computation*, **82**(1), 93–133. (Cited on p. 427.)

[434] Sirugo, G., Williams, S. M., and Tishkoff, S. A. 2019. The Missing Diversity in Human Genetic Studies. *Cell*, **177**(1), 26–31. (Cited on p. 28.)

[435] Smith, R. C. 2013. *Uncertainty Quantification: Theory, Implementation, and Applications*. Computational Science and Engineering Series. Philadelphia, PA: Society for Industrial and Applied Mathematics. (Cited on p. 395.)

[436] Snijders, T. A. B., Pattison, P. E., Robins, G. L., and Handcock, M. S. 2006. New Specifications for Exponential Random Graph Models. *Sociological Methodology*, **36**(1), 99–153. (Cited on p. 369.)

[437] Snijders, T. A., and Nowicki, K. 1997. Estimation and Prediction for Stochastic Blockmodels for Graphs with Latent Block Structure. *Journal of Classification*, **14**(1), 75–100. (Cited on pp. 353, 374.)

[438] Sood, V., and Redner, S. 2005. Voter Model on Heterogeneous Graphs. *Physical Review Letters*, **94**(17), 178701. (Cited on p. 235.)

[439] Soranzo, N., Bianconi, G., and Altafini, C. 2007. Comparing Association Network Algorithms for Reverse Engineering of Large-Scale Gene Regulatory Networks: Synthetic versus Real Data. *Bioinformatics*, **23**(13), 1640–1647. (Cited on p. 134.)

[440] Soundarajan, S., Eliassi-Rad, T., and Gallagher, B. 2014. A Guide to Selecting a Network Similarity Method. Pages 1037–1045 of: *Proceedings of the 2014 SIAM International Conference on Data Mining (SDM)*. Philadelphia, PA: Society for Industrial and Applied Mathematics. (Cited on p. 232.)

[441] Spielman, D. A. 2007. Spectral Graph Theory and Its Applications. Pages 29–38 of: *48th Annual IEEE Symposium on Foundations of Computer Science (FOCS'07)*. Providence, Rhode Island: IEEE. (Cited on p. 427.)

[442] Sporns, O. 2016. *Networks of the Brain*. Cambridge, MA: MIT Press. (Cited on pp. 25, 82.)

[443] Staiano, J., Lepri, B., Aharony, N., Pianesi, F., Sebe, N., and Pentland, A. 2012. Friends Don't Lie: Inferring Personality Traits from Social Network Structure. Pages 321–330 of: *Proceedings of the 2012 ACM Conference on Ubiquitous Computing*. UbiComp '12. New York, NY: Association for Computing Machinery. (Cited on p. 37.)

[444] Stewart, J. 2017. *Python for Scientists*. 2nd edn. Cambridge; New York, NY: Cambridge University Press. (Cited on p. 61.)

[445] Stopczynski, A., Sekara, V., Sapiezynski, P. et al. 2014. Measuring Large-Scale Social Networks with High Resolution. *PLOS ONE*, **9**(4), e95978. (Cited on p. 473.)

[446] Strang, G. 2019. *Linear Algebra and Learning from Data*. Wellesley, MA: Wellesley-Cambridge Press. (Cited on pp. 62, 277, 427.)

[447] Strang, G. 2023. *Introduction to Linear Algebra*. 6th edn. Wellesley, MA: Wellesley-Cambridge. (Cited on pp. 61, 427.)

[448] Stumpf, M. P. H., Thorne, T., de Silva, E. et al. 2008. Estimating the Size of the Human Interactome. *Proceedings of the National Academy of Sciences*, **105**(19), 6959–6964. (Cited on pp. 390, 395.)

[449] Sugimoto, C. R., Ahn, Y.-Y., Smith, E., Macaluso, B., and Larivière, V. 2019. Factors Affecting Sex-Related Reporting in Medical Research: A Cross-Disciplinary Bibliometric Analysis. *The Lancet*, **393**(10171), 550–559. (Cited on p. 28.)

[450] Sullivan, D. 2020. *A Reintroduction to Our Knowledge Graph and Knowledge Panels*. (Cited on p. 449.)

[451] Sullivan, T. J. 2015. *Introduction to Uncertainty Quantification*. Texts in Applied Mathematics, no. 63. Cham, Switzerland: Springer. (Cited on p. 395.)

[452] Sun, J., Bagrow, J. P., Bollt, E. M., and Skufca, J. D. 2009. Dynamic Computation of Network Statistics via Updating Schema. *Physical Review E*, **79**(3), 036116. (Cited on pp. 459, 465, 466, 469.)

[453] Tang, J., Qu, M., Wang, M., Zhang, M., Yan, J., and Mei, Q. 2015. LINE: Large-scale Information Network Embedding. Pages 1067–1077 of: *Proceedings of the 24th International Conference on World Wide Web*. WWW '15. Geneva, Switzerland: International World Wide Web Conferences Steering Committee. (Cited on pp. 437, 440.)

[454] Taylor, D., Myers, S. A., Clauset, A., Porter, M. A., and Mucha, P. J. 2017. Eigenvector-Based Centrality Measures for Temporal Networks. *Multiscale Modeling & Simulation*, **15**(1), 537–574. (Cited on p. 241.)

[455] Taylor, J. R. 1997. *An Introduction to Error Analysis: The Study of Uncertainties in Physical Measurements*. 2nd edn. Sausalito, CA: University Science Books. (Cited on pp. 134, 392.)

[456] The Gene Ontology Consortium, Carbon, S., Douglass, E. et al. 2021. The Gene Ontology Resource: Enriching a GOld Mine. *Nucleic Acids Research*, **49**(D1), D325–D334. (Cited on p. 116.)

[457] Thiebes, S., Lins, S., and Sunyaev, A. 2021. Trustworthy Artificial Intelligence. *Electronic Markets*, **31**(2), 447–464. (Cited on p. 31.)

[458] Thorndike, R. L. 1953. Who Belongs in the Family? *Psychometrika*, **18**(4), 267–276. (Cited on p. 426.)

[459] Tibshirani, R., Walther, G., and Hastie, T. 2001. Estimating the Number of Clusters in a Data Set via the Gap Statistic. *Journal of the Royal Statistical Society: Series B (Statistical Methodology)*, **63**(2), 411–423. (Cited on p. 426.)

[460] Ting, D. 2016. Towards Optimal Cardinality Estimation of Unions and Intersections with Sketches. Pages 1195–1204 of: *Proceedings of the 22nd ACM SIGKDD International Conference on Knowledge Discovery and Data Mining*. KDD '16. New York, NY: Association for Computing Machinery. (Cited on p. 456.)

[461] Traag, V. A., Waltman, L., and van Eck, N. J. 2019. From Louvain to Leiden:

Guaranteeing Well-Connected Communities. *Scientific Reports*, **9**(1), 5233. (Cited on p. 179.)

[462] Tracy, S. J. 2010. Qualitative Quality: Eight "Big-Tent" Criteria for Excellent Qualitative Research. *Qualitative Inquiry*, **16**(10), 837–851. (Cited on p. 226.)

[463] Trouillon, T., Welbl, J., Riedel, S., Gaussier, É., and Bouchard, G. 2016. Complex Embeddings for Simple Link Prediction. Pages 2071–2080 of: *Proceedings of the 33rd International Conference on International Conference on Machine Learning*. ICML'16, vol. 48. New York, NY: JMLR.org. (Cited on p. 437.)

[464] Tufte, E. R. 2001. *The Visual Display of Quantitative Information*. 2nd edn. Cheshire, CT: Graphics Press. (Cited on p. 473.)

[465] Tukey, J. W. 1977. *Exploratory Data Analysis*. Addison-Wesley Series in Behavioral Science. Reading, MA: Addison-Wesley. (Cited on pp. 71, 162.)

[466] Tutte, W. T. 1963. How to Draw a Graph. *Proceedings of the London Mathematical Society*, **3**(1), 743–767. (Cited on p. 221.)

[467] Ugander, J., Backstrom, L., Marlow, C., and Kleinberg, J. 2012. Structural Diversity in Social Contagion. *Proceedings of the National Academy of Sciences*, **109**(16), 5962–5966. (Cited on p. 201.)

[468] Ulrich, W., Almeida-Neto, M., and Gotelli, N. J. 2009. A Consumer's Guide to Nestedness Analysis. *Oikos*, **118**(1), 3–17. (Cited on p. 202.)

[469] van Buuren, S. 2018. *Flexible Imputation of Missing Data*. 2nd edn. Chapman and Hall/CRC Interdisciplinary Statistics Series. Boca Raton, FL: CRC Press, Taylor and Francis Group. (Cited on pp. 116, 134.)

[470] van den Heuvel, M. P., and Sporns, O. 2011. Rich-Club Organization of the Human Connectome. *Journal of Neuroscience*, **31**(44), 15775–15786. (Cited on p. 202.)

[471] van der Maaten, L., and Hinton, G. 2008. Visualizing Data Using T-SNE. *Journal of Machine Learning Research*, **9**(86), 2579–2605. (Cited on p. 259.)

[472] van Mieghem, P. 2010. *Graph Spectra for Complex Networks*. Cambridge: Cambridge University Press. (Cited on pp. 406, 427.)

[473] Vanderplas, J. T. 2016. *Python Data Science Handbook: Essential Tools for Working with Data*. Sebastopol, CA: O'Reilly Media. (Cited on p. 102.)

[474] Vaswani, A., Shazeer, N., Parmar, N. et al. 2017. Attention Is All You Need. In: *Advances in Neural Information Processing Systems*, vol. 30. Long Beach, CA: Curran Associates, Inc. (Cited on p. 273.)

[475] Veličković, P., Cucurull, G., Casanova, A., Romero, A., Liò, P., and Bengio, Y. 2018. Graph Attention Networks. In: *International Conference on Learning Representations*. Vancouver, BC: OpenReview.net. (Cited on pp. 273, 442.)

[476] Vitter, J. S. 1985. Random Sampling with a Reservoir. *ACM Transactions on Mathematical Software*, **11**(1), 37–57. (Cited on p. 453.)

[477] Voitalov, I., van der Hoorn, P., van der Hofstad, R., and Krioukov, D. 2019. Scale-Free Networks Well Done. *Physical Review Research*, **1**(3), 033034. (Cited on p. 346.)

[478] von Luxburg, U. 2007. A Tutorial on Spectral Clustering. *Statistics and Computing*, **17**(4), 395–416. (Cited on pp. 205, 232, 423, 425, 428.)

[479] Vreeman, R. C., and Carroll, A. E. 2008. Festive Medical Myths. *BMJ*, **337**, a2769. (Cited on p. 86.)

[480] Wagner, C., Strohmaier, M., Olteanu, A., Kıcıman, E., Contractor, N., and Eliassi-Rad, T. 2021. Measuring Algorithmically Infused Societies. *Nature*, **595**(7866), 197–204. (Cited on p. 473.)

[481] Walpole, R. E., Myers, R. H., Myers, S. L., and Ye, K. (eds.). 2017. *Probability & Statistics for Engineers & Scientists*. 9th edn. Boston, MA: Pearson. (Cited on p. 62.)

[482] Wang, C. C., Prather, K. A., Sznitman, J. et al. 2021. Airborne Transmission of Respiratory Viruses. *Science*, **373**(6558), eabd9149. (Cited on p. 86.)

[483] Wang, Y. X. R., and Bickel, P. J. 2017. Likelihood-Based Model Selection for Stochastic Block Models. *The Annals of Statistics*, **45**(2), 500–528. (Cited on p. 354.)

[484] Wasserman, L. A. 2004. *All of Statistics: A Concise Course in Statistical Inference*. New York, NY: Springer Science & Business Media. (Cited on pp. 62, 375, 472.)

[485] Wasserman, S., and Faust, K. 1994. *Social Network Analysis: Methods and Applications*. Structural Analysis in the Social Sciences. Cambridge: Cambridge University Press. (Cited on pp. 25, 163, 201, 380, 472.)

[486] Watts, D. J., and Strogatz, S. H. 1998. Collective Dynamics of 'Small-World' Networks. *Nature*, **393**(6684), 440–442. (Cited on pp. 15, 201, 343, 346, 349.)

[487] Weisfeiler, B., and Leman, A. 1968. The Reduction of a Graph to Canonical Form and the Algebra Which Appears Therein. *Nauchno-Technicheskaya Informatsia, Seriya*, **2**(9), 12–16. (Cited on p. 443.)

[488] Wernicke, S. 2006. Efficient Detection of Network Motifs. *IEEE/ACM Transactions on Computational Biology and Bioinformatics*, **3**(4), 347–359. (Cited on p. 175.)

[489] White, H. C., Boorman, S. A., and Breiger, R. L. 1976. Social Structure from Multiple Networks. I. Blockmodels of Roles and Positions. *American Journal of Sociology*, **81**(4), 730–780. (Cited on p. 374.)

[490] Wickham, H. 2014. Tidy Data. *Journal of Statistical Software*, **59**(Sept.), 1–23. (Cited on pp. 109, 116, 217.)

[491] Wickham, H., and Grolemund, G. 2016. *R for Data Science: Import, Tidy, Transform, Visualize, and Model Data*. Sebastopol, CA: O'Reilly Media. (Cited on pp. 61, 102.)

[492] Wilf, H. S. 2006. *Generatingfunctionology*. Wellesley, MA: A K Peters, Ltd. (Cited on p. 349.)

[493] Wilson, G., Aruliah, D. A., Brown, C. T. et al. 2014. Best Practices for Scientific Computing. *PLOS Biology*, **12**(1), e1001745. (Cited on p. 300.)

[494] Wilson, G., Bryan, J., Cranston, K., Kitzes, J., Nederbragt, L., and Teal, T. K.

2017. Good Enough Practices in Scientific Computing. *PLOS Computational Biology*, **13**(6), e1005510. (Cited on p. 300.)

[495] Wong, D. M. 2013. *The Wall Street Journal Guide to Information Graphics: The Do's And Don'ts Of Presenting Data Facts And Figures*. Washington, DC: National Geographic Books. (Cited on p. 473.)

[496] Xin, R. S., Gonzalez, J. E., Franklin, M. J., and Stoica, I. 2013. GraphX: A Resilient Distributed Graph System on Spark. Pages 1–6 of: *First International Workshop on Graph Data Management Experiences and Systems*. GRADES '13. New York, NY: Association for Computing Machinery. (Cited on p. 453.)

[497] Xu, K., Hu, W., Leskovec, J., and Jegelka, S. 2019. How Powerful Are Graph Neural Networks? In: *International Conference on Learning Representations*. New Orleans, LA: OpenReview.net. (Cited on p. 443.)

[498] Xu, M. 2021. Understanding Graph Embedding Methods and Their Applications. *SIAM Review*, **63**(4), 825–853. (Cited on p. 445.)

[499] Xu, R., and Wunsch, D. 2005. Survey of Clustering Algorithms. *IEEE Transactions on Neural Networks*, **16**(3), 645–678. (Cited on p. 232.)

[500] Yanai, I., and Lercher, M. 2020. A Hypothesis Is a Liability. *Genome Biology*, **21**(1), 231. (Cited on p. 70.)

[501] Yang, B., Yih, W.-t., He, X., Gao, J., and Deng, L. 2015. *Embedding Entities and Relations for Learning and Inference in Knowledge Bases*. arXiv:1412.6575. (Cited on p. 437.)

[502] Young, J.-G., Desrosiers, P., Hébert-Dufresne, L., Laurence, E., and Dubé, L. J. 2017. Finite-Size Analysis of the Detectability Limit of the Stochastic Block Model. *Physical Review E*, **95**(6), 062304. (Cited on p. 357.)

[503] Young, J.-G., Cantwell, G. T., and Newman, M. E. J. 2021. Bayesian Inference of Network Structure from Unreliable Data. *Journal of Complex Networks*, **8**(6), cnaa046. (Cited on p. 394.)

[504] Yule, G. U. 1924. II.—A Mathematical Theory of Evolution, Based on the Conclusions of Dr. J. C. Willis, F. R. S. *Philosophical Transactions of the Royal Society of London. Series B, Containing Papers of a Biological Character*, **213**(402–410), 21–87. (Cited on pp. 344, 349.)

[505] Zachary, W. W. 1977. An Information Flow Model for Conflict and Fission in Small Groups. *Journal of Anthropological Research*, **33**(4), 452–473. (Cited on pp. 23, 82, 355, 418.)

[506] Zhang, C., Bengio, S., Hardt, M., Recht, B., and Vinyals, O. 2021. Understanding Deep Learning (Still) Requires Rethinking Generalization. *Communications of the ACM*, **64**(3), 107–115. (Cited on p. 277.)

[507] Zhang, H., Goel, A., Govindan, R., Mason, K., and Van Roy, B. 2004. Making Eigenvector-Based Reputation Systems Robust to Collusion. Pages 92–104 of: Leonardi, S. (ed.), *Algorithms and Models for the Web-Graph*, vol. 3243. Berlin, Heidelberg: Springer. (Cited on p. 189.)

[508] Zhou, J., Cui, G., Hu, S. et al. 2020. Graph Neural Networks: A Review of Methods and Applications. *AI Open*, **1**(Jan.), 57–81. (Cited on p. 445.)

[509] Zhou, S., and Mondragon, R. 2004. The Rich-Club Phenomenon in the Internet Topology. *IEEE Communications Letters*, **8**(3), 180–182. (Cited on p. 202.)

[510] Zhou, Z.-H. 2021. *Machine Learning*. Singapore: Springer Singapore. (Cited on p. 472.)

[511] Ziemann, M., Eren, Y., and El-Osta, A. 2016. Gene Name Errors Are Widespread in the Scientific Literature. *Genome Biology*, **17**(1), 177. (Cited on p. 36.)

[512] Zuboff, S. 2020. *The Age of Surveillance Capitalism: The Fight for a Human Future at the New Frontier of Power*. New York, NY: PublicAffairs. (Cited on pp. 32, 34.)

[513] Zuckerman, E. W., and Jost, J. T. 2001. What Makes You Think You're so Popular? Self-Evaluation Maintenance and the Subjective Side of the "Friendship Paradox". *Social Psychology Quarterly*, **64**(3), 207–223. (Cited on p. 324.)

[514] Zügner, D., Akbarnejad, A., and Günnemann, S. 2018. Adversarial Attacks on Neural Networks for Graph Data. Pages 2847–2856 of: *Proceedings of the 24th ACM SIGKDD International Conference on Knowledge Discovery & Data Mining*. KDD '18. New York, NY: Association for Computing Machinery. (Cited on p. 444.)

Index